FOOD-BORNE PARASITIC ZOONOSES

World Class Parasites

VOLUME 11

Volumes in the *World Class Parasites* book series are written for researchers, students and scholars who enjoy reading about excellent research on problems of global signifi cance. Each volume focuses on a parasite, or group of parasites, that has a major impact on human health, or agricultural productivity, and against which we have no satisfactory defense. The volumes are intended to supplement more formal texts that cover taxonomy, life cycles, morphology, vector distribution, symptoms and treatment. They integrate vector, pathogen and host biology and celebrate the diversity of approach that comprises modern parasitological research.

Series Editors
Samuel J. Black, *University of Massachusetts, Amherst, MA, U.S.A.*
J. Richard Seed, *University of North Carolina, Chapel Hill, NC, U.S.A.*

FOOD-BORNE PARASITIC ZOONOSES

Fish and Plant-Borne Parasites

edited by

K. Darwin Murrell

and

Bernard Fried

Springer

K. Darwin Murrell
Centre for Experimental Parasitology
Department of Veterinary Pathobiology
Faculty of Life Sciences
University of Copenhagen
Denmark
kdmurrell@comcast.net

Bernard Fried
Department of Biology
Lafayette College
Easton, PA
USA
friedb@lafayette.edu

Series Editors

Samuel J. Black
University of Massachusetts
Amherst, MA
USA

Richard J. Seed
University of North Carolina
Chapel Hill, NC
USA

Library of Congress Control Number: 2007926436

ISBN-13: 978-0-387-71357-1 e-ISBN-13: 978-0-387-71358-8

Printed on acid-free paper.

9 8 7 6 5 4 3 2 1

springer.com

Preface

Humans suffer from numerous parasitic foodborne zoonoses, many of which are caused by helminths. The helminth zoonoses of concern in this book are normally limited to diseases of animals that have now become transmissible to humans. In the past, these diseases were limited to populations living in low- and middle-income countries, but the geographical limits and populations at risk are expanding and changing because of growing international markets, improved transportation systems, and demographic changes (such as population movements). The World Health Organization (WHO) has estimated the number of people currently infected with just foodborne trematodes exceeds 41 million, and the number of people at risk worldwide, including those in developed countries, is 750 million. The increasing recognition of the public health significance of these zoonoses, their complicated epidemiology, and their links to poverty, agricultural intensification, environmental degradation, and lack of appropriate tools for control has been welcome. However, because the development of priorities in a national public health system is often a competitive exercise, the claim for more attention and resources for foodborne parasitic zoonoses is usually handicapped by a lack of reliable health and economic impact data. The genesis of this book, then, was a desire to draw attention to the problem of these zoonoses and, hopefully, to inspire greater efforts to acquire a reliable global impact assessment which would strengthen the efforts to develop improved prevention and control actions for these zoonoses.

The list of potential helminth zoonoses that might be discussed in a book such as this is large, and could include all those transmitted by ingestion of any food such as meat, fish, invertebrates and plants. However, we have chosen to focus on those zoonoses that are the least under appreciated and recognized of the foodborne helminths, the fish, plant and invertebrate-borne helminths. While people, especially those living in developed countries, are commonly aware of meat-borne zoonoses such as trichinellosis and cysticercosis, fewer are acquainted with fish-borne parasitic diseases like opisthorchiasis, intestinal trematodiasis or capillariasis. Yet these zoonoses are responsible for large numbers of human infections. For example, at least 10 million people in China are infected with the fish-borne liver fluke *Clonorchis sinensis*, and at least 7 million in Thailand are infected with the species *Opisthorchis viverrini, both of which are associated with liver cancer.* The intestinal flukes are even more common throughout Asia, Russia, and the Middle East.

Compared to other parasitic diseases such as malaria, filariasis, and schistosomiasis, these parasitic zoonoses are public health "orphans" in the world of research funding, due in no small measure to insufficient appreciation of a crucial fact: that most of them exist as a complex of parasites whose transmission often depends on well-entrenched cultural behaviors that are difficult to change. Because the transmission routes to human infection are similar, collectively these zoonoses may have a much greater effect in the aggregate than as single infections. The difficulties of diagnosis, the complexities of human cultural traits and agricultural practices and the lack of realistic assessments of their real or potential economic costs, have made this field simultaneously daunting, scientifically obscure and, therefore, unattractive to investigators. The challenge of developing a prevention and control strategy that accommodates strong cultural and agricultural traditions, however, will test the imaginations and skills of researchers, an intellectual challenge that could provide the stimulation needed to build a more concerted international effort toward control.

This book reviews not only the prevalence and distribution of these zoonoses, including available health and economic impact data, but will highlights gaps in knowledge that must be filled in order to gain the assessment needed to depict the overall importance of a particular zoonosis. This is critical for comparisons to other pressing public health and development needs in resource allocations. The topics on epidemiology, diagnosis, and clinical aspects emphasize the knowledge gaps that limit a full understanding of these zoonoses, and target where greater research investments on these parasitic diseases should be focused.

Contents

Contributors

Takeshi Agatsuma
Department of Environmental Health Science
Kochi Medical School
Oko, Nankoku City
Kochi 783-8505
Japan

Maria Dolores Bargues
Department of Parasitology
Faculty of Pharmacy
University of Valencia
Burjassot, Valencia
Spain

V. Belizario
Department of Parasitology
College of Public Health
University of the Philippines Manila
Manila, Philippines

David Blair
School of Marine and Tropical Biology
James Cook University
Townsville, Queensland 4811
Australia

Jong-Yil Chai
Department of Parasitology and Tropical Medicine
Seoul National University College of Medicine
Institute of Endemic Diseases
Seoul National University Medical Research Center
Seoul 110-799, Korea

S. Paz Diaz Chamacho
Department of Public Health Research "Louis Pasteur"
Faculty of Chemical and Biological Sciences
Universidad Autonoma de Sinaloa
Culiacán, Sinaloa, Mexico

F.Y. Cheah
Fish Health Unit
School of Veterinary and Biomedical Sciences
Murdoch, Western Australia 6150
Australia

E.R. Chen
Department of Parasitology
Kaohsiung Medical University
Kaohsiung, Taiwan, R.O.C.

J. H. Cross
Department of Preventive Medicine and Biometrics
Uniformed Services University of the Health Sciences
Bethesda, MD

Terry A. Dick
Department of Zoology
University of Manitoba
Winnipeg, Manitoba
Canada

Bernard Fried
Department of Biology
Lafayette College
Easton, PA USA

A.J. Lymbery
Fish Health Unit
School of Veterinary and Biomedical Sciences
Murdoch, Western Australia 6150
Australia

Haruhiko Maruyama
Division of Parasitology
Department of Infectious Diseases
Faculty of Medicine
University of Miyazaki
Kiyotake 889-1692, Japan

Santiago Mas-Coma
Department of Parasitology
Faculty of Pharmacy
University of Valencia
Burjassot, Valencia
Spain

K.Darwin Murrell
Danish Centre for Experimental Parasitology
Department of Veterinary Pathobiology
Faculty of Life Sciences
Copenhagen University
Denmark

Yukifumi Nawa
Division of Parasitology
Department of Infectious Diseases
Faculty of Medicine
University of Miyazaki
Kiyotake 889-1692, Japan

Chawalit Pairojkul
Department of Pathology
Liver Fluke and Cholangiocarcinoma Research Center
Faculty of Medicine
Khon Kaen University
Khon Kaen, 40002
Thailand

Nevi Parameswaran
WHO Collaborating Centre for the Molecular Epidemiology of Parasitic Infections
School of Veterinary and Biomedical Sciences
Murdoch University, WA 6150, Australia

Paiboon Sithithaworn
Department of Parasitology
Liver Fluke and Cholangiocarcinoma Research Center
Faculty of Medicine
Khon Kaen University
Khon Kaen, 40002
Thailand

Smarn Tesana
Department of Parasitology
Liver Fluke and Cholangiocarcinoma Research Center
Faculty of Medicine
Khon Kaen University
Khon Kaen, 40002
Thailand

RC Andrew Thompson
WHO Collaborating Centre for the Molecular Epidemiology of Parasitic Infections
School of Veterinary and Biomedical Sciences
Murdoch University, WA 6150, Australia

Rebecca J Traub
School of Veterinary Science
University of Queensland
St Lucia, QLD 4072
Australia

Maria Adela Valero
Department of Parasitology
Faculty of Pharmacy
University of Valencia
Burjassot, Valencia
Spain

Wenlin Wang
Department of Parasitology
Kunming Medical College
Kunming, Yunnan 650031
P.R. China

J. Waikagul
Department of Helminthology
Faculty of Tropical Medicine
Mahidol University
Bangkok, Thailand

Puangrat Yongvanit
Department of Biochemistry
Liver Fluke and Cholangiocarcinoma Research Center
Faculty of Medicine
Khon Kaen University
Khon Kaen, 40002
Thailand

I
Fish- and Invertebrate-Borne Parasites

1
Liver Flukes

Paiboon Sithithaworn, Puangrat Yongvanit, Smarn Tesana,
and Chawalit Pairojkul

Food-borne trematodes (FBTs) contribute to parasitic zoonoses such as liver, lung, and intestinal flukes of humans and they are contracted by the consumption of larval stages in food-related products. Food-borne trematodiasis has been recognized as an important health problem in Asia (WHO, 1995, 2004). The infections are prevalent in developing countries and are closely linked to poverty, pollution, and population growth, and are also associated with cultural determinants, that is, food behavior and tradition. Recent changes in FBT infections in some countries are thought to have been brought about by social and economic development (Cross, 1984). In many areas, greater frequency and transmission such as in *Clonorchis sinensis* in China have been reported (Keiser and Utzinger, 2005; Lun et al., 2005). One possible explanation of this phenomenon is an increase in the production of fresh water aquaculture in the endemic areas where FBTs exist in Asia and more importantly a large proportion of world productions (>90%) were farmed in this area (WHO, 2004). This could facilitate the transmission to consumers both in the domestic and overseas markets. Moreover, a serious consequence of clonorchiasis and opisthorchiasis is intrahepatic cholangiocarcinoma (ICC); therefore, in addition to various aspects of the biology of liver flukes, their roles in carcinogenesis are addressed in this chapter.

Geographical Distribution

The major human liver flukes—*Opisthorchis viverrini, Clonorchis sinensis*, and *Opisthorchis felineus*—are endemic in Asia and Eastern Europe. A rough estimate of the global number of infections is about 17 million, comprising 7 million with *C. sinensis*, 9 million with *O. viverrini*, and 1.2 million with *O. felineus* infections (Preuksaraj, 1984; Rim, 1986; WHO, 1995). *O. viverrini* is prevalent mainly in Thailand, Laos, Cambodia, and Vietnam, and recently 67 million people were estimated to be at risk (Keiser and Utzinger, 2005). *C. sinensis* is widespread in Korea, China, Taiwan, Vietnam, and previously in Japan. Recent estimates suggested that about 35 million humans are infected by *C. sinensis* globally, and in China alone there could be up to 15 million human infections (Lun et al., 2005). *O. felineus* is found in Russia and possibly Eastern Europe. In addition to these endemic areas, regional and global migration of people has expanded the parasite's distribution. Since life cycles cannot be established in places where no intermediate hosts are available, this has limited epidemiologic relevance.

The geographical pattern of liver fluke infection is not uniform. For example, in Thailand, *O. viverini* has marked regional variation. It is highly prevalent in the northeast in contrast to other regions of the country (Preuksaraj, 1984; Jongsuksuntigul, 2002). Within the northeast, the high variability occurs at the provincial, district, and village levels; therefore, the average prevalence and intensity of infection are often accompanied by large standard deviations. In Laos, *O. viverrini* was found mainly in southern areas, that is, Saravan, Suvannakhet, and Khammuan (Rim et al., 2003). A similar uneven distribution is true for *C. sinensis* in China, where south China, particularly Guangdong, has long been known as a heavily endemic area; other major areas include Guangxi, Helongjang, Hubei, and Sichun (Lun et al., 2005).

Biology and Genetic Variation

The liver flukes are hermaphroditic trematodes, dorsoventrally flattened, and the body armed with two muscular suckers: the oral sucker situated anteriorly and the ventral sucker one fifth at the mid-body anteriorly. Differentiation of species is based on morphology; the adult worms differ mainly in the shape and position of testes and the arrangement of vitelline glands. *C. sinensis* can be separated from the other two species by the presence of branched testes in a tandem position and the continuously distributed vitelline glands. *O. viverrini* is similar to *O. felineus* in having lobed testes and a cluster vitelline gland, but it differs from *O. felineus* in the deeper lobulation of and greater extremity of the testes and also no transversely compressed patterns of vitelline follicles. The size of adult flukes varies according to the species. *O. viverrini* is the smallest measuring 5.5–10 × 0.77–1.65 mm, while *C. sinensis* is larger, measuring 10–25 × 3–5 mm. *O. felineus* is smaller measuring 7–12 × 2–3 mm (Beaver et al., 1984) (Fig. 1.1). Variation in the size of adults depends on the intensity of infection and the diameters of the bile ducts they inhabit (Flavell et al., 1983).

Based on nuclear ribosomal DNA and mitochondrial DNA sequences, there was little genetic diversity in *C. sinensis* isolates from China and Korea (Lee and Huh, 2004; Park, 2006). The difference in sequence of mitochondrial DNA between *C. sinensis* and *O. viverrini* from Laos was relatively low (3.9%), although they are considered different genera. By application of allozyme electrophoresis, it is possible to use α-glycerophosphate dehydrogenase (GPD) as a genetic marker to differentiate *C. sinensis* isolates from Korea and China (Park et al., 2000). The evidence of genetic polymorphism of *O. viverrini* by using random amplified polymorphic DNA was recently shown among *O. viverrni* isolates from Thailand and Laos (Sithithaworn et al., 2006). The genetic polymorphism evidence was confirmed by using systematic allozyme analyses, and the existence of cryptic species in *O. viverrini* was demonstrated and correlated with the wetlands of river

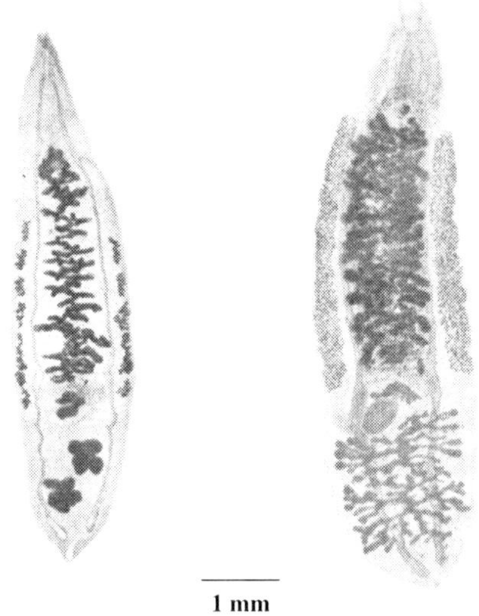

(a) Opisthorchis viverrini (b) *Clonorchis sinensis*

FIGURE 1.1. Adult worms. (A) *O. viverrini*. (B) *C. sinensis*.

systems and *Bithynia* snails in northeast Thailand and Laos (Saijuntha et al., 2006a,b, 2007). The coevolution between host and parasite and the biological significance of this observed genetic variation of *O. viverrini* need further study.

It was reported that *C. sinensis* may survive up to 26 years in a human host (Attwood and Chou, 1978). This observation may represent an extreme survival condition in an individual host. Since there has been no direct estimate of the life expectancy of *O. viverrini*, it is expected that the average survivorship may be much shorter. Based on the pattern of the age intensity profile, it is anticipated that *O. viverrini* may survive in humans for approximately 10 years.

Life Cycle

After reaching sexual maturity, the adult worms cross-fertilize and produce yellow-brown egg that are released in feces. The eggs measure 25 to 35 μm in length and 15 to 17 μm in width; all three species of liver fluke eggs have prominent shoulders, with an operculum at one end and a small knob at the other. The egg shell surface is rough or seen as musk-melon patterns by electron microscopy (Kaewkes et al., 1991; Scholz et al., 1992) (Fig. 1.2). Eggs of the three species of liver flukes are similar and difficult to differentiate. The number of eggs produced per worm depends on the worm burden, that is, a density-dependent effect.

When eggs reach a body of freshwater (small ponds, streams and rivers, flooded rice fields, and large reservoirs) and are ingested by an appropriate snail,

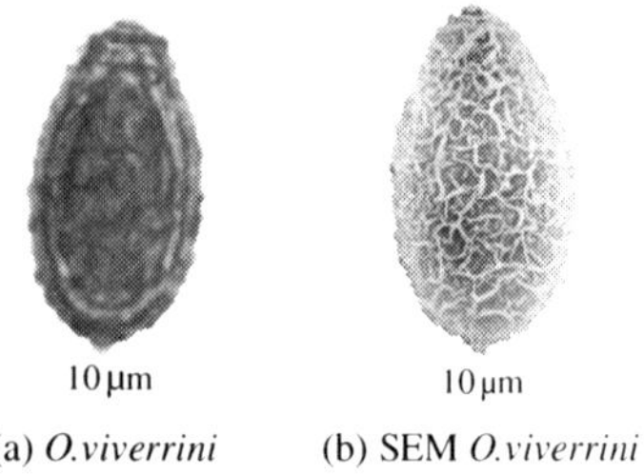

FIGURE 1.2. Eggs of liver flukes. (A) Light micrograph. (B) Scanning electron micrograph.

they develop into sporocysts. Different species of *Bithynia* (Table 1.1 and Fig. 1.3) serve as intermediate hosts for these liver flukes (Brockelman et al., 1986; Ditrich et al., 1990b; Chen et al., 1994). The sporocysts produce rediae, which gave rise to cercariae, which are released daily approximately 2 months after snail infection. The free-swimming cercariae shed their tails and penetrate into the tissues of fish and encyst, becoming a fully infective metacercariae after 21 days. The life cycle requires at least 4 weeks to complete, and in temperate areas the cycle may be prolonged by winter hibernation of snails (Rim, 1986).

TABLE 1.1. Snail intermediate hosts of liver flukes.

	Region, country	References
Opisthorchis viverrini		
Bithynia siamensis siamensis	Central Thailand	Brandt,1974; Upatham and Sukhapanth, 1980; Chitramvong, 1992
B. siamensis goniomphalos	Northeast Thailand, Laos	Brandt,1974; Brockelman et al., 1986; Ditrich et al., 1990; Giboda et al., 1991a; Chitramvong,1992
B. funiculata	North Thailand	Brandt, 1974; Chitramvong, 1992
Clonorchis sinensis		
Parafossarulus manchouricus	China; Han, Kum, Mankyung, Naktong, and Yeongsan, South Korea	Bae et al., 1983; Choi, 1984
P. anomalospiralis	China	Li, 1985
Bulinus striatulus	Han, Kum, Mankyung, Naktong, and Yeongsan, Korea; China; Taiwan	Li, 1964; Choi, 1984; Chen et al., 1997
Bithynia fuchsiana	China, Taiwan	Zhou, 1985
Alocinma longicornis	Hubei, China	Li, 1964; Zou et al., 1994; Chen et al., 1997
O. felineus		
Bithynia inflata	Ukraine	Beer and German, 1987
B. tentaculata	East and central Europe, Middle East, India	Bowman, 1999

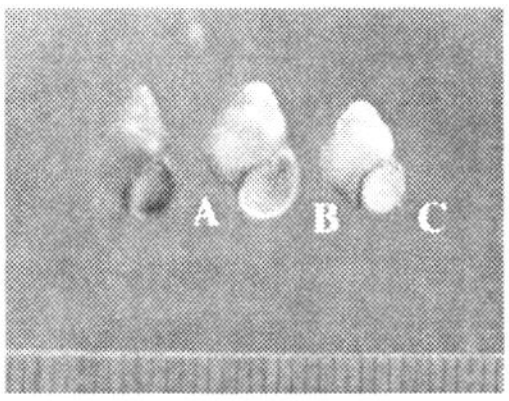

(a) Photographs of Bithynia snails A: *B. siamensis goniomphalos*, B: *B. funiculata*, C: *B. siamensis siamensis* (scale= 1 mm)

(b) Photographs of *Parafossarulus manchouricus* (scale= 1 mm)

FIGURE 1.3. Snail intermediate hosts. (A) *Bithynia*. Left: *B. siamensis goniomphalos*. Middle: *B. funiculata*. Right: *B. siamensis siamensis*. Scale = 1 mm. (B) *Parafossarulus manchouricus*. Scale = 1 mm.

Similar to other trematodes, the prevalence of liver fluke infection in snail intermediate hosts is typically low and a range of 0.07% to 0.63% for *O. viverrini* in *Bithynia* sp. and 0.03% to 43.7% for *C. sinensis* in *P. manchouricus* were reported (Joo, 1980; Harinasuta and Harinasuta, 1984; Brockelman et al., 1986). In temperate countries such as Korea, it is believe that cercariae and rediae within the snail are unable to survive during winter, and new infections are needed to initiate cercarial shedding in the spring (Rim, 1986). The snail population shows strong seasonality and is apparently dependent on rainfall, being highly abundant in the rainy season and distributed extensively in shallow water and paddy fields, but disappear rapidly in the dry season (Brockelman et al., 1986). Due to the nature of such high fluctuation in number, liver fluke control via the reduction of snail population is not a feasible approach.

In contrast to the infection in snail, the prevalence of infection in the fish intermediate host is much higher. As many as 90% to 95% of several species of cyprinoid fish had been reported to habor *O. viverrini* metacercariae (Harinasuta and Vajrasthira, 1960; Vichasri et al., 1982). The most common species of cyprinoid fish were in the genera *Puntius, Cyclocheilichthys*, and *Hampala* (Wykoff et al., 1966) (Table 1.2 and Fig. 1.4). In spite of complex cercarial host finding mechanisms (Haas et al., 1990), free swimming cercariae are very efficient in locating the appropriate species of fish in reservoirs with large volumes of water. The intensity of liver fluke infections in fish varies by season, species, individuals, and types of water bodies (Vichasri et al., 1982; Rim, 1986; Sithithaworn et al., 1997). Most metacercariae were found in the body (Vichasri et al., 1982) or head of the fish (Tesana et al., 1985) (Fig. 1.5). The observed discrepancy probably depends

TABLE 1.2. Fish intermediate hosts of liver flukes.

	Areas/country	References
Opisthorchis viverrini		
Cyclocheilicthys apogon	Northeast, Thailand	Vichasri et al., 1982
Cyclocheilichthys siaja	Northeast, Thailand	Harinasuta and Harinasuta, 1984
Cyclocheilichthys repasson	Laos	Ditrich et al., 1990
Puntius leiacanthus	Northeast, Thailand	Vichasri et al., 1982; Sithithaworn et al., 1997
P. orphoides	North, Thailand	Sukontason et al., 1999
Puntius gonionotus,	Laos	Ditrich et al., 1990
Hampala dispar	Northeast, Thailand; Laos	Ditrich et al., 1990; Sithithaworn et al., 1997
Hampala macrolepidota	Laos	Ditrich et al., 1990
Clonorchis sinensis		
Parabramis bransula	Guangdong, China	Zou et al., 1994
Ctenopharyngodon idellus	Guangdong, Hubei, China	Zou et al., 1994
Ctenopharyngodon idellus	Hubei, China	Chen et al., 1997
Pseudorabora parva	A river in Kyungpook Province, Nam-river, Han, Kum, Mankyung, Naktong, and Yeongsan rivers, Korea; Hubei, China	Choi, 1976; Bae et al., 1983; Choi, 1984; Chen et al., 1997
Zacco platypus	Nam-river, Korea	Bae et al., 1983
Hemibarbus sp.	Nam-river, Korea	Bae et al., 1983
Gnathopogon sp.	Nam-river, Korea	Bae et al., 1983
Ischikauia steenackeri	Nam-river, Korea	Bae et al., 1983
Pseudogobio esocinus	A river in Kyungpook Province, Nam-river, Korea	Choi, 1976; Bae et al., 1983
Hypophthalmichthys nobilis	Hubei, China	Chen et al., 1997
Sarcocheilichthys sinensis	A river in Kyungpook Province,	Choi, 1976
Hemibarbus labeo	Han, Kum, Mankyung, Naktong, and Yeongsan rivers, a river in Kyungpook Province, Korea	Choi, 1976; Choi, 1984
Pungtungia herzi	A river in Kyungpook Province,	Choi, 1976
Erythroculter erythropterus	A river in Kyungpook Province,	Choi, 1976
Gnathopogon atromaculatus	Han, Kum, Mankyung, Naktong, and Yeongsan, Korea;	Choi, 1984;
Hemiculter leucisculus	Sun Moon lake in Central Taiwan	Ooi et al., 1997
Zacco platyus	Soyang lake, Korea	Park et al, 2004
Hypomesus olidus	Soyang lake, Korea Daechung lake, Korea	Park et al, 2004
O. felineus		
Leuciscus idus	Western Siberian region, Ukraine and Kazahstan.	Bronstein, 1998
Leuciscus leuciscus	Western Siberian region, Ukraine and Kazahstan.	Bronstein, 1998
Rutilus rutilus	Western Siberian region, Ukraine and Kazahstan.	Bronstein, 1998

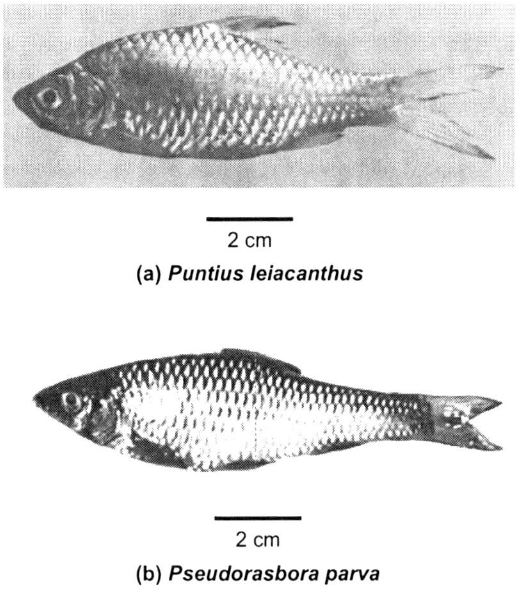

FIGURE 1.4. Fish intermediate hosts. (A) *Puntius leiacanthus*. (B) *Pseudorasbora parva*.

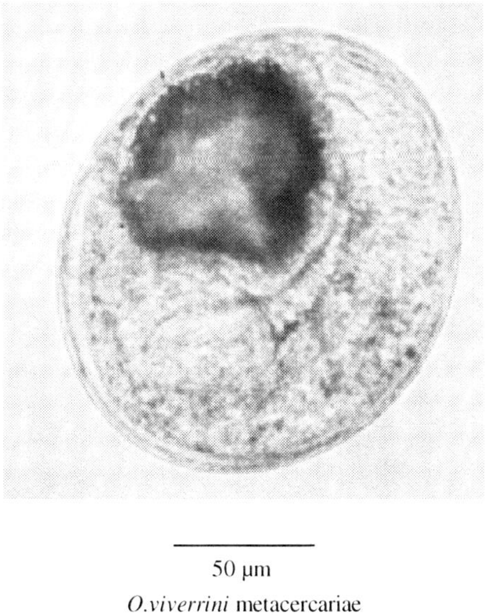

FIGURE 1.5. Metacercaria of *O. viverrini*.

on the technique used for metacercarial extraction; the pepsin digestion technique is believed to be the most reliable (WHO 1995; Sithithaworn et al., 1997; Waikagul 1998). The metacercarial burden peaks in winter (around October to February), and becomes low in the rainy season and summer (Vichasri et al., 1982; Sithithaworn et al., 1997); for *C. sinensis* the peaks occur in spring and summer (Kang et al., 1985). The seasonal patterns of prevalence and intensity of infection in fish intermediate hosts for *O. viverrini* and *C. sinensis* are shown in Figure 1.6.

The number of metacercariae reported in fish generally range from 1 to hundreds. However, large numbers have been reported in some studies; for example, over 30,000 per fish and >6000 per gram were found in *Pseudorasbora parva* in both China and Korea (Kang et al., 1985; Rim, 1986; Chen et al., 1994).

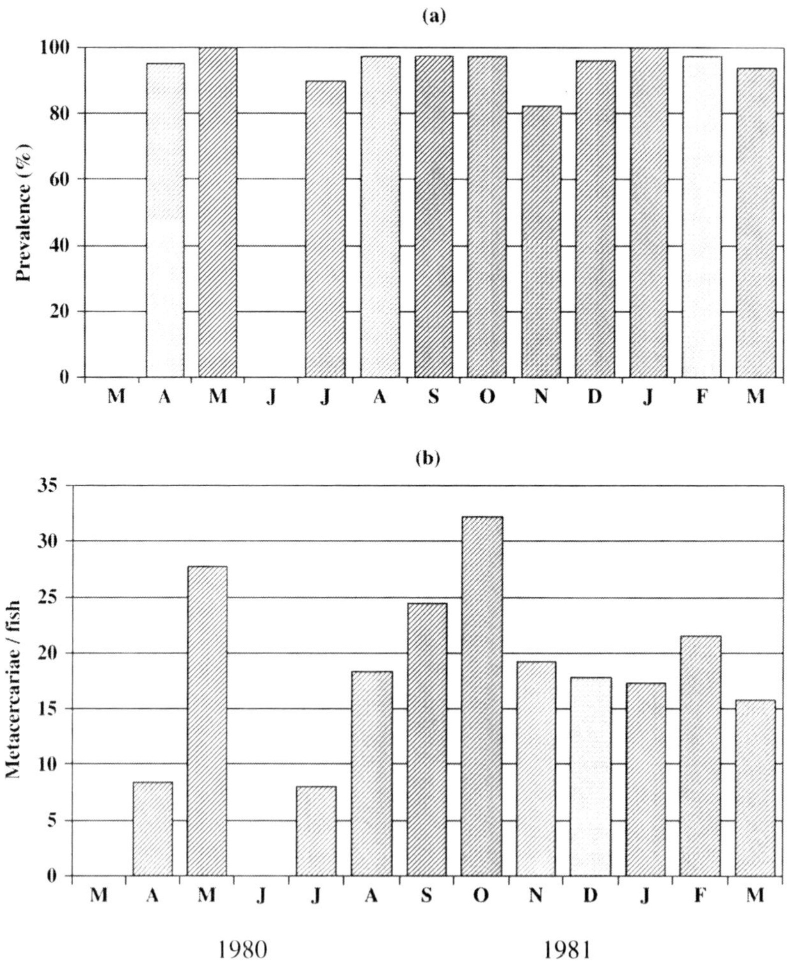

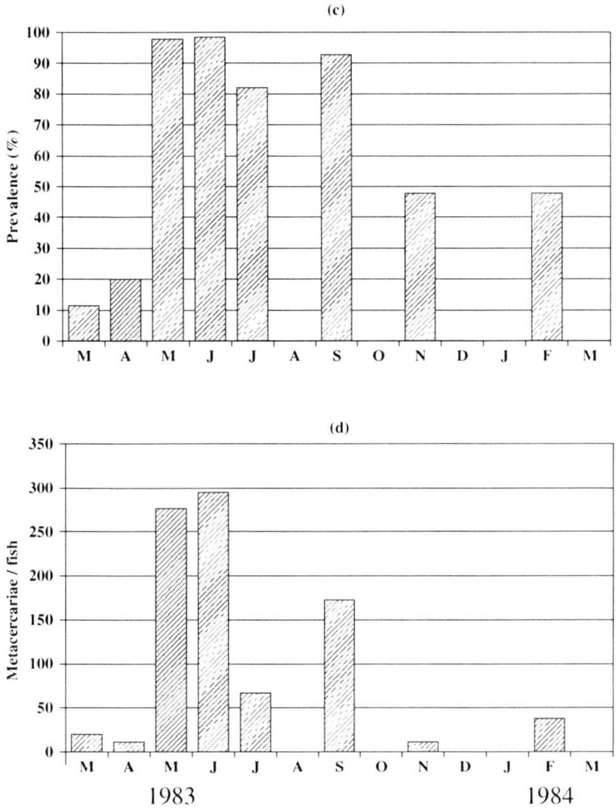

FIGURE 1.6. Seasonal variation of liver fluke metacerariae burden in cyprinid fish from natural water bodies. (A) Prevalence; (B) intensity of *O. vierrini* metacercariae in *P. leiacanthus* in Thailand (Vichasri et al., 1982). (C) Prevalence; (D) intensity of *C. sinensis* metacercariae in *P. parva* in Korea (Kang et al., 1985).

The pattern of frequency distributions of metacercariae for both *O. viverrini* and *C. sinensis* are not uniform; most fish have few metacercariae, while a few fish harbor a heavy metacercarial load as shown in Figure 1.7 (Kim et al., 1979; Vichasri et al., 1982; Kang et al., 1985). The pattern of frequency distribution is best described by a negative binomial distribution model. Recent findings of trematode metacercariae other than *O. viverrini* in cyprinoid fish in the northeast (Srisawangwong et al., 1997; Waikagul, 1998) and northern Thailand (Sukontason et al., 1999) indicated that the occurrence of mixed species of trematodes in a given fish species is common. Careful identification of metacercariae from fish is needed.

The frequency of infection in reservoir hosts, for example, pigs, cats, rats, and dogs, varies considerably by area and is not closely associated with human

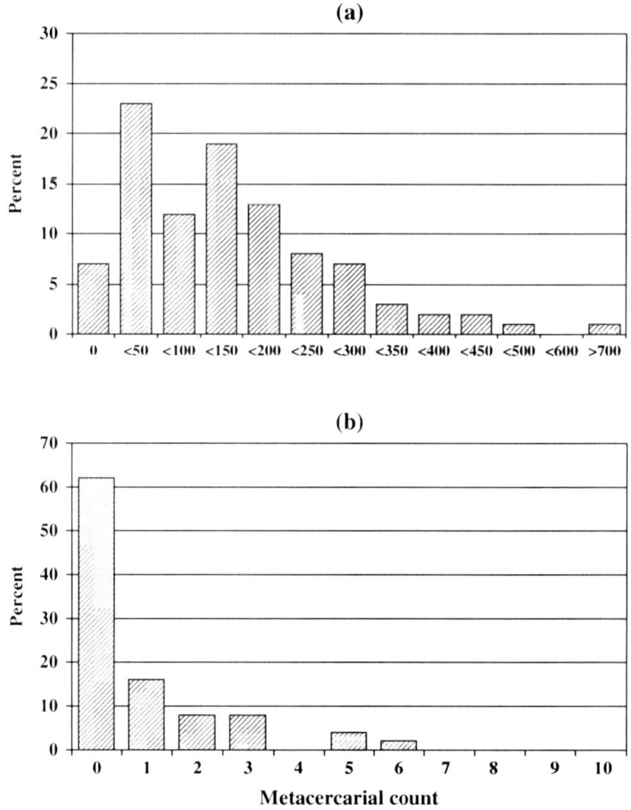

FIGURE 1.7. Frequency distribution of liver fluke metacercariae in cyprinoid fish. (A) Metacerariae of *C. sinensis* in *P. parva* in Korea (Kang et al., 1985). (B) Metacercaria of *O. vierrini* in *P. leiacanthus* in Thailand (Sithithaworn, unpublished).

infection patterns. Relatively low prevalence of liver fluke infection are observed in dogs in northeast Thailand, Taiwan, and some parts of Korea where human infection is common; similar or higher prevalence have been reported in these animals in parts of Thailand and China, even in areas where humans are not infected (Sadun, 1955; Chen et al., 1994). Although fecal contamination from infected animals undoubtedly contributes to parasite transmission to snails, consideration of its actual importance relative to higher egg release from humans, human eating behavior, and sanitation may be minor (Sadun, 1955; Rim, 1986). However, in a situation where fecal contamination from humans is eliminated by mass treatment and proper sanitation, the consumption of raw fish continues and infection among reservoir hosts may serve to maintain the source of reinfection.

Source of Human Infection

Fresh water fish are second intermediate hosts in the life cycle of *O. viverrini*. In Thailand, at least 15 species of native fish serve as intermediate hosts and sources of human infection (WHO, 1995). Although most published descriptions of social habits regarding raw fish consumption are anecdotal and careful sociological investigation is needed, it is obvious that raw or undercooked fish are the major sources of liver fluke infection. Raw or undercooked fish are prepared in many different ways, and these dishes are of considerable cultural and nutritional significance, making change in food habits difficult. Infection may also occur through contamination of utensils, hands, and surfaces used to prepare the fish for cooking.

In northeast Thailand, at least three types of preparations contain uncooked, usually small and medium-sized fish, namely *koi pla*, eaten soon after preparation, moderately fermented (*pla som*; stored for a few days to weeks), or extensively fermented (*pla ra;* highly salted, stored for 2 to 3 months to over 1 year) (Sadun, 1955). *Pla ra* is consumed in different forms on a daily basis. In the past, reported consumption frequencies of *koi pla* were very high; up to 80% in some communities ate the dish on a weekly basis (Changbumrung et al., 1989). A comparison of rural and urban dwellers (Kurathong et al., 1987) reported a higher prevalence of liver fluke infection among rural, compared to urban, residents from the northeast, and among those who reported having eaten *koi pla* (87%), compared to those who did not (61%). A closer relationship was found with *koi pla* consumption within a heavily infected village, with just 19% of those uninfected reporting consumption, compared to 79% of infected and >90% of heavily infected people (Upatham et al., 1984).

More recent surveys suggest that the frequencies of *koi pla* consumption are reduced and generally confined to special social occasions, while other under cooked fish preparations, for example, *pla som* and other moderately preserved fish, are generally eaten several times a week (Changbumrung et al., 1989). Although most people believe that liver fluke infection comes from these dishes, the infectivity of various preparations remains unclear. Several studies have indicated that survival of the infective stages depends on the concentration of salt and degree of fermentation (Vichasri et al., 1982; Tesana et al., 1983). *Koi pla* is probably the most infective, followed by fish preserved for less than 7 days, while viable metacercariae are probably very rare in *pla ra*.

In southern China and Korea, *C. sinensis* may be acquired by consuming whole raw fish directly or slices of raw fish marinated with vinegar, soya sauce, or chili or prepared as pickle fish or raw fish in hot bean paste (Rim, 1986; Chen et al., 1994).

It is important to note that small fish such as *P. parva* and *P. leiacanthus* contained more metacercariae of liver flukes per unit weight than did large fish (Chen et al., 1994; Sithithaworn et al., 1997). Because of the highly overdispersed distribution of metacercariae in fish, the probability of exposure to heavy infection from a single fish meal should be rare. Thus, infection should occur in a repeated trickle manner of infection, and the heavily infected people accumulate worms over time (Sithithaworn and Haswell-Elkins, 2003).

Fecundity

Approximately 1 month after ingestion of metacerariae, adult worms begin releasing eggs, which pass down the bile duct and are released in the feces. Egg can also be found in the gallbladder bile. In *O. viverrini*, estimates of daily fecal egg output per worm are variable. Wykoff et al. (1966) described nine autopsy cases with a mean worm burden of 2588 with an average egg output of 3000 egg per gram feces (epg). Sithithaworn et al. (1991a) performed 181 autopsy cases and found a mean worm burden of 157.5 and an average egg output of 53.3 epg per worm. By expulsion chemotherapy, Elkins et al. (1991) reported the average egg output of 180 epg per worm from a mean worm burden of 42.6. The observed wide variation in estimates between these studies in human infection may be partially explained by a density-dependent fecundity similar to those documented for other helminths (Anderson and May, 1985). The estimated fecundity in infected animals also varied depending on the worm burden, but was in the range of 80 to 300 epg per worm (Flavell et al., 1983; Sripa and Kaewkes, 2000b).

Prevalence of Infection

The most recent national survey by the Ministry of Public Health of Thailand in 2001 showed that of a total helminth infection of 22.5%, hookworm is the most common (11.4%), while *O. viverrini* is second in rank with an average prevalence of 9.6% (Jongsuksuntigul, 2002). Regarding the liver fluke, it is distributed mainly in the north (19.3%) and the northeast (15.7%), and has a low distribution in the central reglon (3.8%) and the south (0%). The decline in the prevalence of infection in northeast Thailand of 34.6% in 1981 (Jongsuksuntigul et al., 1992) to the current level in 2001 is to a large degree attributable to intensive and continuous control activities (Jongsuksuntigul, 2002). The relatively high prevalence observed in northern Thailand probably represents a mixture of small-sized intestinal flukes as well as *O. viverrini* due to the limitation of the diagnostic technique used (Kato-thick smear technique) (Radomyos et al., 1994, 1998). Within northeast Thailand, a high variation of prevalence among provinces, ranged from 4% to 33%, and is a common pattern of infection (Jongsuksuntigul, 2002).

O. viverrini is common in the lowlands of Laos among people with close ethnic ties to the majority of the northeast Thai population; however, national surveys are not available to provide the total numbers of infections (Giboda et al., 1991b). More recent surveys revealed that the prevalence in certain areas, in the range of 36% to 60%, is much higher than previous records indicate (Kobayashi et al., 1996, 2000). Again the presence of mixed infections with both heterophyid and lecithodendriid flukes, in addition to *O. viverrini,* were demonstrated in Vientiane and Saravan along the Mekong River (Chai et al., 2005).

Incidence and Reinfection

In contrast to recorded data on the prevalence of infection, data on the incidence of *O. viverrini* infections are not frequently reported. The available data are reported by a few investigators (Sornmani et al., 1984; Upatham et al., 1988; Saowakontha et al., 1993). In an endemic community in Chonnabot, Khon Kaen, northeast Thailand, the incidence of infection per year was 19.4% to 46% and in children <5 year old, the incidence was 2.1% to 6.2%; males tended to have a higher incidence of infection than females (Upatham et al., 1988). In a more recent study in three villages in Khon Kaen, the incidence was 1.7% to 25% for 6 months (Saowakontha et al., 1993). The very high incidence of infection in some communities explains why a high prevalence of infection was observed. For example, with an incidence of 40% per year, only 6 years are required for the prevalence of an originally uninfected cohort to exceed 95%.

Studies on reinfection posttreatment in northeast Thailand are well correlated to the high incidence of infection. A study in irrigated areas in Khon Kaen revealed that with a pretreatment prevalence of *O. viverrini* of 55.1%, at 1 year after treatment the prevalence returned to 54.8% (Sornmani et al., 1984). Similarly, Upatham et al., (1988) reported that in an area with heavy infection of *O. viverrini* in Chonnabot, Khon Kaen, where 97.4% of villagers were infected, the prevalence was 94% at 1 year post–praziquantel treatment. In addition, those with high pretreatment intensity tended to have a high intensity of reinfection, suggesting that there may be a predisposition to heavy infection in some individuals. Studies using a similar approach based on the quantification of worm load at pre- and posttreatment, also provided supportive evidence of a predisposition to heavy infection with other parasites such as *Ascaris lumbricoides* (Elkins et al., 1986), *Necator americanus* (Schad and Anderson, 1985), *Trichuris trichiura* (Bundy and Golden, 1987), and *Schistosoma mansoni* (Bented-Smith et al., 1987). The rapid reinfection after treatment in opisthorchiasis indicated little evidence for protective immunity, but did not rule it out.

Age- and Sex-Related Patterns of Infection

Although the rates of *O. viverrini* infection vary considerably between villages, patterns of infection are similar. In general, the youngest age groups (0–5 years) have a low prevalence and intensity. Prevalence and intensity of infection increase in the pre- and early teens, often reaching a plateau in the late teenage years (e.g., 15–19). In some areas, the intensity of egg release showed an increased trend with age (Upatham et al., 1984), while the worm burden declined (Haswell-Elkins et al., 1991; Sithithaworn et al., 1991b). Possible reasons suggested for this decline include late-developing immune responses, lower parasite survival in more heavily fibrosed bile ducts, death of parasite in heavily infected people, or reduced exposure to infection in elderly groups (Fig. 1.8).

Mothers feeding raw fish to their infants are a possible source of infection, as young infants have been observed infected (Sadun, 1955; Upatham et al., 1982, 1984). However, the reported intensities of infection under age 4 are invariably low, and there is little evidence that young children frequently experience intensive exposure to infection. To date, this practice is rare.

In typical endemic communities, the prevalence and average intensity of *O. viverrini* infection usually does not differ, or is slightly higher, among males compared to females (Wykoff et al., 1965; Upatham et al., 1982, 1984; Haswell-Elkins et al., 1991). However, higher frequencies of heavy infections may be more frequent among males than females (Wykoff et al., 1965; Upatham et al., 1982, 1984; Haswell-Elkins et al., 1991). Since the risk of clinical manifestations,

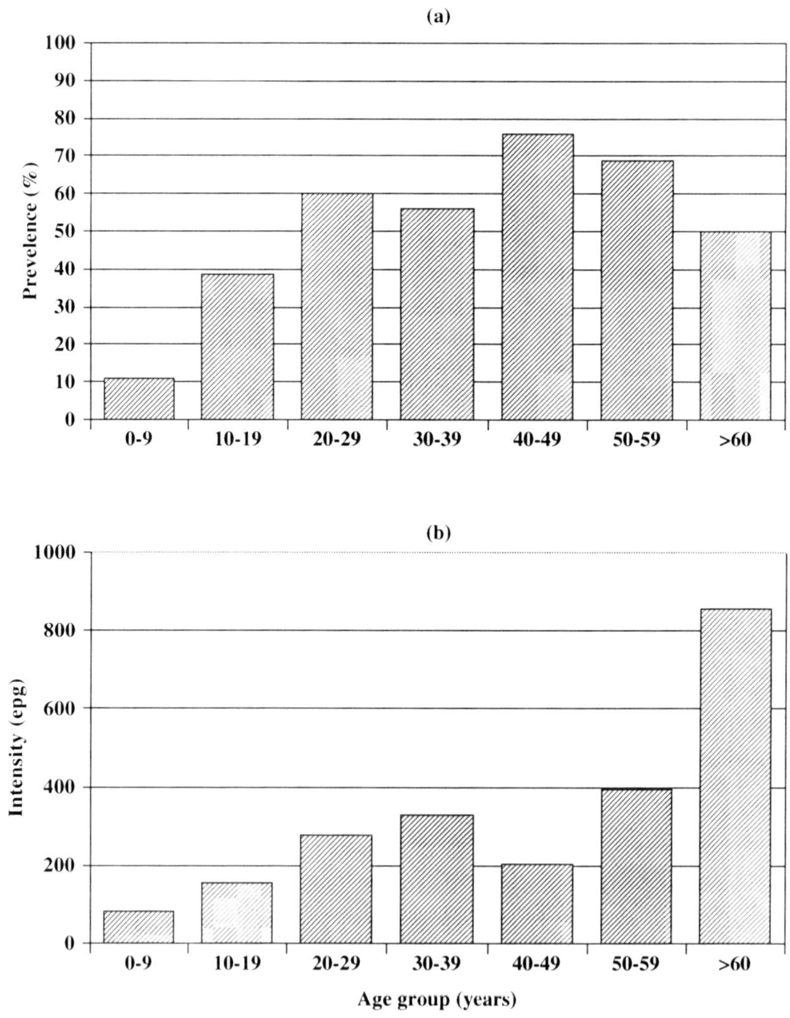

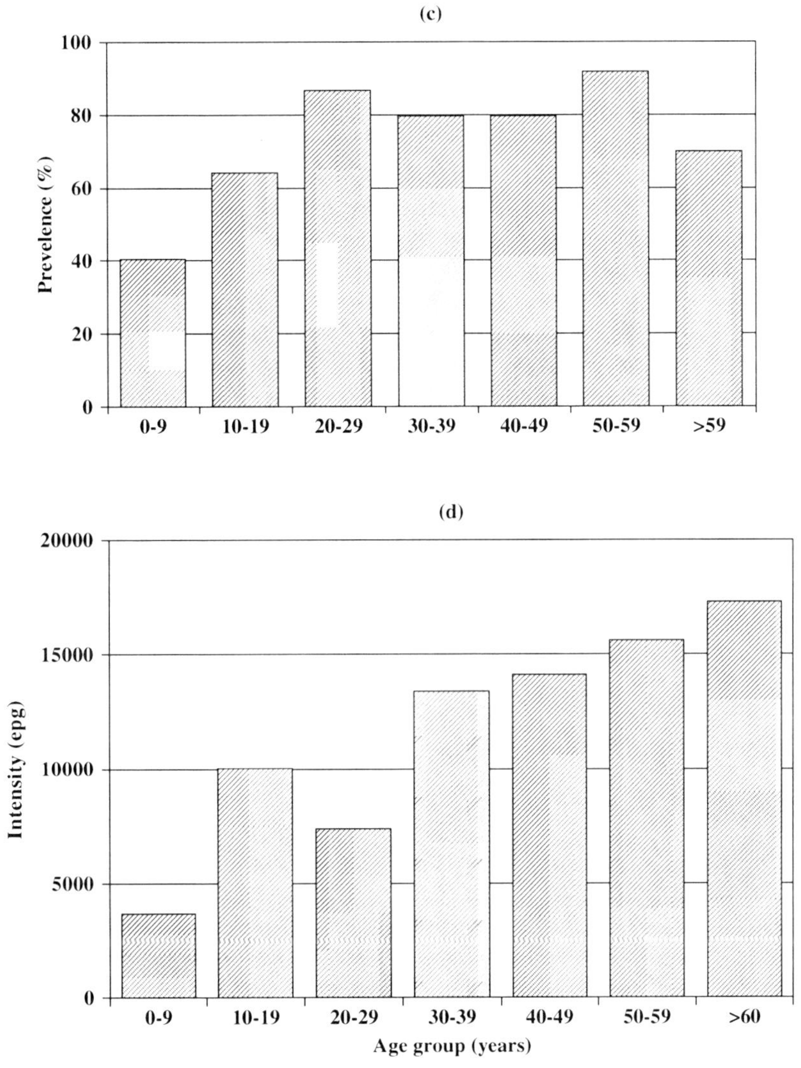

FIGURE 1.8. Age-prevalence and intensity of liver fluke infection in endemic areas. (A) Prevalence; (B) intensity of *O. viverrini* in Laos (Sithithaworn et al, 2006). (C) Prevalence; (D) intensity of *C. sinensis* in Korea (Song et al., 1982).

including cancer, may increase in a nonlinear fashion with infection, the sex of the host may be one of the important disease determinants (Haswell-Elkins et al., 1994a; Elkins et al., 1996).

The age-prevalence profiles in clonorchiasis are similar to that of opisthorchiasis; as the infection rate increases with age, the infection rate in the group of 60 years or a higher in age is greater than in the younger aged groups (Rim, 1986).

Frequency Distribution in Humans

As for other helminths, the population infected with *O. viverrini*, and probably with all three liver flukes, is highly aggregated within a small minority of the heavily infected people. Ramsay et al. (1989) performed expulsion chemotherapy in 33 villagers in Khon Kaen, northeast Thailand, and found that the frequency distribution of *O. viverrini* is highly clumped in a small group of subjects. The highest worm load was 565 and the mean number of flukes was 85 [standard deviation (SD) = 154]. Haswell-Elkins et al. (1991), using a larger sample size, observed that 81% of 11,000 worms recovered after treatment of 246 village residents were expelled by just 25 individuals (10% of the sample population), with burdens of over 100 worms. In addition, no worm was seen in some egg-positive individuals when using expulsion chemotherapy. In an autopsy study in Khon Kaen in which the worm burden was accurately measured, Sithithaworn et al. (1991a) reported that 30 of a total of 181 cadavers examined contained 66% of all the worms recovered at autopsy and only 13 people (7%) had worm burdens greater than 400 (Fig. 1.9).

Similar to opisthorchiasis, patterns of distribution of intensity of infection (epg) in clonorchiasis are overdispersed. The negatives and light intensity class (epg <1000) consist of many people but only relatively few cases with heavy infections (1.5–2.4%) (Seo et al., 1981).

Pathology and Pathogenesis

Humans

The pathological changes in liver fluke infection are confined mainly to the bile duct, liver, and gallbladder in both human and animal models. The magnitude of the pathology depends on the intensity, duration, and susceptibility of the host. In light infections, the liver appears grossly normal. In massive infections, a localized dilation of the thickened peripheral bile ducts can be seen on the surface beneath the fibrotic capsule of the liver (Glisson's capsule) (Rim, 2005). When the infection is well established, all the large and medium-size bile duct are prominent on the cut surface (Fig. 1.10).

A classical histopathological change in clonorchiasis was described by Hou (1955), which is also applicable in opisthorchiasis (Bhamarapravati et al., 1978). In the early stage of infection, the biliary epithelium frequently becomes edematous, and desquamation may be seen in the areas of tissue in close proximity to the flukes. Periductal infiltrates of mononuclear cells are frequently found; however, inflammation of the bile duct walls is generally only slight in uncomplicated cases. Metaplasia of the biliary epithelial cells into mucin-producing cells (goblet cells) occurs during early infection, and these cells may proliferate to produce many small gland-like structures in the mucosa (adenomatous hyperplasia), leading to a persistent and excessively high mucus content in the bile. Chronic and

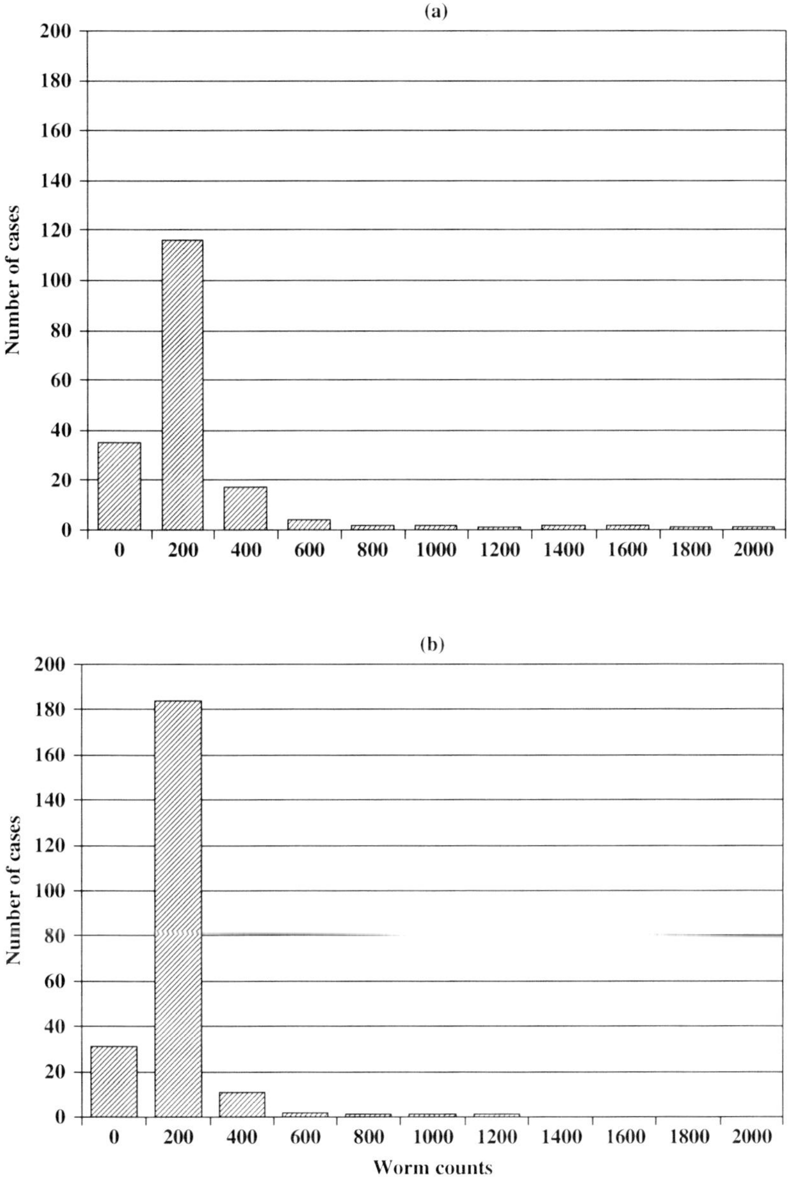

FIGURE 1.9. Frequency distribution of adult *O. viverrini* in humans in endemic communities in Thailand. (A) Worm recovered from autopsy subjects (n = 181) with the mean of 157.5 (data from Sithithaworn et al., 1991a). (B) Worm recovery from expulsion chemotherapy with praziquantel, with the mean of 44.8 (data from Elkins et al., 1991).

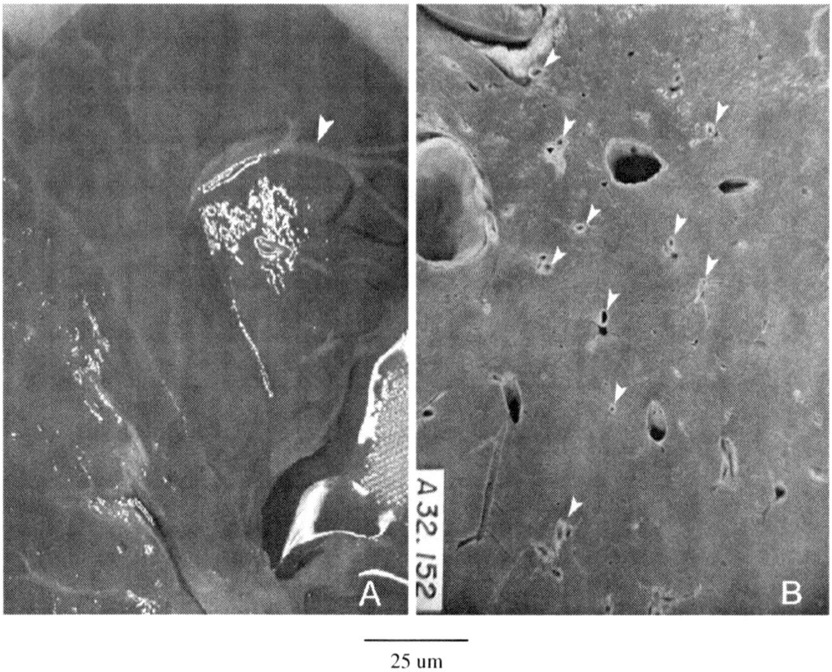

FIGURE 1.10. Gross morphology of the liver with chronic *O. viverrini* infection. (A) Prominent thickening of subcapsular bile duct (arrow). (B) Diffuse fibrotic intrahepatic bile duct wall (arrows).

persistent infections result in a gradual increase in the amount of fibrous tissues, which may eventually engulf some of the proliferating glands, giving the appearance of cholangiofibrosis. As this fibrosis proceeds, the epithelial proliferation becomes milder. In such chronic cases, fecal egg counts may drop markedly. These histopathological changes are distinctive features of clonorchiasis; therefore, when proliferation of the ductal epithelium with metaplastic cells and periductal fibrosis are observed in patients in an area of endemic infection, they are highly suggestive of liver fluke infection on histological grounds (Hou, 1955; Gibson and Sun, 1971).

In opisthorchiasis, during early infections, there was no epithelial hyperplasia or fibrous proliferation. In chronic infections, there was proliferation of epithelial cells with formation of glandular acini, similar to the adenomatous changes in clonorchiasis, and there were varying degrees of periductal fibrosis (Fig. 1.11). The major microscopic changes are confined to the large and medium-sized bile duct where the flukes live. The gross and microscopic characteristics of human opisthorchiasis are well established within 7 to 15 years after *O. viverrini* infection. In chronic and heavy infections, various degrees of cellular infiltration are caused by superimposed bacterial infection. This may result in suppurative

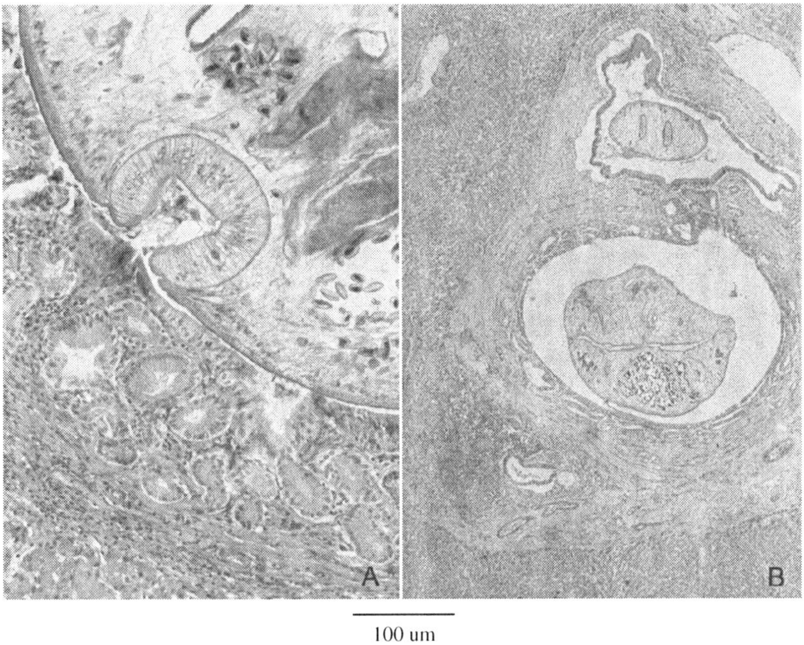

FIGURE 1.11. Histopathology of the bile duct in chronic liver fluke infection. (A) Epithelial and adenomatous hyperplasia of the intrahepatic bile duct. The worm uses its ventral sucker to attach to the surface epithelium and inflammatory reaction at injury site and periductal tissue. (B) Periductal fibrosis and focal mononuclear cell infiltration and cross section of worms in dilated ducts.

cholangitis, and the infection may extend into parenchyma of the liver tissue, causing cholangiohepatitis with abscess formation (Fig. 1.12). In heavy infections with *O. viverrini*, adult parasites are always found in the gallbladder, the common bile duct, and the pancreatic duct. In the large and medium-sized bile ducts, the parasites give rise to chronic cholecystitis. When there is superimposed bacterial infection, empyema of the gallbladder may result. No stone formation was seen, however, either in the bile ducts or in the gallbladder in one series of 70 cases at autopsy (Tansurat, 1971) or in another series of 154 cases (Sonakul et al., 1978). This finding is in contrast to that seen in clonorchiasis, in which cholelithiasis is one of the most serious complications (Rim, 1986). These complications are the result of biliary obstruction. Parasite-induced mucin-secreting cells produce bile with a high mucin content, which, combined with adult flukes and eggs, serves as a nidus for bacterial superinfection and intrahepatic stone formation (Riganti et al., 1988; Sripa et al., 2004). The mucin-rich bile and the presence of worms and eggs in the bile duct cause cholestasis and provide a favorable environment for secondary bacterial infection, which are of enteric origin, with *Escherichia coli* being identified most frequently as a pathogen. The dilatation of intrahepatic bile

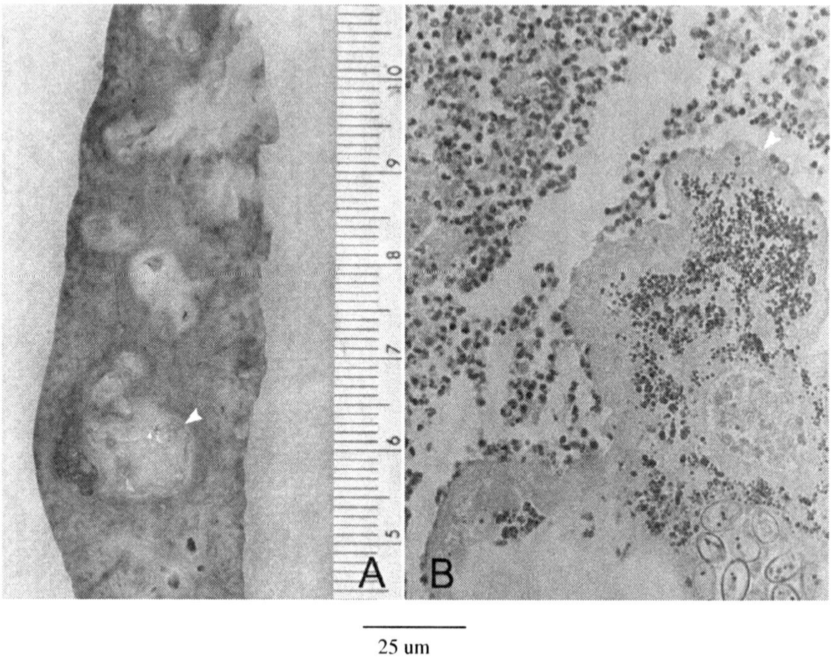

FIGURE 1.12. Acalculus suppurative cholangitis in opisthorchiasis. (A) Cut surface of liver segment reveals thickening bile ducts with necrotic debris in the lumen (upper part) and a cholangitic abscess (arrow). (B) Histology of abscess showing acute inflammatory exudates and dead worm (arrow).

ducts may progress to pyogenic cholangitis, liver abscess, and hepatitis (Sun, 1984; Rim, 2005). Enlargement of the gallbladder is commonly found in opisthorchiasis both at autopsy (Riganti et al., 1989) and during ultrasonographic studies in patients with *O. viverrini* infection (Dhiensiri et al., 1984; Elkins et al., 1990; Mairiang et al., 1992).

Community-based studies in northeast Thailand, using ultrasonography and cholecystography, demonstrated a dramatic increase in the frequency and severity of gallbladder disease, specifically wall irregularity, enlargement, bile sludge, and impaired function among apparently healthy individuals with moderate to heavy *O. viverrini* infections (Haswell-Elkins et al., 1991; Mairiang et al., 1992; Elkins et al., 1996). Following anthelmintic treatment, most of the gallbladder abnormalities were eliminated, as indicated by the reduction of gallbladder size and its regained contractility (Mairiang et al., 1993). Cholelithiasis is not particularly frequent in opisthorchiasis; however, biliary sludge is often seen in the gallbladder in heavy *O. viverrini* infections (Elkins et al., 1990; Mairiang et al., 1992). Granulomatous reactions around the entrapped parasite eggs are only occasionally seen in the gallbladder wall (Viranuvatti and Stitnimarnkarn, 1972).

A radiological finding in heavy infections with *O. viverrini* and early intrahepatic cholangiocarcinoma (ICC) patients that may be associated with pathogenesis is a frequent enhancement of portal vein radicle echoes (Haswell-Elkins et al., 1991; Mairiang et al., 1992). Although the origin is unclear, these echoes may represent fibrotic change resulting from chronic inflammation along the small intrahepatic bile ducts where the flukes reside, as is sometimes observed in the histopathology of liver specimens (Pairojkul et al., 1991).

Clinical Manifestation

The frequency and types of clinical disease seem to differ between the three human liver flukes. Most notably, there were many reports in the Russian literature detailing specific signs and symptoms accompanying well-defined clinical stages of opisthorchiasis, from acute to chronic (Bronshtein, 1986). Acute infection, characterized by high fever, hepatitis-like symptoms, and eosinophilia, is frequently reported for *O. felinius*, but has been documented only once for clonorchiasis (Bronshtein, 1986) and never for *O. viverrini*. This may be due to the large number of migrants into the endemic area of *O. felinius* who become infected as adults, which is quite unusual in the other two liver fluke infections.

Ascending cholangitis and obstructive jaundice are frequently listed as common complications of opisthorchiasis. Pungpak et al. (1989), however, reported only 88 cases with severe manifestations among 15,243 infected people who presented at a hospital in Bangkok. Of these, 41 had obstructive jaundice, 26 had cholangitis, at least 16 had cholangiocarcinoma, and six had other cancers. Since radiological investigations were not performed, cholangiocarcinoma as a cause of these manifestations could not be ruled out.

Two large studies by Upatham et al. (1982, 1984) and Bronshtein (1986), within a heavily infected community, reported significantly increased frequencies of abdominal pain in the upper right quadrant, flatulence or dyspepsia, and weakness associated with increasing intensity of infection. They estimated that 5% to 10% of the population had symptoms attributable to the infection. The findings of this classic study would be better supported by ruling out the presence of cholangiocarcinoma among those with symptoms, controlling for age and sex, and performing posttreatment examinations to demonstrate recovery in the absence of liver flukes.

Studies using ultrasonography have determined strong relationships between gallbladder enlargement, wall irregularities, and sludge, and enhanced echogenicity of portal vein radicles and the intensity of infection (Dhiensiri et al., 1984; Mairiang et al., 1992). In a large-scale study using a sample size of 1807 individuals, the results confirmed the above findings and showed the relationships in more quantitative terms. With the intensity of infection of greater than 3000 egg per gram feces (epg), the adjusted odd ratio for distended gallbladder and portal vein radicles was between 14–20 and 10–28, respectively (Elkins et al., 1996). Reversibility of these abnormalities was observed 11 months after praziquantel treatment (Dhiensiri et al., 1984; Mairiang et al., 1993).

Experimental Animals

Hamsters (*Mesocricetus auratus*), jirds (*Meriones unguiculatus*), guinea pigs, and rabbits are reported to be susceptible laboratory hosts for *O. viverrini* (Wykoff et al., 1966). In *C. sinensis*, mice and rats are additional experimental hosts (Sohn et al., 2006) Extensive study has been done mainly in Syrian golden hamsters starting from the pioneer pathological description of *O. viverrini* infection by Bhamarapravati et al. (1978); they noted that several pathological features were similar to those in human.

The early pathological changes consisted of an acute inflammatory reaction involving the large intrahepatic bile ducts (branch of hepatic duct) and portal connective tissue. Focal hemorrhagic and coagulation necrosis of the liver lobules were also noted. In the chronic phase, when the flukes developed into the adult stage (about 4 weeks postinfection), hyperplasia and adenomatous formations of the bile duct epithelium occurred. Granulomatous responses to adult flukes and eggs were more common. Resolution of the granulomas around the eggs leads to periductal and periportal fibrosis and scarring, which later becomes the most prominent feature in the chronic infection stage (Bhamarapravati et al., 1978). The fibrosis correlates with a marked increase in synthesis and hepatic content of type I and, particularly, type III collagen as demonstrated in long-term infection (Hutradilok et al., 1983; Chotigeat and Ruenwongsa, 1986). The inflammatory responses become less severe in chronic infection than in the acute one, suggesting that immunomodulation may occur (Sripa and Kaewkes, 2000a).

The gross appearance of the gallbladders and extrahepatic bile ducts in infected animals was unremarkable during the first 14 days postinfection (Sripa and Kaewkes, 2000a). From day 30 on, the gallbladder and extrahepatic bile duct wall were slightly opaque (thickened wall), especially in the heavy infected group (100 metacercariae). In chronic infection, particularly from day 90 onward, the most striking lesions were observed in the extrahepatic bile ducts. These included ductal dilatation, increased opacity (thickened wall), and periductal nodule formation. The worms were commonly seen in the gallbladders and extrahepatic bile ducts, and in heavy infection the worms may present in the pancreatic duct (Thamavit et al., 1988). Histologically, acute inflammatory reactions, including congestion and neutrophil and eosinophil infiltration, occur in the gallbladder as early as day 7 of infection; the extrahepatic bile ducts show the similar changes on day 3 postinfection (Sripa and Kaewkes, 2002). Mononuclear cell infiltration, mucus hypersecretion, and fibrosis are gradually observed thereafter. Active inflammation reaches a plateau at approximately day 60 of a 180-day experiment in all infected animals. The well-established chronic histological changes of the gallbladder and extrahepatic bile ducts are fibrosis and mononuclear cell infiltration with lymphoid aggregation and, additionally, ductal dilatation of the bile ducts. Overall, the pathological changes in the extrahepatic bile ducts are more severe than those in the gallbladder for the same dose and period of infection. Figure 1.13 shows histopathological changes in the acute and chronic phase of *O. viverrini* infection in hamsters.

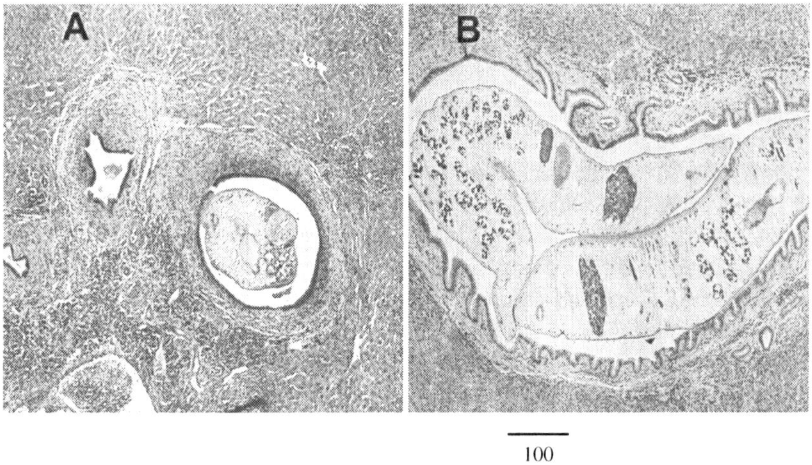

FIGURE 1.13. Histopathology of hamster liver infected with *O. viverrini* infection showing inflammatory reaction around an infected bile duct. (A) Intense inflammatory reaction in the early phase of infection (before day 21). (B) Chronic phase of infection showing periductal fibrosis and less inflammation (90 days postinfection). See also color insert.

O. viverrini infection in hamsters was also found to be associated with renal disease. Six weeks after infection, tegumental and antitegumental membrane immune-complexes and amyloid fibrils were found in the glomeruli. Acute proliferative glomerulonephritis due to immune-complex deposits developed a few weeks later. The intensity of immune-complexes in all the glomeruli was reduced gradually thereafter and was replaced by amyloid. Progressive obsolescence of the glomeruli, tubular atrophy, interstitial inflammation, and fibrosis associated with massive proteinuria and deterioration of renal function appeared at 10 weeks postinfection and beyond (Boonpucknavig et al., 1992).

Host Immune Response

Although *O. viverrini* lives in a biliary system and does not invade into the host's tissue, the infection elicits a systemic immune response. At present, relatively little is known about specific immune responses to infection, especially cell-mediated immunity and the roles of T cells and cytokines in protective immunity and the pathogenesis of opisthorchiasis. Most immunological research has focused on the development of serodiagnosis and antigen characterization (Wongratanacheewin et al., 2003).

There is marked humoral immune response [immunoglobulin (Ig) G, A, and E] to parasite-antigens in the serum and bile of human and animals infected by the liver fluke (Sirisinha et al., 1983b; Srivatanakul et al., 1985; Wongratanacheewin et al., 1987; Itoh et al., 1994; Akai et al., 1995). The level of antibody IgG against crude somatic antigen correlated with the hepatobiliary abnormalities diagnosed by ultrasonography, but there was little correlation with the intensity of infection (Elkins et al., 1996). The 89-kd antigen, mainly appeared in excretory-secretory antigen, was specific to *O. viverrini*, and has a potential diagnostic application (Wongratanacheewin et al., 1988a,c). In infected individuals, the level of serum antibodies (IgG, IgA, IgM) changed slowly and remained elevated several months after praziquantel treatment (Ruangkunaporn et al., 1994). It is not clear whether this reflects long-lasting immunological memory or cross-antigenic stimulations. In the hamster model, antibody responses were first detected as early as 14 days after infection and increased rapidly to a plateau at around 2 months postinfection and were relatively stable thereafter (Sripa and Kaewkes, 2000b). Reinfection with *O. viverrini* metacercariae elicits higher levels of IgG, which correlated with the degree of periductal fibrosis (Pinlaor et al., 2004c).

As opposed to humoral responses, proliferative responses of mononuclear cells to mitogen stimulation were unchanged after drug treatment (Wongratanacheewin et al., 1988b). The role of cellular response in biliary pathology in *O. viverrini* infection is evident since the severity of bile duct inflammation diminished in T-cell–deprived hamsters (Flavell and Flavell, 1986). A reduction of an in vitro lymphoproliferative response was also observed in hamsters infected with *O. viverrini* (Wongratanacheewin et al., 1987).

In the acute phase of infection in an animal model, evidence of cellular immune response was seen as abundant inflammatory cells at 3 weeks postinfection with *O. viverrini* and a marked infiltration of cells, included eosinophils, mononuclear cells, and neutrophils (Bhamarapravati et al., 1978; Sripa and Kaewkes, 2000b). In the chronic phase from 1 to 6 months of infection, there was an increase in mononuclear cells and a decline of eosinophils. Following reinfection, a smaller number of infiltrated inflammatory cells was seen but mainly in the lymphoid follicles in association with the progression of periductal fibrosis (Pinlaor et al., 2004c).

Although evidence presented above clearly showed that infection by *O. viverrini* stimulated both systemic humoral and cell-mediated immune responses during the course of infection, the significance of these immune responses on protective immunity is not convincing. The fact that some individuals from the endemic area may harbor several thousand worms (Bunnag et al., 1981; Haswell-Elkins et al., 1994a) suggests that reinfection does occur and that immune responses fail to prevent reinfection by the same parasite. This conclusion was confirmed when no significant reduction in worm burden was seen in hamsters receiving immune spleen cells and serum compared to the control groups despite a substantial reduction in the fecal egg count (Flavell et al., 1980). It was found that the primary infection of hamsters with *O. viverrini*, whether or not eliminated with praziquantel before challenge, failed to confer any protective immunity to reinfection (Sirisinha et al., 1983a). However, both groups of investigators demonstrated that the previous exposure to *O. viverrini* depressed the egg output in subsequent infections. Egg reduction could be due to biliary obstruction and not necessarily due to immunological damage of the reproductive system of the flukes.

The lack of protective immunity is perhaps due to immunosuppression of both cellular and humoral immune responses by *O. viverrini* infection to unrelated antigens as shown by reduced phytohemagglutinin-induced lymphoproliferation and response to sheep red blood cell stimulation (Wongratanacheewin et al., 1987). Moreover, when specific antibody was measured in the infected animals, the titer was found to be depressed during the late stage of infection, particularly in the heavily infected group (Sirisinha et al., 1983a). Moreover, the parasites may be resistant to immune damage or able to evade the host defense system. The inability to kill the worms by exposure to serum from infected animals or from patients with opisthorchiasis and the resistance to immune damage may be related to tegumental shedding and repair of the parasite (Sirisinha and Wongratanacheewin, 1986).

In the case of experimental clonorchiasis, protective response against reinfection was observed in rats and mice, which are good animal models to investigate the mechanism of resistance to reinfection with *C. sinensis* (Chung et al., 2004; Sohn et al., 2006). In contrast, primary infection of hamsters with *O. viverrini* showed no resistance against reinfection with the same parasite (Sirisinha et al., 1983a). There is evidence that both humoral and cellular responses may play a role in the observed resistance in the animal model (Quan et al., 2004).

Liver Fluke Infection and Cholangiocarcinoma

Infection and chronic inflammation are proposed to contribute to carcinogenesis through inflammation-related mechanisms. Infection with hepatitis C virus, *Helicobacter pylori*, and the liver flukes *O. viverrini* and *C. sinensis* is an important risk factor for hepatocellular carcinoma, gastric cancer, and cholangiocarcinoma, respectively. In 1994 the International Agency for Research on Cancer (IARC, 1994) classified *O. viverrini* as a class 1 carcinogen and *C. sinensis* as a class 2A carcinogen to humans. On the other hand, the evidence in regard to *O. felineus* is insufficient to assess its role in carcinogenesis.

Intrahepatic cholangiocarcinoma (ICC), also called peripheral cholangiocarcinoma, is a malignant tumor arising from the biliary epithelium of the intrahepatic biliary tree; it accounts for 10% to 15% of primary liver cancer and overlaps in its geographical distribution with endemic areas of *O. viverrini* and *C. sinensis* (Parkin et al., 1993). Although the etiology is not known, chronic inflammation of the bile ducts and conditions associated with bile stasis are predisposing factors to the development of ICC. These conditions include primary sclerosing cholangitis, liver fluke infection, and recurrent cholangitis with hepatolithiasis. Intrahepatic cholangiocarcinoma is an intrahepatic bile duct cancer. In Thailand, ICC was 12 times more common in the northeast, the endemic area of *O. viverrini*, than in regions with low prevalence of *O. viverrini* (Srivatanakul et al., 1988). The age-standardized incidence of ICC in Khon Kaen of 84.6 and 36.8 per 100,000 for male and female, respectively, was the highest in the world (Vatanasapt et al., 1993). Several epidemiological studies revealed that evidence of current or past experience of liver fluke infections, in terms of intensity of infection, seropositivity for liver fluke infection, radiological evidence of *C. sinensis* infection, is the major risk factor of ICC (Honjo et al., 2005; Choi et al., 2006; Lim et al., 2006). A linear trend of the frequency of suspected ICC and fecal egg count was also observed where the odds ratio of 14.1 was found in a group with an epg >6000, which is equivalent to >120 worms (Haswell-Elkins et al., 1994b). Additional risk factors reported in ICC were area of residence, alcohol consumption, age, and sex; host genetic polymorphism of glutathione S-transferase enzyme (GSTM1) in association with seropositivity for opisthorchiasis was found to be an important risk factor of ICC. This evidence suggests a classic gene–environment interaction that plays a crucial role in individual susceptibility to ICC.

An ICC in the endemic area of liver fluke infection is similar to that described in nonendemic regions. In ICC, the tumor arises from the peripheral bile duct and can directly invade the sinusoids of the adjacent liver parenchyma to form a tumor nodule; in most clinical cases, the disease is in an advanced stage and usually appears as a large, single, white firm tumor with a distinct margin in the noncirrhotic liver (Nakajima et al., 1988). The most common histopathology of ICCs (>95%) is adenocarcinoma showing a glandular or papillary structure with a variable fibrous stroma (Fig. 1.14). There is no dominant histologic type of ICC in the

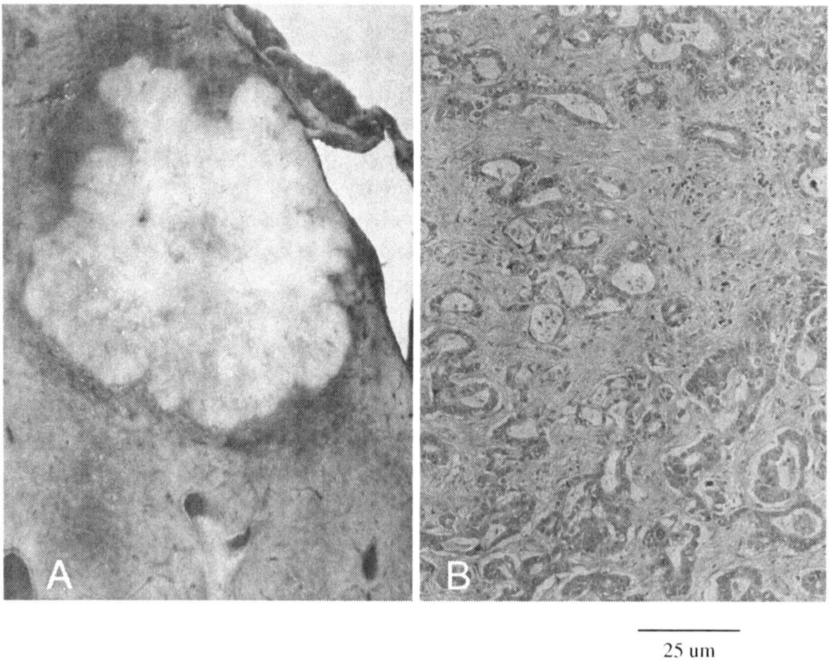

FIGURE 1.14. Intrahepatic cholangiocarcinoma (ICC) in association with *O. viverrini* infection. (A) Gross pathology of ICC appears in focal liver mass in a noncirrhotic liver. (B) Histology of ICC showing well-differentiated adenocarcinoma with fibrotic stroma. See also color insert.

case of liver flukes or hepatolithiasis when compared to those from non-endemic areas (Shirai et al., 1992). Intrahepatic cholangiocarcinoma is originated from the surface epithelial cells lining bile ducts. Atypical hyperplasia and dysplasia that lie adjacent to ICC are potential precancerous lesions in both fluke and non-fluke-related ICC (Hou, 1955; Kim, 1984). This observation supports the fact that ICCs with different etiology backgrounds may have a common histogenesis.

Chronic infection results in chronic inflammation of the bile ducts, with epithelial changes including hyperplasia, proliferation with acini formation (adenomatous hyperplasia), goblet cell metaplasia, and periductal fibrosis (Hou, 1955; Tansurat, 1971; Riganti et al., 1989; Pairojkul et al., 1991). The acini formation of the adenomatous hyperplasia in liver fluke infection is intramural peribiliary gland hyperplasia (Terada and Nakanuma, 1992). Morphological studies in Hong Kong and Korea indicate that carcinoma usually arises in association with preexisting epithelial changes and neoplastic transformation from adenomatous hyperplasia in bile ducts to ICC through dysplastic changes of lining cell or from cholangiofibrosis and is clearly demonstrable (Fig. 1.15). Histopathologic characteristics indicated that *O. viverrini* and *C. sinensis* are predisposing conditions for pathogenesis of ICC in Asia. Strong supporting evidence was demonstrated in

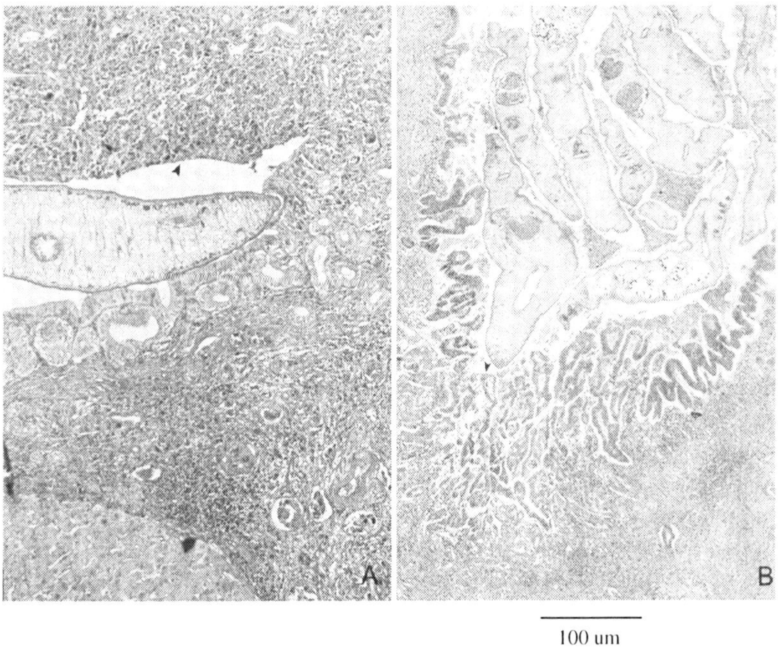

FIGURE 1.15. Microscopic view of ICC with preexisting liver fluke-associated pathology. (A) Early ICC (arrow) in liver fluke–associated adenomatous hyperplasia of the intrahepatic bile duct. (B) Early ICC (arrow) in epithelial hyperplasia of periductal fibrotic lining of the bile duct. Note several worms in the duct lumen. See also color insert.

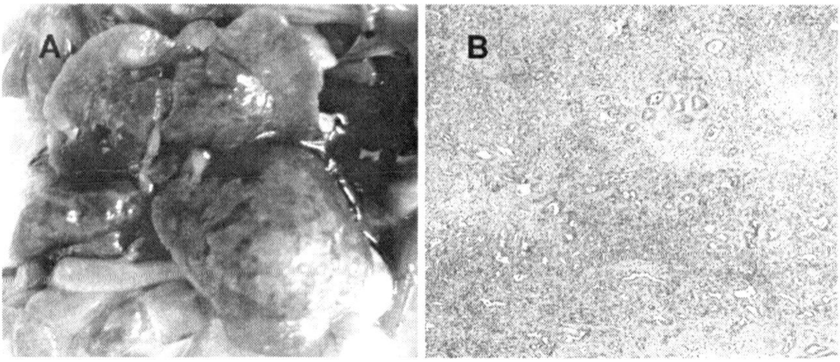

FIGURE 1.16. Intrahepatic cholangiocarcinoma (ICC) in hamster model from its induction with *O. viverrini* and dimethylnitrosamine. (A) Gross appearance, showing the enlarged liver with its nodular surface at the right lateral lobe. (B) Microscopic feature of well-differentiated tubular adenocarcinoma. See also color insert.

an animal model in which hamsters could be induced to develop ICC by infection with *O. viverrini* or *C. sinensis* and administration of a subcarcinogenic dose of dimethylnitrosamine, as shown in Figure 1.16.

Inflammation-Related Cholangiocarcinogenesis

Most authors of the earlier study suggested that the liver flukes mediate tissue damage directly by mechanical and chemical irritation (Tansurat, 1971; Viranuvatti and Stitnimarnkarn, 1972; Harinasuta and Harinasuta, 1984; Kim, 1984; Thammapalerd et al., 1988), but some studies postulated that parasite-specific immune responses may play a major role (Bhamarapravati et al., 1978; Flavell, 1981; Flavell and Flavell, 1986; Haswell-Elkins et al., 1991). At present, although the underlying enhancement of neoplasia by liver flukes is not fully understood, carcinogenesis is a multistage process in which many factors are likely to be involved. However, the mechanistic explanation is gradually becoming clearer, particularly since Ohshima et al., (1994) have hypothesized that the chronic infection and inflammation is a risk factor of carcinogenesis due to the possible role of nitric oxide.

During liver fluke infection, macrophages and other cell types (e.g., mast cells, eosinophils, epithelial cells), activated by parasite-specific T cells and cytokines, synthesize nitric oxide (NO) from L-arginine via the induction of inducible nitric oxide synthase (iNOS) in order to kill the parasite. Nitric oxide is not only cytotoxic but also genotoxic at physiological concentrations. Excess NO production plays a crucial role in a variety of pathological processes, including cancer (Hussain et al., 2003; Ohshima et al., 2003). Nitric oxide reacts with superoxide ($O_2{\bullet}-$) to form peroxynitrite ($ONOO-$), a highly reactive species causing nitrative and oxidative

DNA damage. ONOO- can mediate the formation of 8-oxodG (see below) (Inoue and Kawanishi, 1995) and 8-nitroguanine, a marker of nitrative DNA damage.

Endogenous nitrosation caused by liver fluke infection has been studied in both animal models and humans. In *O. viverrini*–infected people, there were significant increases in plasma and urinary nitrate and *N*-nitrosoproline (NPRO) in urine after proline loading, nitrate, and NPRO, the indicators of endogenous NO synthesis and nitrosation reactions, respectively (Srivatanakul et al., 1991; Haswell-Elkins et al., 1994b). The site of the endogenous reaction was proven to be extragastric, that is, within the inflamed bile duct (Satarug et al., 1996a). Direct measurement of the carcinogenic product of nitrosation reaction, *N*-nitrosodimethylamine (NDMA), excreted in the urine, appeared to be associated with in vitro lymphoproliferative responses to liver fluke antigens; after the flukes were removed following praziquantel treatment, the association disappeared (Satarug et al., 1998). In multivariate analyses, NDMA levels were related to urinary nitrates, stimulation indices for two T-cell responses to two parasite antigens (MW 37 kd and 110 kd) and gallbladder dimensions. In active infection with *O. viverrini*, NDMA is further activated by some isoforms of cytochrome P-450 (CYP) enzymes, mainly CYP2E1 and CYP2A6, to become ultimately carcinogenic as shown in human and animal models (Kirby et al., 1994; Satarug et al., 1996b). This increase in CYP2A6-related enzyme activity may represent an important mechanistic link between inflammatory products of chronic liver fluke infection and the high risk of cholangiocarcinoma faced by infected individuals.

To mimic the pattern of infection in humans, the role of reinfection with *O. viverrini* on carcinogenesis in a hamster model was examined (Pinlaor et al., 2004c). The degree of pathological changes, including periductal fibrosis, bile duct dilation, small bile duct formation, and *O. viverrini*–specific IgG level, correlated with the number of reinfections, that is, triple infection > double infection > single infection groups. Liver enzyme activity was related to the degree of inflammatory cell infiltration. It was concluded that reinfection with *O. viverrini* induced a more rapid inflammatory reaction and more severe pathological changes in association with parasite-specific antibody level through chronic inflammation. Further evidence at the molecular level was that infected hamsters expressed iNOS; hence, more NO was produced and the increased reactive oxygen species (ROS) mediated the formation of DNA adducts, namely 8-oxo-7,8-dihydro-2′-deoxyguanosine (8-oxodG) and 8-nitroguanine through chronic inflammation (Pinlaor et al., 2003). These two DNA adducts increased more prominently in the bile duct epithelium in triple > double > single infections (Pinlaor et al., 2004b).

On the other hand, glutathione (GSH) content was significantly decreased on day 21 and rebounded to a significantly higher level than that of the noninfected hamsters, leading to the notion that GSH takes part in cytoprotection against oxidative stress. It is noteworthy that GSH levels and alanine aminotransferase (ALT) activity tended to return to the normal level in long-term chronic infections, while nucleic acid damage, especially 8-oxodG, remained in the epithelium of the bile ducts. Sustained formation of 8-oxodG and separated localization of 8-oxodG and 8-nitroguanine may be explained by the fact that lipid peroxidation products cause additional 8-oxodG formation in human DNA (Kaneko and Tahara, 2000). Therefore, NO and ROS may mediate 8-oxodG and 8-nitroguanine formation triggered by *O. viverrini* infections through chronic inflammation (Pinlaor et al., 2004a).

O. viverrini causes the formation of oxidative and nitrative DNA adducts as proven through the Toll-like receptor 2 (TLR2)-mediated pathway leading to the activation of nuclear factor κB (NF-κB), which is a key player in tumor promotion in inflammation-associated cancer (Pikarsky et al., 2004; Pinlaor et al., 2005). Additionally, proliferating cell nuclear antigen (PCNA), a cofactor for DNA polymerase, is associated with DNA replication and long-patch base excision repair, accumulated in the epithelium of bile ducts after repeated *O. viverrini* infection, supporting the hypothesis that cell proliferation was promoted by inflammation-mediated DNA damage (Pinlaor et al., 2004b). It was shown that the responses of a fibroblast cell line, NIH-3T3, to excretory/secretory (ES) product(s) released from *O. viverrini* in a noncontact coculture condition had a marked increase in cell proliferation (Thuwajit et al., 2004).

Treatment with praziquantel not only reduced bile duct inflammation via the TLR–2 mediated pathway but also oxidative and nitrative DNA damage when assessed during 1 to 3 weeks posttreatment, and thus may help prevent carcinogenesis (Pinlaor et al., 2006). However, a short-term effect of treatment on an antigenic burst of the dead worms and the possibility to trigger an inflammatory reaction as seen in reinfection experiments remain to be determined.

In conclusion, a series of experiments conducted with single and repeated infections of *O. viverrini* in hamsters showed that iNOS expression in the bile duct epithelial cells and inflammatory cells via the TLR2-mediated pathway, resulting in nitrative and oxidative damage to nucleic acids, enabling the simultaneous formation of 8-oxodG and 8-nitroguanine in the nucleus of inflammatory cells and epithelium of bile ducts in the first phase of single infection. In the second phase, both 8-oxodG and 8-nitroguanine were mostly observed in small inflammatory cells, bile duct epithelium, and small bile ducts, which may participate in every step of cholangiocarcinoma development, including initiation, promotion, and progression. Repeated infection with *O. viverrini* induces more prominent iNOS expression in the bile duct epithelial cells, resulting in a higher degree of nitrative and oxidative damage to nucleic acids. Praziquantel treatment inhibits iNOS-dependent DNA damage through not only the elimination of the parasites but also via a potential antiinflammatory effect.

The possible mechanism of liver fluke-associated CCA development based on evidence of a series of experimental studies has been proposed and divided into four stages similar to those in colon cancer (Fig. 1.17) (Ohshima et al., 1994; Holzinger et al., 1999). First, exposure to a risk factor may lead to chronic inflammation or cholestasis through the biochemical or mechanical processes. Second, genotoxic events may be the consequence of NO overproduction by inflammatory cells. Nitric oxide and oxygen radicals can inactivate and injure the lipid, protein, DNA, and carbohydrate within the cells, via direct oxidative damage. Free radicals caused by inflammatory cytokines can induce many subsequent events, such

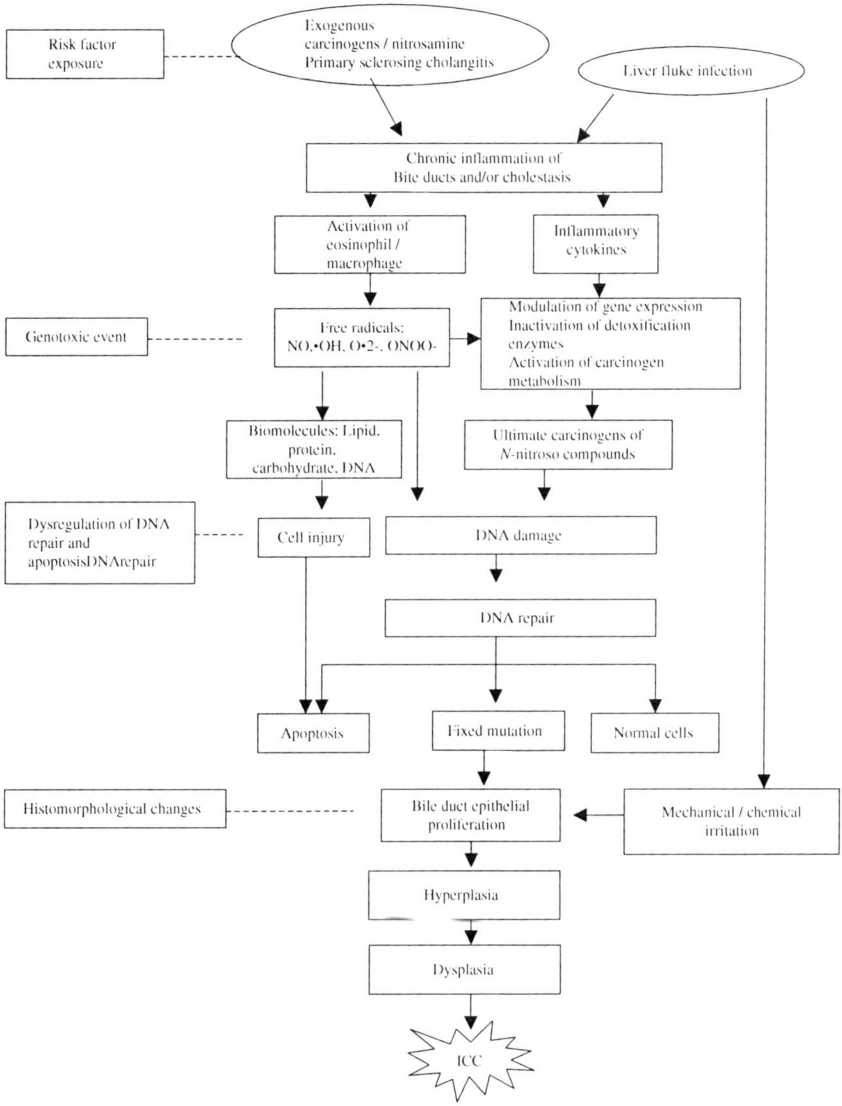

FIGURE 1.17. The possible carcinogenesis of *O. viverrini*–associated ICC. (Modified from Oshima and Bartsch, 1994; Holzinger et al., 1999.)

as the activation of gene expression, alteration of detoxification gene expression, and activation of carcinogen metabolism to form ultimate carcinogens, which enhance DNA damage. Moreover, overproduction of NO has been reported to be immunosuppressive for lymphocyte proliferation. Third, dysregulation of DNA repair and apoptosis, genotoxic events with DNA damage, usually leads to either DNA mismatch repair mechanism or, if the damage is beyond repair, cell death occurs through apoptosis. Finally, the histological changes to malignancy could occur in the following step of the "hyperplasia-dysplasia-carcinoma" model, which has been proven for colon and gastric cancers (Holzinger et al., 1999).

Diagnosis

Recent studies in clonorchiasis and opisthorchiasis have taken advantage of considerable advances in radiology imaging to aid in the diagnosis of clonorchiasis and hepatobiliary diseases, and among several radiological diagnoses, abdominal ultrasonography seems to be the most practical and noninvasive (Mairiang and Mairiang, 2003; Choi et al., 2004). Characteristic sonographic findings in active liver fluke infection include increased periductal echogenicity, floating echogenic foci in gallbladder or gallbladder sludge, diffuse dilatation of the intrahepatic bile duct, and gallbladder enlargement. Occasionally, biliary stones including gallbladder stones, common bile duct stones, and intrahepatic stones may be found.

Parasitological Diagnosis

The presence of egg in fecal samples or duodenal fluids or the presence of adult flukes during laparotomy remains the gold standard diagnosis for the liver flukes. The most convenient clinical sample for egg examination is the feces. The sensitivity and reliability of the diagnosis depends on the fecal examination technique and the experience of the microscopist.

The fecal examination techniques frequently employed for diagnoses of liver fluke infection are the formalin-ether concentration technique, the Kato thick smear, and Stoll's dilution egg count technique. Other standard techniques such as a simple direct smear or simple sedimentation techniques are also used in certain situations based on the available facilities. Single examination by these techniques seems to have a high false-negative rate, especially with light infections or those with a history of recent treatment. Repeated stool examinations are necessary to improve the sensitivity of the examination. Although repeated examination, increases sensitivity, repeated or multiple stool sample collections is difficult in practice, particularly in a rural community.

The formalin-ether concentration technique is considered to be relatively sensitive compared to other techniques, since the specimen is filtered and lipid is removed; the resulting sediment is smaller in quantity and the eggs are easily seen microscopically. Although this technique gives qualitative results, it can be a modified to be a quantitative technique, if the weight or volume of the stool sample is known, to obtain a number of egg counts. The other advantage is that the stool samples can be kept for several months prior to processing and examining once fixed in formalin. The typical amount of stool sample take for processing by this technique is 1 to 2 g, which is 1000 to 2000 more than the direct simple smear technique and hence is more sensitive. In light infection cases with low worm burdens, egg detection in the stool is problematic. The influence of the worm

TABLE 1.3. The sensitivity of stool examination for *O. viverini* eggs in autopsy subjects grouped according to the number of worms recovered from the liver. The methods for egg counts were Stoll's dilution and quantitative formalin-ether concentration methods (data from Sithithaworn et al., 1991a).

		Number positive (%)	
Worm burden	Sample size	Stoll's	Formalin-ether
0	26	0 (0)	0 (0)
1–9	25	8 (32)	7 (28)
10–19	13	10 (76.9)	8 (61.5)
20–39	19	19 (100)	16 (84.2)
40–99	13	13 (100)	13 (100)
100–199	20	20 (100)	19 (95)
>200	23	23 (100)	23 (100)
Total	139	93 (66.9)	86 (61.9)

burden of *O. viverrini* on fecal examination by formalin-ethyl acetate concentration and Stoll's egg count technique is shown in Table 1.3. Based on recovery of adult *O. viverrini* directly from the liver at autopsy ($n = 139$), the prevalence (81.29%) was significantly greater than both the formalin-ether concentration (61.87%) and the Stoll's dilution egg count (66.91%) technique (Sithithaworn et al., 1991a). A detailed analysis revealed that at a worm burden of less than 10 worms per liver, as much as 68% to 72% were egg negative in their stools. At a worm load of 10 to 19 worms, the proportion of false negatives became 23% to 29%, while at a higher worm burden >20, the rate of egg detection was comparable with worm recovery.

The Kato thick smear technique is commonly used for routine diagnosis of opisthorchiasis in Thailand, Laos, and also is widely used in the field for clonorchiasis in Korea and China. It is believed to be a sensitive and reliable technique comparable to the formalin-ether concentration technique and more sensitive than the simple sedimentation technique particularly for clonorchiasis (Hong et al., 2003). When the amount of stool specimen is measured or weighed, it is then known as the Kato-Katz technique. However, the identification of eggs in the Kato thick smear needs experience to distinguish the transparency of egg shells.

The Stoll's dilution egg count technique uses sodium hydroxide to digest lipid and clear the fecal suspension. The detection rate was slightly inferior to Kato-Katz but it is believed to be suitable for the measurement of the intensity of *O. viverini* (Viyanant et al., 1983).

Since there are several species of food-borne trematodes belonging to the Opisthorchiidae, Heterohyidae, and Lecithodendriidae families, which have similar egg morphology, recognition of the eggs is essential for correct identification. In Thailand, for example, the latter two families are collectively referred to as minute intestinal flukes (MIFs) because of their small size compared to *O. viverrini* (Kaewkes et al., 1991). In Thailand, the Lecithodendriidae contains *Phaneropsolus bonnei* and *Prostodendrium molenkampi*, while the Heterohyidae contains

Haplorchis taichui, H. pumilio, and *Stellanchasmus falcatus* (Radomyos et al., 1998). Within the heterophyid, there are *Metagonimus* spp., *Hetrophyid* spp., *Haplorchis* spp., and *Pygidiopsis* spp. found in Korea (Chai et al., 2005). The eggs of liver flukes, namely *O. viverrini, C. sinensis*, and *O. felineus*, are indistinguishable; therefore, background information on the geographical source of stool samples is helpful. In addition, eggs of heterophyids are very similar in all three liver flukes; thus, adult worms are required for identification (Rim, 2005). Under light microscopy, eggs of *O. viverrini* are characterized by rough and thick egg shells; and by scanning electron microscopy the egg shell shows a musk-melon pattern. This can be differentiated from *P. bonnei* and *P. molenkampi*, which have smooth and thin egg shells, and, when stained with iodine, iodophilic bodies are seen (Kaewkes et al., 1991). No iodophilic staining appears in *O. viverrini* eggs. Attempts to used potassium permanganate staining may help to distinguish the musk-melon pattern of the egg shell of *O. viverrini* from *H. taichuii* and *P. bonnei*.

Immunodiagnosis

Although stool examination is the gold standard for the diagnosis of liver fluke infection, it has become more difficult to do because of the lack of compliance in developing countries or the lack of large field studies. Serodiagnosis to detect antiparasite antibodies is widely used and may replace stool examination in the future. In the past decade, serological tests have been developed so that the sensitivities and specificities of the tests have greatly improved. Currently, there are several techniques available for serodiagnosis such as an intradermal test, an immunoelectrophoresis (IEP), an indirect hemagglutination assay (IHA), an indirect fluorescent antibody test (IFAT), and an indirect enzyme-linked immunosorbent assay (indirect ELISA) (Rim, 1986; Chen et al., 1994; Wongratanacheewin et al., 2003). Among these tests, ELISA is the favorite, but its sensitivity and specificity vary depending on the nature of the antigen used in the system. Different parasite components are used and can be classified into adult somatic extracts, excretory and secretory (ES) substance, and surface and egg antigens.

In opisthorchiasis, adult somatic extracts were used for IEP and ELISA with a sensitivity of 76% to 100% for IgG and lower sensitivities for IgA and IgE (Janechaiwat et al., 1980; Srivatanakul et al., 1985; Wongratanacheewin et al., 1987). A partially purified somatic extract of surface tegument yielded a higher sensitivity and specificity (Poopyruchpong et al., 1990). The 89-kd protein most prominent in the ES antigen of *O. viverrini* was claimed to be specific with immunodiagnostic potential; however, it also cross-reacted with *C. sinensis* (Wongratanacheewin et al., 1988a; Sirisinha et al., 1990). The affinity-purified egg antigen using specific monoclonal antibody to detect serum antibody, gives a high sensitivity and specificity (Wongsaroj et al., 2001). Extensive immunodiagnostic studies were performed on clonorchiasis mainly in Korea and also China (Rim, 1986; Chen et al., 1994). Purification of native antigens of *C. sinensis* have been attempted and different antigenic bands reacting to IgG appeared to give variable seropositive rates, and IgG4 antibody reactive to certain antigen was more specific but not sensitive for diagnostic purposes (Hong et al., 1997, 1999; Choi et al., 2003). Recently, several novel genes of *C. sinensis* were reported, and recombinant proteins were produced for a diagnostic test with varying degrees of success. One candidate recombinant protein is the cathepsin B-like cysteine proteinase enzyme, which showed a promising diagnostic potential with a high sensitivity and specificity for clonorchiasis (Na et al., 2002; Nagano et al., 2004). However, one drawback of these antibody detection methods is the inability to differentiate past and current infections since antibodies in opisthorchiasis and clonorchiasis persist in the infected hosts for months or years even after curative treatment (Thammapalerd et al., 1988; Chen et al., 1994). However, with more advanced molecular biology techniques, more recombinant antigens are becoming available; therefore, there are possibilities that the serological tests could be improved to be used not only for better diagnostic efficacy but also in the future for quantification of the worm burdens and past experience of infection as well as the risk of morbidity.

In contrast to antibody detection, the detection of antigens is informative in determining the current state of infection. For this approach, several monoclonal antibodies to *O. viverrini* antigens were produced against different antigenic proteins such as 16 kd of tegumental protein, 89 kd glycoprotein, and 90 kd somatic protein, and were used for parasite antigen detection in the stool with moderate success, with a sensitivity of 31% to 57% and a specificity of 70% to 100% (Chaicumpa et al., 1991; Sirisinha et al., 1995). Although this is a promising approach to develop further into a simple diagnostic kit, improvement is needed to increase the sensitivity and reduce the cross-reactivity of the test.

Molecular Diagnosis

As an alternative to conventional parasitological and immunological diagnostic methods, which have weak points in terms of their specificities, the molecular methods are of particular advantage. In opisthorchiasis, a specific DNA probe designed from repeated DNA elements (satellite DNA) of 334 base pairs was used for the detection of egg DNA (Sermswan et al., 1991; Sirisinha et al., 1991). Recently, a polymerase chain reaction (PCR)-based detection of *O. viverrini* in human stools, based on a pair of primers complementary to the same target DNA of *O. viverrini*, was used to detect egg DNA in infected animals and human stool samples a with high sensitivity and specificity compared to the parasitological methods (Wongratanacheewin et al., 2001, 2002). From these studies, a specificity of 97.8% and a sensitivity of 100% were achieved in moderate to severe infections (eggs per gram >1000), but in light infections (eggs per gram <200) the sensitivity was reduced to 68.2%. Application of this PCR-based method for the detection of *O. viverrini* DNA in stool samples from Laos yielded the sensitivity of approximately 50% in samples with egg count >1000 epg (Stensvold et al., 2006). It is suggested that the presence of a PCR inhibitor in the fecal specimen may influence the sensitivity of the diagnosis rather than the diagnostic reference–related issue. Nevertheless, because these molecular diagnostic tests are extremely specific, they will play significant roles in the accurate assessment of cure as well as the rate of reinfection.

In addition, species-specific PCR tests to identify the species of liver fluke are now available for *O. viverrini* (Ando et al., 2001; Wongratanacheewin et al., 2001), *O. viverrini* and *C. sinensis* (Le et al., 2006), and *O. felineus* (Pauly et al., 2003). It is possible to identify the parasite species from various stages of the parasite, including the egg, the metacercaria, as well as the adult worm.

Treatment

Praziquantel is the drug of choice for treatment of several species of trematodes including liver flukes, except *Fasciola* spp. The recommended dose for mass treatment in humans is 40 mg/kg body weight, and a dose of 25 mg/kg three times a day gives 100% cure rate for *O. viverrini* infection; however, it gives an 80% to 85% cure rate in *C. sinensis* (WHO, 1995). Two consecutive days of treatment of humans infected with *C. sinensis* increases the cure rate to nearly 100%. To date, there has been no evidence of praziquantel resistance by liver flukes. The mechanism of action of praziquantel is not fully known, but it induces muscle contraction and vacuolization of the tegumental syncytium of the flukes within a short time after drug exposure; eventually the worm tegument is disrupted and it bursts, thus allowing host phagocytic cells to infiltrate (Sirisinha et al., 1984). The expelled worms posttreatment are often deformed or broken. Since praziquantel is absorbed rapidly, the peak serum level occurs in 1 to 3 hours posttreatment, and it is excreted in the bile and urine within 24 hours, the praziquantel treatment has little effect on subsequent exposure to infection.

Although praziquantel is well tolerated, it occasionally causes side-effect reactions, such as abdominal discomfort, nausea, vomiting, headache, dizziness, and, rarely, pyrexia and urticaria. Drowsiness and tachycardia have also been recorded. Praziquantel is not recommended for use with pregnant woman to prevent any mutagenic, teratogenic, or embryotoxic effects on the fetus (WHO, 1995).

Prevention and Control

According to the WHO recommendation for food-borne trematode control (WHO, 1995), morbidity can be prevented or controlled by treatment, health education, improved sanitary conditions, and implementation of food safety measures.

The most effective measure is chemotherapy using a single dose of praziquantel at 40 mg/kg body weight because it reduces the worm burden and also the morbidity rate. In theory, it is conceivable that liver fluke control, that is, opisthorchiasis control, should also contribute to the reduction in the incidence of cholangiocarcinoma in Thailand; however, because of the complex nature of carcinogenesis processes involving risk factors other than *O. viverrini* infection

(Haswell-Elkins et al., 1992; Sithithaworn et al., 1997), a reduction in the trend in the incidence of ICC in Thailand is far from conclusive.

Chemotherapy alone, with or without supported health education, is not effective in every endemic area. That the history of liver fluke control in Thailand has proven to be successful as it is today can be attributed not only to the availability of praziquantel treatment but also to the extensive improvement of the health care system and to socioeconomic development (Jongsuksuntigul and Imsomboon, 2003). The pilot parasite control project in Laos recently conducted by providing yearly treatment with praziquantel for 2 years seems to show little impact on *O. viverrini* transmission (Strandgaard, 2006). Based on the sample size of 768 to 1363 people, the baseline prevalence of *O. viverrini* was 40%, with an intensity of 150 epg at pretreatment. After the first and the second treatment, the prevalence became 16% and 24%, while the intensities were 58 and 168 epg, respectively. Obviously, there are several explanations of these observed data; reinfection is one likely factor due to continued parasite transmission (Kobayashi et al., 2000). Timing the treatment in the period when transmission potential is low (low metacercarial load in fish) may help to reduce reinfection and increase the effectiveness of the program (Hinz et al., 1994). Therefore, without suitable and appropriate health promotion, together with drug treatment, reinfection may occur rapidly. Emphasis on health education should be placed on the younger generation in school as a part of the conventional curriculum.

To ensure the success of parasite control, collaboration with different sectors are suggested, for example, with the fisheries and aquaculture, with the food industries, educational sectors, and nongovernment organizations. An initative of the Food and Agriculture Organization (FAO) has stressed the need to assess the relative importance of aquaculture vis-à-vis captured fisheries as a source of food-borne trematode infection (WHO, 1995). Of particular interest is the need for food safety assurance and products from aquaculture for both domestic consumption and international trade. In addition, food control and inspection techniques such as Hazard Analysis and Critical Control Point approach (HACCP) are available for the effective control of the food risk at the production stage. Preliminary application of this approach in the culture of fresh water carp (*Puntius gonionotus*) was tested for the first time (Khamboonruang et al., 1997). This may lead to the identification of good aquaculture practice at all levels of the production system as a means of controlling trematode cross-contamination. However, more studies are needed to optimize additional production costs of food safety to meet local conditions and constraints in applying the HACCP approach.

Conclusion

O. viverrini, C. sinensis and *O. felineus* are three major food-borne trematodes of public health importance in several countries in Asia and in Russia. The distribution of the live fluke, *O. viverrini*, in Thailand is highly concentrated in northeast Thailand and corresponds well with the availability of snail and fish intermediate hosts and the food consumption habits of local the population. The transmission of these flukes to humans and to the intermediate hosts has a strong seasonal pattern; peak tramsmission to humans occurs after the rainy season or after the winter, based on the countries where the fish have the highest metacecarial burden. Several traditional dishes prepared from fresh water fish serve as the source of live fluke infection. The prevalence of infection in humans occurs early in life and plateaus after the teenage years and into older age. The worm burden increases with age with a slight drop in the older age groups, thus suggesting that immunity to infection is probably minimal. The distribution of the worm burden in humans is highly aggregated; a few people harbor heavy worm loads while most people have light or no infection. In heavy transmission areas, a high incidence as well as rapid reinfection after treatment is common. The infection induces little apparent clinical manifestation, but a considerable proportion of heavily infected people may have hepatobiliary diseases including cancer of the bile duct or ICC. However, these hepatobiliary pathologies could be reversed after chemotherapy. There is substantial evidence that inflammation caused by liver fluke infection induced ROS and NO, which give rise to nitrative and oxidative DNA damage via iNOS. Nitric oxide could act as a nitrosating agent to form NDMA, which is activated by xenobiotic metabolizing enzymes to be the ultimate carcinogen. The excretory-secretory products of *O. viverrini* were found to induce cell proliferation in hosts. These data imply that liver fluke infection, at least with *O. viverrini*, plays a role as promoter as well as initiator in a multistage carcinogenesis of ICC. Moreover, praziquantel treatment, in the long term, eliminates inflammation, and a subsequent induction of oxidative and nitrative DNA damage of the bile duct epithelium and eventually prevents carcinogenesis. However, the short-term impact of an antigenic burst from a dead worm requires more detailed investigation.

For parasite control, the rationale is that treatment is required to eliminate the long-lived parasites immediately, sanitation interrupts transmission from human feces to snails, and health education stops people from eating raw fish and becoming reinfected after treatment. A number of studies have suggested that control programs using treatment plus health education are more effective at suppressing reinfection than programs using treatment alone. However, the vehicle and type of health education utilized is rarely explained, and the possibility that seasonal variation or sampling problems bias the results is rarely considered. Control campaigns that promote messages that are foreign to local beliefs and demand too many changes may not be successful. The long-term and sustainable control program should be focused on schoolchildren who will then be a liver fluke–free generation.

References

Akai, P.S., Pungpak, S., Kitikoon, V., Bunnag, D., and Befus, D. (1995). Egg-negative residents of an Opisthorchiasis-endemic area possess high levels of anti-Opisthorchis antibodies. Adv Exp Med Biol 983–986.

Anderson, R.M., and May, R.M. (1985). Helminth infections in humans: mathematical models, population dynamics and control. Adv Parasitol 24, 1–101.

Ando, K., Sithithaworn, P., Nuchjungreed, C., Tesana, S., Srisawangwong, T., Limviroj, W., and Chinzei, Y. (2001). Nucleotide sequence of mitochondrial CO I and ribosomal ITS II genes of *Opisthorchis viverrini* in northeast Thailand. Southeast Asian J Trop Med Public Health 32, 17–22.

Attwood, H.D., and Chou, S.T. (1978). The longevity of *Clonorchis sinensis*. Pathology 10, 153–156.

Bae, K.H., Ahn, Y.K., Soh, C.T., and Tsutsumi, H. (1983). [Epidemiological studies on *Clonorchis sinensis* infection along the Nam-River in Gyeongnam Province, Korea]. Kisaengchunghak Chapchi 21, 167–186.

Beaver, P.C., Jung, R.C., and Cupp, E.W. (1984). Clinical Parasitology. Philadelphia: Lea & Febiger.

Beer, S.A., and German, S.M. (1987). [Susceptibility of Bithynia inflata mollusks from discrete populations for *Opisthorchis felineus* infestation from different foci of opisthorchiasis]. Parazitologiia 21, 585–588.

Bensted-Smith, R., Anderson, R.M., Butterworth, A.E., Dalton, P.R., Kariuki, C.H., Koech, D., Mugambi, M., Ouma, J.H., Arap Siongok, T.K., and Sturrock, R.F. (1987). Evidence for predisposition of individual patients to reinfections with *Schistosoma mansoni* after treatment. Trans R Soc Trop Med Hyg 81, 651–656.

Bhamarapravati, N., Thammavit, W., and Vajrasthira, S. (1978). Liver changes in hamsters infected with a liver fluke of man, *Opisthorchis viverrini*. Am J Trop Med Hyg 27, 787–794.

Boonpucknavig, S., Boonpucknavig, V., Tanvanich, S., Doungchawee, G., and Thamavit, W. (1992). Development of immune-complex glomerulonephritis and amyloidosis in Syrian golden hamsters infected with *Opisthorchis viverrini*. J Med Assoc Thai 75 (suppl 1), 7–19.

Bowman, D. (1999). Georgis' Parasitology for Veterinarians. Philadelphia. WB Sauders, 119–120

Brandt, R. (1974). The non-marine aquatic Mollusca of Thailand. Arch Mollusk, 105 1–423.

Brockelman, W.Y., Upatham, E.S., Viyanant, V., Ardsungnoen, S., and Chantanawat, R. (1986). Field studies on the transmission of the human liver fluke, *Opisthorchis viverrini*, in northeast Thailand: population changes of the snail intermediate host. Int J Parasitol 16, 545–552.

Bronshtein, A.M. (1986). Morbidity from opisthorchiasis and diphyllobothriasis in the aboriginal population of the Kyshik village in the Khanty-Mansy Autonomous Region. Med Parazitol 3, 44–48.

Bronstein, A.M., Luchshev VI. (1998). Liver-fluke inflections: opisthorchiasis, clonorchiasis. The Russian Med J 6, 140–148.

Bundy, D.A., and Golden, M.H. (1987). The impact of host nutrition on gastrointestinal helminth populations. Parasitology 95, 623–635.

Bunnag, D., Harinasuta, T., and Desakorn, V. (1981). Studies on the chemotherapy of human opisthorchiasis: II. Clinical trial of niclofolan. Southeast Asian J Trop Med Public Health 12, 107–109.

Chai, J.Y., Park, J.H., Han, E.T., Guk, S.M., Shin, E.H., Lin, A., Kim, J.L., Sohn, W.M., Yong, T.S., Eom, K.S., Min, D.Y., Hwang, E.H., Phommmasack, B., Insisiengmay, B., and Rim, H.J. (2005). Mixed infections with *Opisthorchis viverrini* and intestinal flukes in residents of Vientiane Municipality and Saravane Province in Laos. J Helminthol 79, 283–289.

Chaicumpa, W., Ruangkunaporn, Y., Kalambaheti, T., Limavongpranee, S., Kitikoon, V., Khusmith, S., Pungpak, S., Chongsa-nguan, M., and Sornmani, S. (1991). Specific monoclonal antibodies to *Opisthorchis viverrini*. Int J Parasitol 21, 969–974.

Changbumrung, S., Tungtrongchitr, R., Hongtong, K., Supawan, V., Kwanbunjan, K., Prayurahong, B., Sitabutra, P., Srithip, U., Theerachai, A., Vudhivai, N., Migasena, P., and Sornamani, S. (1989). Food patterns and habits of people in and endemic area for liver fluke infection. J Nutr Assoc 23, 133–146.

Chen, M., Lu, Y., Hua, X., and Mott, K.E. (1994). Progress in assessment of morbidity due to *Clonorchis sinensis* infection: a review of recent literature. Trop Dis Bull 91, R7-R65.

Chen, S., Wu, F., Ke, S., and Pan, Z. (1997). Epidemiological survey on clonorchiasis sinensis in Yangxin County of Hubei Province of PR China. Southeast Asian J Trop Med Public Health 28 (suppl 1), 51–53.

Chitramvong, Y. (1992). The Bithyniidae (Gastropoda : Prosobranchia) of Thailand. Comparative external morphology. Malacol Rev 25, 23–38.

Choi, B.I., Han, J.K., Hong, S.T., and Lee, K.H. (2004). Clonorchiasis and cholangiocarcinoma: etiologic relationship and imaging diagnosis. Clin Microbiol Rev 17, 540–552.

Choi, D., Lim, J.H., Lee, K.T., Lee, J.K., Choi, S.H., Heo, J.S., Jang, K.T., Lee, N.Y., Kim, S., and Hong, S.T., (2006). Cholangiocarcinoma and *Clonorchis sinensis* infection: a case-control study in Korea. J Hepatol 44, 1066–1073.

Choi, D.W. (1976). *Clonorchis sinensis* in Kyungpook province, Korea: 2. Demonstration of metacercaria of *Clonorchis sinensis* from fresh-water fish. Kisaengchunghak Chapchi 14, 10–16.

Choi, D.W. (1984). *Clonorchis sinensis*: life cycle, intermediate hosts, transmission to man and geographical distribution in Korea. Arzneimittelforschung 34, 1145–1151.

Choi, M.H., Ryu, J.S., Lee, M., Li, S., Chung, B.S., Chai, J.Y., Sithithaworn, P., Tesana, S., and Hong, S.T. (2003). Specific and common antigens of *Clonorchis sinensis* and *Opisthorchis viverrini* (Opisthorchidae, Trematoda). Korean J Parasitol 41, 155–163.

Chotigeat, W., and Ruenwongsa, P. (1986). Types of collagen in *Opisthorchis viverrini* infected hamster liver. Mol Biochem Parasitol 18, 377–387.

Chung, B.S., Zhang, H., Choi, M.H., Jeon, D., Li, S., Lee, M., and Hong, S.T. (2004). Development of resistance to reinfection by *Clonorchis sinensis* in rats. Korean J Parasitol 42, 19–26.

Cross, J.H. (1984). Changing patterns of some trematode infections in Asia. Arzneimittelforschung 34, 1224–1226.

Dhiensiri, T., Eua-Ananta, Y., Bunnag, D., Harinasuta, T., and Schelp, P.F. (1984). Roentgenographically controlled healing of gallbladder lesions in opisthorchiasis after praziquantel treatment. Arzneimittelforschung 34, 1175–1177.

Ditrich, O., Scholz, T., and Giboda, M. (1990). Occurrence of some medically important flukes (Trematoda: Opisthorchiidae and Heterophyidae) in Nam Ngum water reservoir, Laos. Southeast Asian J Trop Med Public Health 21, 482–488.

Elkins, D.B., Haswell-Elkins, M., and Anderson, R.M. (1986). The epidemiology and control of intestinal helminths in the Pulicat Lake region of Southern India. I. Study design and pre- and post- treatment observations on Ascaris lumbricoides infection. Trans R Soc Trop Med Hyg 80, 774–792.

Elkins, D.B., Haswell-Elkins, M.R., Mairiang, E., Mairiang, P., Sithithaworn, P., Kaewkes, S., Bhudhisawasdi, V., and Uttaravichien, T. (1990). A high frequency of hepatobiliary disease and suspected cholangiocarcinoma associated with heavy *Opisthorchis viverrini* infection in a small community in north-east Thailand. Trans R Soc Trop Med Hyg 84, 715–719.

Elkins, D.B., Mairiang, E., Sithithaworn, P., Mairiang, P., Chaiyakum, J., Chamadol, N., Loapaiboon, V., and Haswell-Elkins, M.R. (1996). Cross-sectional patterns of hepatobiliary abnormalities and possible precursor conditions of cholangiocarcinoma associated with *Opisthorchis viverrini* infection in humans. Am J Trop Med Hyg 55, 295–301.

Elkins, D.B., Sithithaworn, P., Haswell-Elkins, M., Kaewkes, S., Awacharagan, P., and Wongratanacheewin, S. (1991). *Opisthorchis viverrini*: relationships between egg counts, worms recovered and antibody levels within an endemic community in northeast Thailand. Parasitology 102 (pt 2), 283–288.

Flavell, D.J. (1981). Liver-fluke infection as an aetiological factor in bile-duct carcinoma of man. Trans R Soc Trop Med Hyg 75, 814–824.

Flavell, D.J., and Flavell, S.U. (1986). *Opisthorchis viverrini*: pathogenesis of infection in immunodeprived hamsters. Parasite Immunol 8, 455–466.

Flavell, D.J., Flavell, S.U., and Field, G.F., (1983). *Opisthorchis viverrini*: the relationship between egg production, worm size and intensity of infection in the hamster. Trans R Soc Trop Med Hyg 77, 538–545.

Flavell, D.J., Pattanapanyasat, K., and Flavell, S.U. (1980). *Opisthorchis viverrini*: partial success in adoptively transferring immunity with spleen cells and serum in the hamster. J Helminthol 54, 191–197.

Giboda, M., Ditrich, O., Scholz, T., Viengsay, T., and Bouaphanh, S. (1991a). Current status of food-borne parasitic zoonoses in Laos. Southeast Asian J Trop Med Public Health 22 (suppl), 56–61.

Giboda, M., Viengsay, M., Bouaphan, S., and Ditrich, O. (1991b). [Epidemiology of intestinal parasitosis in Laos (with anti-amoebic antibody levels)]. Bull Soc Pathol Exot 84, 184–193.

Gibson, J.B., and Sun, T. (1971). Clonochiasis. Baltimore: Williams & Wilkins.

Haas, W., Granzer, M., and Brockelman, C.R. (1990). *Opisthorchis viverrini*: finding and recognition of the fish host by the cercariae. Exp Parasitol 71, 422–431.

Harinasuta, C., and Harinasuta, T. (1984). *Opisthorchis viverrini*: life cycle, intermediate hosts, transmission to man and geographical distribution in Thailand. Arzneimittelforschung 34, 1164–1167.

Harinasuta, C., and Vajrasthira, S. (1960). *Opisthorchis viverrini*: life cycle, intermediate hosts, transmission to man and geographical distribution in Thailand. Ann Trop Med Parasitol 54, 100–105.

Haswell-Elkins, M.R., Elkins, D.B., Sithithaworn, P., Treesarawat, P., and Kaewkes, S. (1991). Distribution patterns of *Opisthorchis viverrini* within a human community. Parasitology 103 (pt 1), 97–101.

Haswell-Elkins, M.R., Mairiang, E., Mairiang, P., Chaiyakum, J., Chamadol, N., Loapaiboon, V., Sithithaworn, P., and Elkins, D.B. (1994a). Cross-sectional study of *Opisthorchis viverrini* infection and cholangiocarcinoma in communities within a high-risk area in northeast Thailand. Int J Cancer 59, 505–509.

Haswell-Elkins, M.R., Satarug, S., Tsuda, M., Mairiang, E., Esumi, H., Sithithaworn, P., Mairiang, P., Saitoh, M., Yongvanit, P., and Elkins, D.B. (1994b). Liver fluke infection and cholangiocarcinoma: model of endogenous nitric oxide and extragastric nitrosation in human carcinogenesis. Mutat Res 305, 241–252.

Haswell-Elkins, M.R., Satarug, S., and Elkins, D.B. (1992). *Opisthorchis viverrini* infection in northeast Thailand and its relationship to cholangiocarcinoma. J Gastroenterol Hepatol 7, 538–548.

Hinz, E., Saowakontha, S., and Pipitgool, V. (1994). Opisthorchiasis control in northeast Thailand: proposal for a new approach. Appl Parasitol 35, 118–124.

Holzinger, F., Z'Graggen, K., and Buchler, M.W. (1999). Mechanisms of biliary carcinogenesis: a pathogenetic multi-stage cascade towards cholangiocarcinoma. Ann Oncol 10 (suppl 4), 122–126.

Hong, S.T., Choi, M.H., Kim, C.H., Chung, B.S., and Ji, Z. (2003). The Kato-Katz method is reliable for diagnosis of *Clonorchis sinensis* infection. Diagn Microbiol Infect Dis 47, 345–347.

Hong, S.T., Kho, W.G., Lee, M., Lee, J.S., and Lee, S.H. (1997). Immunoblot patterns of clonorchiasis. Korean J Parasitol 35, 87–93.

Hong, S.T., Lee, M., Sung, N.J., Cho, S.R., Chai, J.Y., and Lee, S.H. (1999). Usefulness of IgG4 subclass antibodies for diagnosis of human clonorchiasis. Korean J Parasitol 37, 243–248.

Honjo, S., Srivatanakul, P., Sriplung, H., Kikukawa, H., Hanai, S., Uchida, K., Todoroki, T., Jedpiyawongse, A., Kittiwatanachot, P., Sripa, B., Deerasamee, S., and Miwa, M. (2005). Genetic and environmental determinants of risk for cholangiocarcinoma via *Opisthorchis viverrini* in a densely infested area in Nakhon Phanom, northeast Thailand. Int J Cancer 117, 854–860.

Hou, P.C. (1955). The pathology of *Clonorchis sinensis* infestation of the liver. J Pathol Bacteriol 70, 53–64.

Hussain, S.P., Hofseth, L.J., and Harris, C.C. (2003). Radical causes of cancer. Nat Rev Cancer 3, 276–285.

Hutradilok, N., Ruenwongsa, P., Thamavit, W., and Upatham, E.S. (1983). Liver collagen in *Opisthorchis viverrini* infected hamsters following praziquantel treatment. Southeast Asian J Trop Med Public Health 14, 290–293.

IARC. (1994). Schistosomes, liver flukes and Helicobacter pylori. IARC Working Group on the Evaluation of Carcinogenic Risks to Humans. Lyon, 7–14 June 1994. IARC Monogr Eval Carcinog Risks Hum 61, 1–241.

Inoue, S., and Kawanishi, S. (1995). Oxidative DNA damage induced by simultaneous generation of nitric oxide and superoxide. FEBS Lett 371, 86–88.

Itoh, M., Pairojkul, C., Thamawit, W., Sithithaworn, P., Tiwawech, D., Uttaravicien, T., Shirai, T., and Ito, N. (1994). Association of antibodies to *Opisthorchis viverrini* with hepatobiliary disease in northeastern Thailand. Am J Trop Med Hyg 51, 424–429.

Janechaiwat, J., Tharavanij, S., Vajrasthira, S., and Chaicumpa, W. (1980). The immunological diagnosis of human opisthorchiasis and the humoral immune response to opisthorchis infection in the hamster. J Med Assoc Thai 63, 439–447.

Jongsuksuntigul, P. (2002). Parasitic diseases in northeast Thailand. In: Seminar in Parasitic Diseases in Northeast Thailand. Khon Kaen, Thailand.

Jongsuksuntigul, P., Chaychumsri, W., Techamontrikul, P., Cheeradit, P., and Suratawanich, P. (1992). Studies on prevalence and intensity of intestinal helminthiasis and liver fluke in Thailand in 1991. J Trop Med Assoc Thai 2, 80–95.

Jongsuksuntigul, P., and Imsomboon, T. (2003). Opisthorchiasis control in Thailand. Acta Trop 88, 229–232.

Joo, C.Y. (1980). Epidemiological studies of Clonorchis sinesis in vicinity of river Taewha, Kyungnam province, Korea. Korean J Parasitof 18,199–214.

Kaewkes, S., Elkins, D.B., Sithithaworn, P., and Haswell-Elkins, M.R. (1991). Comparative studies on the morphology of the eggs of *Opisthorchis viverrini* and lecithodendriid trematodes. Southeast Asian J Trop Med Public Health 22, 623–630.

Kaneko, T., and Tahara, S. (2000). Formation of 8-oxo-2′-deoxyguanosine in the DNA of human diploid fibroblasts by treatment with linoleic acid hydroperoxide and ferric ion. Lipids 35, 961–965.

Kang, S.Y., Kim, S.I., and Cho, S.Y. (1985). Seasonal variations of metacercarial density of *Clonorchis sinensis* in fish intermediate host, Pseudorasbora parva. Kisaengchunghak Chapchi 23, 87–94.

Keiser, J., and Utzinger, J. (2005). Emerging foodborne trematodiasis. Emerg Infect Dis 11, 1507–1514.

Khamboonruang, C., Keawvichit, R., Wongworapat, K., Suwanrangsi, S., Hongpromyart, M., Sukhawat, K., Tonguthai, K., and Lima dos Santos, C.A. (1997). Application of hazard analysis critical control point (HACCP) as a possible control measure for *Opisthorchis viverrini* infection in cultured carp (Puntius gonionotus). Southeast Asian J Trop Med Public Health 28, 65–72.

Kim, Y.I. (1984). Liver carcinoma and liver fluke infection. Arzneimittelforschung 34, 1121–1126.

Kim, Y.K., Kang, S.Y., and Lee, S.H. (1979). [Study on the frequency distribution of the metacercarial density of *Clonorchis sinensis* in fish host, Pseudorasbora parva]. Kisaengchunghak Chapchi 17, 127–131.

Kirby, G.M., Pelkonen, P., Vatanasapt, V., Camus, A.M., Wild, C.P., and Lang, M.A. (1994). Association of liver fluke (*Opisthorchis viverrini*) infestation with increased expression of cytochrome P450 and carcinogen metabolism in male hamster liver. Mol Carcinog 11, 81–89.

Kobayashi, J., Vannachone, B., Sato, Y., Manivong, K., Nambanya, S., and Inthakone, S. (2000). An epidemiological study on *Opisthorchis viverrini* infection in Lao villages. Southeast Asian J Trop Med Public Health 31, 128–132.

Kobayashi, J., Vannachone, B., Xeutvongsa, A., Manivang, K., Ogawa, S., Sato, Y., and Pholsena, K. (1996). Prevalence of intestinal parasitic infection among children in two villages in Lao PDR. Southeast Asian J Trop Med Public Health 27, 562–565.

Kurathong, S., Lerdverasirikul, P., Wongpaitoon, V., Pramoolsinsap, C., and Upatham, E.S. (1987). *Opisthorchis viverrini* infection in rural and urban communities in northeast Thailand. Trans R Soc Trop Med Hyg 81, 411–414.

Le, T.H., Van De, N., Blair, D., Sithithaworn, P., and McManus, D.P. (2006). *Clonorchis sinensis* and *Opisthorchis viverrini:* development of a mitochondrial-based multiplex PCR for their identification and discrimination. Exp Parasitol 112, 109–114.

Lee, S.U., and Huh, S. (2004). Variation of nuclear and mitochondrial DNAs in Korean and Chinese isolates of *Clonorchis sinensis*. Korean J Parasitol 42, 145–148.

Li, B. (1985). Discovery of Parafossarulus anomaloispiralis as the first intermediate host of *Clonorchis sinensis*. Zhongguo Ji Sheng Chong Xue Yu Ji Sheng Chong Bing Za Zhi 6, 199.

Li, X. (1964). Survey on clonorchiasis in Fuyang, Anhui. Chinese J Hyg 9, 98.

Lim, M.K., Ju, Y.H., Franceschi, S., Oh, J.K., Kong, H.J., Hwang, S.S., Park, S.K., Cho, S.I., Sohn, W.M., Kim, D.I., Yoo, K.Y., Hong, S.T., and Shin, H.R. (2006). *Clonorchis sinensis* infection and increasing risk of cholangiocarcinoma in the Republic of Korea. Am J Trop Med Hyg 75, 93–96.

Lun, Z.R., Gasser, R.B., Lai, D.H., Li, A.X., Zhu, X.Q., Yu, X.B., and Fang, Y.Y. (2005). Clonorchiasis: a key foodborne zoonosis in China. Lancet Infect Dis 5, 31–41.

Mairiang, E., Elkins, D.B., Mairiang, P., Chaiyakum, J., Chamadol, N., Loapaiboon, V., Posri, S., Sithithaworn, P., and Haswell-Elkins, M. (1992). Relationship between intensity of *Opisthorchis viverrini* infection and hepatobiliary disease detected by ultrasonography. J Gastroenterol Hepatol 7, 17–21.

Mairiang, E., Haswell-Elkins, M.R., Mairiang, P., Sithithaworn, P., and Elkins, D.B. (1993). Reversal of biliary tract abnormalities associated with *Opisthorchis viverrini* infection following praziquantel treatment. Trans R Soc Trop Med Hyg 87, 194–197.

Mairiang, E., and Mairiang, P. (2003). Clinical manifestation of opisthorchiasis and treatment. Acta Trop 88, 221–227.

Na, B.K., Lee, H.J., Cho, S.H., Lee, H.W., Cho, J.H., Kho, W.G., Lee, J.S., Song, K.J., Park, P.H., Song, C.Y., and Kim, T.S. (2002). Expression of cysteine proteinase of *Clonorchis sinensis* and its use in serodiagnosis of clonorchiasis. J Parasitol 88, 1000–1006.

Nagano, I., Pei, F., Wu, Z., Wu, J., Cui, H., Boonmars, T., and Takahashi, Y. (2004). Molecular expression of a cysteine proteinase of *Clonorchis sinensis* and its application to an enzyme-linked immunosorbent assay for immunodiagnosis of clonorchiasis. Clin Diagn Lab Immunol 11, 411–416.

Nakajima, T., Kondo, Y., Miyazaki, M., and Okui, K. (1988). A histopathologic study of 102 cases of intrahepatic cholangiocarcinoma: histologic classification and modes of spreading. Hum Pathol 19, 1228–1234.

Ohshima, H., Bandaletova, T.Y., Brouet, I., Bartsch, H., Kirby, G., Ogunbiyi, F., Vatanasapt, V., and Pipitgool, V. (1994). Increased nitrosamine and nitrate biosynthesis mediated by nitric oxide synthase induced in hamsters infected with liver fluke (*Opisthorchis viverrini*). Carcinogenesis 15, 271–275.

Ohshima, H., Tatemichi, M., and Sawa, T. (2003). Chemical basis of inflammation-induced carcinogenesis. Arch Biochem Biophys 417, 3–11.

Ooi, H.K., Chen, C.I., Lin, S.C., Tung, K.C., Wang, J.S., and Kamiya, M. (1997). Metacercariae in fishes of Sun Moon lake which is an endemic area for *Clonorchis sinensis* in Taiwan. Southeast Asian J Trop Med Public Health 28 (suppl 1), 222–223.

Pairojkul, C., Shirai, T., Hirohashi, S., Thamavit, W., Bhudhisawat, W., Uttaravicien, T., Itoh, M., and Ito, N. (1991). Multistage carcinogenesis of liver-fluke-associated cholangiocarcinoma in Thailand. Princess Takamatsu Symp 22, 77–86.

Park, G.M. (2007). Genetic comparison of liver flukes, *Clonorchis sinensis* and *Opisthorchis viverrini,* based on rDNA and mtDNA gene sequences. Parasitol Res 100, 351–357.

Park, G.M., Yong, T.S., Im, K., and Lee, K.J. (2000). Isozyme electrophoresis patterns of the liver fluke, *Clonorchis sinensis* from Kimhae, Korea and from Shenyang, China. Korean J Parasitol 38, 45–48.

Park, J.H., Guk, S.M., Kim, T.Y., Shin, E.H., Lin, A., Park, J.Y., Kim, J.L., Hong, S.T., and Chai, J.Y. (2004). *Clonorchis sinensis* metacercarial infection in the pond smelt Hypomesus olidus and the minnow Zacco platypus collected from the Soyang and Daechung Lakes. Korean J Parasitol 42, 41–44.

Parkin, D.M., Ohshima, H., Srivatanakul, P., and Vatanasapt, V. (1993). Cholangiocarcinoma: epidemiology, mechanisms of carcinogenesis and prevention. Cancer Epidemiol Biomarkers Prev 2, 537–544.

Pauly, A., Schuster, R., and Steuber, S. (2003). Molecular characterization and differentiation of opisthorchiid trematodes of the species *Opisthorchis felineus* (Rivolta, 1884) and Metorchis bilis (Braun, 1790) using polymerase chain reaction. Parasitol Res 90, 409–414.

Pikarsky, E., Porat, R.M., Stein, I., Abramovitch, R., Amit, S., Kasem, S., Gutkovich-Pyest, E., Urieli-Shoval, S., Galun, E., and Ben-Neriah, Y. (2004). NF-kappaB functions as a tumour promoter in inflammation-associated cancer. Nature 431, 461–466.

Pinlaor, S., Hiraku, Y., Yongvanit, P., Tada-Oikawa, S., Ma, N., Pinlaor, P., Sithithaworn, P., Sripa, B., Murata, M., Oikawa, S., and Kawanishi, S. (2006). iNOS-dependent DNA damage via NF-kappaB expression in hamsters infected with *Opisthorchis viverrini* and its suppression by the antihelminthic drug praziquantel. Int J Cancer 119, 1067–1072.

Pinlaor, S., Hiraku, Y., Ma, N., Yongvanit, P., Semba, R., Oikawa, S., Murata, M., Sripa, B., Sithithaworn, P., and Kawanishi, S. (2004a). Mechanism of NO-mediated oxidative and nitrative DNA damage in hamsters infected with *Opisthorchis viverrini:* a model of inflammation-mediated carcinogenesis. Nitric Oxide 11, 175–183.

Pinlaor, S., Ma, N., Hiraku, Y., Yongvanit, P., Semba, R., Oikawa, S., Murata, M., Sripa, B., Sithithaworn, P., and Kawanishi, S. (2004b). Repeated infection with *Opisthorchis viverrini* induces accumulation of 8-nitroguanine and 8-oxo-7,8-dihydro-2′-deoxyguanine in the bile duct of hamsters via inducible nitric oxide synthase. Carcinogenesis 25, 1535–1542.

Pinlaor, S., Sripa, B., Sithithaworn, P., and Yongvanit, P. (2004c). Hepatobiliary changes, antibody response, and alteration of liver enzymes in hamsters re-infected with *Opisthorchis viverrini.* Exp Parasitol 108, 32–39.

Pinlaor, S., Sripa, B., Ma, N., Hiraku, Y., Yongvanit, P., Wongkham, S., Pairojkul, C., Bhudhisawasdi, V., Oikawa, S., Murata, M., Semba, R., and Kawanishi, S. (2005). Nitrative and oxidative DNA damage in intrahepatic cholangiocarcinoma patients in relation to tumor invasion. World J Gastroenterol 11, 4644–4649.

Pinlaor, S., Yongvanit, P., Hiraku, Y., Ma, N., Semba, R., Oikawa, S., Murata, M., Sripa, B., Sithithaworn, P., and Kawanishi, S. (2003). 8-nitroguanine formation in the liver of hamsters infected with *Opisthorchis viverrini.* Biochem Biophys Res Commun 309, 567–571.

Poopyruchpong, N., Viyanant, V., Upatham, E.S., and Srivatanakul, P. (1990). Diagnosis of opisthorchiasis by enzyme-linked immunosorbent assay using partially purified antigens. Asian Pac J Allergy Immunol 8, 27–31.

Preuksaraj, S. (1984). Public health aspects of opisthorchiasis in Thailand. Drugs Res 34, 1119–1120.

Pungpak, S., Sornmani, S., Suntharasamai, P., and Vivatanasesth, P. (1989). Ultrasonographic study of the biliary system in opisthorchiasis patients after treatment with praziquantel. Southeast Asian J Trop Med Public Health 20, 157–162.

Quan, F.S., Matsumoto, T., Shin, Y.O., Min, Y.K., Yang, H.M., Othman, T., and Lee, J.B. (2004). Relationships between IgG, IgM, IgE and resistance to reinfection during the early phase of infection with *Clonorchis sinensis* in rats. Immunol Invest 33, 51–60.

Radomyos, B., Wongsaroj, T., Wilairatana, P., Radomyos, P., Praevanich, R., Meesomboon, V., and Jongsuksuntikul, P. (1998). Opisthorchiasis and intestinal fluke infections in northern Thailand. Southeast Asian J Trop Med Public Health 29, 123–127.

Radomyos, P., Radomyos, B., and Tungtrongchitr, A. (1994). Multi-infection with helminths in adults from northeast Thailand as determined by post-treatment fecal examination of adult worms. Trop Med Parasitol 45, 133–135.

Ramsay, R.J., Sithithaworn, P., Prociv, P., Moorhouse, D.E., and Methaphat, C. (1989). Density-dependent fecundity of *Opisthorchis viverrini* in humans, based on faecal recovery of flukes. Trans R Soc Trop Med Hyg 83, 241–242.

Riganti, M., Pungpak, S., Punpoowong, B., Bunnag, D., and Harinasuta, T. (1989). Human pathology of *Opisthorchis viverrini* infection: a comparison of adults and children. Southeast Asian J Trop Med Public Health 20, 95–100.

Riganti, M., Pungpak, S., Sachakul, V., Bunnag, D., and Harinasuta, T. (1988). *Opisthorchis viverrini* eggs and adult flukes as nidus and composition of gallstones. Southeast Asian J Trop Med Public Health 19, 633–636.

Rim, H.J. (1986). The current pathobiology and chemotherapy of clonorchiasis. Kisaengchunghak Chapchi 24 (suppl), 1–141.

Rim, H.J. (2005). Clonorchiasis: an update. J Helminthol 79, 269–281.

Rim, H.J., Chai, J.Y., Min, D.Y., Cho, S.Y., Eom, K.S., Hong, S.J., Sohn, W.M., Yong, T.S., Deodato, G., Standgaard, H., Phommasack, B., Yun, C.H., and Hoang, E.H. (2003). Prevalence of intestinal parasite infections on a national scale among primary schoolchildren in Laos. Parasitol Res 91, 267–272.

Ruangkunaporn, Y., Akai, P.S., Chongsa-nguan, M., Sri-Yeythong, S., Kitikoon, V., and Chaicumpa, W. (1994). Changes in serum antibodies to *Opisthorchis viverrini* in humans and hamsters following treatment of opisthorchiasis. Asian Pac J Allergy Immunol 12, 83–84.

Sadun, E.H. (1955). Studies on *Opisthorchis viverrini* in Thailand. Am J Hyg 62, 81–115.

Saijuntha, W., Sithithaworn, P., Wongkham, S., Laha, T., Pipitgool, V., Petney, T.N., and Andrews, R.H. (2006a). Genetic markers for the identification and characterization of *Opisthorchis viverrini,* a medically important food borne trematode in Southeast Asia. Acta Tropica 100, 246–251.

Saijuntha, W., Sithithaworn, P., Wongkham, S., Laha, T., Pipitgool, V., Petney, T.N., Chilton, N.B., and Andrews, R.H. (2006b). Enzyme markers to identify and characterise *Opisthorchis viverrini* in Thailand and Laos. Southeast Asian J Trop Med Public Health 37(suppl), 43–47.

Saijuntha, W., Sithithaworn, P., Wongkham, S., Laha, T., Pipitgool, V., Petney, T.N., Chilton, N.B., Tesana, S., and Andrews, R.H. (2007). Evidence of a species complex within the food-borne trematode *Opisthorchis viverrini* and possible coevolution with their first intermediate hosts. Int J Parasitol 37, 695–703.

Saowakontha, S., Pipitgool, V., Pariyanonda, S., Tesana, S., Rojsathaporn, K., and Intarakhao, C. (1993). Field trials in the control of *Opisthorchis viverrini* with an integrated programme in endemic areas of northeast Thailand. Parasitology 106, 283–288.

Satarug, S., Haswell-Elkins, M.R., Sithithaworn, P., Bartsch, H., Ohshima, H., Tsuda, M., Mairiang, P., Mairiang, E., Yongvanit, P., Esumi, H., and Elkins, D.B. (1998). Relationships between the synthesis of N-nitrosodimethylamine and immune responses to chronic infection with the carcinogenic parasite, *Opisthorchis viverrini,* in men. Carcinogenesis 19, 485–491.

Satarug, S., Haswell-Elkins, M.R., Tsuda, M., Mairiang, P., Sithithaworn, P., Mairiang, E., Esumi, H., Sukprasert, S., Yongvanit, P., and Elkins, D.B. (1996a). Thiocyanate-independent nitrosation in humans with carcinogenic parasite infection. Carcinogenesis 17, 1075–1081.

Satarug, S., Lang, M.A., Yongvanit, P., Sithithaworn, P., Mairiang, E., Mairiang, P., Pelkonen, P., Bartsch, H., and Haswell-Elkins, M.R. (1996b). Induction of cytochrome P450 2A6 expression in humans by the carcinogenic parasite infection, opisthorchiasis viverrini. Cancer Epidemiol Biomarkers Prev 5, 795–800.

Schad, G.A., and Anderson, R.M. (1985). Predisposition to hookworm infection in humans. Science 228, 1537–1540.

Scholz, T., Ditrich, O., and Giboda, M. (1992). Study on the surface morphology of the developmental stages of the liver fluke, *Opisthorchis viverrini* (Trematoda: Opisthorchiidae). Ann Parasitol Hum Comp 67, 82–90.

Seo B.S., Lee, S.H., Cho, S.Y., Chai, J.Y., Hong, S.T., Han, I.S., Sohn, J.S., Cho, B.H., Ahh, S.R., Lee, S.K., Chung, S.C., Kang, K.S., Shin, H.S. and Hwang, I.S. (1981). An Epidemiologic Study On Clonarchiasis And Metagonimiasis In Reverside Areas In Korea. Korea J Parasitol 19, 137–150.

Sermswan, R., Mongkolsuk, S., Panyim, S., and Sirisinha, S. (1991). Isolation and characterization of *Opisthorchis viverrini* specific DNA probe. Mol Cell Probes 5, 399–407.

Shirai, T., Pairojkul, C., Ogawa, K., Naito, H., Thamavit, W., Bhudhisawat, W., and Ito, N. (1992). Histomorphological characteristics of cholangiocellular carcinomas in northeast Thailand, where a region infection with the liver fluke, *Opisthorchis viverrini* is endemic. Acta Pathol Jpn 42, 734–739.

Sirisinha, S., Chawengkirttikul, R., Haswell-Elkins, M.R., Elkins, D.B., Kaewkes, S., and Sithithaworn, P. (1995). Evaluation of a monoclonal antibody-based enzyme linked immunosorbent assay for the diagnosis of *Opisthorchis viverrini* infection in an endemic area. Am J Trop Med Hyg 52, 521–524.

Sirisinha, S., Chawengkirttikul, R., Sermswan, R., Amornpant, S., Mongkolsuk, S., and Panyim, S. (1991). Detection of *Opisthorchis viverrini* by monoclonal antibody-based ELISA and DNA hybridization. Am J Trop Med Hyg 44, 140–145.

Sirisinha, S., Puengtomwatanakul, S., Sobhon, P., Saitongdee, P., Wongpayabal, P., Mitranonde, V., Radomyos, P., Bunnag, D., and Harinasuta, T. (1984). Alterations of the surface tegument of *Opisthorchis viverrini* exposed to praziquantel in vitro and in vivo. Southeast Asian J Trop Med Public Health 15, 95–103.

Sirisinha, S., Sahassananda, D., Bunnag, D., and Rim, H.J. (1990). Immunological analysis of *Opisthorchis* and *Clonorchis* antigens. J Helminthol 64, 133–138.

Sirisinha, S., Tuti, S., Tawatsin, A., Vichasri, S., Upatham, E.S., and Bunnag, D. (1983a). Attempts to induce protective immunity in hamsters against infection by a liver fluke of man (*Opisthorchis viverrini*). Parasitology 86, 127–136.

Sirisinha, S., Tuti, S., Vichasri, S., and Tawatsin, A. (1983b). Humoral immune responses in hamsters infected with *Opisthorchis viverrini*. Southeast Asian J Trop Med Public Health 14, 243–251.

Sirisinha, S., and Wongratanacheewin, S. (1986). Immunization of hamsters against *Opisthorchis viverrini* infection. Southeast Asian J Trop Med Public Health 17, 567–573.

Sithithaworn, P., and Haswell-Elkins, M. (2003). Epidemiology of *Opisthorchis viverrini*. Acta Trop 88, 187–194.

Sithithaworn, P., Nuchjungreed, C., Srisawangwong, T., Ando, K., Petney, T.N., Chilton, N.B., and Andrews, R.H. (2006). Genetic variation in *Opisthorchis viverrini* (Trematoda: Opisthorchiidae) from northeast Thailand and Laos PDR based on random amplified polymorphic DNA analyses. Parasitol Res 100, 613–617.

Sithithaworn, P., Pipitgool, V., Srisawangwong, T., Elkins, D.B., and Haswell-Elkins, M.R. (1997). Seasonal variation of *Opisthorchis viverrini* infection in cyprinoid fish in north-east Thailand: implications for parasite control and food safety. Bull WHO 75, 125–131.

Sithithaworn, P., Tesana, S., Pipitgool, V., Kaewkes, S., Pairojkul, C., Sripa, B., Paupairoj, A., and Thaiklar, K. (1991a). Relationship between faecal egg count and worm burden of *Opisthorchis viverrini* in human autopsy cases. Parasitology 102 (pt 2), 277–281.

Sithithaworn, P., Tesana, S., Pipitgool, V., Kaewkes, S., Thaiklar, K., Pairojkul, C., Sripa, B., Paupairoj, A., Sanpitak, P., and Aranyanat, C. (1991b). Quantitative post-mortem study of *Opisthorchis viverrini* in man in north-east Thailand. Trans R Soc Trop Med Hyg 85, 765–768.

Sohn, W.M., Zhang, H., Choi, M.H., and Hong, S.T. (2006). Susceptibility of experimental animals to reinfection with *Clonorchis sinensis*. Korean J Parasitol 44, 163–166.

Sonakul D., Koompirochana, C., China, K., and Stitnimakarn, T. (1978). Hepatic carcinoma with opisthorchiasis. Southeast Asian J Trop Med Public Health 9, 215–219.

Song S.B. (1982). Epidemiological studies of Clonorchis sinersis in lower area of Nag Dong river nearby Busan City i Korea. Korea J Parasitol 20,133–141.

Sornmani, S., Schelp, F.P., Vivatanasesth, P., Patihatakorn, W., Impand, P., Sitabutra, P., Worasan, P., and Preuksaraj, S. (1984). A pilot project for controlling *O. viverrini* infection in Nong Wai, Northeast Thailand, by applying praziquantel and other measures. Arzneimittelforschung 34, 1231–1234.

Sripa, B., and Kaewkes, S. (2000a). Localisation of parasite antigens and inflammatory responses in experimental opisthorchiasis. Int J Parasitol 30, 735–740.

Sripa, B., and Kaewkes, S. (2000b). Relationship between parasite-specific antibody responses and intensity of *Opisthorchis viverrini* infection in hamsters. Parasite Immunol 22, 139–145.

Sripa, B., and Kaewkes, S. (2002). Gall bladder and extrahepatic bile duct changes in *Opisthorchis viverrini*-infected hamsters. Acta Trop 83, 29–36.

Sripa, B., Kanla, P., Sinawat, P., and Haswell-Elkins, M.R. (2004). Opisthorchiasis-associated biliary stones: light and scanning electron microscopic study. World J Gastroenterol 10, 3318–3321.

Srisawangwong, T., Sithithaworn, P., and Tesana, S. (1997). Metacercariae isolated from cyprinoid fishes in Khon Kaen District by digestion technic. Southeast Asian J Trop Med Public Health 28, 224–226.

Srivatanakul, K., Sontipong, S., Chotiwan, P., and Parkin, D.M. (1988). Liver cancer in Thailand: temporal and geographic variations. J Gastroenterol Hepatol 3, 413–420.

Srivatanakul, P., Ohshima, H., Khlat, M., Parkin, M., Sukarayodhin, S., Brouet, I., and Bartsch, H. (1991). Endogenous nitrosamines and liver fluke as risk factors for cholangiocarcinoma in Thailand. IARC Sci Publ 105, 88–95.

Srivatanakul, P., Viyanant, V., Kurathong, S., and Tiwawech, D. (1985). Enzyme-linked immunosorbent assay for detection of *Opisthorchis viverrini* infection. Southeast Asian J Trop Med Public Health 16, 234–239.

Stensvold, C.R., Saijuntha, W., Sithithaworn, P., Wongratanacheewin, S., Strandgaard, H., Ornbjerg, N., and Johansen, M.V. (2006). Evaluation of PCR based coprodiagnosis of human opisthorchiasis. Acta Trop 97, 26–30.

Strandgaard, H. (2006). Applied Aspects of Opisthorchiasis Control in Lao PDR. Faculty of Science. Denmark: University of Copenhagen, 64.

Sukontason, K., Piangjai, S., Muangyimpong, Y., Methanitikorn, R., and Chaithong, U. (1999). Prevalence of trematode metacercariae in cyprinoid fish of Ban Pao district, Chiang Mai Province, northern Thailand. Southeast Asian J Trop Med Public Health 30, 365–370.

Sun, T. (1984). Pathology and immunology of *Clonorchis sinensis* infection of the liver. Ann Clin Lab Sci 14, 208–215.

Tansurat, P. (1971). Opisthorchiasis. In Marcial-Rojas RA, ed. Pathology of Protozoal and Helminthic Diseases. Baltimore: Williams & Wilkins, 536–545.

Terada, T., and Nakanuma, Y. 1992. Pathological observations of intrahepatic peribiliary glands in 1,000 consecutive autopsy livers. Hyperplasia of intramural and extramural glands. Hum Pathof 23, 438–490.

Tesana, S., Bhuripanyo, K., Sanpitak, P., and Sithithaworn, P. (1983). Intestinal capillariasis from Udon Thani province, northeastern part of Thailand: report of an autopsy case. J Med Assoc Thai 66, 128–131.

Tesana, S., Kaewkes, S., Srisawangwong, T., and Pinlaor, P. (1985). The distribution and density of *Opisthorchis viverrini* metacercariae in cyprinoid fish ni Khon Kaen Province. J Parasitol Trop Med Med Ass 8, 36–39.

Thamavit, W., Moore, M.A., Hiasa, Y., and Ito, N. (1988). Enhancement of DHPN induced hepatocellular, cholangiocellular and pancreatic carcinogenesis by *Opisthorchis viverrini* infestation in Syrian golden hamsters. Carcinogenesis 9, 1095–1098.

Thammapalerd, N., Tharavanij, S., Nacapunchai, D., Bunnag, D., Radomyos, P., and Prasertsiriroj, V. (1988). Detection of antibodies against *Opisthorchis viverrini* in patients before and after treatment with praziquantel. Southeast Asian J Trop Med Public Health 19, 101–108.

Thuwajit, C., Thuwajit, P., Kaewkes, S., Sripa, B., Uchida, K., Miwa, M., and Wongkham, S. (2004). Increased cell proliferation of mouse fibroblast NIH-3T3 in vitro induced by excretory/secretory product(s) from *Opisthorchis viverrini*. Parasitology 129, 455–464.

Upatham, E.S., and Sukhapanth, N. (1980). Field studies on the bionomics of Bithynia siamensis siamensis and the transmission of *Opisthorchis viverrini* in Bangna, Bangkok, Thailand. Southeast Asian J Trop Med Public Health 11, 355–358.

Upatham, E.S., Viyanant, V., Brockelman, W.Y., Kurathong, S., Lee, P., and Kraengraeng, R. (1988). Rate of re-infection by *Opisthorchis viverrini* in an endemic northeast Thai community after chemotherapy. Int J Parasitol 18, 643–649.

Upatham, E.S., Viyanant, V., Kurathong, S., Brockelman, W.Y., Menaruchi, A., Saowakontha, S., Intarakhao, C., Vajrasthira, S., and Warren, K.S. (1982). Morbidity in relation to intensity of infection in *Opisthorchiasis viverrini*: study of a community in Khon Kaen, Thailand. Am J Trop Med Hyg 31, 1156–1163.

Upatham, E.S., Viyanant, V., Kurathong, S., Rojborwonwitaya, J., Brockelman, W.Y., Ardsungnoen, S., Lee, P., and Vajrasthira, S. (1984). Relationship between prevalence and intensity of *Opisthorchis viverrini* infection, and clinical symptoms and signs in a rural community in north-east Thailand. Bull WHO 62, 451–461.

Vatanasapt, V., Kosuwon, W., and Pengsaa, P. (1993). Unit cost analysis in a university hospital: an example from Srinagarind Hospital, Khon Kaen. J Med Assoc Thai 76, 647–653.

Vichasri, S., Viyanant, V., and Upatham, E.S. (1982). *Opisthorchis viverrini* : intensity and rates of infection in cyprinoid fish from an endemic focus in Northeast Thailand. Southeast Asian J Trop Med Public Health 13, 138–141.

Viranuvatti, V., and Stitnimarnkarn, T. (1972). Liver fluke infection and infestation in Southeast Asia. In Schaffner F, ed. Progress in Liver Disease. New York: Grune & Stratton, 537–547.

Viyanant, V., Brockelman, W.Y., Lee, P., Ardsungnoen, S., and Upatham, E.S. (1983). A comparison of a modified quick-Kato technique and the Stoll dilution method for field examination for *Opisthorchis viverrini* eggs. J Helminthol 57, 191–195.

Waikagul, J. (1998). *Opisthorchis viverrini* metacercaria in Thai freshwater fish. Southeast Asian J Trop Med Public Health 29, 324–326.

WHO. (1995). Control of Foodborne Trematode Infection. WHO Technical Report Series 849.

WHO. (2004). Report of Joint WHO/FAO workshop on food-borne trematode infections in Asia. 1–58.

Wongratanacheewin, S., Bunnag, D., Vaeusorn, N., and Sirisinha, S. (1988a). Characterization of humoral immune response in the serum and bile of patients with opisthorchiasis and its application in immunodiagnosis. Am J Trop Med Hyg 38, 356–362.

Wongratanacheewin, S., Charupatana, C., Bunnag, D., and Sirisinha, S. (1988b). Effect of praziquantel treatment on antibody levels and lymphoproliferative responses in patients with opisthorchiasis. Southeast Asian J Trop Med Public Health 19, 109–116.

Wongratanacheewin, S., Chawengkirttikul, R., Bunnag, D., and Sirisinha, S. (1988c). Analysis of *Opisthorchis viverrini* antigens by immunoprecipitation and polyacrylamide gel electrophoresis. Parasitology 96, 119–128.

Wongratanacheewin, S., Pumidonming, W., Sermswan, R.W., and Maleewong, W. (2001). Development of a PCR-based method for the detection of *Opisthorchis viverrini* in experimentally infected hamsters. Parasitology 122, 175–180.

Wongratanacheewin, S., Pumidonming, W., Sermswan, R.W., Pipitgool, V., and Maleewong, W. (2002). Detection of *Opisthorchis viverrini* in human stool specimens by PCR. J Clin Microbiol 40, 3879–3880.

Wongratanacheewin, S., Rattanasiriwilai, W., Priwan, R., and Sirisinha, S. (1987). Immunodepression in hamsters experimentally infected with *Opisthorchis viverrini*. J Helminthol 61, 151–156.

Wongratanacheewin, S., Sermswan, R.W., and Sirisinha, S. (2003). Immunology and molecular biology of *Opisthorchis viverrini* infection. Acta Trop 88, 195–207.

Wongsaroj, T., Sakolvaree, Y., Chaicumpa, W., Maleewong, W., Kitikoon, V., Tapchaisri, P., Chongsa-nguan, M., and Cross, J.H. (2001). Affinity purified oval antigen for diagnosis of *Opisthorchiasis viverrini*. Asian Pac J Allergy Immunol 19, 245–258.

Wykoff, D.E., Chittayasothorn, K., and Winn, M.M. (1966). Clinical manifestations of *Opisthorchis viverrini* infections in Thailand. Am J Trop Med Hyg 15, 914–918.

Wykoff, D.E., Harinasuta, C., Juttijudata, P., and Winn, M.M. (1965). *Opisthorchis viverrini* in Thailand—the life cycle and comparison with *O. felineus*. J Parasitol 51, 207–214.

Zhou, W. (1985). Seasonal fluctuation of C. sinensis infection in Bithynia Fuchsiana. J Parasitol Parasit Dis 3, 159.

Zou, H., Peng, Y., Cai, W., and Lu, S. (1994). [Studies on Clonorchiasis sinensis control in Sanshui City, Guangdong province]. Zhongguo Ji Sheng Chong Xue Yu Ji Sheng Chong Bing Za Zhi 12, 294–296.

2
Intestinal Flukes

Jong-Yil Chai

Around the world, 40 to 50 million people are currently estimated to be infected with food-borne intestinal trematodes (Fried et al., 2004), including at least 18 million people infected by fish-borne trematodes (Chai et al., 2005a). However, this may be an underestimate of the total number of humans infected. The number of trematode species, currently known to be involved, is 70 (Yu and Mott, 1994; Chai and Lee, 2002). Morphologically they are diverse, belonging to the families Heterophyidae, Echinostomatidae, Plagiorchiidae, Lecithodendriidae, Neodiplostomidae, Nanophyetidae, Paramphistomatidae, Cathaemaciidae, Fasciolidae, Gastrodiscidae, Gymnophallidae, Microphallidae, Strigeidae, and Brachylaimidae (Yu and Mott, 1994; Chai and Lee, 2002; Fried et al., 2004). Life cyles and geographical distributions are also diverse and characteristic for each species. This chapter, briefly describes the characteristics of each species of intestinal fluke involved, in terms of the biology, epidemiology, host–parasite relationships, pathogenicity, clinical aspects, diagnosis, and treatment.

Brachylaimidae Joyeux and Foley, 1930

Species Infecting Humans

Brachylaima cribbi Butcher and Grove, 2001

The first human infection with this fluke was reported in South Australia (Butcher et al., 1998), and subsequently 10 adults and children in South Australia (Butcher et al., 2003) were reported as infected. Birds, reptiles, and mammals were found to be infected with this fluke (Butcher et al., 1998; Butcher and Grove, 2001, 2005). The first intermediate host is a helicid land snail, *Theba pisana,* and cercariae begin to emerge 8 weeks after exposure to the eggs (Butcher and Grove, 2001). Cercariae encyst in other species of helicid land snails, such as *Cernuella virgata*, which serve as the source of human infections (Butcher and Grove, 2005). Symptoms due to this fluke infection vary depending on the worm burden and include diarrhea, abdominal pain, low-grade fever, and fatigue (Butcher et al., 2003).

Cathaemaciidae Fuhrmann, 1928

Species Infecting Humans

Cathaemacia cabrerai Jueco and Monzon, 1984

The first human infection with this fluke was reported from a patient in the Philippines (Jueco and Monzon, 1984). No information is available on the life cycle and the source of infection.

Echinostomatidae Poche, 1926

Species Infecting Humans

Acanthoparyphium tyosenense Yamaguti, 1939 (Fig. 2.1)

This species was originally found in the small intestines of ducks *Melanitta fusca stejnegeri* and *M. nigra americana* caught in the Republic of Korea (Yamaguti, 1939a). It is characterized by 23 collar spines on the oral sucker, a long cirrus sac reaching beyond the posterior margin of the acetabulum, and the vitellaria extending to the level of the cirrus sac or the Mehlis' gland (Chai et al., 2001b). Human infections were first identified in 10 patients residing in two coastal villages in Chollabuk-do Province (Chai et al., 2001b). The patients had consumed various species of brackish water mollusks caught in an estuary near their villages; two species of bivalves, *Mactra veneriformis* and *Solen grandis*, and a gastropod *Neverita bicolor* were found to have the metacercariae (Chai et al., 2001b). The first intermediate hosts include the marine megagastropods *Lunatia fortuni* and *Glassaulax didyma* (Kim et al., 2004). The adult flukes were confirmed after experimental infection of metacercariae in chicks (Chai et al., 2001b; Han et al., 2003) and sea gulls *Larus crassiostris* (Kim et al., 2004).

Artyfechinostomum malayanum (Leiper, 1911), Railliet, 1925
[syn. *Artyfechinostomum surfrartyfex* Lane, 1915, *Paryphostomum surfrartyfex* Bhalerao, 1931, *Artyfechinostomum mehrai* Faruqui, 1930]

This fluke (under the name *A. surfrartyfex*) was first found in an Assamese girl in India and then found in pigs in India (Beaver et al., 1984). The source of infection is a snail, *Digoniostoma pulchella*, and the dog and rat are other definitive hosts (Yu and Mott, 1994). The taxonomic position of this species, in relation to related genera and species, has been confusing (Yamaguti, 1958; Lie, 1963; Beaver et al., 1984). However, a review by Kostadinova et al. (2002) suggested *A. surfrartyfex* to be conspecific with *A. malayanum*, which precedes *A. surfrartyfex*. In the meantime, *Artyfechinostomum mehrai* was reported from human infections in India (Beaver et al., 1984), but synonymized with *A. surfrartyfex* (Ahluwalia, 1962).

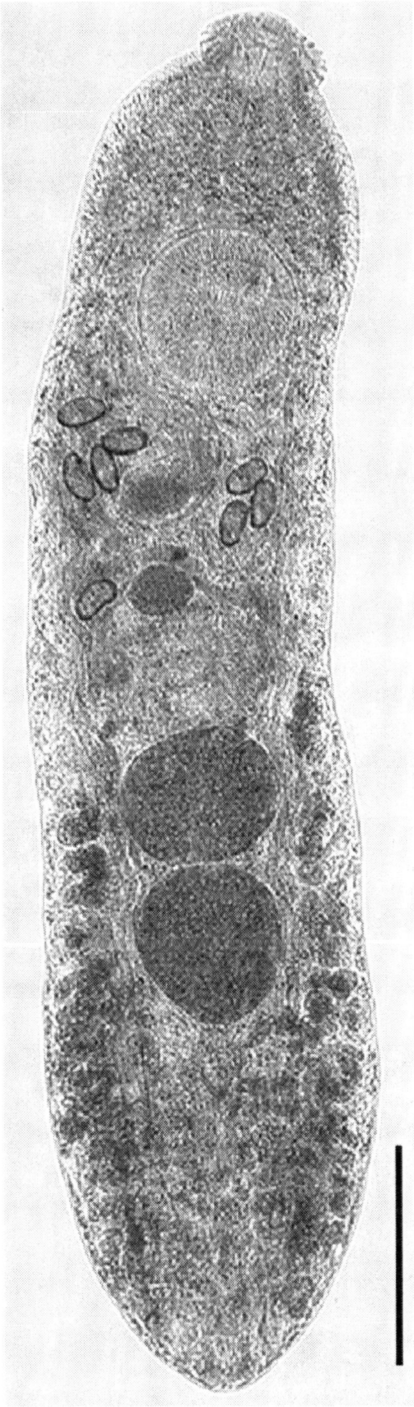

FIGURE 2.1. *Acanthoparyphium tyosenense* adult from an experimentally infected chick necropsied at day 20 postinfection. This echinostome species has 23 collar spines. Acetocarmine stain. Scale bar = 0.5 mm.

Artyfechinostomum oraoni Bandyopadhyay et al., 1989

This species was reported from human infections in a tribal community in India (Bandyopadhyay et al., 1989). The freshwater snail, *Lymnaea* sp., was experimentally proven to be a first intermediate host (Maji et al., 1995). In naturally infected pigs, this parasite provoked fatal diarrhea (Bandyopadhyay et al., 1995).

Echinochasmus fujianensis Cheng et al., 1992

This is a species reported from humans, dogs, cats, pigs, and rats in Fujian Province, China (Cheng et al., 1992a). The prevalence among residents in five areas of southern Fujian Province was 3.2% (1.6–7.8%); two thirds of the infected people were 3 to 15 years of age. Its first intermediate host is *Bellamya aeruginosa* and the second intermediate hosts include *Pseudorasbora parva* and *Cyprinus carpio* (Yu and Mott, 1994).

Echinochasmus japonicus Tanabe, 1926

This species was first reported from experimental animals such as dogs, cats, rats, mice, and birds that have been fed the metacercariae encysted in fresh water fish in Japan (Tanabe, 1926). The characteristic features of the adult worm include a small, plump body, the presence of total 24 collar spines, which are interrupted dorsally, two large, tandem testes, and a very small number of eggs in the uterus (usually less than five) (Chai and Lee, 2002). This species is known to exist mainly in countries in the Far East (Chai and Lee, 2002). An experimental human infection was reported in Japan (Ujiie, 1936a); natural human infections have been found in China (Zhu et al., 1986), and the Republic of Korea (Seo et al., 1985b). The first intermediate host is a fresh water snail, *Parafossarulus manchouricus* (Lee et al., 1983; Choi et al., 2006). Eighteen species of fresh water fish have been found to be the second intermediate host, including *Pseudorasbora parva, Hypomesus olidus*, and *Gnathopogon strigatus* (Lee et al., Kim, 1984a; Chai et al., 1985b; Choi et al., 2006). Natural infections in avian species, such as ducks (Eom and Rim, 1984), and mammalian species, such as cats (Sohn and Chai, 2005) have been confirmed by the recovery of adult flukes.

Echinochasmus jiufoensis Yu and Mott, 1994

This species was reported from a 6-month-old girl who died from pneumonia and dehydration in Guangzhou, China (Liang and Ke, 1988). The life cycle and route of infection are unknown (Yu and Mott, 1994).

Echinochasmus liliputanus (Looss, 1896); Odhner, 1910

This species was originally described from cats and dogs in Egypt, Syria, and Palestine (Yamaguti, 1958). Later, human infections were first discovered in Anhui Province, China, in 1991, with the prevalence rate of 13.4% among 2426 examined people (Xiao et al., 1992). Higher infection rates were observed in age groups 3 to 15 years (22.7%) and 16 to 30 years (16.4%) than in others. In the same place, the infection rates in dogs and cats were 60% and 45%, respectively (Xiao et al., 1992). The freshwater snail *Parafossarulus striatulus* (Yu and Mott, 1994), and the freshwater fish *Pseudorasbora parva* (Yu and Mott, 1994) and goldfish (Xiao et al., 2005) were the first and second intermediate hosts, respectively. Interestingly, it was postulated that humans can be infected with this parasite through drinking untreated water containing the cercariae (Xiao et al., 2005). As a mechanism for human oral infections with cercariae, the phenomenon of cercarial encystment in the presence of human gastric juice was proposed (Xiao et al., 2005).

Echinochasmus perfoliatus (Ratz, 1908); Dietz, 1910

This is a common parasite of the small intestine of dogs and cats in Hungary, Italy, Rumania, Russia, Japan, China, and Taiwan (Yu and Mott, 1994; Shimalov and Shimalov, 2002), and of red foxes in Denmark (Saeed et al., 2006). The rat, dog, and wild boar are also found infected (Beaver et al., 1984). Human infections were reported from China (Guangdong, Fujian, Anhui, and Hubei Provinces), with the prevalence rate of 1.8% (34/1846), including a child who died from the infection; about 14,000 worms were found in the child at autopsy (Yu and Mott, 1994). Many species of freshwater fish such as *Carassius* sp. harbor the metacercariae, which are encysted only on the gills (Yu and Mott, 1994).

Echinoparyphium recurvatum von Linstow, 1873 [syn. *Echinoparyphium koidzumii* Tsuchimochi, 1924]

This is a common intestinal parasite of birds and mammals, including wild rats, in Egypt (Yu and Mott, 1994) and Poland (Betlejewska and Jorol, 2002). This parasite was also reported from house rats in the Republic of Korea (Lee et al., 1990c). Human infections were recorded in Taiwan, Indonesia, and Egypt (Beaver et al., 1984; Yu and Mott, 1994). The metacercariae encyst in tadpoles and frogs of *Rana temporaria* and also in snails, *Planorbis planorbis*, *Lymnaea* sp. (Yu and Mott, 1994), and *Lymnaea stagnalis* (Yurlova et al., 2006).

Echinostoma angustitestis Wang, 1977

The genus *Echinostoma* (Rudolphi, 1809) is characterized by an elongated body and presence of a head collar with dorsally uninterrupted crown of spines around the oral sucker (Yamaguti, 1958). More than 95 species are known (Yamaguti, 1958), and seven to eight species infect humans (Yu and Mott, 1994). *Echinostoma angustitestis* was first described in 1977 from dogs experimentally infected with metacercariae isolated from the freshwater fish (Yu and Mott, 1994). Two human infections were reported in Fujian, China (Cheng et al., 1992b).

Echinostoma cinetorchis Ando and Ozaki, 1923 (Fig. 2.2)

This species was first reported in house rats in Japan (Ando and Ozaki, 1923), and also in rats in the Republic of Korea (Seo et al., 1964, 1981a). Characteristic

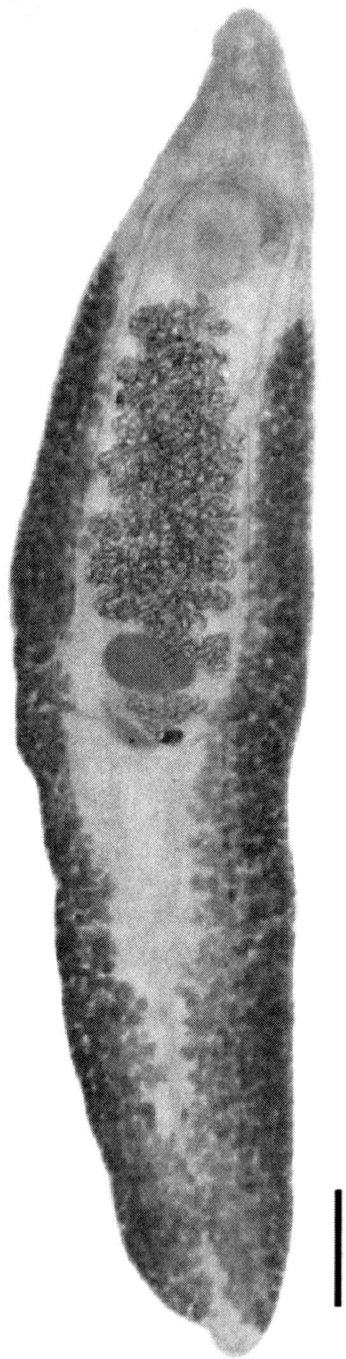

FIGURE 2.2. An adult fluke of *Echinostoma cinetorchis* from an experimentally infected rat. As a characteristic feature in this species, both testes disappeared. Acetocarmine stain. Scale bar = 1 mm.

features include the abnormal location or disappearance of one, or both, of the two testes, and the presence of 37 or 38 collar spines around the oral sucker (Chai and Lee, 2002). Human infections were reported in Japan (Kawahara and Tamamoto, 1933), the Republic of Korea (Seo et al., 1980a; Ryang et al., 1986; Lee et al., 1988a). The fresh water snail *Hippeutis cantori* was experimentally confirmed as the first as well as the second intermediate host (Lee et al., 1990a). Other fresh water snails including *Radix auricularia coreanus, Physa acuta,* and *Cipangopaludina chinensis malleata* (Ahn et al., 1989; Chung and Jung, 2000), and fresh water fish, especially the loach *Misgurnus anguillicaudatus* (Seo et al., 1984c), were proven to harbor the metacercarial stage. House rats (Ando and Ozaki, 1923; Seo et al., 1964, 1981a) and dogs (Cho et al., 1981) were found to be natural definitive hosts. Laboratory rats and mice are highly susceptible to experimental infections (Lee et al., 1988c).

Echinostoma echinatum (Zeder, 1803)
[syn. *Echinostoma lindoense* Sandground and Bonne, 1940]

This parasite has 37 collar spines and resembles *Echinostoma revolutum. Echinostoma lindoense* has been synonymized with *E. echinatum* (Huffman and Fried, 1990; Fried and Graczyk, 2004). During 1937 and 1956, a high prevalence of 24% to 96% and heavy infections among people were known in Celebes, Indonesia, under the name of *E. lindoense* (Yu and Mott, 1994). The mode of human infection was eating raw or insufficiently cooked mussels *Corbicula lindoensis, Corbicula sucplanta,* and *Idiopoma javanica* (Beaver et al., 1984). This fluke was found in Brazil where *Biomphalaria glabrata* snail was the source of infection (Lie, 1968). Rats and mice are experimental definitive hosts (Beaver et al., 1984).

Echinostoma hortense Asada, 1926 (Fig. 2.3)

This parasite was first described from house rats in Japan (Asada, 1926), and then reported from Korea (Park, 1938; Seo et al., 1964, 1981a, 1983; Chai and Lee, 2002) and China (Fan and Sun, 1989). Human infections have been found in Japan, Korea, and China. In Japan, more than 20 human infections have been reported based on the recovery of the adult flukes (Miyamoto et al., 1983). In the Republic of Korea, an infection rate of 22.4% was reported among residents of Cheongsong-gun (Lee et al., 1988b). In a survey in Liaoning Province of northeast China, six of 10 hospitalized hepatitis patients who had eaten raw loach were found infected (Chen et al., 1993). Morphological characters include a laterally located ovary, and 27 to 28 collar spines around the oral sucker (Chai and Lee, 2002). The molluscan intermediate hosts are the fresh water snail, *Lymnaea pervia* and *Radix auricularia coreana* (Chai and Lee, 2002). The second intermediate hosts are the loaches, *Misgurnus anguillicaudatus* and *Misgurnus mizolepis*, and other fresh water fish, including *Odontobutis obscura interrupta, Moroco oxycephalus, Coreoperca kawamebari*, and *Squalidus coreanus* (Chai et al., 1985c; Ryang, 1990; Chai and Lee, 2002). In a survey in China, 69.7% of the loach *Misgurnus anguillicaudatus* from a market in Liaoning Province was

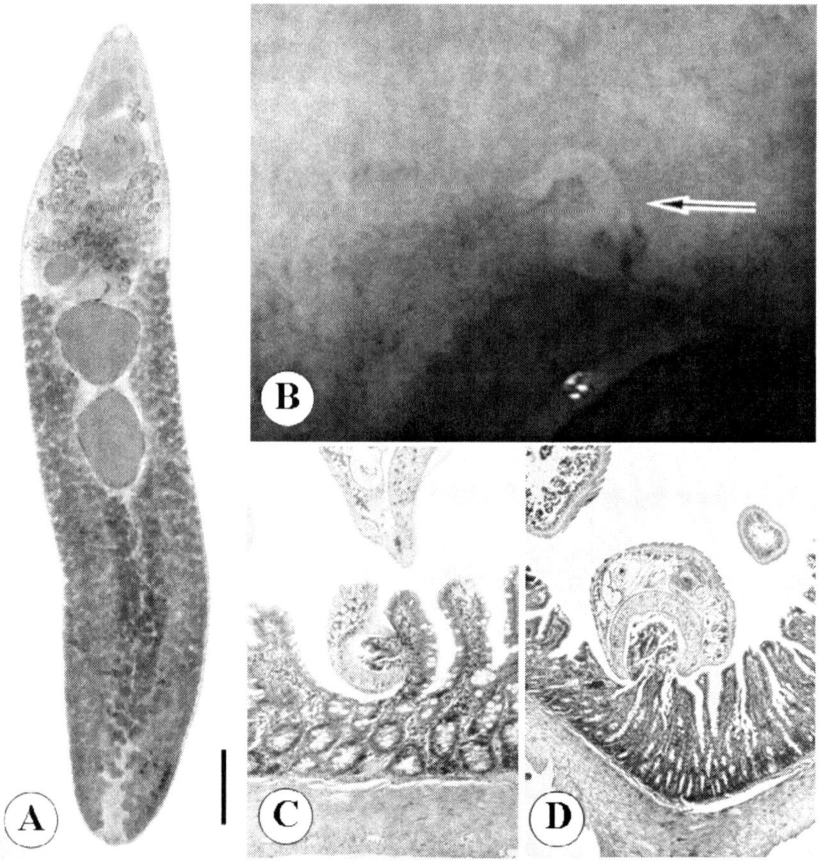

FIGURE 2.3. (A) An adult *Echinostoma hortense* worm recovered from an experimental rat. Acetocarmine stain. Scale bar = 1 mm. (B) Gastroendoscopic view of an *E. hortense* worm attached to the ulcerated lesion of the stomach wall of a Korean patient. (From Chai et al., 1994c, with permission.) (C, D) Duodenal sections of experimental rats infected with *E. hortense*. Two worms are sucking the duodenal villi with their oral (C, ×100) and ventral suckers (D, ×100). Hematoxylin and eosin (H&E) stain. (From Lee et al., 1990b, with permission.)

infected (Yu and Mott, 1994). Rats (Park, 1938; Seo et al., 1964, 1981a), dogs (Cho et al., 1981), and cats (Sohn and Chai, 2005) have been found to be natural definitive hosts. Mice, rats, and humans have all been determined, experimentally, susceptible to *E. hortense* infection (Seo et al., 1985a; Lee et al., 2004a).

Echinostoma ilocanum (Garrison, 1908); Odhner, 1911

The eggs of this species were first found in the feces of a man in Manila, Philippines, in 1907, and later 21 adult flukes were recovered after anthelmintic treatment (Beaver et al., 1984). The Norway rat and the dog are reservoir hosts

(Beaver et al., 1984). Human infections were reported from Celebes, Java, Indonesia, China, Thailand, and India (Radomyos et al., 1982; Yu and Mott, 1994; Grover et al., 1998). The prevalence among the Ilocano population in northern Luzon, Philippines, was 5% on average (range 0–11%) (Cross and Bassaca-Sevilla, 1981). Characteristic morphological features include the presence of 49 to 51 collar spines and deeply lobed testes. The first intermediate hosts are *Gyraulus* or *Hippeutis* snails (Yu and Mott, 1994). The sources of human infections are the large snails *Pila conica* (Philippines) and *Viviparus javanicus* (Java) (Beaver et al., 1984). Infected humans may experience intestinal colic and diarrhea (Beaver et al., 1984).

Echinostoma macrorchis Ando and Ozaki, 1923

This parasite was first described from naturally infected rats in Japan (Ando and Ozaki, 1923). Subsequently, human infections were reported in Japan (Majima, 1927). Metacercarial cysts were found from snails *Cipangopaludina malleata, Cipangopaludina japonica, Segmentina nitiella, Viviparus malleatus*, and the frog *Rana* sp. (Yu and Mott, 1994).

Echinostoma malayanum Leiper, 1911

This fluke was first found in human infections in Singapore and Kuala Lumpur, Malaysia, in 1911, and then reported from Thailand, Indonesia, and India (Beaver et al., 1984; Maji et al., 1993; Radomyos et al., 1998; Yu and Mott, 1994). This fluke is now known to be distributed in the Philippines (Yu and Mott, 1994). The dog, rat, mouse, and hamster are experimental definitive hosts (Yu and Mott, 1994). The first intermediate host is a freshwater snail *Indoplanorbis exustus* and *Gyraulus convexiusculus* and the cercariae encyst in various species of snails, that is, *Pila scutata* and *Lymnaea (Bullastra) cumingiana* (Yu and Mott, 1994).

Echinostoma revolutum (Froelich, 1802); Looss, 1899 [syn. *Echinoparyphium paraulum* Diez, 1909]

This fluke is an intestinal fluke of the duck, goose, muskrat, and human in Asia, Europe, New Zealand, and Brazil (Yu and Mott, 1994; Fried and Graczyk, 2004). In Russia, domestic and wild animals were found to be infected with this fluke (Shimalov and Shimalov, 2002). The first human infection was reported from Taiwan in 1929, and the prevalence was estimated to be between 2.8% and 6.5% (Yu and Mott, 1994). This fluke was also reported from human infections in Yunnan and Guangdong Provinces, China, Indonesia, Thailand, and Russia (Beaver et al., 1984; Ashford and Crewe, 2002). This parasite was also reported from house rats and cats in the Republic of Korea (Lee et al., 1990c; Sohn and Chai, 2005). The snail host includes *Lymnaea* sp., *Physa* sp., *Paludina* sp., *Segmentina* sp., and *Heliosoma* sp. (Beaver et al., 1984). Cercariae penetrate into tadpoles, snails, or the clam *Corbicula producta*, which is the source of infection for definitive hosts (Beaver et al., 1984). *Echinoparyphium paraulum*, regarded as a synonym of *E. revolutum* (Beaver et al., 1984), was described from dogs in India (Yamaguti, 1958) and birds and a human in Russia (Ashford and Crewe, 2002).

Episthmium caninum (Verma, 1935); Yamaguti, 1958

This fluke was described from dogs in Calcutta, India (Yamaguti, 1958). Human cases were reported from northeast Thailand (Radomyos et al., 1985, 1991), and the source of infection was freshwater fish (Radomyos et al., 1991). The genus *Episthmium* was suggested to be tentatively retained as a synonym of *Echinochasmus* (Kostadinova and Gibson, 2001).

Himasthla muehlensi Vogel, 1933

Five adult worms were found from a German patient who lived in Colombia and traveled to New York City where he had eaten raw clams *Venus mercenaria* (Beaver et al., 1984). A species of marine operculate snail *Littorina littoria* serves as the first intermediate host and the bivalve mollusks *Mytilus* and *Mya* spp. as the host for metacercariae (Beaver et al., 1984). Birds are natural definitive hosts (Beaver et al., 1984).

Hypoderaeum conoideum (Block, 1872); Diez, 1909

This species was originally described from naturally infected birds in Europe, Japan and Siberia (Yamaguti, 1958). It is also a parasite of humans and birds in Thailand (Yokogawa et al., 1965a). In an area of northeast Thailand, 55% of 254 residents were found infected (Yokogawa et al., 1965a). The duck and fowl are reservoir hosts (Harinasuta et al., 1987). The first intermediate hosts are freshwater snails, *Planorbis corneus, Indoplanorbis exustus, Lymnaea stagnalis, Lymnaea limosa, Lymnaea ovata,* and *Lymnaea rubiginosa*, and snails and tadpoles are the second intermediate hosts (Yamaguti, 1958; Harinasuta et al., 1987).

Isthmiophora melis (Schrank, 1788); Lühe, 1909 [syn. *Euparyphium melis* Railliet, 1919, *Euparyphium jassyense* Leon and Ciurea, 1922]

This fluke has 27 collar spines, with four corner ones on each side, and was recovered from the diarrheic stools of a Romanian patient and at autopsy from a Chinese patient (Beaver et al., 1984). The taxonomic position of this parasite has been unstable, placing it into *Euparyphium* (Beaver et al., 1984) or *Echinostoma* (Harinasuta et al., 1987); however, it was later designated as a species of *Isthmiophora* (Fried, 2001; Kostadinova and Gibson, 2002). Human infections were also reported in Taiwan (Yu and Mott, 1994). Domestic and wild animals were found to be infected with this fluke in Russia (Shimalov and Shimalov, 2002). In the region of Douglas Lake, Michigan, the snail *Stagnicola emarginata angulata* is the first intermediate host and tadpoles serve as the second intermediate host (Beaver et al., 1984).

Psilorchis hominis Kifune and Takao, 1973

This fluke was described from a 48-year-old Japanese patient, who had a mixed infection with *E. macrorchis*, after anthelmintic medication (Kifune and Takao, 1973). The genus *Psilorchis* was assigned to the family Psilostomatidae (Ashford and Crewe, 2002). No other information is available on this species.

Pathogenicity and Host–Parasite Relationships of Echinostomes

The pathogenicity of echinostomes has not been well studied (Yu and Mott, 1994); however, it may be closely related to individual worm burdens (Rim, 1982). The intestinal histopathology was studied in a few species, including *E. revolutum* (Bindseil and Christensen, 1984; Huffman et al., 1986), *E. hortense* (Lee et al., 1990b; Chai and Lee, 2002) and *Echinostoma trivolvis* (Fujino et al, 1993). The worms were located in the lumen of the upper small intestine, and the pathological changes were observed chiefly in the mucosal layer (Chai and Lee, 2002). Villous atrophy, crypt hyperplasia, inflammation of the stroma, and decreased villus/crypt ratios were observed in the small intestine of experimentally infected animals (Huffman et al., 1986; Lee et al., 1990b) (Fig. 2.3). Compared with heterophyid infections, the mucosal damages were more severe, and in focal areas massive destruction and detachment of the villi and at times complete loss of the mucosal integrity and ulcerations were observed (Huffman et al., 1986; Chai and Lee, 2002). It was also reported that naturally infected pigs with *Artyfechinostomum oraoni* developed fatal diarrhea; at autopsy a massive infection with hemorrhagic and edematous mucosa of the jejunum and duodenum, extending up to the pyloric end of the stomach, was observed (Bandyopadhyay, 1995).

Spontaneous expulsion of worms from the small intestine of experimentally infected mice was observed in different species of echinostome infections; *E. trivolvis*, *E. hortense*, and *Echinostoma caproni* (Fujino et al., 1993; Kim et al., 2000; Brunet et al., 2000). Mucosal goblet cells (Fujino et al., 1993) and mucosal mast cells (Kim et al., 2000) were suggested to play important roles in the worm expulsion, although further studies are required to clarify the precise role of these cells. Roles of T-helper-1 (Th1) and Th2 cytokines were also suggested; injection of mice with anti–interferon-γ (anti–IFN-γ) monoclonal antibodies significantly lowered the worm burden, suggesting that host Th1 responses are related to establishment of a chronic infection (Brunet et al., 2000).

Clinical Symptoms, Diagnosis, and Treatment of Echinostomiases

Abdominal pain, diarrhea, and easy fatigability are the major symptoms due to echinostome infections (Rim, 1982; Chai and Lee, 2002; Fried et al., 2004). The symptoms are thought to be more severe than those seen in heterophyid infections, considering the more severe mucosal damages and even ulcerations of the mucosa seen in experimental rats infected with *E. hortense*, for instance (Lee et al., 1990b). A patient with an *E. hortense* infection complained of lower abdominal pain, diarrhea and tenesmus, easy fatigability, and urinary incontinence (Lee et al., 1986). Interesting to note are reports of patients with *E. hortense* infection suffering from severe epigastric discomfort and ulcerative lesions in the duodenum and diagnosed by discovery of worms at gastroduodenal endoscopy (Chai et al., 1994c; Lee and Hong, 2002; Cho et al., 2003; Chang et al., 2005). One of the patients was admitted to a hospital because of epigastric pain and hematemesis; through gastroduodenoscopy an adult fluke was seen attached at the lesion (Fig. 2.3) and it was removed by an endoscopic clipper; three more adult flukes were recovered from him after praziquantel treatment (Chai et al., 1994c). The levels of peripheral blood eosinophilia due to the *E. hortense* infection were dependent on individual worm burdens; 11% to 24% (average, 17%) among the patients with more than 100 worms, 4% to 21% (average, 10%) among those with 51 to 100 worms, and 2% to 14% (average, 5%) among those with less than 50 worms (Lee et al., 1988b). The clinical symptoms in *E. cinetorchis, E. japonicus*, and *A. tyosenense* infections are not well known (Seo et al., 1980a, 1985b; Chai et al., 2001b), but may be similar to those of *E. hortense* infection.

The diagnosis is based on recovery of echinostomatid eggs in the feces. Specific diagnosis can be made through careful observations and measurements of the eggs. However, a recovery and identification of the adult flukes is strongly recommended, if a definite diagnosis is preferred. The detectability of eggs in the feces is remarkably different among the species of echinostomes; it is high in *E. hortense* (Seo et al., 1985a) and *E. cinetorchis* (Seo et al., 1984c) infections, but very low in *E. japonicus* (Chai et al., 1985b) and *A. tyosenense* (Chai et al., 2001b) infections. The difference is greatly due to remarkably different numbers of the intrauterine eggs and the different egg-laying capacity of each species.

Echinostome infections can be treated successfully using praziquantel 10 to 20 mg/kg in a single oral dose (Seo et al., 1985a,b; Ryang et al., 1986; Lee et al., 1988a; Chai et al., 1994c, 2001b). Albendazole may also be effective. For prevention, eating raw or improperly cooked flesh of fresh water fish and fresh or brackish water snails should be avoided.

Fasciolidae Railliet, 1895

Three species of the family Fasciolidae (Railliet, 1895) are known to infect humans (Beaver et al., 1984; Mas-Coma et al., 2005). Two of them, *Fasciola hepatica* and *F. gigantica*, parasitize the liver of livestock animals, and accidentally infect humans. Only the remaining species, *Fasciolopsis buski*, is the species infecting the intestinal tract of animals and humans. Their cercariae encyst on the surface of aquatic plants, on debris, and also on the water surface (Fried et al., 2004).

Species Infecting Humans

Fasciolopsis buski (Landkester, 1857); Odhner, 1902 (Fig. 2.4)

This species was first discovered in the duodenum of an Indian man who died in London (Beaver et al., 1984). It is the largest fluke parasitizing humans (Kuntz and Lo, 1967; Mas-Coma et al., 2005), and a common intestinal parasite of humans and pigs in central and south China, Taiwan, Thailand, Vietnam, Laos, Cambodia, Bangladesh, India, Indonesia, and Malaysia (Yu and Mott, 1994;

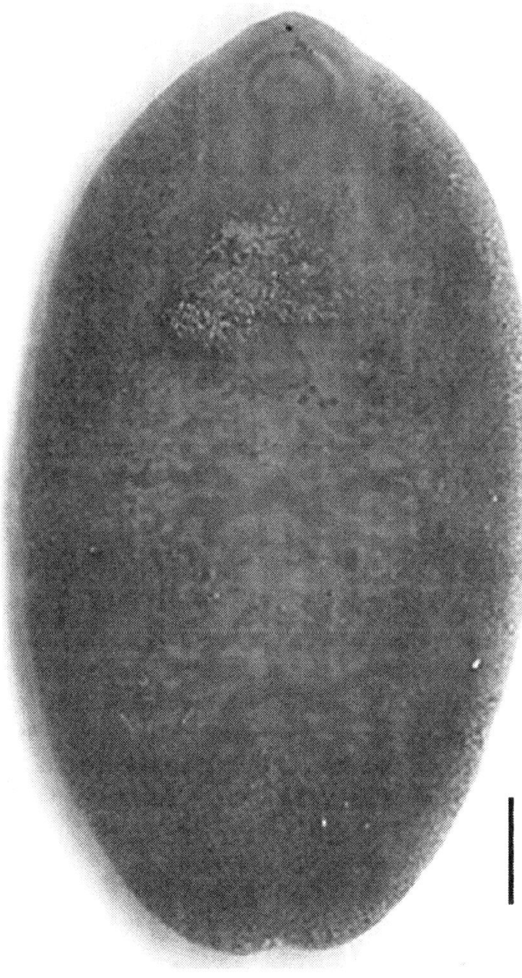

FIGURE 2.4. An adult fluke of *Fasciolopsis buski* recovered from a patient. Acetocarmine stain. Scale bar = 5 mm.

Mas-Coma et al., 2005; Rohela et al., 2005). The prevalence in children varies according to countries, 10% in Thailand (Bunnag et al., 1983), 25% in Taiwan (Shyu et al., 1984), 57% in China (Lee, 1972) and 60% in India (Muttalib and Islam, 1975). Its first intermediate hosts are the freshwater snails *Segmentina* sp., *Hippeutis* sp., and *Gyraulus* sp. Metacercarial encystment occurs on the surface of edible aquatic plants, such as water chestnut *Eliocharis tuberose*, water caltrop *Trapa natans*, water hyacinth *Eichhornia* sp., roots of the lotus, water bamboo *Zizania* sp., other aquatic vegetation, or in the water (Beaver et al., 1984; Yu and Mott, 1994; Fried et al., 2004). People are infected through consuming raw or improperly cooked aquatic plants, or peeling off the hull or skin of the plants by mouth before eating the raw nut (Yu and Mott, 1994). The drainage of pig excreta in farms is an important factor for maintaining high endemicity (Yu and Mott, 1994).

Pathogenicity and Host–Parasite Relationships of Fasciolopsiasis

The disease can be fatal depending on worm burden (Bunnag et al., 1983; Mas-Coma et al., 2005). In light infections, anemia, eosinophilia, headache, dizziness, gastric pain, and loose stools can occur (Gilman et al., 1982). In moderate and heavy infections, severe epigastric and abdominal pain, diarrhea, bowel obstruction, nausea, acute ileus, anasarca, and marked eosinophilia and leukocytosis may occur (Gilman et al., 1982). Adult flukes damage the intestinal mucosa and cause extensive intestinal and duodenal erosions, ulceration, hemorrhages, abscess, and catarrhal inflammation (Marty and Andersen, 2000; Fried et al., 2004).

Symptoms, Diagnosis, and Treatment of Fasciolopsiasis

Absorption of toxic and allergic worm metabolites can cause ascites, general edema, and facial and orbital edema (Jaroonvesama et al., 1986). Diagnosis can be made by recovery of eggs in the feces. Praziquantel, in a single dose of 15 mg/kg, has been reported successful for treatment of fasciolopsiasis (Bunnag et al., 1983; Fried et al., 2004). Fasciolopsiasis is aggravated by socioeconomic factors, such as poverty, malnutrition, a lack of food inspection, poor sanitation, other helminthiases, and declining economic conditions (Yu and Mott, 1994).

Gastrodiscidae Stiles and Goldberger, 1910

Species Infecting Humans

Gastrodiscoides hominis (Lewis and McConnell, 1876); Leiper, 1913

This species was first found and described from the cecum of an Indian patient (Beaver et al., 1984). It has been known to be a common human parasite in India,

Pakistan, Myanmar, Vietnam, the Philippines, Thailand, China, Kazakstan, Indian immigrants in Guyana, and the Volga Delta in Russia (Beaver et al., 1984; Mas-Coma et al., 2005). The pig is a common reservoir host, and the napu mouse deer, field rat, and rhesus monkey are local reservoir hosts (Beaver et al., 1984; Yu and Mott, 1994). The first intermediate host is the planorbid snail *Helicorbis coenosus* (Beaver et al., 1984) and cercariae encyst on aquatic plants, or in tadpoles, frogs, and crayfish (Yu and Mott, 1994). In human infections, the worms attach to the cecum and ascending colon and may produce a mucous diarrhea (Beaver et al., 1984).

Gymnophallidae Morozov, 1955

Species Infecting Humans

Gymnophalloides seoi Lee, Chai and Hong, 1993 (Fig. 2.5)

This fluke was first discovered in 1988 in a Korean woman suffering from acute pancreatitis and gastrointestinal discomforts (Lee et al., 1993b; Chai et al., 2003). Her home village in a southwestern coastal village (Aphaedo, Shinan-gun) was found to be a highly endemic area (Lee et al., 1994). Subsequently, 24 villages on western and southern coastal islands (Chai et al., 1997, 1998a, 2001c, Lee et al., 1996) and three nonisland coastal villages (Guk et al., 2006) were identified as endemic areas. This parasite has never been reported from other countries. The adult parasite is very small, and characterized by a large oral sucker, a small ventral sucker, short ceca, two compact masses of vitellaria, and a unique ventral pit (Lee et al., 1993). The first intermediate host is yet unknown (Lee and Chai, 2001; Chai et al., 2003), but the second intermediate host was confirmed to be the oyster *Crassostrea gigas* (Lee et al., 1995b; Sohn et al., 1998). Other than humans (Lee et al., 1993, 1994), the Palearctic oystercatcher *Haematopus ostralegus* (Ryang et al., 2000) has been shown to be a natural definitive host. Wading birds, such as the Kentish plover *Charadrius alexandrinus*, Mongolian plover *Charadrius mongolus*, and gray plover *Pluvialis squatarola*, were highly susceptible to experimental infection with this fluke (Ryang et al., 2001). Mammals, such as gerbils, hamsters, cats, and several strains of mice, were also found susceptible to experimental infections (Lee et al., 1997b). In vitro cultivation of the metacercariae into adults was successful using National Cancer Tissue Culture (NCTC) 109 medium (Kook et al., 1997).

Pathogenicity and Host–Parasite Relationships of Gymnophallids

In experimental mice, *G. seoi* parasitizes the small intestine, chiefly the duodenum and jejunum, pinching and sucking the intestinal villi with their large oral suckers (Chai et al., 2001a) (Fig. 2.5). Histopathologically, the infected

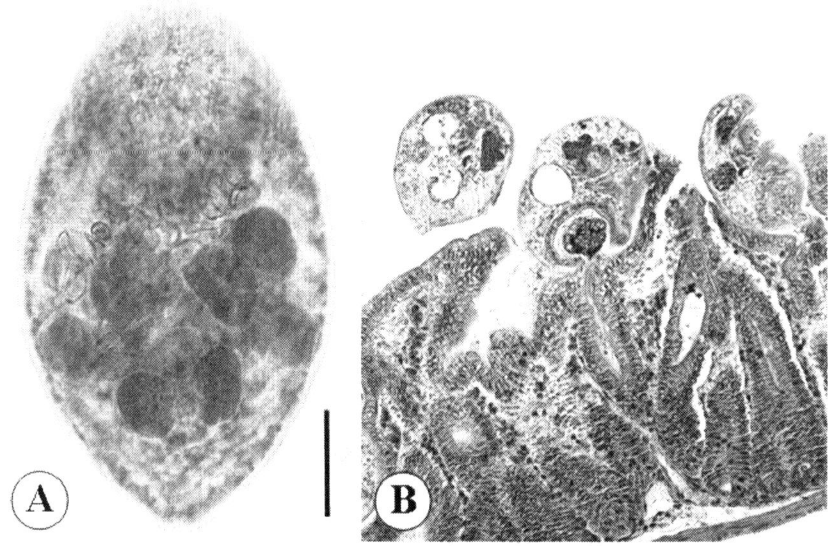

FIGURE 2.5. (A) *Gymnophalloides seoi* adult recovered from a patient. Acetocarmine stain. (B) Intestinal section of an experimentally infected mouse showing adult worms sucking the mucosal layer. H&E stain, ×200. Scale bar = 0.1 mm.

intestinal mucosa showed villous atrophy and crypt hyperplasia, with inflammatory reactions in the villous stroma and the crypt (Chai et al., 2001a). The histopathological changes were generally not so severe, and the mucosal integrity was restored around day 14 to 21 postinfection (Chai et al., 2001a). The worms did not invade the submucosal layer in immunocompetent hosts. However, in immunosuppressed hosts, worms were found to invade the submucosa (Chai et al., 2001a). It is of interest to note that, in a colon cancer patient who received anticancer chemotherapy, a *G. seoi* worm was found to have penetrated into the colonic lymphoid tissue (Seo et al., 2006). With regard to the mechanisms of the pathogenicity, mechanical irritation by the flukes was considered to be important (Chai et al., 2001a). Cysteine proteinases were isolated from the metacercariae and adults, and functionally characterized (Choi 1998a,b).

After experimental infections with metacercariae, Institute for Cancer Research (ICR) and BALB/c mice retained many worms by day 3 postinfection; however, most worms were expelled before day 7 postinfection (Lee et al., 1997). This had probably been caused by an innate immune response of the host. In this respect, it was of note that goblet cell proliferation was marked in the small intestines of infected mice, in particular on the villous epithelia of the jejunum during days 3 to 7 postinfection (Chai et al., 2001a). The importance of goblet cells in worm expulsion was further suggested by only a small number of flukes retained in the intestines at days 7 to 14 postinfection (Chai et al., 1999; Seo et al., 2003). This was also suggested by the minimal degree of goblet cell hyperplasia in immunocompromised mice, with many more flukes surviving in the intestine (Chai et al., 1999; Seo et al., 2003). This is in good agreement with other workers studying other intestinal helminths, in which also goblet cells were found to be an important effector for the worm expulsion from the host, for example, *Nippostrongylus brasiliensis* (Nawa et al., 1994; Onah and Nawa, 2000) and *Echinostoma trivolvis* (Fujino et al., 1993).

The susceptibility of mice to an experimental infection with *G. seoi* was variable in different species of animals as well as different strains of mice (Lee et al., 1997). Among mouse strains, C3H/HeN mice were the most highly susceptible and the best for the growth and development of the worms, although the worm recovery at days 7 to 21 postinfection was not sufficiently high (Lee et al., 1997; Chai et al., 1999). It is likely that the difference in the susceptibility is caused by the genetic backgrounds of the host that regulate the immune responses of the host, including the mucosal goblet cell responses. In fact, when C3H/HeN mice were immunosuppressed by injection with prednisolone, the survival and the recovery of the worms were greatly enhanced (Lee et al., 1997; Chai et al., 1999), and the enhancement of the worm recovery had a strong correlation with the duration of the immunosuppression of mice. Immunosuppression of the C3H/HeN mice also enhanced the fecundity of the worms (Lee et al., 1997; Chai et al., 1999).

The habitat of *G. seoi* in the human host is presumed to be the small intestine, as seen in rodents (Lee et al., 1997; Chai et al., 2001a). However, the first patient suffered from acute pancreatitis (Lee et al., 1993), and two other patients were accompanied by diabetes mellitus (Lee et al., 1995a). It is thus suspected that *G. seoi* could infect the pancreatic duct in humans. In this respect, it is of note that other gymnophallids were found in the bursa Fabricii and gallbladder of shore birds as well as in the intestine (Yamaguti, 1975). However, the possibility of involvement of extraintestinal organs by *G. seoi* has not been experimentally proved; experiments using larger animals, such as monkeys, seem to be necessary to verify this possibility. Since *G. seoi* was able to invade the submucosa of immunosuppressed mice (Chai et al., 2001a), there is also a possibility that the eggs may be transferred to remote organs to cause erratic parasitisms in immunocompromised hosts, as reported in heterophyid infections (Africa et al., 1940).

Clinical Symptoms, Diagnosis, and Treatment of Gymnophalloidiasis

People infected with *G. seoi* complained of variable degrees of gastrointestinal troubles and indigestion (Lee et al., 1994; Chai et al., 2003). Fever, anorexia, weight loss, easy fatigability, and weakness may accompany the infection. However, the degree of symptoms seems to be variable among patients; the first patient underwent acute pancreatitis or acute cholecystitis, with episodes of epigastric discomforts, indigestion, and diarrhea (Lee et al., 1993), whereas other patients complained of only mild gastrointestinal troubles such as indigestion (Lee et al., 1994; Chai et al., 2003). In the first patient, laboratory studies revealed elevated serum and urine amylase levels, increased serum alkaline phosphatase activity, and slight to moderate degrees of eosinophilia (Lee et al., 1993). However, 5 days after treatment with praziquantel, epigastric pain and diarrhea completely disappeared, and serum and urine amylase levels returned to their normal levels (Lee et al., 1993). In the case of two patients, *G. seoi* infection was accompanied by diabetes mellitus (Lee et al., 1995a). Hence, some relations between the *G. seoi* infection and diabetes were suspected. It is of note that some of the infected patients in a highly endemic area in Aphaedo (Lee et al., 1994) complained of symptoms such as thirst, polydipsia, and polyuria (Chai et al., 2000a), that may occur among the patients with diabetes mellitus. Their blood and urine glucose levels, however, were within normal limits (Chai et al., 2000a).

In patients infected with *G. seoi*, the diagnosis can be made by detection of eggs in the feces; however, an expert is needed to identify the eggs (Lee and Chai, 2001). The eggs are very small, only 0.020 to 0.025 mm in length, smaller than those of *Clonorchis sinensis, Metagonimus yokogawai*, or other heterophyids, except for those of *P. summa*, and have a very thin and transparent shell (Lee et al., 1993, 1994). The problem is that the eggs are not readily detected in routine fecal examinations performed by formalin-ether sedimentation or cellophane thick smear techniques. They may be overlooked or misdiagnosed as an air bubble or other artifacts (Lee et al., 1993). Another problem is a very low egg-laying capacity of *G. seoi*, compared to other intestinal parasites (Chai et al., 2000a; Chai and Lee, 2002). The daily egg output was estimated to be only 2 to 84 eggs per adult fluke in the human host (Chai et al., 2000a). Unless more than 100 worms are present, less than 8400 eggs would be discharged in a whole-day stool; the eggs per gram of feces (epg) would be only 42 (daily stool weight; 200 g). This value means the appearance of only one to two eggs on the whole field of a fecal smear made by the Kato-Katz technique (41.7 mg of feces/smear) (Chai and Lee, 2002). When *G. seoi* (gymnophallid) eggs are detected in the feces, differential diagnosis is needed, because the egg morphology between different species of gymnophallids is similar to each other.

Praziquantel in a single oral dose of 10 mg/kg is highly effective for treatment of *G. seoi* infection in humans (Lee et al., 1993, 1994; Chai et al., 2000a). Albendazole may also be effective against *G. seoi* infection, but this needs confirmation. The best way to prevent *G. seoi* infection is the avoidance of consuming infected oysters, under raw or improperly cooked conditions. As control measures, oyster irradiation (Chai et al., 1996), and repeated chemotherapy of the people in endemic areas (Chai et al., 2000) were tried, with considerable success.

Heterophyidae Odhner, 1914

Species Infecting Humans

Apophallus donicus (Skrjabin and Lindtrop, 1919); Price, 1931

An experimental human infection with this species was successful in the U.S. (Niemi and Macy, 1974), and there were other reports of infection with this species in humans where fish are eaten raw (Schell, 1985). Heterophyid cercariae shed by the stream snail *Flumenicola virens* were found to encyst in hatchery-reared coho salmon, *Oncorhynchus kisutch* (Niemi and Macy, 1974). Many other kinds of fish, including blackside dace, suckers, squawfish, redside shiners, and rainbow trout, were found naturally infected with the metacercariae (Niemi and Macy, 1974). Reservoir hosts are dogs, cats, rats, foxes, and rabbits (Yamaguti, 1958).

Ascocotyle (Phagicola) longa Ransom, 1920

Flukes of *Ascocotyle* (Looss, 1899) [subgenus *Phagicola* Faust, 1920] are intestinal parasites of fish-eating birds or mammals in Europe, Asia, Africa, and the Americas (Yu and Mott, 1994). Human infections presumably with this species (described as a *Phagicola* sp.) were reported in Brazil (Chieffi et al., 1992). A dog was also found infected with this fluke (Chieffi et al., 1992). Freshwater fish are the second intermediate hosts (Chieffi et al., 1992). The taxonomic status of this species in relation to other related species was extensively studied (Scholz, 1999).

Centrocestus armatus (Tanabe, 1922b); Price, 1932

This fluke was first reported in dogs, cats, rabbits, rats, and mice experimentally fed on cyprinoid fish infected with the metacercariae (Tanabe, 1922b). Characteristic features of this species are the presence of 42 to 48 circumoral spines on the oral sucker, a small number of intrauterine eggs, the median location of the ovary, and the side-by-side location of the two testes. With regard to human infection, a successful experimental infection was reported in Japan (Tanabe, 1922b), and a case of natural human infection was reported in the Republic of Korea (Hong et al., 1988). The first intermediate host is the fresh water snail, *Semisulcospira* sp. (Takahashi, 1929b). The second intermediate hosts are fresh water fish, such as, *Zacco platypus, Zacco temminckii, Rhodeus ocellatus, Gobius similis, Pseudorasbora parva*, and *Pelteobagrus fulvidraco* (Lee et al., 1984a; Hong et al., 1989). The large egret *Egretta alba modesta* (Ryang et al., 1991) and the cat (Sohn and Chai, 2005) have been reported to be natural definitive hosts.

Centrocestus caninus (Leiper, 1913); Yamaguti, 1958 [syn. *Stephanopirumus longus* Onji and Nishio, 1916]

This fluke was first reported from dogs and foxes in Taiwan (Yamaguti, 1958), and two human infections in Thailand (Waikagul et al., 1997). The adult worm has 26 to 30 circumoral spines (Waikagul et al., 1997). Cercariae emerge from

Melanoides tuberculata and infect freshwater fish, such as *Cyprinus carpio, Hampala dispar, Puntius* sp., *Cyclocheilichthys* sp., and *Tilapia nilotica* (Wakagul et al., 1997). Adult worms were recovered from the posterior half of the small intestine of rats, 18 days after the infection (Waikagul et al., 1997).

Centrocestus cuspidatus (Looss, 1896), 1899

This species was described from a naturally infected dog in Egypt (Ransom, 1920). The adult worm has 36 circumoral spines (Ransom, 1920). Rats were experimentally infected after feeding them with the fish *Astatotilapia desfontainesi*, and adult worms were obtained in Tunis, Tunisia (Yamaguti, 1958). Human infections were reported in Egypt (Yu and Mott, 1994), although the literature background is uncertain.

Centrocestus formosanus (Nishigori, 1924); Price, 1932

This species was described from an experimental dog fed freshwater fish infected with the metacercariae and found also in a naturally infected fox (Nishigori, 1924). An experimental human infection was successful (Nishigori, 1924). The presence of natural human infections in Taiwan and Japan was mentioned by various authors (Ito, 1964; Premvati and Pande, 1974). The adult worm has 32 circumoral spines (Chen, 1942). This fluke is now known to be present in Taiwan, China, Japan, the Philippines, India, Hawaii, and Mexico (Chen, 1942; Martin, 1958; Yamaguti, 1975; Madhavi, 1986; Yu and Mott, 1994; Scholz and Salgado-Maldonado, 1999). Cercariae were found in the snail, *Stenomelania newcombi* (Martin, 1958).

Centrocestus kurokawai (Kurokawa, 1935); Yamaguti, 1958

This fluke was originally described from a naturally infected man in Hiroshima Prefecture, Japan (Kurokawa, 1935). The adult worm has 38 or 40 circumoral spines (Kurokawa, 1935). No information is available on the intermediate and reservoir hosts.

Cryptocotyle lingua (Creplin, 1825); Fischoeder, 1903

This species has been described from cats, dogs, rats, birds, and wild animals in Europe, North America, Russia, Denmark, and Japan (Yamaguti, 1958; Saeed et al., 2006). Human infection with this fluke was reported in Greenland (Yu and Mott, 1994). Cercariae develop in littorina snails, *Littorina littoria*, and encyst in the fish *Gobius ruthensparri* and *Labrus bergylta*; adults can be grown in gulls fed metacercarial cysts (Yamaguti, 1958).

Haplorchis pleurolophocerca (Sonsino, 1896); Yamaguti, 1958

The genus *Haplorchis* is characterized by the presence of only one testis and a ventrogenital-sucker complex armed with gonotyl and chitinous spines (Chen, 1936).

The genus consists of more than 20 species, and five of them, *Haplorchis pleurolophocerca, H. pumilio, H. taichui, H. vanissimus,* and *H. yokogawai* (one more species, *H. microrchis* is a synonym of *H. taichui*; Chen, 1936), are acknowledged as the species responsible for human infections (Yu and Mott, 1994). *H. pleurolophocerca* was described based on adults from cats in Egypt (Yamaguti, 1958). Cercariae were found in the snails, *Melania tuberculata* and *Cleopatra bulimoides* (Yamaguti, 1958). The fish *Gambusia affinis* is the second intermediate host (Yu and Mott, 1994). Human infections are known from Egypt (Yu and Mott, 1994).

Haplorchis pumilio (Looss, 1986), 1899

This species was originally described based on adult flukes obtained from birds and mammals in Egypt (Yamaguti, 1958). Later the same species was discovered in Taiwan and described under the name of *Monorchotrema taihokui* (Chen, 1936). It has also been found in the Philippines, Thailand, Laos, South China, Taiwan, and Egypt (Velasquez, 1982; Yu and Mott, 2002; Chai et al., 2005b). Human infections were reported in all of the above countries (Yu and Mott, 1994). The snail host is *Melania reiniana* var. *hitachiens* (Faust and Nishigori, 1926; Velasquez, 1982). The fish hosts are freshwater species belonging to the Cyprinidae, Siluridae, and Cobitidae, similar to *H. tachui* (Velasquez, 1982). Dogs and cats are the natural definitive hosts (Yamaguti, 1958).

Haplorchis taichui (Nishigori, 1924); Chen, 1936 (Fig. 2.6)
[syn. *Monorchotrema microrchia* Matsuda, 1932; *Haplorchis microrchis* Yamaguti, 1958]

This species was originally described from adult flukes in birds and mammals caught in the middle part of Taiwan (Faust and Nishigori, 1926; Ujiie, 1936b). It is also distributed widely in the Philippines, Bangladesh, India, Palestine, Egypt, Malaysia, Thailand, Laos, Vietnam, and South China (Velasquez, 1982, Yu and Mott, 1994; Chai et al., 2005b). Human infections are not uncommon in Thailand, Laos, the Philippines, and South China (Velasquez, 1982; Yu and Mott, 2002; Chai et al., 2005b; Belzario et al., 2005). The snail host is *Melania obliquegranosa, Melania juncea*, or *Melanoides tuberculata* (Faust and Nishigori, 1926; Velasquez, 1982). The fish hosts are mainly freshwater species, including *Cyprinus carpio, Cyprinus auratus, Zacco platypus, Pseudorasbora parva, Rhodeus ocellatus, Gambusia affinis, Puntius orphoides, Puntius leicanthus, Puntius gonionotus, Puntius binotatus*, and *Puntius palata* (Velasquez, 1982) and *Raiamas guttatus, Mystacoleucus marginatus*, and *Henichoryhnchus siamensis* (Kumchoo et al., 2005). Dogs, cats, and birds are the natural definitive hosts (Yamaguti, 1958).

Haplorchis vanissimus Africa, 1938

This species was originally described from adult flukes obtained in a naturally infected man in the Philippines (Africa et al., 1940). Later, this fluke was reported

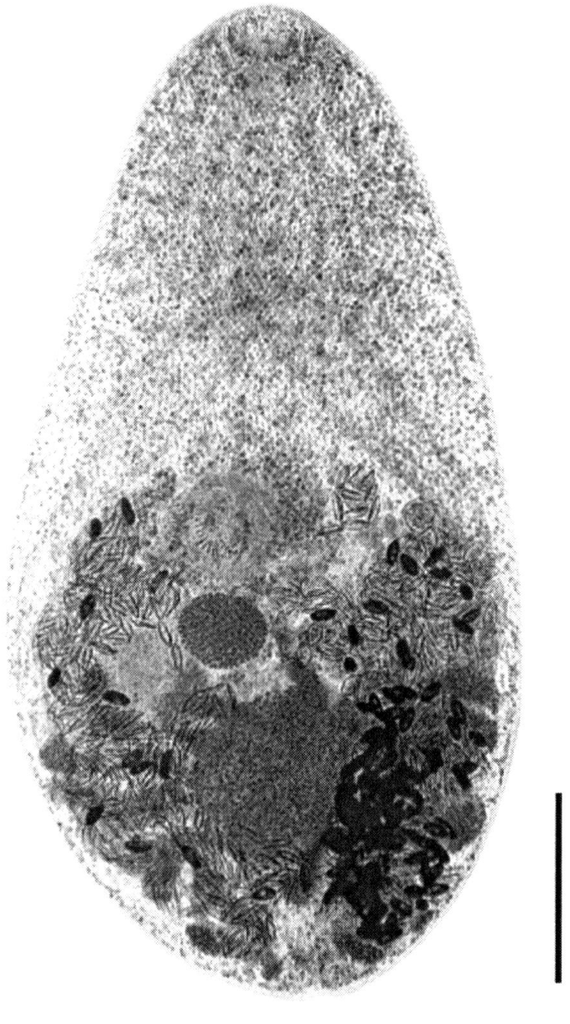

FIGURE 2.6. An adult fluke of *Haplorchis taichui* from a human infection. Acetocarmine stain. Scale bar = 0.1 mm.

from pelicans and wild mammals in Australia (Pearson and Ow-Yang, 1982). The snail host is unknown. Freshwater fish are second intermediate hosts (Yu and Mott, 1994).

Haplorchis yokogawai (Katsuta, 1932); Chen, 1936

This species was originally described from adult flukes obtained in dogs and cats experimentally fed the metacercariae encysted in the mullet *Mugil cephalus* in Taiwan (Katsuta, 1932b). Later this fluke was reported in the Philippines, South China, Malaysia, Indonesia, Thailand, Laos, India, Australia, and Egypt (Velasquez, 1982;

Yu and Mott, 1994). Human infections were reported in most of the above countries (Yu and Mott, 1994; Chai et al., 2005b). The snail host is *Melanoides tuberculata* or *Stenomelania newcombi* (Velasquez, 1982). The fish hosts are freshwater species including *Mugil* spp., *Puntius* spp., *Misgurnus* sp., and *Ophicephalus striatus* (Velasquez, 1982). Dogs, cats, cattle, and other mammals are natural definitive hosts (Yamaguti, 1958).

Heterophyes dispar Looss, 1902

The genus *Heterophyes* is characterized by the median location of the ventral sucker and the presence of a genital sucker armed with gonotyl (Chai and Lee, 2002). The genus consists of about 10 species, and three of them, *Heterophyes dispar, H. heterophyes*, and *H. nocens* (syn. *H. katsuradai*), are the species responsible for human infections (Yu and Mott, 1994; Chai and Lee, 2002). *Heterophyes dispar* was first discovered in the intestines of dogs and cats in Egypt, and then from mammals including the fox and wolf in the northern Africa and eastern Mediterranean (Yu and Mott, 1994). Brackish water fish are second intermediate hosts (Taraschewski, 1984). Human infections were reported from two Korean men who returned from Saudi Arabia (Chai et al., 1986a) and from Thailand (Yu and Mott, 1994).

Heterophyes heterophyes (v. Siebold, 1852); Stiles and Hassall, 1900

This species was first discovered by Bilharz in 1851 at autopsy of an Egyptian in Cairo, and is now known to cause human infections along the Nile Delta of Egypt and Sudan (Yu and Mott, 1994; Fried et al., 2004; Chai et al., 2005a). It is also present in Greece, Iran, Turkey, Italy, and Tunisia (Himonas, 1964; Yu and Mott, 1994; Pica et al., 2003). In Asia, several foci have been reported (Yu and Mott, 1994); however, this parasite species might have been confused with *H. nocens* and should be verified. Imported human infections were reported in Japan (Kagei et al., 1980) and the Republic of Korea (Chai et al., 1986a; Chai and Lee, 2002), from people who returned from Egypt to Japan and from Saudi Arabia and Sudan to Korea. The snail host is *Pirenella conica* in Egypt (Taraschewski, 1984). Important second intermediate hosts are brackish water fishes including *Mugil cephalus, Tilapia nilotica, Aphanius fasciatus*, and *Acanthogobius* sp. Humans become infected by eating infected fish raw or inadequately cooked. A variety of mammals other than humans takes the role of the reservoir host, for example, dogs in India (Beaver et al., 1984; Harinasuta et al., 1987). In Egypt, human infections are prevalent among the inhabitants of the northern part of the Nile Delta, particularly around the Lakes Manzala, Borollos, and Edco where fishermen and domestic animals frequently consume fish (Yu and Mott, 1994). During 1987–1991, the prevalence of heterophyiasis in five governorates of the Nile Delta ranged between 0.01% and 1% (Yu and Mott, 1994). A review of 299 cases in Dakahlia Governorate indicated that the disease is common in both urban and rural localities owing to the habit of consuming salted or insufficiently baked fish (Sheir and Aboul-Enein, 1970). The mean prevalence of heterophyid infections in the villages of Khuzestan, Islamic Republic, was found to be 8% (range 2–24%) (Yu and Mott, 1994). In postmortem examination of carnivores in the same areas, 14.2% of jackals, 33.3% of foxes, and 2.5% of dogs were infected with heterophyid flukes including *H. heterophyes, M. yokogawai*, and *H. katsuradai* (a synonym of *H. nocens*) in order of frequency (Massoud et al., 1981). Metacercariae of *H. heterophyes* can survive up to 7 days in salted fish. The pathogenesis and intestinal pathology, clinical disease, diagnosis, chemotherapy, prevention, and control are the same as those for metagonimiasis.

Heterophyes nocens Onji and Nishio, 1916 (Fig. 2.7) [syn. *Heterophyes katsuradai* Ozaki and Asada, 1926]

This species was first reported in Japan from experimental dogs and cats fed the metacercariae encysted in the mullet *Mugil cephalus* (Onji and Nishio, 1916). It is now known to occur as human infections in Japan, China, and the Republic of Korea (Yokogawa et al., 1965b; Xu and Li, 1979; Seo et al., 1981b; Chai et al., 1984a, 1985a). In China, the species was described as *H. heterophyes* (Xu and Li, 1979), but is now presumed to be *H. nocens. H. nocens* is distinguished from *H. heterophyes* by the morphology of the genital sucker, especially the smaller number of rodlets on the gonotyl; 50 to 62 in *H. nocens* and 70 to 85 in *H. heterophyes* (Taraschewski, 1984; Chai and Lee, 2002). The first intermediate host is a brackish water snail *Cerithidea cingulata* (= *Tympanotonus microptera*). The second intermediate hosts are brackish water fish such as the mullet *Mugil cephalus* or goby *Acanthogobius flavimanus* (Chai and Lee, 2002). Domestic or feral cats were found naturally infected with this fluke (Eom et al., 1985; Sohn and Chai, 2005). In the Republic of Korea, the metacercariae were found in the mullet *Mugil cephalus* captured in three southern coastal areas (Seo et al., 1980b). Over 40% prevalences were detected in several southwestern coastal areas (Chai et al., 1994b, 1997; Chai and Lee, 2002). Individual worm burdens ranged from 1 to 1338, Average 263 per person (Chai et al., 1994a, 1997, 1998a). Many western and southern coastal islands were added to the list of endemic areas (Chai et al., 2004). In Japan, human *H. nocens* infections were reported from Kochi, Chiba, Yamaguchi, Chugoku, and Hiroshima Prefectures (Suzuki et al., 1982). Recently, two lakeside villages of Mikkabi-cho, north end of Hamana Lake, Shizuoka Prefecture, were added as new endemic areas, with prevalence rates of 7.5% and 10.5% (Kino et al., 2002).

Heterophyopsis continua (Onji and Nishio, 1916); Yamaguti, 1958

This species was first discovered from experimental cats fed the mullet *Mugil cephalus* that harbored the metacercariae in Japan (Onji and Nishio, 1916). *H. continua* differs from other heterophyid species in its elongate body, genital sucker located separately from the ventral sucker, and two obliquely tandem testes (Chai and Lee, 2002). The presence of human infections was first mentioned in Japan (Yamaguti, 1939b). Subsequently, in the Republic of Korea, two natural human infections were discovered (Seo et al., 1984a). Including these two

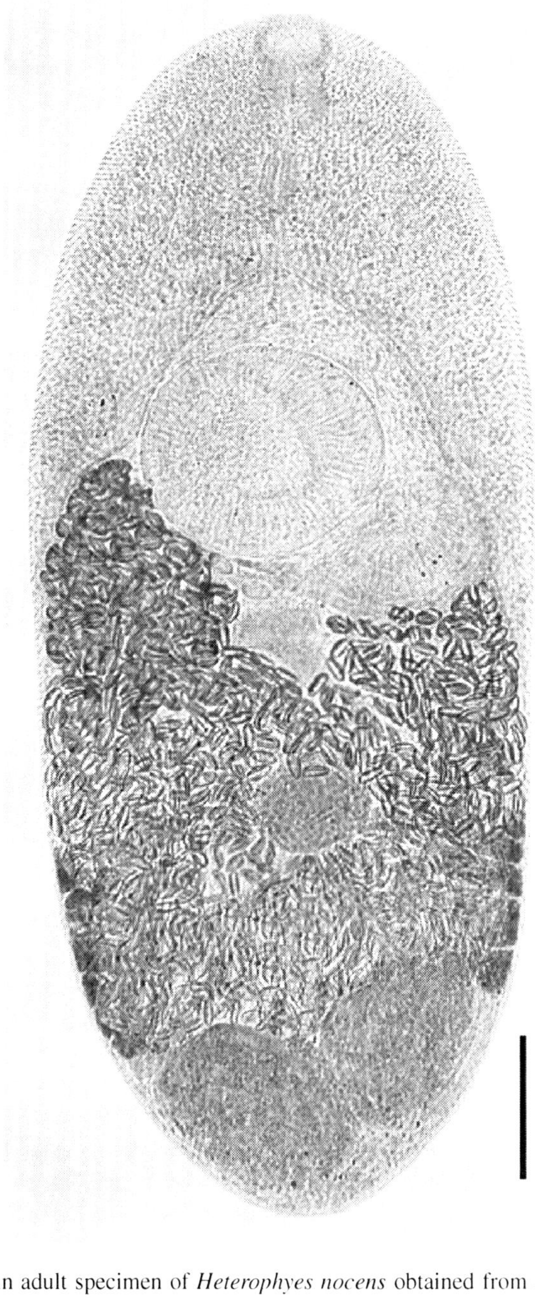

FIGURE 2.7. An adult specimen of *Heterophyes nocens* obtained from an experimentally infected rat. Acetocarmine stain. Scale bar = 0.1 mm.

cases, eight human cases in total have been confirmed by the recovery of adult flukes in the Republic of Korea (Chai et al., 1997, 1998a; Hong et al., 1996a). The first intermediate host is unknown. Metacercariae were found in the perch *Lateolabrax japonicus* and goby *Acanthogobius flavimanus* (Chun, 1960b). Other fish hosts include shad *Clupanodon punctatus* (Chun, 1960b; Sohn et al., 1994b), conger eel *Conger myriaster* (Kim et al., 1996), and sweetfish *Plecoglossus altivelis* (Cho and Kim, 1985). Domestic or feral cats (Eom et al., 1985; Sohn and Chai, 2005), ducks (Onji and Nishio, 1916), and sea gulls (Yamaguti, 1939a) were reported to be natural definitive hosts. Experimental definitive hosts include cats (Onji and Nishio, 1916), dogs (Chun, 1960b; Seo et al., 1984a), and domestic chicks (Hong et al., 1990a, 1991).

Metagonimus minutus Katsuta, 1932

Flukes of *Metagonimus* are characterized by the small body size, laterally deviated ventral sucker, and absence of the ventrogenital apparatus or genital sucker, which is present in other genera including *Heterophyes, Heterophyopsis, Haplorchis*, and *Stellantchasmus* (Yu and Mott, 1994; Chai and Lee, 2002). A total of seven species have been reported (Saito et al., 1997), and four of them, namely *M. yokogawai* (Korea, China, Taiwan, Japan, Indonesia, and Russia), *M. takahashii* (Korea and Japan), *M. minutus* (Taiwan), and *M. miyatai* (Korea and Japan), have been reported from human infections (Yu and Mott, 1994; Chai and Lee, 2002). *Metagonimus minutus*, characterized by small sized uterine eggs, was reported as adult flukes recovered from experimental mice and cats fed mullets infected with the metacercariae in Taiwan (Katsuta, 1932a). This parasite is listed among the human-infecting intestinal trematodes (Beaver et al., 1984; Yu and Mott, 1994), but no literature background is traceable.

Metagonimus miyatai Saito et al., 1997

This parasite was first found by I. Miyata, in 1941 in Japan, but its taxonomic significance was not established until 1997, when it was reported as a distinct species in Japan and Korea (Saito et al., 1997). The description was based on adult flukes collected from dogs and hamsters experimentally fed the metacercariae from sweetfish, dace, common fat-minnow *Morocco steindachneri*, pale chub *Zacco platypus*, dark chub *Zacco temmincki*, and also on specimens collected from naturally infected humans. This fluke is morphologically different from *M. yokogawai* and *M. takahashii* in the position of the posterior testis (separated considerably from the anterior one), the distribution of vitelline follicles (never crossing over the posterior testis), and the intermediate size of eggs (28–32 μm) (Saito et al., 1997). This species is genetically distinct from *M. takahashii* and *M. yokogawai*, as shown by the polymerase chain reaction–based restriction fragment length polymorphism (PCR-RFLP) patterns (Yu et al., 1997a), karyotypes (Lee et al., 1999), simple sequence repeat (SSR)-PCR patterns (Yang et al., 2000), random amplification of polymorphic DNA patterns (Yu et al., 1997b), and 28S ribosomal DNA and cytochrome C oxydase subunit I patterns (Lee et al., 2004b). The snail intermediate hosts are *Semisulcospira globus* (Kim et al., 1987), *Semisulcospira libertina*, or

Semisulcospira dolorosa (Shimazu, 2002). Mice, rats, hamsters, and dogs are experimental definitive hosts (Chai and Lee, 2002; Guk et al., 2005). Reservoir hosts are unknown. In the Republic of Korea, the presence of this species was first reported in 1980 (under the name of *Metagonimus* sp.) from people along the Gum River, which is uninhabited by the sweetfish (Kim, 1980). A high prevalence of this fluke infection was reported among people residing around the Daechong Reservoir and its upper reaches (Kim et al., 1987), the upper reaches of the Namhan River (Chai et al., 1993a), the Hantaan River basin (Park et al., 1993), and the western inland of Gangwon-do (Ahn, 1993). A western tributary of the Nakdong River in Keochang-gun was also reported as a low-grade endemic area (erroneously under the name of *M. takahashii*) (Son et al., 1994). Small rivers of Shizuoka Prefecture, Japan, were found to be endemic areas of this fluke (Kino et al., 2006).

Metagonimus takahashii Suzuki, 1930

This fluke was first reported in Japan from mice and dogs fed metacercariae encysted in several species of fresh water fish other than the sweetfish (Suzuki, 1930), and is also distributed in Korea and Japan (Chai and Lee, 2002). It differs from *M. yokogawai* in the position of two testes (anterior testis separated from the posterior testis), the distribution of vitelline follicles (more abundant and crossing over the posteriormost end), and by the larger size of the eggs (*M. yokogawai*, 28–30 μm; *M. takahashii*, 32–36 μm) (Chai and Lee, 2002). The two species are genetically differentiated by the PCR-RFLP patterns, karyotypes, SSR anchored-PCR (SSR-PCR) patterns (Chai and Lee, 2002), and 28S ribosomal DNA and cytochrome C oxydase subunit I patterns (Lee et al., 2004b). The Koga type of *Metagonimus* encysting in the dace *Tribolodon* spp. (Saito, 1984) is regarded as a synonym of *M. takahashii* (Chai et al., 1991). The snail hosts involved may be *Semisulcospira coreana* or *Koreanomelania nodifila* (Cho et al., 1984), but this has to be confirmed. The fish hosts are the crussian carp *C. carassius* (Chun, 1960a), carp *C. carpio* (Saito, 1984), dace *Tribolodon taczanowskii* (Chai et al., 1991), and perch *Lateolabrax japonicus* (Kim et al., 2006). Numerous strains of mice could be experimentally infected with fluke (Guk et al., 2005). There are no reports on reservoir hosts. In the Republic of Korea, the presence of this species was documented based on adult flukes recovered from experimental rabbits in 1960 (Chun, 1960a). In humans, adult flukes were confirmed in 1993 from inhabitants of Umsong-gun, Chunchungnam-do, along the upper reaches of the Namhan River (Chai et al., 1993a). The inhabitants had mixed infections with *M. miyatai*, with an egg positive rate of 9.7% for both species (Chai et al., 1993a). People in the western inland of Gangwon-do were also found to be infected with the two species (Ahn, 1993). *M. takahashii* is distributed along small streams in various inland areas of the Republic of Korea (Chai and Lee, 2002).

Metagonimus yokogawai (Katsurada, 1912) (Fig. 2.8)

This parasite is probably the most common intestinal fluke infecting humans in the Far East (Chai and Lee, 2002). Human infections were reported also from the

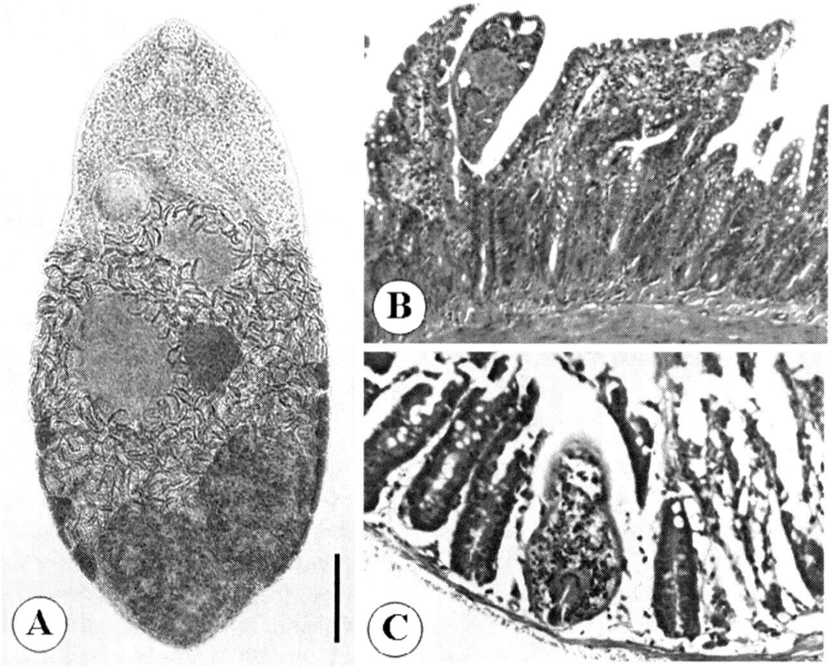

FIGURE 2.8. (A) *Metagonimus yokogawai* adult recovered from an experimentally infected rat. Acetocarmine stain. Scale bar = 0.1 mm. (B) Intestinal section of a rat experimentally infected with *M. yokogawai*, 4 weeks after infection. The mucosal pathology is characterized by villous atrophy and crypt hyperplasia. H&E stain, ×200. (C) An adolescent worm of *M. yokogawai*, which invaded beyond the submucosa and facing the muscular layer of the duodenum of an immunosuppressed mouse, day 5 postinfection. H&E stain, ×400. (From Chai et al., 1995a, with permission).

northern provinces of Siberia, Israel, the Balkan states, and Spain (Yu and Mott, 1994). The eggs of this species are confused with many other heterophyid species, thus the accuracy of the data on the prevalence of this infection is limited. The molluscan first intermediate host is the fresh water snail, *Semisulcospira coreana* or *Semisulcospira libertina* (Cho et al., 1984). The most important second intermediate host is the sweetfish *Plecoglossus altivelis* (Chai and Lee, 2002), but the dace *Tribolodon* sp. (Chai et al., 1991) and the perch *Lateolabrax japonicus* (Ahn, 1983) also serve as second intermediate hosts. Dogs, rats, and cats were reported as natural definitive hosts (Cho et al., 1981; Seo et al., 1981a; Huh et al., 1993), although their significance as the source of the human infection (i.e., as reservoir hosts) has not been established. Variable strains of mice were found to be susceptible to *M. yokogawai* infection but less susceptible to *M. miyatai* and *M. takahashii* infections (Guk et al., 2005). In the Republic of Korea, almost all the large and small streams in eastern and southern coastal areas are endemic foci of metagonimiasis (Seo et al., 1981d; Song et al., 1985). The Sumjin, Tamjin, and Boseong Rivers, Geoje Island, and Osip Stream (Gangwon-do) were the highest endemic areas with 20% to 70% egg positive rates in the villagers (Chai et al., 1977, 2000c; Seo et al., 1981d; Chai and Lee, 2002). The nationwide egg positive rate of people residing in river basins was once estimated at 4.8% (Seo et al., 1981d). Human infection has been recorded in Guangdong, Anhui, Hubei, and Zhejiang Provinces of China and Taiwan (Xu and Li, 1979). In Japan, the prevalence rate of *M. yokogawai* infection was low or negligible after the 1970s, except a few foci such as areas surrounding the Hamana Lake (Ito et al., 1991). However, small rivers of Shizuoka Prefecture still were prevalent with *M. yokogawai* in the fish intermediate hosts (Kino et al., 2006). In Russia, *M. yokogawai* is endemic in the Amur and Ussuri valleys of Khabarovsk territory where the prevalence in ethnic minority groups varies between 20% and 70% (Yu and Mott, 1994). In the north of Sakhalin Island the infection rate was 1.5% in Russians and 10% in ethnic minorities. Sporadic cases were also reported in Amur district and Primorye territory (Yu and Mott, 1994).

Procerovum calderoni (Africa and Garcia, 1935); Price, 1940

This species was reported from dogs and then two native people in the Philippines (Africa and Garcia, 1935). Later, it was reported from China and Africa (Harinasuta et al., 1987). Second intermediate hosts are the freshwater fish, *Ophiocephalus striatus, Glossogobius giurus, Mollienesia latipinna, Mugil* sp., and *Creisson validus* (Velasquez, 1973a,b; Yu and Mott, 1994). The first intermediate host is the brackish water snail, *Thiara riquetti* (Velasquez, 1973b).

Procerovum varium Onji and Nishio, 1916

This parasite was described from experimentally infected dogs with the metacercariae encysted in the mullet *Mugil cephalus* in Japan (Onji and Nishio, 1916). Experimental human infections were reported (Aokage, 1956), but there have been no reports of natural human infections. It is now known to be distributed in China, the Philippines, Australia, India (Umadevi and Madhavi, 2000), and Korea (Sohn and Chai, 2005). Natural infection of cats has been found (Sohn and Chai, 2005).

Pygidiopsis summa Onji and Nishio, 1916 (Fig. 2.9)

This species was first found in dogs fed brackish water fish infected with the metacercariae in Japan (Onji and Nishio, 1916). It is now known to be present in the Republic of Korea (Chai and Lee, 2002). Human infections were first reported in Japan by detection of eggs in feces in 1929 (Takahashi, 1929a), and adult flukes in human infections were identified in 1965 (Yokogawa et al., 1965b). The worms are characterized by a small concave body, median location of the ventral sucker, unique morphology of the ventrogenital apparatus, and side-by-side location of the two testes (Chai et al., 1986b). The metacercariae were detected in the gills and muscles of the mullet *Mugil cephalus* and goby *Acanthogobius flavimanus* (Chun, 1963; Seo et al., 1981c; Sohn et al., 1994b). Human infections in the Republic of Korea were first reported in eight residents of a seaside salt-farm village of

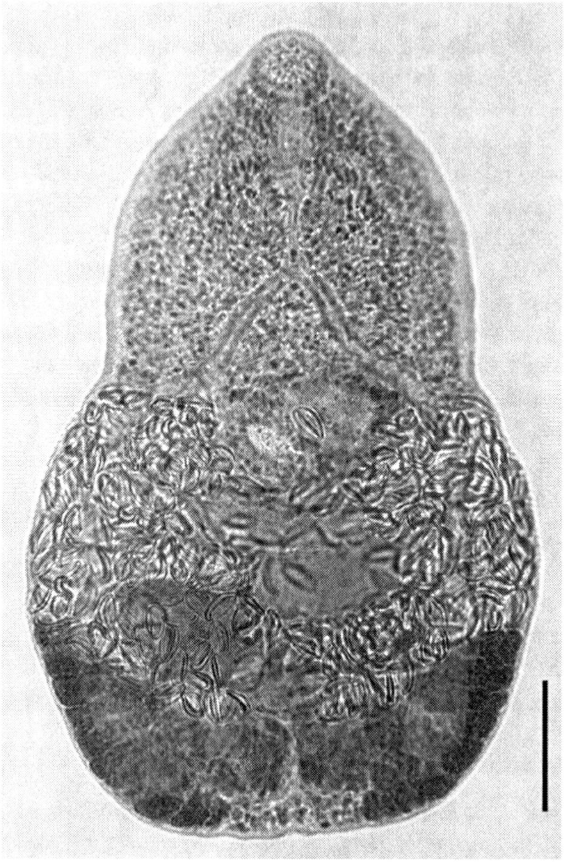

FIGURE 2.9. An adult worm of *Pygidiopsis summa* from an experimentally infected rat. Acetocarmine stain. Scale bar = 0.1 mm.

Okku-gun, Chollabuk-do, who habitually ate the raw flesh of the mullet (Seo et al., 1981b). In another coastal area of the Republic of Korea, 18 of 20 heterophyid egg–positive people were found to be infected with this fluke (Chai et al., 1997). Five infected people were detected in Buan-gun, Chollabuk-do (Chai et al., 1998a). It is now known to be distributed widely along the western and southern coastal islands (Chai et al., 2004). The first intermediate host in the Republic of Korea is the brackish water snail *Cerithidea* sp. or *Tympanotonus* sp. (personal observation). The natural infection of domestic or feral cats has also been reported (Eom et al., 1985; Sohn and Chai, 2005).

Stellantchasmus falcatus Onji and Nishio, 1916

This species was first reported from cats experimentally fed the mullet harboring the metacercariae, in Japan (Onji and Nishio, 1916). Morphological characteristics of this fluke include the ventral sucker, which is slightly deviated to the right side of the body, and the presence of an elongated sac-like seminal vesicle on the opposite side of the ventral sucker (Chai and Lee, 2002). Human infections were reported first (Takahashi, 1929a) and then repeatedly reported in Japan (Ito, 1964; Kagei et al., 1964). Thereafter, human infections have been reported in various Asian-Pacific countries: the Philippines, Hawaii, Japan, Palestine, Thailand, and Korea (Yamaguti, 1958; Radomyos et al., 1990; Seo et al., 1984b; Chai and Lee, 2002). A successful life cycle study was performed in Hawaii; the first intermediate host was confirmed to be the brackish water snail *Stenomelania newcombi* or *Thiara granifera* (Martin, 1958; Noda, 1959), and the second intermediate host was shown to be the mullet (Chai and Sohn, 1988; Chai and Lee, 2002), and half-beaked fish *Dermogenus pusillus* (Wongsawad et al., 1998; Sripalwit et al., 2003). For experimental hosts, rats were better than mice (Saenphet et al., 2003). Natural infections in cats are known (Takahashi, 1929a; Sohn and Chai, 2005).

Stellantchasmus formosanus Katsuta, 1931

This parasite was described from experimentally infected cats, dogs, and mice with the metacercariae encysted in the mullet *Mugil cephalus* in Taiwan (Katsuta, 1931). An experimental human infection was reported in Taiwan (Katsuta, 1931), but there have been no reports of natural human infections.

Stellantchasmus pseudocirratus (Witenberg, 1929); Yamaguti, 1958 [syn. *Stellantchasmus amplicaecalis* Katsuta, 1932]

This parasite was described from naturally infected dogs and cats in Palestine (Witenberg, 1929) and cats, dogs, and mice fed mullet in Taiwan (Katsuta, 1932c). The mullet, *Mugil* sp., is the second intermediate host (Witenberg, 1929; Yamaguti, 1958). Human infections were reported in the Philippines and Hawaii (Africa et al., 1940; Yamaguti, 1958).

Stictodora fuscata (Onji and Nishio, 1916); Yamaguti, 1958

This species was originally described from cats experimentally fed on infected mullet in Japan (Onji and Nishio, 1916). The worm is morphologically characterized by the presence of a gonotyl, which is superimposed on the ventral sucker and armed with 12 chitinous spines, a metraterm, and two testes located obliquely in the middle field of the body. Human infection with this fluke (reported under the name of *Stictodora* sp.) was found in a young Korean man, who regularly ate raw mullets and gobies (Chai et al., 1988). Thirteen additional human cases were subsequently detected in a seashore village in the southwestern coastal area (Chai and Lee, 2002). The metacercariae were found in gobies, *Acanthogobius flavimanus*, collected from a market in Chollanam-do Province (Sohn et al., 1994a, 1994b). The domestic cat (*Felis catus*) has been used as an experimental definitive host (Sohn et al., 1994b). Feral cats were found naturally infected with this fluke (Sohn and Chai, 2005).

Stictodora lari Yamaguti, 1939

This fluke was first found in the small intestine of the sea gull *Larus crassirostris* in Japan (Yamaguti, 1939a). Morphological characters include a gonotyl armed with 70 to 80 minute spines (Chai et al., 1989a). Adult flukes were first recovered from six Korean people who resided in two southern coastal villages (Chai et al., 2002). The first intermediate host is the brackish water gastropod *Velacumantus australis* in Australia (Howell, 1973). The metacercariae of this fluke were found in a species of brackish water fish, that is, the goby *Acanthogobius flavimanus*, in the Republic of Korea (Chai et al., 1989a). Other fish hosts include a number of species of estuarine fish (Howell, 1973). In gobies, metacercariae were observed mainly in the head of the fish (Chai et al., 1989a). Cats and dogs were used as experimental definitive hosts (Chai et al., 1989a). Reservoir hosts include feral cats (Sohn and Chai, 2005).

Pathogenicity and Host–Parasite Relationships of Heterophyid Flukes

At the site of attachment in the host intestinal mucosa, *H. heterophyes* adults can cause mild inflammatory reactions, ulcers, irritation, and superficial necrosis of the mucosa (Yu and Mott, 1994; Fried et al., 2004). The intestinal histopathology was studied in *M. yokogawai* (Chai, 1979; Lee et al., 1981; Kang et al., 1983) (Fig. 2.8), *P. summa* (Seo et al., 1986), *H. heterophyes* (Marty and Andersen, 2000), and *C. armatus* (Hong et al., 1997) using experimental animals, including mice, rats, cats, and dogs. The adult flukes of *M. yokogawai* were found to parasitize the middle part of the small intestines; within the crypts of Lieberkühn in early stages of the infection (by day 2 to 3 postinfection), and between the villi in later stages (Chai, 1979; Lee et al., 1981; Kang et al., 1983).

The pathological features were characterized by villous atrophy and crypt hyperplasia, with variable degrees of inflammatory reactions. The infected mucosa showed blunting and fusion of the villi, edema of the villus tips, congestion and inflammatory cell infiltrations in the villous stroma, and decreased villus/crypt height ratios (Chai and Lee, 2002). In a naturally infected human with *M. yokogawai*, similar intestinal histopathology was reported (Chi et al., 1988). In immunocompetent animals, the location of worms was confined to the intestinal mucosa (Kang et al., 1983; Rho et al., 1984; Jang et al., 1985). However, immunosuppression of mice by prednisolone injection allowed a deeper invasion of the worms into the submucosa (Chai et al., 1995a) (Fig. 2.8). In addition, immunosuppression enhanced the survival of worms and prolonged their life spans in the same mouse strain (Chai et al., 1984b, 1995a). In *M. miyatai*–infected mice, similar intestinal histopathology was observed, although the degree of mucosal damage was less severe than in *M. yokogawai*–infected mice, as represented by stronger expression patterns of the proliferating cell nuclear antigen (PCNA) in the intestinal mucosa (Yu et al., 1997c). Similar features were also observed in rats and mice experimentally infected with *P. summa*; the middle part of their small intestines was most frequently affected, and like *M. yokogawai*, the worms caused severe villous atrophy and crypt hyperplasia, with inflammation of the villous stroma (Seo et al., 1986). In experimental *C. armatus* infection in rats, the worms caused mechanical irritation and mucosal inflammations in the small intestines from as early as 3 days after the infection (Hong et al., 1997).

Intestinal histopathology due to *M. yokogawai* infection was normalized at 3 to 4 weeks after the infection (Chai et al., 1995a). Hence, there may be host protective mechanisms against *M. yokogawai* and other heterophyid fluke infections. However, the immunophysiology and pathogenesis of the intestinal pathology and symptoms due to intestinal fluke infections have seldom been studied, in contrast to other intestinal helminth infections including nematode infections such as trichinosis (Castro, 1989). One of the possible immune effectors for the spontaneous recovery of the histopathology includes intestinal intraepithelial lymphocytes that increase remarkably along the villous epithelial layer of infected rats (Chai et al., 1994a). Mucosal mast cells were suggested as another effector responsible for the worm expulsion from infected rats (Chai et al., 1993b). Goblet cells were suggested to be a third effector for the expulsion of worms (Chai and Lee, 2002). However, intensive studies are required to understand the precise roles of mucosal mast cells and goblet cells in the host defense against heterophyid infections. Immunogold studies revealed that the antigenicity of *M. yokogawai* originated from the syncytial tegument, tegumental cell cytoplasms, vitelline cells, and epithelial lamellae of the cecum (Ahn et al., 1991; Rim et al., 1992). A sodium dodecyl sulfate–polyacrylamide gel electrophoresis (SDS-PAGE)/immunoblot analysis of crude extracts of the metacercariae showed that out of 14 protein bands found, 11 reacted with infected human sera, and among them 66-kd and 22-kd proteins were the specific antigens (Lee et al., 1993a).

In terms of pathogenicity of heterophyid flukes, it is of interest to note that several species, namely *S. falcatus*, *Haplorchis* spp., and *Procerovum* spp., can cause erratic parasitism in humans, which is often fatal (Africa et al., 1940). The most frequently affected sites of the erratic parasitism were the heart valve, brain, and spinal cord, where eggs and adult flukes originating from the intestinal mucosa embolized in the blood vessels (Africa et al., 1940). Eggs of *H. heterophyes* (Collomb et al., 1960) and probably of *H. nocens* (under the name *H. heterophyes*) (Zhang and Fan, 1990) were found encapsulated in the brain of patients with nerological symptoms. Presumably such erratic parasitism could occur in immunocompromised patients, rather than immunocompetent individuals. In this respect, it is worthwhile mentioning that *M. yokogawai* worms were found to have invaded the submucosa of the small intestine in the immunosuppressed mice (Chai et al., 1995a). Further studies are required to elucidate the pathogenesis of this erratic parasitism by heterophyid flukes.

Clinical Symptoms, Diagnosis, and Treatment of Heterophyidiases

Clinical symptoms due to intestinal heterophyid infections are generally mild and transient, unless patients are heavily infected (Chai and Lee, 2002) or immunocompromised. In patients with *M. yokogawai* infection, for example, the most frequent symptoms reported are mild to moderate degrees of abdominal pain, diarrhea, lethargy, anorexia, and weight loss (Cho et al., 1984; Seo et al., 1985c). Decreased enzyme activities were also observed in association with diarrhea and malabsorption, which are commonly encountered in acute infections with *M. yokogawai* (Hong et al., 1990b). From a biochemical study on the watery content in the small intestines of dogs experimentally infected with *M. yokogawai*, it was suggested that the watery content might be a result of a poor absorption of the intestinal secretions from the secretory crypt cells (Cho et al., 1985).

The degree of clinical symptoms seems to be related to the individual worm burdens; heavier infection cases tend to suffer from more severe illness. However, the severity of the symptoms may also be related to the susceptibility as well as the degree of acquired immunity of the individual patient. A new visitor to an endemic area, for example, suffered from a severe illness after a primary infection (Chai et al., 1989b). On the other hand, long-term residents in endemic areas generally complained of milder symptoms than those expected (Seo et al., 1985c). Even in the most heavily infected case in the Korean literature, a man who resided in a highly endemic area and harbored as many as 63,587 worms, complained of only minor gastrointestinal trouble—indigestion and epigastric pain (Seo et al., 1985c). Clinical symptoms due to other heterophyid fluke infections were not much different from those seen in *M. yokogawai* infection (Chai et al., 1994b, 1997, 1998a).

The diagnosis can be made by the recovery of heterophyid eggs in fecal examinations, and is expressed using the term *heterophyid fluke infection*, because the eggs of different heterophyid species closely resemble each other (Lee et al., 1984c). The confirmatory diagnosis should be carried out usually after the recovery of adult flukes following anthelmintic treatment and purgation. The specific diagnosis using only eggs is difficult in areas of no previously known endemicity, as well as in endemic areas with mixed heterophyid infections. Close observations and measurements of the heterophyid eggs in the feces are useful for differential diagnosis (Lee et al., 1984c). For example, the eggs of *M. yokogawai* can be differentiated from other heterophyid eggs by their length of 26.9 to 31.6 μm, elliptical shape with length/width ratio of 1.5 to 2.1, clean shell surface, less prominent operculum, no shoulder rims, and dark yellow or brown color (Lee et al., 1984c). The eggs of *M. takahashii* and *M. miyatai* have similar morphology to those of *M. yokogawai*, with the exception of the larger egg sizes of the two former species; hence, the measurement of egg size is essential. The eggs of *H. nocens* are similar to those of *M. yokogawai*, but the former is a little smaller and has slight attenuations at one or both ends (Lee et al., 1984c). The eggs of *H. continua* are broadly oval in shape, and difficult to differentiate from those of

M. yokogawai (Lee et al., 1984c). The eggs of *P. summa* are characteristically small and pyriform in shape (Lee et al., 1984c); they resemble the eggs of *Clonorchis sinensis*, a liver fluke, but lack musk-melon patterns on the shell surface, and shoulder rims around the operculum are inconspicuous. Similarly, eggs of *H. taichui, H. pumilio*, and *H. yokogawai*, closely resemble each other, and need differentiation from eggs of *Opisthorchis viverrini* and lecithodendriid flukes, including *Phaneropsolus bonnei* and *Prosthodendrium molenkampi* (Tesana et al., 1991). There could be false egg-negative cases among the light infection cases with *M. yokogawai*, for example, with less than 100 worms in an infected person. The number of eggs produced per day per worm for *M. yokogawai* was reported to be only 14 to 64 in the human host (Ahn, 1993), so the detectability of eggs in feces from such a case is negligible. Serological tests such as enzyme-linked immunosorbent assay (ELISA) are helpful in false egg-negative cases (Chai et al., 1989b; Cho et al., 1987).

Praziquantel is the drug of choice for all infections by heterophyids. A single oral dose of 10 to 20 mg/kg praziquantel is satisfactory, with a 95% to 100% cure rate for *M. yokogawai* infection (Rim et al., 1978; Lee et al., 1984b). Irradiation of the sweetfish, *Plecoglossus altivelis*, by 200 Gy was highly effective in controlling infectivity of the *M. yokogawai* metacercariae (Chai et al., 1995b). The heterophyid fluke infections could be prevented by the avoidance of eating uncooked fresh water or brackish water fish.

Lecithodendriidae Odhner, 1911

Species Infecting Humans

Phaneropsolus bonnei Lie Kian Joe, 1951 (Fig. 2.10)

This species was first reported from a single human autopsy in Jakarta, Indonesia, and after that from 15 human autopsies in Udornthani Provincial Hospital, Thailand (Manning et al., 1971). This fluke was also discovered in Malaysia and India (Manning et al., 1971). Later, high prevalences of this fluke infection were found in Thailand (Radomyos et al., 1998). In Laos, a total of 366 adult specimens were recovered from four people residing in Mekong riverside areas of Saravane Province (Chai et al., 2005b). Metacercariae were discovered in naiads and adult dragonflies and damselflies in Thailand (Manning and Lertprasert, 1973). Local people in northeast Thailand and Laos are known to eat naiads of these insects (Manning and Lertprasert, 1973; Chai et al., 2005b). The egg morphology is very similar to that of heterophyid fluke and of *Opisthorchis viverrini* (Kaewkes et al., 1991b; Tesana et al., 1991).

Phaneropsolus spinicirrus Kaewkes et al., 1991a

This species was reported from only one human infection in northeast Thailand (Kaewkes et al., 1991a). No further information on this parasite is available.

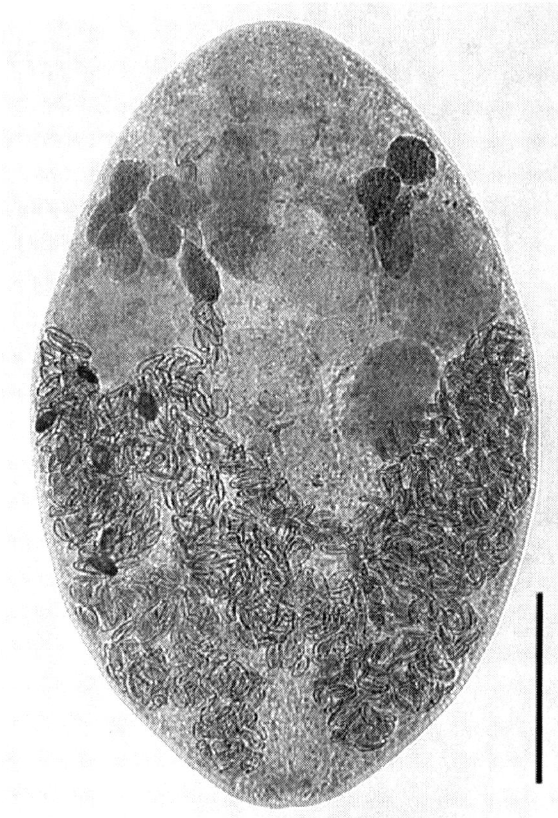

FIGURE 2.10. *Phaneropsolus bonnei* adult recovered from a human infection. Acetocarmine stain. Scale bar = 0.1 mm.

Prosthodendrium molenkampi Lie Kian Joe, 1951 (Fig. 2.11)

This species was first reported from a single human autopsy at Jakarta, Indonesia, and then from 14 human autopsies in Udornthani Provincial Hospital, Thailand (Manning et al., 1971). Later, high prevalences were reported in Thailand (Radomyos et al., 1998). The prevalence for *P. molenkampi* was 19.4% among 681 small trematode egg–positive individuals (including *Opisthorchis viverrini*) treated with praziquantel and purged in northeast Thailand (Radomyos et al., 1998). In Laos, a total of 502 adult specimens were recovered from 14 infected people residing along the Mekong riverside areas of Vientian Minicipality and Saravane Province (Chai et al., 2005b). Metacercariae of *P. molenkampi* were discovered in naiads and adult dragonflies and damselflies in Thailand (Manning and Lertprasert, 1973). Local people in northeast Thailand and Laos are known to eat naiads of these insects (Manning and Lertprasert, 1973).

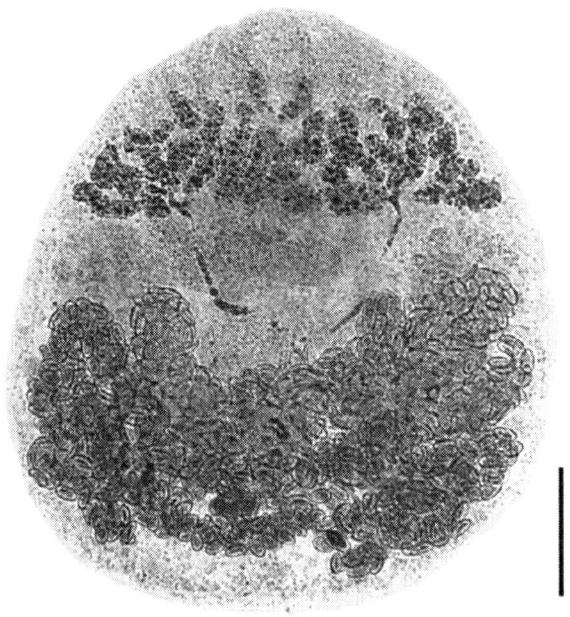

FIGURE 2.11. *Prosthodendrium molenkampi* adult recovered from a human infection. Acetocarmine stain. Scale bar = 0.1 mm.

Microphallidae Travassos, 1920

Species Infecting Humans

Spelotrema brevicaeca (Africa and Garcia, 1935); Tubangui and Africa, 1939

This fluke was originally reported under the name *Heterophyes brevicaeca* from human infections in the Philippines (Africa and Garcia, 1935). Fluke eggs were found in the heart, brain, and spinal cord of persons who died of acute cardiac dilatation (Africa and Garcia, 1935). The metacercariae were found in the crab *Carcinus maenas* and the shrimp *Macrobrachium* sp. in the Philippines (Beaver et al., 1984).

Nanophyetidae Dollfus, 1939

Species Infecting Humans

Nanophyetus salmincola Chapin, 1926, 1927 [syn. *Troglotrema salmincola* Witenberg, 1932]

This fluke infects the intestine of mammals including humans, dogs, cats, raccoons, and fox, and three species of birds in the Pacific coast of North America and Canada, and Eastern Siberia (Millemann and Knapp, 1970; Beaver et al.,

1984; Chai et al., 2005a). It has a minute pyriform body, and is characterized by the presence of two large testes in the posterior half of the body. Its snail host is *Oxytrema silicula*, and second intermediate hosts are a wide variety of fish, including salmon, trout, and nonsalmonid fish (Millemann and Knapp, 1970; Yu and Mott, 1994). Nanophyetiasis is endemic in the far-eastern part of Russia including Amur and Ussuri valleys of Khabarovsk territory and north Sakhalin, where the average prevalence is 5% (Yu and Mott, 1994). In local ethnic minorities, the prevalence is higher, 20%, and reaches up to 60% in some localities. In the U.S., 20 human cases were reported after 1974 (Eastburn et al., 1987). People acquire the infection by ingestion of improperly cooked salmon or trout. Infected people may experience diarrhea, abdominal discomfort, and eosinophilia, but the symptoms are generally mild. In animals such as dogs, foxes, and coyotes, however, the fluke has been proven to be the vector of a rickettsia, *Neorickettsia helmintheca*, which causes a serious and often fatal systemic infection known as salmon poisoning. Salmon poisoning has not been reported in humans. A new species, *Nanophyetus schikhobalowi*, was described from natives of far-eastern Siberia by Skrjabin and Podjapolskaja (1931) (Yamaguti, 1958). However, it is regarded as a subspecies, *Nanophyetus salmincola schikhobalowi* (Milleman and Knapp, 1970). Its major difference from *Nanophyetus salmincola* is that *Nanophyetus salmincola schikhobalowi* is apparently not a vector for the rickettsial organism (Milleman and Knapp, 1970).

Neodiplostomidae Shoop, 1989

Species Infecting Humans

Fibricola cratera (Barker and Noll, 1915); Dubois, 1932

This species is a parasite of wild mammals in North America (Shoop, 1989). Frogs are the second intermediate hosts, and snakes are paratenic hosts (Shoop, 1989). An experimental human infection was proved to be successful with recovery of eggs in the feces; the worms lived longer than 3 years in the human (Shoop, 1989).

Neodiplostomum seoulense (Seo et al., 1964); Hong and Shoop, 1995 (Fig. 2.12)

This species was first reported from naturally infected house rats in the Republic of Korea (Seo et al., 1964) and then repeatedly reported from house rats (Seo et al., 1981a, 1988). This parasite is now known to be distributed countrywide in the Republic of Korea, but predominantly in mountainous areas (Seo, 1990). This species has never been reported in other countries (Chai and Lee, 2002), except in a northeastern part of China (Quan et al., 1995). Its characteristic morphology includes a bisegmented body; the tribocytic organ, which is for dissolving the host tissues; two butterfly-shaped testes; and a wide distribution of vitellaria in the anterior body to the level of the ventral sucker (Seo, 1990). The first human infection was found in 1982 in a young man suffering from acute abdominal pain

FIGURE 2.12. An adult fluke of *Neodiplostomum seoulense* recovered from an experimentally infected rat. Acetocarmine stain. Scale bar = 0.1 mm.

and fever (Seo et al., 1982). He had a history of consuming improperly cooked snakes 7 days prior to admission to a hospital (Seo et al., 1982). Subsequently, the grass snake *Rhabdophis* (= *Natrix*) *tigrina* was found carrying the metacercariae (Hong et al., 1982). Further, 25 human cases were found among soldiers who had eaten raw snakes during their survival training (Hong et al., 1984, 1986). Later, an egg-positive case was found in a soldier (Huh et al., 1994). Studies on the life cycle revealed that the first intermediate hosts are freshwater snails, *Hippeutis cantori* (Seo et al., 1988) and *Segmentina (Polypylis) hemisphaerula* (Chung et al., 1996); the second intermediate hosts are tadpoles and frogs of *Rana* sp. (Seo et al., 1988). The snake *Rhabdophis tigrina* is regarded as a paratenic host (Seo, 1990). Mice, rats, and guinea pigs have been found to be susceptible laboratory hosts (Seo, 1990).

Pathogenicity and Host–Parasite Relationships of Neodiplostomes

The duodenum is the most favored site of *N. seoulense* in experimental rodents (Seo, 1990). Villous atrophy, crypt hyperplasia, mucosal inflammation, and bleeding are the major histopathological features of the affected mucosa (Lee et al., 1985). The worms entrapped the host villi using their concave ventral curvature of the anterior body, with their tribocytic organs piercing into the villous stroma (Lee et al., 1985). The affected villi underwent severe destruction with hemorrhages, and finally the intestinal mucosa lost its integrity (Lee et al., 1985). The histopathological changes due to *N. seoulense* were more severe compared with those observed in other intestinal trematode infections including *Metagonimus* (Chai, 1979), *Pygidiopsis* (Seo et al., 1986), and *E. hortense* (Lee et al., 1990b) infections.

A unique feature in host–parasite relationships of *N. seoulense* infection is high pathogenicity and lethality to laboratory mice (Huh et al., 1988; Kook et al., 1998). A 100% fatality of experimentally infected mice was observed by day 23 postinfection, with 200 metacercariae per animal (Kook et al., 1998). The whole intestine of the infected mice was severely contracted, and the contraction was irreversible, which was strongly suggestive of an intestinal paralysis (Kook et al., 1998). The fatality of the host animal varied according to genetic backgrounds of the mice (Chai et al., 2000b). The tribocytic organ of *N. seoulense* was suspected to be an important body structure responsible for the host mucosal damage (Huh et al., 1990). In a histochemical study, it was shown that the tribocytic organ, which entrapped and pierced into the host villi, secreted alkaline phosphatases, which could lyse the host villi and help the mucosal invasion of worms (Huh et al., 1990). The secretory function of this organ was explained by demonstrating the presence of microvilli on the surface of the organ (Huh and Song, 1993). The tribocytic organ was shown to contain neutral mucopolysaccharides, and thus the organ is suggested to play a protective role against host digestive enzymes (Huh et al., 1990). A cysteine protease, with the molecular weight of 54 kDa, was purified from the crude extract of *N. seoulense* adults, although its function was suggested to aid nutrient uptake, rather than host tissue lysis (Choi et al., 1999). Other proteases should be purified and their functions be elucidated to understand their roles in eliciting the pathogenicity to the host.

It is interesting to note that the survival of worms in the host intestine was variable depending on the strain of mice (Chai et al., 1998b). BALB/c mice revealed a consistently higher recovery of worms than C3H mice, based on 28 days' observation after an experimental infection. In experimental mice and rats, mucosal mast cells and goblet cells were shown to increase markedly (Chai et al., 1998b; Kho et al., 1990). Despite a suggestion that proliferation of these host cells is a result of local immune response due to the presence of worms (Kho et al., 1990), they may be more or less responsible for the different susceptibility of the different strains of mice. It was speculated that binding of histamine from mast cells to its receptor on intestinal smooth muscles would be more important than the level of histamine alone, or mastocytosis (Shin et al., 2003). Serum and mucosal tissue immunoglobulin A (IgA) were increased after an experimental infection in mice, but the increase was not directly related to the worm expulsion (Huh et al., 1995). Immunogold studies revealed that the tribocytic organ, seminal vesicle, ceca, and vitelline follicles were the major origins of worm antigens (Lee et al., 1997).

Clinical Symptoms, Diagnosis, and Treatment of Neodiplostomiases

In patients infected with *N. seoulense*, acute abdominal pain, diarrhea, lethargy, fever, and weight loss may occur. However, clinical symptoms and signs due to *N. seoulense* infection are not well documented except in the first patient, who experienced severe abdominal pain that led to admission to an emergency room of a university hospital (Seo et al., 1982). The severity of symptoms may be dependent on the individual worm burdens, as well as on the acquired immunity of each individual. Repeatedly infected patients may complain of milder symptoms than primarily infected patients, as seen in the asymptomatic soldiers infected during survival training (Hong et al., 1984, 1986).

Humans or animals infected with *N. seoulense* can be diagnosed by the recovery of typical eggs in the feces (Seo, 1990). The eggs are ellipsoid to elliptical, thin-shelled, with an inconspicuous operculum, and frequently asymmetrical (Seo, 1990). They differ from the eggs of *E. hortense* or *E. cinetorchis*, in that they have a clean shell surface and, unlike the latter, they do not have abopercular wrinkles at the posterior end. Praziquantel in a single oral dose of 10 to 20 mg/kg is a highly effective treatment for *N. seoulense* infection (Hong et al., 1984, 1986). For prevention, ingestion of raw or improperly cooked flesh of snakes or frogs should be avoided.

Paramphistomatidae Fischoeder, 1901

Species Infecting Humans

Fischoederius elongatus (Poirier, 1883); Stiles and Goldberger, 1910

This species is a parasite of ruminants infected by ingesting aquatic plants having the metacercariae (Yu and Mott, 1994). The first human infection was reported from Guandong, China (Yu and Mott, 1994). The patient complained of epigastric pain for several months (Yu and Mott, 1994).

Watsonius watsoni (Conyngham, 1904); Stiles and Goldberger, 1910

This species, an aquatic plant-borne trematode, was discovered only once at the autopsy of a West African Negro who died of severe diarrhea (Beaver et al., 1984). Many worms were recovered from the intestine, some attached to the duodenal and jejunal wall, others free in the lumen of the colon (Beaver et al., 1984). Various species of primates are natural hosts of this parasite in eastern Asia and Africa (Beaver et al., 1984).

Plagiorchiidae Ward, 1917

Species Infecting Humans

Plagiorchis harinasutai Radomyos et al., 1989

Four humans infected with this fluke were discovered, and the worm was described as a new species (Radomyos et al., 1989). The life cycle is unknown.

Plagiorchis javensis Sandground, 1940

This species was reported from human infections on several occasions in Indonesia (Sandground, 1940). Larval insects are the source of infection, and birds and bats are reservoir hosts (Yu and Mott, 1994).

Plagiorchis muris (Tanabe, 1922); Shul'ts and Skvortsov, 1931

This species was described in Japan from worms recovered from the small intestines of mice experimentally infected with the metacercariae (Tanabe, 1922a). It has been found in house rats in Japan (Tanabe, 1922a), and rats (Seo et al., 1964, 1981a) and cats (Sohn and Chai, 2005) in the Republic of Korea. Its morphological characteristics include a laterally located ovary, two tandem testes, an extensive distribution of the vitellaria, and large eggs. Experimental human infection has been reported in the U.S. (McMullen, 1937), and natural ones in both Japan (Asada et al., 1962) and the Republic of Korea (Hong et al., 1996b). The molluscan intermediate host in Japan is the freshwater snail, *Lymnaea pervia* (Tanabe, 1922a), and *Stagnicola emarginata angulata* in the U.S. (McMullen, 1937). The snail host in the Republic of Korea is unknown. The second intermediate hosts include a wide range of animals such as aquatic insects (mosquito larvae), insect naiads, fresh water snails, and fresh water fish (Tanabe, 1922a, McMullen, 1937; Hong et al., 1996b; Hong et al., 1999). Albino rats are an experimental definitive host (Hong et al., 1999). The reservoir hosts are unknown.

Plagiorchis philippinensis Sandground, 1940

Adult flukes were recovered at the autopsy of a resident in Manila, the Philippines (Yamaguti, 1958; Yu and Mott, 1994). Infection was acquired by eating insect larvae. Birds and rats are reservoir hosts (Yu and Mott, 1994).

Strigeidae (Railliet, 1919)

Species Infecting Humans

Cotylurus Japonicus (Ishii, 1932)

The first human infection with this fluke was reported from a 13-year-old girl in Hunan Province, China (Chen and Cai, 1985). Ducks were found to be infected with this fluke (Yu and Mott, 1994). The first intermediate hosts are freshwater snails belonging to the genera *Stagnicola, Lymnaea, Physa*, and *Heligsoma*, and cercariae encyst in the same snail hosts to become specialized metacercariae known as tetracotyles (Fried et al., 2004). Infection may occur when birds or mammals ingest tetracotyles in infected snails (Fried et al., 2004).

Summary

A total of 70 species (14 families and 36 genera) of food-borne human intestinal flukes are known around the world. The largest family is the Heterophyidae, which constitutes 29 species in 12 genera (*Apophallus, Ascocotyle, Centrocestus, Cryptocotyle, Haplorchis, Heterophyes, Heterophyopsis, Metagonimus, Procerovum, Pygidiopsis, Stellantchasmus*, and *Stictodora*). The next is the Echinostomatidae, in which 22 species in 10 genera (*Artyfechinostomum, Acanthoparyphium, Echinochasmus, Echinoparyphium, Echinostoma, Episthmium, Euparyphium, Himasthla, Hypoderaeum*, and *Psilorchis*) are involved. The Lecithodendriidae includes three species in two genera (*Phaneropsolus* and *Prosthodendrium*), and the Paramphistomatidae two species in two genera (*Fischoederius* and *Watsonius*). For the other families, one to four species in one genus each is involved; Brachylaimidae (*Brachylaima*) (one species), Cathaemaciidae (*Cathaemacia*) (one species), Fasciolidae (*Fasciolopsis*) (one species), Gastrodiscidae (*Gastrodiscoides*) (one species), Gymnophallidae (*Gymnophalloides*) (one species), Microphallidae (*Spelotrema*) (one species), Nanophyetidae (*Nanophyetus*) (one species), Neodiplostomidae (*Neodiplostomum*) (two species), Plagiorchiidae (*Plagiorchis*) (four species), and Strigeidae (*Cotylurus*) (one species). Among these trematodes, heterophyids and echinostomes are the two major groups, in terms of the number of species involved, the number of people infected, and the distribution of endemic areas. Various types of foods are sources of human infections. They include freshwater fish, brackish water fish, fresh water snails, brackish water snails (including the oyster), amphibians, terrestrial snakes, aquatic insects, and aquatic plants. The reservoir hosts are various species of mammals or birds.The host–parasite relationships have been studied extensively in several species, including *Heterophyes heterophyes, Metagonimus yokogawai, Echinostoma hortense, Echinostoma trivolvis, Fasciolopsis buski, Neodiplostomum seoulense*, and *Gymnophalloides seoi*; however, more information is needed. The pathogenicity of each parasite species and host mucosal defense mechanisms are poorly understood. Clinical aspects of each parasite species need more clarification. Diagnosis of intestinal fluke infections can be done by fecal examination, but differential diagnosis is difficult because of morphological similarity of eggs. Praziquantel is an effective anthelmintic for most of the intestinal flukes. Epidemiological surveys and detection of further human cases are required for a better understanding of the distribution and endemicity of each trematode species.

TABLE 2.1. Taxonomic classifications of food-borne intestinal flukes.

Family	Genus	Species
Brachylaimidae	*Brachylaima*	*B. cribbi*
Cathaemaciidae	*Cathaemacia*	*C. cabrerai*
Echinostomatidae	*Acanthoparyphium*	*A. tyosenense*
	Artyfechinostomum	*A. malayanum, A. oraoni*
	Echinochasmus	*E. fujianensis, E. japonicus, E. jiufoensis, E. liliputanus, E. perfoliatus*
	Echinoparyphium	*E. recurvatum*
	Echinostoma	*E. angustitestis, E. cinetorchis, E. echinatum, E. hortense, E. ilocanum, E. macrorchis, E. malayanum, E. revolutum*
	Episthmium	*E. caninum*
	Himasthla	*H. muehlensi*
	Hypoderaeum	*H. conoideum*
	Isthmiophora	*I. melis*
	Psilorchis	*P. hominis*
Fasciolidae	*Fasciolopsis*	*F. buski*
Gastrodiscidae	*Gastrodiscoides*	*G. hominis*
Gymnophallidae	*Gymnophalloides*	*G. seoi*
Heterophyidae	*Apophallus*	*A. donicus*
	Ascocotyle	*A. (Phagicola) longa*
	Centrocestus	*C. armatus, C. caninus, C. cuspidatus, C. formosanus, C. kurokawai*
	Cryptocotyle	*C. lingua*
	Haplorchis	*H. pleurolophocerca, H. pumilio, H. taichui, H. vanissimus, H. yokogawai*
	Heterophyes	*H. dispar, H. heterophyes, H. nocens*
	Heterophyopsis	*H. continua*
	Metagonimus	*M. minutus, M. miyatai, M. takahashii, M. yokogawai*
	Procerovum	*P. calderoni, P. varium*
	Pygidiopsis	*P. summa*
	Stellantchasmus	*S. falcatus, S. formosanus, S. pseudocirratus*
	Stictodora	*S. fuscata, S. lari*
Lecithodendriidae	*Phaneropsolus*	*P. bonnei, P. spinicirrus*
	Prosthodendrium	*P. molencampi*
Microphallidae	*Spelotrema*	*S. brevicaeca*

Nanophyetidae	*Nanophyetes*	*N. salmincola*
Neodiplostomidae	*Fibricola*	*F. cratera*
	Neodiplostomum	*N. seoulense*
Paramphistomatidae	*Fischoederius*	*F. elongatus*
	Watsonius	*W. watsoni*
Plagiorchiidae	*Plagiorchis*	*P. harinasutai, P. javensis, P. muris, P. philippinensis,*
Strigeidae	*Cotylurus*	*C. japonicus*

TABLE 2.2. Fish-borne intestinal flukes.

Parasite species	Source of human or animal infections
Apophallus donicus	Freshwater fiish, blackside dace, sucker, squawfish, redside shiner, rainbow trout, coho salmon
Ascocotyle (Phagicola) longa	Freshwater fish
Centrocestus armatus	Fresh water fish, *Zacco platypus, Zacco temminckii, Rhodeus ocellatus, Gobius similis, Pseudorasbora parva, Pelteobagrus fulvidraco*
Centrocestus caninus	Freshwater fish, *Cyprinus carpio, Hampala dispar, Puntius* spp., *Cyclocheilichthys* sp., *Tilapia nilotica*
Centrocestus cuspidatus	Freshwater fish, *Astatotilapia desfontainesi*
Centrocestus formosanus	Freshwater fish
Cryptocotyle lingua	Freshwater fish, *Gobius ruthensparri, Labrus bergylta*
Echinochasmus fujianensis	Freshwater fish, *Pseudorasbora parva, Cyprinus carpio*
Echinochasmus japonicus	Fresh water fish, *Pseudorasbora parva, Hypomesus olidus, Gnathopogon strigatus*
Echinochasmus jiufoensis	Unknown
Echinochasmus liliputanus	Freshwater fish, *Pseudorasbora parva,* goldfish
Echinochasmus perfoliatus	Freshwater fish, *Carassius* sp.
Echinostoma angustitestis	Freshwater fish
Echinostoma cinetorchis	Freshwater fish, *Misgurnus anguillicaudatus*
Echinostoma hortense	Freshwater fish, *Misgurnus anguillicaudatus, Misgurnus mizolepis, Odontobutis obscura interrupta, Moroco oxycephalus, Coreoperca kawamebari, Squalidus coreanus*
Episthmium caninum	Freshwater fish
Haplorchis pleurolophocerca	Freshwater fish, *Gambusia affinis*
Haplorchis pumilio	Freshwater fish, Cyprinidae, Siluridae, Cobitidae
Haplorchis taichui	Freshwater fish, *Cyprinus carpio, Carassius auratus, Zacco platypus, Pseudorasbora parva, Rodeus ocellatus, Gambusia affinis, Puntius orphoides, Puntius spp., Raiamas guttatus, Mystacoleucus marginatus, siamensis*
Henichoryhnchus	
Haplorchis vanissimus	Freshwater fish
Haplorchis yokogawai	Freshwater fish, *Mugil* spp., *Puntius* spp., *Misgurnus* sp., *Ophicephalus striatus*
Heterophyes dispar	Brackish water fish
Heterophyes heterophyes	Brackish water fish, *Mugil cephalus, Tilapia nilotica, Aphanius fasciatus, Acanthogobius* sp.
Heterophyes nocens	Brackish water fish, *Mugil* sp., *Acanthogobius* sp.
Heterophyopsis continua	Brackish water fish, *Acanthogobius* sp., *Lateolabrax* sp., *Clupadon punctatus*

(Continued)

TABLE 2.2. *(Continued)*

Parasite species	Source of human or animal infections
Metagonimus minutus	Mullet, *Mugil cephalus*
Metagonimus miyatai	Sweetfish, dace, common fat-minnow *Morocco steindachneri*, pale chub *Zacco platypus*, dark chub *Zacco temmincki*,
Metagonimus takahashii	Crussian carp *C. carassius*, carp *C. carpio*, dace *Tribolodon taczanowskii*, and perch *Lateolabrax japonicus*
Metagonimus yokogawai	Sweetfish *P. altivelis*, dace *Tribolodon* sp., perch *Lateolabrax japonicus*
Nanophyetes salmincola	Freshwater fish, salmon, trout, nonsalmonid fish
Plagiorchis muris	Freshwater fish, various species
Procerovum calderoni	Freshwater fish, *Ophiocephalus striatus*, *Glossogobius giurus*, *Mollienesia latipinna*, *Mugil* sp., and *Creisson validus*
Procerodum varium	Mullet *Mugil cephalus*
Pygidiopsis summa	Mullet *Mugil cephalus* and goby *Acanthogobius flavimanus*
Stellantchasmus falcatus	Mullet, half-beaked fish
Stellantchasmus formosanus	Mullet *Mugil cephalus*
Stellantchasmus pseudocirratus	Mullet *Mugil cephalus*
Stictodora fuscata	Goby *Acanthogobius flavimanus*
Stictodora lari	Goby *Acanthogobius flavimanus*, and other estuarine fish

TABLE 2.3. Snail-borne intestinal flukes.

Parasite species	Source of human or animal infections
Acanthoparyphium tyosenense	Bivalve, *Mactra veneriformis*, *Solen grandis*, gastropod, *Neverita bicolor*
Artyfechinostomum malayanum	Snail, *Digoniostoma pulchella*
Brachylaima cribbi	Helicid land snail, *Cernuella virgata*
Cotylurus japonicus	Freshwater snail, *Stagnicola*, *Lymnaea*, *Physa*, *Heligsoma* spp.
Echinoparyphium recurvatum	Freshwater snail, *Planorbis planorbis*, *Lymnaea* sp., *Lymnaea stagnalis*
Echinostoma cinetorchis	Freshwater snail, *Radix auricularia coreanus*, *Physa acuta*, *Cipangopaludina chinensis malleata*
Echinostoma echinatum	Mussel, *Corbicula lindoensis*, *Corbicula sucplanta*, *Idiopoma javanica*, freshwater snail, *Biomphalaria glabrata*
Echinostoma ilocanum	Large snail, *Pila conica*, *Viviparus javanicus*
Echinostoma macrorchis	Large snail, *Cipangopaludina malleata*, *Cipangopaludina japonica*, *Segmentina nitiella*, *Viviparus malleatus*
Echinostoma malayanum	Large snail, *Pila scutata*, *Lymnaea (Bullastra) cumingiana*
Echinostoma revolutum	Snail or clam, *Corbicula producta*,
Gymnophalloides seoi	Oyster, *Crassostrea gigas*
Himasthla muehlensi	Clams, *Venus mercenaria*, bivalve mollusk, *Mytilus*, *Mya* spp.
Hypoderaeum conoideum	Snail
Plagiorchis muris	Freshwater snail

TABLE 2.4. Amphibia, reptile, crustacean, insect, and aquatic plant-borne intestinal flukes.

Parasite species	Source of human or animal infections
Echinoparyphium recurvatum	Tadpole and frog of *Rana temporaria*
Echinostoma macrorchis	Frog of *Rana* sp.
Echinostoma revolutum	Tadpole
Fasciolopsis buski	Aquatic plant, including water caltrop, water cress, water chestnut, and water bamboo
Fibricola cratera	Snake, frog
Fischoederius elongates	Aquatic plant
Gastrodiscoides hominis	Tadpole, frog, crayfish, aquatic plant
Hypoderaeum conoideum	Tadpole
Isthmiophora melis	Tadpole
Neodiplostomum seoulense	Grass snake, *Rhabdophis tigrina*, Tadpole and frog of *Rana nigromaculata*
Phaneropsolus bonnie	Naiad of dragonfly, damselfly
Plagiorchis javensis	Larval insect
Plagiorchis muris	Larval insect, insect naiad
Plagiorchis philippinensis	Insect larva
Prosthodendrium molenkampi	Naiad of dragonfly, damselfly
Spelotrema brevicaeca	Crab *Carcinus maenas*, shrimp *Macrobrachium* sp.
Watsonius watsoni	Aquatic plant

References

Africa, C.M., and Garcia, E.Y. (1935). Heterophyid trematodes of man and dog in the Philippines with descriptions of three new species. Philippine J Sci 57, 253–267.

Africa, C.M., de Leon, W., and Garcia, E.Y. (1940). Visceral complications in intestinal heterophyidiasis of man. Acta Med Philippina Monogr Series No. 1, 1–132.

Ahluwalia, S.S. (1962). Studies on some helminthes of the domestic pigs *(Sus scrofa domestica)* in western Uttar Pradesh. J Helminthol 36, 347–364.

Ahn, Y.K. (1983). *Lateolaborax japonicus*, a role of the second intemediate host of *Metagonimus yokogawai*. New Med J (Korea), 26, 135–139.

Ahn, Y.K. (1993). Intestinal fluke of genus *Metagonimus* and their second intermediate hosts in Kangwon-do. Korean J Parasitol 31, 331–340 (in Korean).

Ahn, Y.K., Ryang, Y.S., Chai, J.Y., and Sohn, W.M. (1989). Cercarial shedding of *Echinostoma cinetorchis* and experimental infection of the cercariae to several kinds of snails. Korean J Parasitol 27, 203–211 (in Korean).

Ahn, H., Rim, H.J., and Kim, S.J. (1991). Antigenic localities in the tissues of *Metagonimus yokogawai* observed by immunogold labeling method. Korean J Parasitol 29, 245–257 (in Korean).

Ando, R., and Ozaki, Y. (1923). On four new species of trematodes of the family Echinostomatidae. Dobutsugaku Zasshi 35, 108–119 (in Japanese).

Aokage, K. (1956). Studies on the trematode parasites of brackish water fishes in Chugoku coast of Setonaikai. Tokyo Iji Shinshi 73, 217–224 (in Japanese).

Asada, J. (1926). On a new echinostomatid trematode and its life history. Trans Jpn Soc Pathol 16, 293–294 (in Japanese).

Asada, J.I., Otagaki, H., Morita, M., Takeuchi, T., Sakai, Y., Konoshi, T., and Okahashi, K. (1962). A case report on the human infection with *Plagiorchis muris* Tanabe, 1922 in Japan. Jpn J Parasitol 11, 512–516 (in Japanese).

Ashford, R.W., and Crewe, W. (2002). The Parasites of *Homo sapiens*. London: Liverpool School of Tropical Medicine.

Bandyopadhyay, A.K., Manna, B., and Nandy, A. (1989). Human infection of *Artyfechinostomum oraoni* n. sp. (Paryphostominae: Echinostomatidae) in a tribal community, "Oraons' in West Bengal, India. Indian J Parasitol 13, 191–196.

Bandyopadhyay, A.K., Maji, A.K., Manna, B., Bera, D.K., Addy, M., and Nandy, A. (1995). Pathogenicity of *Artyfechinostomum oraoni* in naturally infected pigs. Trop Med Parasitol 46, 138–139.

Belzario, V.Y. Jr., de Leon, W.U., Bsrsaba, M.J.J., Purnomo, Baird, J.K., and Bangs, M.J. (2005). A focus of human infection by *Haplorchis taichui* (Trematoda: Heterophyidae) in the southern Philippines. J Parasitol 90, 1165–1169.

Betlejewska, K.M., and Jorol, E.N. (2002). Taxonomic, topical and quantitative structure of the community of intestinal flukes (Digenea) of mallards, *Anas platyrhynchos* Linnaeus, 1758 from the area of Szcecin. Wiadomosci Parazitologiczne 48, 343–357.

Beaver, P.C., Jung, R.C., and Cupp, E.W. (1984). Clinical Parasitology 9th ed. Philadelphia: Lea & Febiger.

Bindseil, E., and Christensen, N.O. (1984). Thymus-independent crypt hyperplasia and villous atrophy in the small intestine of mice infected with the trematode *Echinostoma revolutum*. Parasitol 88, 431–438.

Brunet, I.R., Joseph, S., Dunne, D.W., and Fried, B. (2000). Immune responses during the acute stage of infection with the intestinal trematode *Echinostoma caproni*. Parasitol 120, 565–571.

Bunnag, D., Radomyos, P., and Harinasuta, C. (1983). Field trial on the treatment of fasciolopsiasis with praziqautnel. Southeast Asian J Trop Med Public Health 14, 216–219.

Butcher, A.R., and Grove, D.I. (2001). Description of life-cycle stages of *Brachylaima cribbi* n. sp. (Digenea: Brachylaimidae) derived from eggs recovered from human feces in Australia. Syst Parasitol 49, 211–221.

Butcher, A.R., and Grove, D.I. (2005). Second intermediate host land snails and definitive host animals of *Brachylaima cribbi* in southern Australia. Parasite 12, 31–37.

Butcher, A.R., Parasuramar, P., Thompson, C.S., and Grove, D.I. (1998). First report of the isolation of an adult worm of the genus *Brachylaima* (Digenea: Brachylaimidae), from the gastrointestinal tract of a human. Int J Parasitol 28, 607–610.

Butcher, A.R., Palethope, H.M., and Grove, D.I. (2003). Response to re-infection with *Brachylaima cribbi* in immunocompetent and immunodeficient mice. Parasitol Int 52, 219–228.

Castro, G.A. (1989). Immunophysiology of enteric parasitism. Parasitol Today, 5, 11–19.

Chai, J.Y. (1979). Study on *Metagonimus yokogawai* (Katsurada, 1912) in Korea V. Intestinal pathology in experimentally infected albino rats. Seoul J Med 20, 104–117.

Chai, J.Y., and Sohn, W.M. (1988). Identification of *Stellantchasmus falcatus* metacercariae encysted in mullets in Korea. Korean J Parasitol 26, 65–68.

Chai, J.Y., and Lee, S.H. (2002). Food-borne intestinal trematode infections in the Republic of Korea. Parasitol Int 51, 129–154.

Chai, J.Y., Cho, S.Y., and Seo, B.S. (1977). Study on *Metagonimus yokogawai* (Katsurada, 1912) in Korea. IV. An epidemiological investigation along Tamjin River Basin, South Cholla Do, Korea. Korean J Parasitol 15, 115–120.

Chai, J.Y., Seo, B.S., and Lee, S.H. (1984a). Studies on intestinal trematodes in Korea XI. Two cases of human infection by *Heterophyes heterophyes nocens*. Korean J Parasitol 22, 37–42.

Chai, J.Y., Seo, B.S., and Lee, S.H. (1984b). Study on *Metagonimus yokogawai* (Katsurada, 1912) in Korea VII. Susceptibility of various strains of mice to *Metagonimus* infection and effect of prednisolone. Korean J Parasitol 22, 153–160.

Chai, J.Y., Hong, S.J., Sohn, W.M., Lee, S.H., and Seo, B.S. (1985a). Further cases of human *Heterophyes heterophyes nocens* infection in Korea. Seoul J Med 26, 197–200.

Chai, J.Y., Hong, S.J., Son, D.W., Lee, S.H., and Seo, B.S. (1985b). Metacercariae of the *Echinochasmus japonicus* encysted in a fresh water fish, *Pseudorasbora parva*, and their development in experimental mice. Korean J Parasitol 23, 221–229.

Chai, J.Y., Hong, S.J., Sohn, W.M., Lee, S.H., and Seo, B.S. (1985c). Studies on intestinal trematode in Korea XVI. Infection status of loaches with the metacercariae of *Echinostoma hortense*. Korean J Parasitol 23, 18–23.

Chai, J.Y., Seo, B.S., Lee, S.H., Hong, S.J., and Sohn, W.M. (1986a). Human infections by *Heterophyes heterophyes* and *H. dispar* imported from Saudi Arabia. Korean J Parasitol 24, 82–88.

Chai, J.Y., Seo, B.S., Lee, S.H., and Hong, S.T. (1986b). Growth and development of *Pygidiopsis summa* in rats and mice with a supplementary note on its morphological characters. Korean J Parasitol 24, 55–62.

Chai, J.Y., Hong, S.J., Lee, S.H., and Seo, B.S. (1988). *Stictodora* sp. (Trematoda: Heterophyidae) recovered from a man in Korea. Korean J Parasitol 26, 127–132.

Chai, J.Y., Park, S.K., Hong, S.J., Choi, M.H., and Lee, S.H. (1989a). Identification of *Stictodora lari* (Heterophyidae) metacercariae encysted in the brackish water fish, *Acanthogobius flavimanus*. Korean J Parasitol 27, 253–259.

Chai, J.Y., Yu, J.R., Lee, S.H., Jung, H.C., Song, I.S., and Cho, S.Y. (1989b). An egg-negative patient of acute metagonimiasis diagnosed serologically by ELISA. Seoul J Med 30, 139–142.

Chai, J.Y., Sohn, W.M., Kim, M.H., Hong, S.T., and Lee, S.H. (1991). Three morphological types of the genus *Metagonimus* encysted in the dace, *Tribolodon taczanowskii*, caught from the Sumjin River. Korean J Parasitol 29, 217–225.

Chai, J.Y., Huh, S., Yu, J.R., Kook, J., Jung, K.C., Park, E.C., Sohn, W.M., Hong, S.T., and Lee, S.H. (1993a). An epidemiological study of metagonimiasis along the upper reaches of the Namhan River. Korean J Parasitol 31, 99–108.

Chai, J.Y., Kim, T.H., Kho, W.G., Chung, S.W., Hong, S.T., and Lee, S.H. (1993b). Mucosal mast cell responses to experimental *Metagonimus yokogawai* infection in rats. Korean J Parasitol 31, 129–134.

Chai, J.Y., Yun, T.Y., Kim, J., Huh, S., Choi, M.H., and Lee, S.H. (1994a). Chronological observation on intestinal histopathology and intraepithelial lymphocytes in the intestine of rats infected with *Metagonimus yokogawai*. Korean J Parasitol 32, 215–221 (in Korean).

Chai, J.Y., Nam, H.K., Kook, J., and Lee, S.H. (1994b). The first discovery of an endemic focus of *Heterophyes nocens* (Heterophyidae) infection in Korea. Korean J Parasitol 32, 157–161.

Chai, J.Y., Hong, S.T., Lee, S.H., Lee, G.C., and Min, Y.I. (1994c). A case of echinostomiasis with ulcerative lesions in the duodenum. Korean J Parasitol 32, 201–204.

Chai, J.Y., Kim, J., and Lee, S.H. (1995a). Invasion of *Metagonimus yokogawai* into the submucosal layer of the small intestine of immunosuppressed mice. Korean J Parasitol 33, 313–321.

Chai, J.Y., Kim, S.J., Kook, J., and Lee, S.H. (1995b). Effects of gamma-irradiation on the survival and development of *Metagonimus yokogawai* metacercariae in rats. Korean J Parasitol 33, 297–303.

Chai, J.Y., Han, M.S., Seo, M., and Lee, S.H. (1996). Effects of gamma-irradiation on the survival and development of *Gymnophalloides seoi* in C3H mice. Korean J Parasitol 34, 21–25.

Chai, J.Y., Kim, I.M., Seo, M., Guk, S.M., Kim, J.L., Sohn, W.M., and Lee, S.H. (1997). A new focus of *Heterophyes nocens, Pygidiopsis summa*, and other intestinal flukes in a coastal area of Muan-gun, Chollanam-do. Korean J Parasitol 35, 233–238.

Chai, J.Y., Song, T.E., Han, E.T., Guk, S.M., Park, Y.K., Choi, M.H., and Lee, S.H. (1998a). Two endemic foci of heterophyids and other intestinal fluke infections in southern and western coastal areas in Korea. Korean J Parasitol 36, 155–161.

Chai, J.Y., Kim, T.K., Cho, W.H., Seo, M., Kook, J., Guk, S.M., and Lee, S.H. (1998b). Intestinal mastocytosis and goblet cell hyperplasia in BALB/c and C3H mice infected with *Neodiplostomum seoulense*. Korean J Parasitol 36, 109–119.

Chai, J.Y., Chung, W.J., Kook, J., Seo, M., Park, Y.K., Guk, S.M., Choi, M.H., and Lee, S.H. (1999). Growth and development of *Gymnophalloides seoi* in immunocompetent and immunosupressed C3H/HeN mice. Korean J Parasitol 37, 21–26.

Chai, J.Y., Lee, G.C., Park, Y.K., Han, E.T., Seo, M., Kim, J., Guk, S.M., Shin, E.H., Choi, M.H., and Lee, S.H. (2000a). Persistent endemicity of *Gymnophalloides seoi* infection in a southwestern coastal village with special reference to its egg laying capacity in the human host. Korean J Parasitol 38, 51–57.

Chai, J.Y., Shin, E.H., Han, E.T., Guk, S.M., Choi, M.H., and Lee, S.H. (2000b). Genetic difference in susceptibility and fatality of three strains of mice experimentally infected with *Neodiplostomum seoulense*. J Parasitol 86, 1140–1144.

Chai, J.Y., Han, E.T., Park, Y.K., Guk, S.M., Kim, J.L., and Lee, S.H. (2000c). High endemicity of *Metagonimus yokogawai* infection among residents of Samchok-shi, Kangwon-do. Korean J Parasitol 38, 33–36.

Chai, J.Y., Lee, H.S., Hong, S.J., Yoo, J.H., Guk, S.M., Seo, M., Choi, M.H., and Lee, S.H. (2001a). Intestinal histopathology and in situ postures of *Gymnophalloides seoi* in experimentally infected mice. Korean J Parasitol 39, 31–41.

Chai, J.Y., Han, E.T., Park, Y.K., Guk, S.M., and Lee, S.H. (2001b). *Acanthoparyphium tyosenense*: the discovery of human infection and identification of its source. J Parasitol 87, 794–800.

Chai, J.Y., Park, J.H., Han, E.T., Shin, E.H., Kim, J.L., Hong, K.S., Rim, H.J., and Lee, S.H. (2001c). A nationwide survey of the prevalence of human *Gymnophalloides seoi* infection on western and southern coastal islands in the Republic of Korea. Korean J Parasitol 39, 23–30.

Chai, J.Y., Han, E.T., Park, Y.K., Guk, S.M., Park, J.H., and Lee, S.H. (2002). *Stictodora lari* (Digenea: Heterophyidae): the discovery of the first human infections. J Parasitol 88, 627–629.

Chai, J.Y., Choi, M.H., Yu, J.R., and Lee, S.H. (2003). *Gymnophalloides seoi*: a new human intestinal trematode. Trends Parasitol 19, 109–112.

Chai, J.Y., Park, J.H., Han, E.T., Shin, E.H., Kim, J.L., Guk, S.M., Hong, K.S., Lee, S.H., and Rim, H.J. (2004). Prevalence of *Heterophyes nocens* and *Pygidiopsis summa* infections among residents of the western and southern coastal islands of the Republic of Korea. Am J Trop Med Hyg 71, 617–622.

Chai, J.Y., Murrell, K.D., and Lymbery, A.J. (2005a). Fish-borne parasitic zoonoses: status and issues. Int J Parasitol 35, 1233–1254.

Chai, J.Y., Park, J.H., Han, E.T., Guk, S.M., Shin, E.H., Lin, A., Kim, J.L., Sohn, W.M., Yong, T.S., Eom, K.S., Min, D.Y., Hwang, E.H., Phommasack, B., Insisiengmai, B., and Rim, H.J. (2005b). Mixed infections with *Opisthorchis viverrini* and intestinal flukes in residents of Vientiane Municipality and Saravane Province in Laos. J Helminthol 79, 283–289.

Chang, Y.D., Sohn, W.M., Ryu, J.H., Kang, S.Y., and Hong, S.J. (2005). A human infection of *Echinostoma hortense* in duodenal bulb diagnosed by endoscopy. Korean J Parasitol 43, 57–60.

Chen, H.T. (1936). A study of the Haplorchinae (Looss, 1899) Poche 1926 (Trematoda: Heterophyidae). Parasitology, 28, 40–55.

Chen, H.T. (1942). The metacercaria and adult of *Centrocestus formosanus* (Nishigori, 1924), with notes on the natural infection of rats and cats with *C. armatus* (Tanabe, 1922). J Parasitol 28, 285–298.

Chen, C.E., and Cai, G.D. (1985). Morphological testimony: demonstration of a case of human infestation with *Cotylurus japonicus* Ishii, 1932 (Trematoda, Strigeidae). Bull Hunan Med Coll 10, 31–34 (in Chinese).

Chen, X.Q., Feng, G.X., and Qian, Z.F. (1993). Survey on infection due to *Echinostoma hortense* in Liaoning Province. Chin J Parasitol Parasitic Dis 11, 226 (in Chinese).

Cheng, Y.Z., Lin, J.X., Chen, B.J., Fang, Y.Y., Guo, Z.F., Xu, G.F., and Wu, Z.Y. (1992a). A new species of *Echinochasmus* parasitic in man and observation of its experimental infection. Wuyi Sci J 9, 43–48 (in Chinese).

Cheng, Y.Z., Lin, J.X., Fang, Y.Y., Lin, A.Q., Chen, B.J., and Zhuang H.J. (1992b). Discovery of human infection with *Echinostoma angustitestis*. Chin J Zoonoses 8, 7–8 (in Chinese).

Chi, J.G., Kim, C.W., Kim, J.R., Hong, S.T., and Lee, S.H. (1988). Intestinal pathology in human metagonimiasis with ultrastructural observations of parasites. J Korean Med Sci 3, 171–177.

Chieffi, P.P., Gorla, M.C.O., Torres, D.M.A.G.V., Dias, R.M.D.S., Maggini, A.C.S., Monteiro, A.V.M., and Woiciechovski, E. (1992). Human infection by *Phagicola* sp. (Trematoda: Heterophyidae) in the municipality of Registro, Sao Paulo State, Brazil. J Trop Med Hyg 95, 346–348.

Cho, S.Y., and Kim, S.I. (1985). *Plecoglossus altivelis* as a new fish intermediate host of *Heterophyopsis continua*. Korean J Parasitol 23, 173–174.

Cho, S.Y., Kang, S.Y., and Ryang, Y.S. (1981). Helminthes infections in the small intestine of stray dogs in Ejungbu City, Kyunggi Do, Korea. Korean J Parasitol 19, 55–59.

Cho, S.Y., Kang, S.Y., and Lee, J.B. (1984). Metagonimiasis in Korea. Session V: Intestinal fluke and *Schistosoma japonicum*. Arzneimittelforschung Drug Res 34, 1211–1213.

Cho, S.Y., Kim, S.I., Earm, Y.E., and Ho, W.K. (1985). A preliminary observation on watery content of small intestine in *Metagonimus yokogawai* infected dogs. Korean J Parasitol 23, 175–177.

Cho, S.Y., Kim, S.I., and Kang, S.Y. (1987). Specific IgG antibody response in experimental cat metagonimiasis. Korean J Parasitol 25, 149–153.

Cho, C.M., Tak, W.Y., Kweon, Y.O., Kim, S.K. Choi, Y.H., Kong, H.H., and Chung, D.I. (2003) A human case of *Echinostoma hortense* (Trematoda: Echinostomatidae) infection diagnosed by gastroduodenal endoscopy in Korea. Korean J Parasitol 41, 117–120.

Choi, M.H., Chai, J.Y., and Lee, S.H. (1998a). Purification and characterization of a 16-kDa cysteine proteinase of *Gymnophalloides seoi* (Gymnophallidae) metacercariae. J Parasitol 84, 350–355.

Choi, M.H., Park, W.J., Park, Y.K., Chai, J.Y., and Lee, S.H. (1998b). Isolation and characterization of a 40 kDa cysteine protease from *Gymnophalloides seoi* adult worms. Korean J Parasitol 36, 133–141.

Choi, M.H., Choe, S.C., and Lee, S.H. (1999). A 54 kDa cysteine protease purified from the crude extract of *Neodiplostomum seoulense* adult worms. Korean J Parasitol 37, 39–46.
Choi, M.H., Kim, S.H., Chung, J.H., Jang, H.J., Eom, J.H., Chung, B.S., Sohn, W.M., Chai, J.Y., and Hong, S.T. (2006) Morphological observations of *Echinochasmus japonicus* cercariae and the in vitro maintenance of its life cycle from cercariae to adults. J Parasitol 92, 236–241.
Chun, S.K. (1960a). A study on the metacercaria of *Metagonimus takahashii* and *Exorchis oviformis* from *Carassius carassius*. Bull Pusan Fisheries Coll 3, 31–39.
Chun, S.K. (1960b). A study on some trematodes whose intermediate hosts are brackish water fish (I). The life history of *Heterophyes continus*, the intermediate host of which is *Lateolabrax japonicus*. Bull Pusan Fisheries Coll 3, 40–44.
Chun, S.K. (1963). A study on some trematodes whose intermediate hosts are brackish water fish (II). The life history of *Pygidiopsis summus*, the intermediate host of which is *Mugil cephalus*. Bull Pusan Fisheries Coll 5, 1–5.
Chung, P.R., and Jung, Y. (2000). *Cipangopaludina chnensis malleata* (Gastropoda: Viviparidae): a new second molluscan intermediate host of a human intestinal fluke *Echinostoma cinetorchis* (Trematoda: Echinostomatidae) in Korea. J Parasitol 85, 963–964.
Chung, P.R., Jung, Y., and Kim, D.S. (1996). *Segmentina (Polypylis) hemisphaerula* (Gastropoda: Planorbidae): a new molluscan intermediate host of a human intestinal fluke *Neodiplostomum seoulense* (Trematoda: Diplostomatidae) in Korea. J Parasitol 82, 336–338.
Collomb, P.H., Deschiens, R., and Demarchi, J. (1960). Sur Deux Cas de Distomatodse Cerebrale a *Heterophyes heterophyes*. Bull Soc Pathol Exotique 53, 144–147.
Cross, J.H., and Bassaca-Sevilla, V. (1981). Intestinal parasitic infections in Southeast Asia. Southeast Asian J Trop Med Public Health 12, 262–274.
Eom, K.S., and Rim, H.J. (1984). A study on the parasitic helminths of domestic duck (*Anas platyrhynchos* var. *domestica* Linnaeus) in Korea. Korean J Parasitol 22, 215–221.
Eom, K.S., Son, S.Y., Lee, J.S., and Rim, H.J. (1985). Heterophyid trematodes (*Heterophyopsis continua, Pygidiopsis summa* and *Heterophyes heterophyes nocens*) from domestic cats in Korea. Korean J Parasitol 23, 197–202.
Eastburn, R.L., Tritsche, T.R., and Terhune, Jr. C.A. (1987). Human intestinal infection with *Nanophyetus salmincola* from salmonid fishes. Am J Trop Med Hyg 36, 586–591.
Faust, E.C., and Nishigori, M. (1926). The life cycles of two new species of Heterophyidae parasitic in manmals and birds. J Parasitol 13, 91–131.
Fan, S.Q., and Sun, M.F. (1989). *Echinostoma hortense.* Bull Harbin Med Univ 23, 161 (in Chinese).
Fried, B. (2001). Biology of echinostomes except *Echinostoma*. Adv Parasitol 49, 163–210.
Fried, B., and Graczyk, T.K. (2004). Recent advances in the biology of *Echinostoma* species in the 'revolutum' group. Adv Parasitol 58, 139–195.
Fried, B., Graczyk, T.K., and Tamang, L. (2004). Food-borne intestinal trematodiases in humans. Parasitol Res 93, 159–170.
Fujino, T., Fried, B., and Tada, I. (1993). The expulsion of *Echinostoma trivolvis*: worm kinetics and intestinal cytopathology in conventional and congenitally athymic BALB/c mice. Parasitology 106, 297–304.
Gilman, R.H., Mondal, G., Maksud, M., Alam, K., Rutherford, E., Gilman, J.B., and Khan, M.U. (1982). Endemic focus of *Fasciolopsis buski* infection in Bangladesh. Am J Trop Med Hyg 31, 796–802.

Grover, M., Dutta, R., Kumar, R., Aneja, S., and Mehta, G. (1998). *Echinostoma ilocanum* infection. Indian Pediatr 35, 549–552.

Guk, S.M., Park, J.Y., Seo, M., Han, E.T., Kim, J.L., and Chai, J.Y. (2005). Susceptibility of inbred mouse strains to infection with three species of *Metagonimus* prevalent in the Republic of Korea. J Parasitol 91, 12–16.

Guk, S.M., Park, J. H, Shin, E.H., Kim, J.L., Lin, A., and Chai, J.Y. (2006). Prevalence of *Gymnophalloides seoi* infection in coastal villages of Haenam-gun and Yeongam-gun, Republic of Korea. Korean J Parasitol 44, 1–5.

Han, E.T., Kim, J.L., and Chai, J.Y. (2003). Recovery, growth and development of *Acanthoparyphium tyosenense* (Digenea: Echinostomatidae) in experimental chicks. J Parasitol 89, 176–179.

Harinasuta, T., Bunnag, D., and Radomyos, P. (1987). Intestinal fluke infections. In Baillieres Clin Trop Med Commun Dis 2(3): 695–721.

Himonas, C.A. (1964). *Heterophyes heterophyes* from dogs in Greece. J Parasitol 50, 799.

Hong, S.T., Hong, S.J., Lee, S.H., Seo, B.S., and Chi, J.G. (1982). Studies on intestinal trematodes in Korea VI. On the metacercaria and the second intermediate host of *Fibricola seoulensis*. Korean J Parasitol 20, 101–111.

Hong, S.T., Cho, T.K., Hong, S.J., Chai, J.Y., Lee, S.H., and Seo, B.S. (1984). Fifteen human cases of *Fibricola seoulensis* infection in Korea. Korean J Parasitol 22, 61–65.

Hong, S.T., Chai, J.Y., and Lee, S.H. (1986). Ten human cases of *Fibricola seoulensis* infection and mixed one with *Stellantchasmus* and *Metagonimus*. Korean J Parasitol 24, 94–96.

Hong, S.J., Seo, B.S., Lee, S.H., and Chai, J.Y. (1988). A human case of *Centrocestus armatus* infection in Korea. Korean J Parasitol 26, 55–60.

Hong, S.J., Woo, H.C., and Kim, I.T. (1989). Study on *Centrocestus armatus* in Korea I. Infection status of *Zacco platypus* and *Z. temminckii* with the metacercariae of *C. armatus*. Korean J Parasitol 27, 41–46 (in Korean).

Hong, S.J., Lee, S.H., Chai, J.Y., and Seo, B.S. (1990a). Recovery rate, growth and development of *Heterophyopsis continua* in experimental chicks. Korean J Parasitol 28, 53–62.

Hong, S.T., Yu, J.R., Myong, N.H., Chai, J.Y., and Lee, S.H. (1990b). Activities of brush border membrane bound enzymes of the small intestine in *Metagonimus yokogawai* infection in mice. Korean J Parasitol 29, 9–20.

Hong, S.J., Chai, J.Y., and Lee, S.H. (1991). Surface ultrastructure of the developmental stages of *Heterophyopsis continua* (Trematoda: Heterophyidae). J Parasitol 77, 613–620.

Hong, S.J., Chang, C.K., Lee, D.H., and Woo, H.C. (1996a). One human case of natural infection by *Heterophyopsis continua* and three other species of intestinal trematodes. Korean J Parasitol 34, 87–90.

Hong, S.J., Woo, H.C., and Chai, J.Y. (1996b). A human case of *Plagiorchis muris* (Tanabe, 1922: Digenea) infection in the Republic of Korea: Freshwater fish as a possible source of infection. J Parasitol 82, 647–649.

Hong, S.J., Han, J.H., Park, C.K., and Kang, S.Y. (1997). Intestinal pathologic findings at early stage infection by *Centrocestus armatus* in albino rats. Korean J Parasitol 35, 135–138.

Hong, S.J., Woo, H.C., Lee, S.U., and Huh, S. (1999). Infection status of dragonflies with *Plagiorchis muris* metacercariae in Korea. Korean J Parasitol 37, 65–70.

Howell, M.J. (1973). The resistance of cysts of *Stictodora lari* (Trematoda: Heterophyidae) to encapsulation by cells of the fish host. Int J Parasitol 3, 653–659.

Huffman, J.E., and Fried, B. (1990). *Echinostoma* and echinostomiasis. Adv Parasitol 29, 215–269.

Huffman, J.E., Michos, C., and Fried, B. (1986). Clinical and pathological effects of *Echinostoma revolutum* (Digenea: Echinostomatidae) in the golden hamster, *Mesocricetus auratus*. Parasitol 93, 505–515.

Huh, S., and Song, H.B. (1993). Transmission electron microscopic findings of the tribocytic organ of *Fibricola seoulensis*. Korean J Parasitol 31, 315–320.

Huh, S., Chai, J.Y., Hong, S.T., and Lee, S.H. (1988). Clinical and histopathological findings in mice heavily infected with *Fibricola seoulensis*. Korean J Parasitol 26, 45–53.

Huh, S., Lee, S.H., and Seo, B.S. (1990). Histochemical findings of the tribocytic organ and tegument of *Fibricola seoulensis*. Korean J Parasitol 28, 155–160.

Huh, S., Sohn, W.M., and Chai, J.Y. (1993). Intestinal parasites of cats purchased in Seoul. Korean J Parasitol 31, 371–373.

Huh, S., Lee, S.U., and Huh, S.C. (1994). A follow-up examination of intestinal parasitic infections of the Army soldiers in Whachon-gun, Korea. Korean J Parasitol 32, 61–63.

Huh, S., Lee, S.U., Won, M.H., Jeong, Y.G., Kwon, Y.H., and Choi, C.S. (1995). IgA response in mice infected with *Neodiplostomum seoulensis*. Korean J Parasitol 33, 55–60.

Ito, J. (1964). *Metagonimus* and other human heterophyid trematodes. Prog Med Parasitol Jpn 1, 314–393.

Ito, J., Mochizuki, H., Ohno, Y., and Ishiguro, M. (1991). On the prevalence of *Metagonimus* sp. among the imhabitants at Hamamatsu basin in Shizuoka Prefecture, Japan. Jpn J Parasitol 40, 274–278 (in Japanese).

Jang, Y.K., Kang, S.Y., Kim, S.I., and Cho, S.Y. (1985). *In situ* posture of anterior body of *Metagonimus yokogawai* in experimentally infected dog. Korean J Parasitol 23, 203–213.

Jaroonvesama, N., Charoenlarp, K., and Areekul, S. (1986). Intestinal absorption studies in *Fasciolopsis buski* infection. Southeast Asian J Trop Med Public Health 17, 582–586.

Jueco, N.L., and Monzon, R.B. (1984). *Cathaemacia cabrerai* sp. n. (Trematoda: Cathaemaciidae) a new oarqasite of man in the Philippines. Southeast Asian J Trop Med Public Health 15, 427–429.

Kagei, N., Oshima, T., Ishikawa, K., and Kihata, M. (1964). Two cases of human infection with *Stellantchasmus falcatus* Onji et Nishio, 1915 (Heterophyidae) in Kochi Prefecture. Jpn J Parasitol 13, 472–478.

Kagei, N., Hayashi, S., and Kato, K. (1980). On the heterophyid trematoda (*Heterophyes heterophyes*) infection cases imported from Egypt to Japan. Jpn J Trop Med 8, 1–7 (in Japanese).

Kaewkes, S., Elkins, D.B., Haswell-Elkins, M.R., and Sithithaworn, P. (1991a). *Phaneropsolus spinicirrus* n. sp. (Digenea: Lecithodendriidae), a human parasite in Thailand. J Parasitol 77, 514–516.

Kaewkes, S., Elkins, D.B., Sithithaworn, P., and Haswell-Elkins, M.R. (1991b) Comparative studies on the morphology of the eggs of *Opisthorchis viverrini* and lecithodendriid trematodes. Southeast Asian J Trop Med Public Health 22, 623–630.

Kang, S.Y., Cho, S.Y., Chai, J.Y., Lee, J.B., and Jang, D.H. (1983). A study on intestinal lesions of experimentally reinfected dogs with *Metagonimus yokogawai*. Korean J Parasitol 21, 53–73.

Katsuta, I. (1931). Studies on the metacercariae of Formosan brackish water fishes. (1) On a new species, *Stellantchamus formosanus* n. sp., parasitic in *Mugil cephalus*. Taiwan Igakkai Zasshi 30, 1404–1417 (in Japanese).

Katsuta, I. (1932a). Studies on the metacercariae of Formosan brackish water fishes. (2) On a new species, *Metagonimus minutus* n. sp., parasitic in *Mugil cephalus*. Taiwan Igakkai Zasshi 31, 26–39 (in Japanese).

Katsuta, I. (1932b). Studies on the metacercariae of Formosan brackish water fishes. (4) A new trematode, *Monorchotrema yokogawai*, which takes *Mugil cephalus* as its intermediate host. Taiwan Igakkai Zasshi 31, 160–175 (in Japanese).

Katsuta, I. (1932c). Studies on the metacercariae of Formosan brackish water fishes. (5) On a new species, *Stellantchasmus amplicaecalis* n. sp., parasitic in *Mugil cephalus*. Taiwan Igakkai Zasshi 31, 457–471 (in Japanese).

Kawahara, S., and Yamamoto, E. (1933). Human cases of *Echinostoma cinetorchis*. Tokyo Iji Shinshi (No. 2840), 1794–1796 (in Japanese).

Kho, W.G., Chai, J.Y., Chun, C.H., and Lee, S.H. (1990). Mucosal mast cell responses to experimental *Fibricola seoulensis* infection. Seoul J Med 31, 191–199.

Kifune, T., and Takao, Y. (1973). Description of *Psilorchis hominis* sp. nov. from man (Trematoda: Echinostomatoidea: Psilostomidae). Jpn J Parasitol 22, 111–115.

Kim, C.H. (1980). Study on the *Metagonimus* sp. in Gum River Basin, Chungchung-nam Do, Korea. Korean J Parasitol 18, 215–228 (in Korean).

Kim, C.H., Kim, N.M., Lee, C.H., and Park, J.S. (1987). Studies on the *Metagonimus* fluke in the Daecheong Reservoir and the upper stream of Geum River, Korea. Korean J Parasitol 25, 69–82 (in Korean).

Kim, K.H., Choi, E.S., and Rim, H.J. (1996). *Conger myriaster*, a new second intermediate host of *Heterophyopsis continua* (Digenea: Heterophyidae). Korean J Parasitol 34, 283–285.

Kim, I., Im, J.A., Lee, K.J., and Ryang, Y.S. (2000). Mucosal mast cell reponses in the small intestine of rats infected with *Echinostoma hortense*. Korean J Parasitol 38, 139–143.

Kim, Y.K., Yu, J.E., Chung, E.Y., and Chung, P.R. (2004). *Acanthoparyphium tyosenense* (Digenea: Echinostomatidae): experimental confirmation of the cercaria and its complete life history in Korea. J Parasitol 90, 97–102.

Kim, D.G., Kim, T.S., Cho, S.H., Song, H.J., and Sohn, W.M. (2006). Heterophyid metacercarial infections in brackish water fishes from Jinju-man (Bay), Kyongsangnam-do, Korea. Korean J Parasitol 44, 7–13.

Kino, H., Oishi, H., Ohno, Y., and Ishiguro, M. (2002). An endemic human infection with *Heterophyes nocens* Onji et Nishio, 1916 at Mikkabi-cho, Shizuoka, Japan. Jpn J Trop Med Hyg 30, 301–304.

Kino, H., Suzuki, Y., Oishi, H., Suzuki, S., Yamagiwa, S., and Ishiguro, M. (2006). Geographical distribution of *Metagonimus yokogawai*, and *M. miyatai* in Shizuoka Prefecture, Japan, and their site preferences in the sweetfish, *Plecoglossus altivelis*, and hamsters. Parasitol Int 55, 201–206.

Kook, J., Lee, S.H., and Chai, J.Y. (1997). *In vitro* cultivation of *Gymnophalloides seoi* metacercariae (Digenea: Gymnophallidae). Korean J Parasitol 35, 25–29.

Kook, J., Nawa, Y., Lee, S.H., and Chai, J.Y. (1998). Pathogenicity and lethality of a minute intestinal fluke, *Neodiplostomum seoulense*, to various strains of mice. J Parasitol 84, 1178–1183.

Kostadinova, A., and Gibson, D.I. (2001). A redescription of *Uroproctepisthmium bursicola* (Creplin, 1837) n. comb. (Digenea: Echinostomatidae), and re-evaluations of the genera *Episthmium* Luhe, 1909 and *Uroproctepisthmium* Fischthal & Kuntz, 1976. Syst Parasitol 50, 63–67.

Kostadinova, A., and Gibson, D.I. (2002). *Isthmiophora* Luhe, 1909 and *Euparyphium* Diet, 1909 (Digenea: Echinostomatidae) re-defined, with comments on their nominal species. Syst Parasitol 52, 205–217.

Kostadinova, A., Vaucher, C., and Gibson, D.I. (2002). Redescriptions of two echinostomes from birds in Paraguay, with comments on *Drepanocephalus* Diez, 1909 and *Paryphostomum* Dietz, 1909 (Digenea: Echinostomatidae). Syst Parasitol 53, 147–158.

Kumchoo, K., Wongsawad, C., Chai, J.Y., Vanittanakom, P., and Rojanapaibul, A. (2005). High prevalence of *Haplorchis taichui* metacercariae in cyprinoid fish from Chiang Mai Province, Thailand. Southeast Asian J Trop Med Public Health 36, 451–455.

Kuntz, R.E., and Lo, C.T. (1967). Preliminary studies on *Fasciolopsis buski* (Lankester, 1857) (giant Asian intestinal fluke) in the United States. Trans Am Microsc Soc 86, 163–166.

Kurokawa, T. (1935). On a new species of *Stamnosoma* found from a man. Tokyo Iji Shinshi (No. 2915), 293–298 (in Japanese).

Lee, H.H. (1972). Survey of *Fasciolopsis buski* infection among children of Liu-ying primary school in Tainan county. Chin J Microbiol 5, 110–114.

Lee, S.H., and Chai, J.Y. (2001). A review of *Gymnophalloides seoi* (Digenea: Gymnophallidae) and human infections in the Republic of Korea. Korean J Parasitol 39, 85–118.

Lee, S., J., and Hong, S.J. (2002). Gastric echinostomiasis diagnosed by endoscopy. Gastrointest Endosc 55, 440–442.

Lee, J.B., Chi, J.G., Lee, S.K., and Cho, S.Y. (1981). Study on the pathology of metagonimiasis in experimentally infected cat intestine. Korean J Parasitol 19, 109–129.

Lee, J.K., Lee, H.I., Baek, B.K., and Kim, P.G. (1983). Survey on encysted cercariae of trematodes from fresh-water fishes in Mangyeong riverside area. Korean J Parasitol 21, 187–192 (in Japanese).

Lee, J.K., Lee, H.I., Baek, B.K., and Kim, P.G. (1984a). Survey on encysted cercariae of trematodes from fresh-water fishes in Tongjin riverside areas in Korea. Korean J Parasitol 22, 190–202 (in Korean).

Lee, S.H., Chai, J.Y., Hong, S.J., Jun, Y.S., and Seo, B.S. (1984b). Anthelmintic effect of praziquantel (Distocide[@]) against *Metagonimus yokogawai* infection. J Korean Soc Chemother 2, 167–171 (in Korean).

Lee, S.H., Hwang, S.W., Chai, J.Y., and Seo, B.S. (1984c). Comparative morphology of eggs of heterophyids and *Clonorchis sinensis* causing human infections in Korea. Korean J Parasitol 22, 171–180.

Lee, S.H., Yoo, B.H., Hong, S.T., Chai, J.Y., Seo, B.S., and Chi, J.G. (1985). A histopathological study on the intestine of mice and rats experimentally infected by *Fibricola seoulensis*. Korean J Parasitol 23, 58–72.

Lee, S.K., Chung, N.S., Ko, I.H., Ko, H.I., and Chai, J.Y. (1986). Two cases of natural human infection by *Echinostoma hortense*. Korean J Parasitol 24, 77–81.

Lee, S.K., Chung, N.S., Ko, I.H., Ko, H.I., and Sohn, W.M. (1988a). A case of natural human infection by *Echinostoma cinetorchis*. Korean J Parasitol 26, 61–64.

Lee, S.K., Chung, N.S., Ko, I.H., Sohn, W.M., Hong, S.T., Chai, J.Y., and Lee, S.H. (1988b). An epidemiological survey of *Echinostoma hortense* infection in Chongsong-gun, Kyongbuk Province. Korean J Parasitol 26, 199–206.

Lee, S.H., Lee, J.K., Sohn, W.M., Hong, S.T., Hong, S.J., and Chai, J.Y. (1988c). Metacercariae of *Echinostoma cinetorchis* encysted in the fresh water snail, *Hippeutis (Helicorbis) cantori*, and their development in rats and mice. Korean J Parasitol 26, 189–197.

Lee, S.H., Chai, J.Y., Hong, S.T., and Sohn, W.M. (1990a). Experimental life history of *Echinostoma cinetorchis*. Korean J Parasitol 28, 39–44.

Lee, S.H., Noh, T.Y., Sohn, W.M., Kho, W.G., Hong, S.T., and Chai, J.Y. (1990b). Chronological observation of intestinal lesions of rats experimentally infected with *Echinostoma hortense*. Korean J Parasitol 28, 45–52 (in Korean).

Lee, S.H., Sohn, W.M., and Chai, J.Y. (1990c). *Echinostoma revolutum* and *Echinoparyphium recurvatum* recovered from house rats in Yangyang-gun, Kangwon-do. Korean J Parasitol 28, 235–240.

Lee, S.C., Chung, Y.B., Kong, Y., Kang, S.Y., and Cho, S.Y. (1993a). Antigenic protein fractions of *Metagonimus yokogawai* reacting with patient sera. Korean J Parasitol 31, 43–48.

Lee, S.H., Chai, J.Y., and Hong, S.T. (1993b). *Gymnophalloides seoi* n. sp. (Digenea: Gymnophallidae), the first report of human infection by a gymnophallid. J Parasitol 79, 677–680.

Lee, S.H., Chai, J.Y., Lee, H.J., Hong, S.T., Yu, J.R., Sohn, W.M., Kho, W.G., Choi, M.H., and Lim, Y.J. (1994). High prevalence of *Gymnophalloides seoi* infection in a village on a southwestern island of the Republic of Korea. Am J Trop Med Hyg 51, 281–285.

Lee, S.H., Chai, J.Y., Seo, M., Choi, M.H., Kim, D.C., and Lee, S.K. (1995a). Two cases of *Gymnophalloides seoi* infection accompanied by diabetes melitus. Korean J Parasitol 33, 61–64.

Lee, S.H., Choi, M.H., Seo, M., and Chai, J.Y. (1995b). Oysters, *Crassostrea gigas*, as the second intermediate host of *Gymnophalloides seoi* (Gymnophallidae). Korean J Parasitol 33, 1–7.

Lee, S.H., Sohn, W.M., Hong, S.J., Huh, S., Seo, M., Choi, M.H., and Chai, J.Y. (1996). A nationwide survey of naturally produced oysters for infection with *Gymnophalloides seoi* metacercariae. Korean J Parasitol 34, 107–112.

Lee, J.C., Kong, Y., Lee, S.U., and Huh, S. (1997a). Localization of worm antigen in *Neodiplostomum seoulense* by immuno-electronmicroscopy. Korean J Parasitol 35, 95–104.

Lee, S.H., Park, S.K., Seo, M., Guk, S.M., Choi, M.H., and Chai, J.Y. (1997b). Susceptibility of various animals and strains of mice to *Gymnophalloides seoi* infection and the effetcs of immunosuppression in C3H/HeN mice. J Parasitol 83, 883–886.

Lee, S.U., Huh, S., Park, G.M., and Chai, J.Y. (1999). A cytogenetic study on human intestinal trematodes of the genus *Metagonimus* (Digenea: Heterophyidae) in Korea. Korean J Parasitol 37, 237–241.

Lee, K.J., Park, S.K., Im, J.A., Kim, S.K., Kim, G.H., Kim, G.Y., Yang, E.J., and Ryang, Y.S. (2004a). Susceptibility of several strains of mice to *Echinostoma hortense* infection. Korean J Parasitol 42, 51–56.

Lee, S.U., Huh, S., Sohn, W.M., and Chai, J.Y. (2004b). Sequence comparisons of 28S ribosomal DNA and mitochondrial cytochrome C oxidase subunit I of *Metagonimus yokogawai, M. takahashii* and *M. miyatai*. Korean J Parasitol 42, 129–135.

Liang, C., and Ke, X.L. (1988). *Echinochasmus jiufoensis* sp. Nov., a human parasite from Guangzhou (Trematoda: Echinostomatidae). Acta Zootaxonom Sinica 13, 4–8 (in Chinese).

Lie, K.J. (1963). Studies on Echinostomatidae in Malaya. III. The adult *Echinostoma malayanum* Leiper, 1911 (Trematoda) and the probable synonymy of *Artyfechinostomum sufrartyfex* Lane, 1915. Z Parasitenkunde 23, 124–135.

Lie, K.J. (1968). Further studies on the life history of *Echinostoma lindoense* Sandground and Bonne, 1940 (Trematoda: Echinostomatidae) with a report of its occurrence in Brazil. Proc Helminthol Soc Washington 35, 74–77.

Madhavi, R. (1986). Distribution of metacercariae of *Centrocestus formosanus* (Trematoda: Heterophyidae) on the gills of *Aplocheilus panchax*. J Fish Biol 29, 685–690.

Maji, A.K., Bera, D.K., Manna, B., Nandy, A., Addy, M., and Bandyopadhyay, A.K. (1993). First record of human infection with *Echinostoma malayanum* in India. Trans R Soc Trop Med Hyg 87, 673.

Maji, A.K., Manna, B., Bandyopadhyay, A.K., Bera, D.K., Addy, M., and Nandy, A. (1995). Studies on the life cycle of *Artyfechinostomum oraoni* Bandyopadhyay, Manna and Nandy, 1989: embryogenesis and development in the intermediate host. Indian J Med Res 102, 124–128.

Majima, M. (1927). On *Echinostoma macrorchis* found in man. Kumamoto Igakkai Zasshi (No. 2252), 2260 (in Japanese).

Manning, G.S., and Lertprasert, P. (1973) Studies on the life cycle of *Phaneropsolus bonnei* and *Prosthodendrium molenkampi* in Thailand. Ann Trop Med Parasitol 67, 361–365.

Manning, G.S., Lertprasert, P., Watanasirmkit, K. and Chetty, C. (1971). A description of newly discovered intestinal parasites endemic to northeastern Thailand. J Med Assoc Thailand 54, 466–474.

Martin, W.E. (1958). The life cycle of some Hawaiian heterophyid trematodes. J Parasitol 44, 305–321.

Marty, A.M., and Andersen, E.M. (2000). Fasciolopsiasis and other intestinal trematodiases. Pathol Infect Dis 1, 93–105.

Mas-Coma, S., Bargues, M.D., and Valero, M.A. (2005). Fascioliasis and other plant-borne trematode zoonoses. Int J Parasitol 35, 1255–1278.

Massoud, J., Jalali, H., and Reza, M. (1981). Studies on trematodes of the family Heterophyidae (Odhner, 1914) in Iran. I. Preliminary epidemiological surveys in man and carnivores in Khuzestan. J Helminthol 55, 255–260.

McMullen, D.B. (1937). An experimental infection of *Plagiorchis muris* in man. J Parasitol 23, 113–115.

Millemann, R.E., and Knapp, S.E. (1970). Biology of *Nanophyetus salmincola* and "salmon poisoning" disease. Adv Parasitol 8, 1–41.

Miyamoto, K., Nakao, M., and Inaoka, T. (1983). Studies on the zoonoses in Hokkaido, Japan. 5. On the epidemiological survey of *Echinostoma hortense* Asada, 1926. Jpn J Parasitol 32, 261–269.

Muttalib, M.A., and Islam, N. (1975). *Fasciolopsis buski* in Bangladesh—a pilot study. J Trop Med Hyg 78, 135–137.

Nawa, Y., Ishikawa, N., Tsuchiya, K., Horii, Y., Abe, T., Khan, A.I., Itoh, H., Ide, H., and Uchiyama, F. (1994). Selective effector mechanisms for the expulsion of intestinal helminths. Parasite Immunol 16, 333–338.

Niemi, D.R., and Macy, R.W. (1974). The life cycle and infectivity to man of *Apophallus donicus* (Skrjabin and Lindtrop, 1919) (Trematoda: Heterophyidae). Proc Helminthol Soc Washington 41, 223–229.

Nishigori, M. (1924). On a new trematode, *Stamnosoma formosanum* n. sp., and its life history. Taiwan Igakkai Zasshi (No. 234), 181–228 (in Japanese).

Noda, K. (1959). The larval development of *Stellantchasmus falcatus* (Trematoda; Heterophyidae) in the first intermediate host. J Parasitol 45, 635–642.

Onah, D.N., and Nawa, Y. (2000). Mucosal immunity against parasitic gastrointestinal nematodes. Korean J Parasitol 38, 209–236.

Onji, Y., and Nishio, T. (1916). On the trematodes whose intermediate host is brackish water fishes. Chiba Igaku Semmon Gakko Zasshi 81&82, 229–249 (in Japanese).

Park, J.T. (1938). A rat trematode, *Echinostoma hortense* Asada, from Korea. Keijo J Med 9, 283–286.

Park, M.S., Kim, S.W., Yang, Y.S., Park, C.H., Lee, W.T., Kim, C.U., Lee, E.M., Lee, S.U., and Huh, S. (1993). Intestinal parasite infections in the inhabitants along the Hantan River, Chorwon-gun. Korean J Parasitol 31, 375–378.

Pearson, J.C., and Ow-Yang, C.K. (1982). New species of *Haplorchis* from Southeast Asia, together with keys to the *Haplorchis*-group of heterophyid trematodes of the region. Southeast Asian J Trop Med Public Health 13, 35–60.
Pica, R., Castellano, C., Cilia, C., and Errico, F.O. (2003). Intestinal fluke infections: the heterophyids. Clin Terapeutica 154, 61–63.
Premvati, G., and Pande, V. (1974). On *Centrocestus formosanus* (Nishigori, 1924) Price, 1932 and its experimental infections in white leghorn chicks. Jpn J Parasitol 23, 79–84.
Quan, F., Jiang, T., Ma, H., Cui, C., and Sun, L. (1995). The first discovery of *Fibricola seoulensis* in China. J Yanbian Med Coll 18, 17–20.
Radomyos, P., Bunnag, D., and Harinasuta, T. (1982). *Echinostoma ilocanum* (Garrison, 1908) Odhner, 1911, infection in man in Thailand. Southeast Asian J Trop Med Public Health 13, 265–269.
Radomyos, P., Bunnag, D., and Harinasuta, T. (1985). Report of *Episthmium caninum* (Verma, 1935) Yamaguti, 1958 (Digenea: Echinostomatidae) in man. Southeast Asian J Trop Med Public Health 16, 508–511.
Radomyos, P., Bunnag, D., and Harinasuta, C. (1989). A new intestinal fluke, *Plagiorchis harinasutai*, n. sp. Southeast Asian J Trop Med Public Health 20, 101–107.
Radomyos, P., Charoenlarp, P., Radomyos, B., and Tungtrongchitr, A. (1990). Two human cases of *Stellantchasmus falcatus* (Trematoda, Heterophyidae) infection in Northeastern Thailand. Jpn J Parasitol 39, 7–11.
Radomyos, P., Charoenlarp, P., and Harinasuta, T. (1991). Human *Episthmium caninum* (Digenea: Echinostomatidae) infection: report of two more cases. J Trop Med Parasitol 14, 48–50.
Radomyos, B., Wongsaroj, T., Wilrairatana, P., Radomyos, P., Praevanich, R., Meesomboon, V., and Jongsuksuntikul, P. (1998). Opisthorchiasis and intestinal fluke infections in northern Thailand. Southeast Asian J Trop Med Public Health 29, 123–127.
Ransom, B.H. (1920). Synopsis of the trematode family Heterophyidae with descriptions of a new genus and five new species. Proc U. S. Natural Museum 57, 527–573.
Rho, I.H., Kim, S.I., Kang, S.Y., and Cho, S.Y. (1984). Observation on the pathogenesis of villous changes in early phase of experimental metagonimiasis. Chung-Ang J Med 9, 67–77 (in Korean).
Rim, H.J. (1982). Echinostomiasis. CRC Handbook Series in Zoonoses, Section C: Parasitic Zoonoses, Vol. III. Trematode Zoonoses. Boca Raton, FL: CRC Press, 53–69.
Rim, H.J., Chu, D.S., Lee, J.S., Joo, K.H., and Won, C.Y. (1978). Anthelmintic effects of various drugs against metagonimiasis. Korean J Parasitol 16, 117–122.
Rim, H.J., Kim, S.J., and Yang, M.G. (1992). Antigenic localities in the tissues of Metagonimus yokogawai in the period of growth. Korean J Parasitol 30, 309–321 (in Korean).
Rohela, M., Jamaiah, I., Menon, J., and Rachel, J. (2005). Fasciolopsiasis: a first case report from Malaysia. Southeast Asian J Trop Med Public Health 36, 456–458.
Ryang, Y.S. (1990). Studies on *Echinostoma* spp. in the Chungju Reservoir and upper streams of the Namhan River. Korean J Parasitol 28, 221–233 (in Korean).
Ryang, Y.S., Ahn, Y.K., Kim, W.T., Shin, K.C., Lee, K.W., and Kim, T.S. (1986). Two cases of human infection by *Echinostoma cinetorchis*. Korean J Parasitol 24, 71–76.
Ryang, Y.S., Ahn, Y.K., and Yoon, M.B. (1991). Trematode infections in the small intestine of *Egretta alba medesta* in Kanhwon-do. Korean J Parasitol 29, 227–233 (in Korean).
Ryang, Y.S., Yoo, J.C., Lee, S.H., and Chai, J.Y. (2000). The Palearctic oystercatcher *Haematopus ostralegus*, a natural definitive host for *Gymnophalloides seoi*. J Parasitol 86, 418–419.

Ryang, Y.S., Yoo, J.C., Lee, S.H., and Chai, J.Y. (2001). Susceptibility of avian hosts to experimental *Gymnophalloides seoi* infection. J Parasitol 87, 454–456.

Saeed, I., Maddox-Hyttel, C., Monrad, J., and Kapel, C.M. (2006). Helminths of red foxes (*Vulpes vulpes*) in Denmark. Veterinary Parasitol 139, 168–179.

Saenphet, S., Wongsawad, C., Saenphet, K., and Chai, J.Y. (2003). Susceptibility of rodents to *Stellantchasmus falcatus* infection. Southeast Asian J Trop Med Public Health 34(suppl), 123–127.

Saito, S. (1984). On the validity of *Metagonimus* spp. Proc Jpn Parasite Taxonomy Morphol Conference, 2, 1–6 (in Japanese).

Saito, S., Chai, J.Y., Kim, K.H., Lee, S.H., and Rim, H.J. (1997). *Metagonimus miyatai* sp. nov. (Digenea: Heterophyidae), a new intestinal trematode transmitted by freshwater fishes in Japan and Korea. Korean J Parasitol 35, 223–232.

Sandground, J.H. (1940). *Plagiorchis javensis* n. sp. a new trematode parasitic in man. Rev Med Trop Parasitol (Habana) 6, 207–211.

Schell, S.C. (1985). Handbook of Trematodes of North America North of Mexico. Idaho: University Press of Idaho.

Scholz, T. (1999). Taxonomic study of *Ascocotyle (Phagicola) longa* Ransom, 1920 (Digenea: Heterophyidae) and related taxa. Syst Parasitol 43, 147–158.

Scholz, T., and Salgado-Maldonado, G. (1999). The introduction and dispersal of *Centrocestus formosanus* (Nishigori, 1924) (Digenea: Heterophyidae) in Mexico: a review. Am Midland Naturalist 143, 185–200.

Seo, B.S. (1990). *Fibricola seoulensis* Seo, Rim and Lee, 1964 (Trematoda) and fibricoliasis in man. Seoul J Med 31, 61–96.

Seo, B.S., Rim, H.J., and Lee, C.W. (1964). Studies on the parasitic helminths of Korea I. Trematodes of rodents. Korean J Parasitol 2, 20–26.

Seo, B.S., Cho, S.Y., and Chai, J.Y. (1980a). Studies on intestinal trematodes in Korea I. A human case of *Echinostoma cinetorchis* infection with an epidemiological investigation. Seoul J Med 21, 21–29.

Seo, B.S., Cho, S.Y., Chai, J.Y., and Hong, S.T. (1980b). Studies on intestinal trematodes in Korea II. Identification of the metacercariae of *Heterophyes heterophyes nocens* in mullets of three southern coastal areas. Seoul J Med 21, 30–38.

Seo, B.S., Cho, S.Y., Hong, S.T., Hong, S.J., and Lee, S.H. (1981a). Studies on parasitic helminths of Korea V. Survey on intestinal trematodes of house rats. Korean J Parasitol 19, 131–136.

Seo, B.S., Hong, S.T., and Chai, J.Y. (1981b). Studies on intestinal trematodes in Korea III. Natural human infection of *Pygidiopsis summa* and *Heterophyes heterophyes nocens*. Seoul J Med 22, 228–235.

Seo, B.S., Hong, S.T., Chai, J.Y., and Cho, S.Y. (1981c). Studies on intestinal trematodes in Korea IV. Geographical distribution of *Pygidiopsis* and *Heterophyes* metacercariae. Seoul J Med 22, 236–242.

Seo, B.S., Lee, S.H., Cho, S.Y., Chai, J.Y., Hong, S.T., Han, I.S., Shon, J.S., Cho, B.H., Ahn, S.R., Lee, S.K., Chung, S.C., Kang, K.S., Shim, H.S., and Hwang, I.S. (1981d). An epidemiologic study on clonorchiasis and metagonimiasis in riverside areas in Korea. Korean J Parasitol 19, 137–150.

Seo, B.S., Lee, S.H., Hong, S.T., Hong, S.J., Kim, C.Y., and Lee, H.Y. (1982). Studies on intestinal trematodes in Korea V. A human case infected by *Fibricola seoulensis* (Trematoda: Diplostomatidae). Korean J Parasitol 20, 93–99.

Seo, B.S., Hong, S.T., Chai, J.Y., and Lee, S.H. (1983). Studies on intestinal trematodes in Korea VII. A human case of *Echinostoma hortense* infection. Korean J Parasitol 21, 219–223.

Seo, B.S., Lee, S.H., Chai, J.Y., and Hong, S.J. (1984a). Studies on intestinal trematodes in Korea XIII. Two cases of natural human infection by *Heterophyopsis continua* and the status of metacercarial infection in brackish water fishes. Korean J Parasitol 22, 51–60.

Seo, B.S., Lee, S.H., Chai, J.Y., and Hong, S.J. (1984b). Studies on intestinal trematodes in Korea XII. Two cases of human infection by *Stellantchasmus falcatus*. Korean J Parasitol 22, 43–50.

Seo, B.S., Park, Y.H., Chai, J.Y., Hong, S.J., and Lee, S.H. (1984c). Studies on intestinal trematodes in Korea XIV. Infection status of loaches with metacercariae of *Echinostoma cinetorchis* and their development in albino rats. Korean J Parasitol 22, 181–189.

Seo, B.S., Chun, K.S., Chai, J.Y., Hong, S.J., and Lee, S.H. (1985a). Studies on intestinal trematodes in Korea XVII. Development and egg laying capacity of *Echinostoma hortense* in albino rats and human experimental infection. Korean J Parasitol 23, 24–32.

Seo, B.S., Lee, S.H., Chai, J.Y., and Hong, S.J. (1985b). Studies on intestinal trematodes in Korea XX. Four cases of natural human infection by *Echinochasmus japonicus*. Korean J Parasitol 23, 214–220.

Seo, B.S., Lee, H.S., Chai, J.Y., and Lee, S.H. (1985c). Intensity of *Metagonimus yokogawai* infection among inhabitants in Tamjin River basin with reference to its egg laying capacity in human host. Seoul J Med 26, 207–212.

Seo, B.S., Cheong, S.K., Chai, J.Y., Lee, S.H., and Lee, J.B. (1986). Histopathology of small intestine of rats and mice experimentally infected with *Pygidiopsis summa*. Seoul J Med 27, 125–134.

Seo, B.S., Lee, S.H., Chai, J.Y., Hong, S.J., and Hong, S.T. (1988). The life cycle and larval development of *Fibricola seoulensis* (Trematoda: Diplostomatidae). Korean J Parasitol 26, 179–188.

Seo, M., Guk, S.M., Han, E.T., and Chai, J.Y. (2003). Role of intestinal goblet cells in the expulsion of *Gymnophalloides seoi* from mice. J Parasitol 89, 1080–1082.

Seo, M., Chun, H., Ahn, G., Jang, K.T., Guk, S.M., and Chai, J.Y. (2006). A case of colonic lymphoid tissue invasion by *Gymnophalloides seoi* in a Korean man. Korean J Parasitol 44, 87–89.

Sheir, Z.M., and Aboul-Enein, M.E. (1970). Demographic, clinical and therapeutic appraisal of heterophyiasis. J Trop Med Hyg 73, 148–152.

Shimalov, V.V., and Shimalov, V.T. (2002). Human parasitizing echinostomatata (Trematoda, fasciolidae) in mammals in Belorussia forest zone. Meditsinskaia Parazitol Parazitamye Bolezni (No. 1), 47–50 (in Russian).

Shimazu, T. (2002). Life cycle and morphology of *Metagonimus miyatai* (Digenea: Heterophyidae). Parasitol Int 51, 271–280.

Shin, E.H., Kim, T.H., Hong, S.J., Park, J.H., Guk, S.M., and Chai, J.Y. (2003). Effects of anti-allergic drugs on intestinal mastocytosis and worm expulsion of rats infected with *Neodiplostomum seoulense*. Korean J Parasitol 41, 81–87.

Shoop, W.L. (1989). Experimental human infection with *Fibricola cratera* (Trematoda: Neodiplostomidae). Korean J Parasitol 27, 249–252.

Shyu, L.Y., Lee, H.H., and Chen, E.R. (1984). A preliminary study on epidemiology of fasciolopsiasis in Tainan Hsien, south Taiwan. Chin J Microbiol Immunol 17, 118–120.

Sohn, W.M., and Chai, J.Y. (2005). Infection status with helminthes in feral cats purchased from a market in Busan, Republic of Korea. Korean J Parasitol 43, 93–100.

Sohn, W.M., Chai, J.Y., and Lee, S.H. (1994a). *Stictodora fuscatum* (Heterophyidae) metacercariae encysted in gobies, *Acanthogobius flavimanus*. Korean J Parasitol 32, 143–148.

Sohn, W.M., Han, G.G., Kho, W.G., Chai, J.Y., and Lee, S.H. (1994b). Infection status with the metacercariae of heterophyid flukes in the brackish water fish from Haenam-gun, Chollanam-do, Korea. Korean J Parasitol 32, 163–169 (in Korean).

Sohn, W.M., Ryang, Y.S., Chai, J.Y., and Lee, S.H. (1998). Discovery of *Gymnophalloides seoi* metacercariae in oysters from islands of the West Sea known as the habitats of paleartic oystercatchers. Korean J Parasitol 36, 163–169.

Son, W.Y., Huh, S., Lee, S.U., Woo, H.C., and Hong, S.J. (1994). Intestinal trematode infections in the villagers in Koje-myon, Kochang-gun, Kyongsangnam-do, Korea. Korean J Parasitol 32, 149–155.

Song, C.Y., Lee, S.H., and Jeon, S.R. (1985). Studies on the intestinal fluke, *Metagonimus yokogawai* Katsurada, 1912 in Korea IV. Geographical distribution of sweetfish and infection status with *Metagonimus* metacercaria in south-eastern area of Korea. Korean J Parasitol 23, 123–138 (in Korean).

Sripalwit, P., Wongsawad, C., Chai, J.Y., Anuntalabhochai, S., and Rojanapaibul, A. (2003). Investigation of *Stellantchasmus falcatus* metacercariae in half-beaked fish, *Dermogenus pusillus* from four districts of Chiang Mai Province, Thailand. Southeast Asian J Trop Med Public Health 34, 281–285.

Suzuki, S. (1930). Yokogawa's *Metagonimus*. List of Publications on Special Animals in Okayama Prefecture. Okayama Prefecture Rep 146–148 (in Japanese).

Suzuki, N., Kumazawa, H., Araki, T., Hara, M., Morita, H., Segawa, T., Shina, T., Yamada, S., and Tsumoto, S. (1982). Occurrence of *Heterophyes heterophyes nocens* Onji et Nishio, 1916 in man in Kochi Prefecture. Jpn J Parasitol 31, 271–274 (in Japanese).

Takahashi, S. (1929a). On the eggs of *Stellantchasmus falcatus* and *Pygidiopsis summus* found in human stools. Okayama Igakkai Zasshi 41, 1502–1513 (in Japanese).

Takahashi, S. (1929b). A contribution to the life-history of *Stamnosoma armatum* Tanabe. Okayama Igakkai Zasshi 4, 1759–1771 (in Japanese).

Tanabe, H. (1922a). A contribution to the study of the life cycle of digenetic trematodes. A study of a new species *Lepoderma muris* n. sp. Okayama Igakkai Zasshi 385, 47–58 (in Japanese).

Tanabe, H. (1922b). Studien uber die Trematoden mit Susswasser-fischen als Zwischenwirt. I. *Stamnosoma armatum* n.g., n. sp. Kyoto Igaku Zasshi 19, 239–252.

Tanabe, H. (1926). Studies on trematodes with fresh water fishes as their intermediate host III. On a new species, *Echinochasmus japonicus* (n.sp.). Nippon Byori Gakkai Kaishi 16, 295–296 (in Japanese).

Taraschewski, H. (1984). Die trematoden der gattung *Heterophyes* taxonomie, biologie, epidemiologie. Ph. D. Dissertation to Universitat Hohenheim.

Tesana, S., Srisawangwonk, Kaewkes, S., Sithithaworn, P., Kanla, P., and Arunyanart, C. (1991). Eggshell morphology of the small eggs of human trematodes in Thailand. Southeast Asian J Trop Med Public Health 22, 631–636.

Ujiie, N. (1936a). On the development, the structure of *Echinochasmus japonicus* and its parasitism in man. Taiwan Igakkai Zasshi 35, 535–546 (in Japanese).

Ujiie, N. (1936b). Differentiation of the species of *Monorchotrema* from Formosa. Taiwan Igakkai Zasshi 35, 938–946 (in Japanese).

Umadevi, K., and Madhavi, R. (2000). Observation on the morphology and life-cycle of *Procerovum varium* (Onji & Nishio, 1916) (Trematoda: Heterophyidae). Syst Parasitol 46, 215–225.

Velasquez, C.C. (1973a). Observations on some Heterophyidae (Trematoda: Digenea) encysted in Philippine fishes. J Parasitol 59, 77–84.

Velasquez, C.C. (1973b). Life cycle of *Procerovum calderoni* (Africa and Garcia, 1935), Price, 1940 (Trematoda: Digenea: Heterophyidae). J Parasitol 59, 813–816.

Velasquez, C.C. (1982). Heterophyidiasis, In: Steele, J.H., ed. CRC Handbook Series in Zoonoses, Section C: Parasitic Zoonoses, Vol. III (Trematode Zoonoses). Boca Raton, FL: CRC Press, 99–107.

Waikagul, J., Wongsaroj, T., Radomyos, P., Meesomboon, V., Praewanich, R., and Jongsuksuntikul, P. (1997). Human infection of *Centrocestus caninus* in Thailand. Southeast Asian J Trop Med Public Health 28, 831–835.

Witenberg, G. (1929). Studies on the trematode—family Heterophyidae. Ann Trop Med Parasitol 23, 131–268.

Wongsawad, C., Chariyahpongpun, P., and Namue, C. (1998). Experimental host of *Stellantchasmus falcatus*. Southeast Asian J Trop Med Public Health 29, 406–409.

Xiao, X., Wang, T.P., Lu, D.B., Gao, J.F., Xu, L.F., Wu, W.D., Mei, J.D., and Xu, F.N. (1992). The frist record of human infection with *Echinochasmus liliputanus*. Chin J Parasitol Parasitic Dis 10, 132–135 (in Chinese).

Xiao, X., Wang, T., Zheng, X., Shen, G., and Tian, Z. (2005). In vivo and in vitro encystment of *Echinochasmus liliputanus* cercariae and biological activity of the metacercariae. J Parasitol 91, 492–498.

Xu, B.K., and Li, D.N. (1979). Rare human parasites in China. Nat Med J China 59, 286–290 (in Chinese).

Yamaguti, S. (1939a). Studies on the helminth founa of Japan. Part 25. Trematodes of birds, IV. Jpn J Zool 8, 131–210.

Yamaguti, S. (1939b). Studies on the helminth fauna of Japan. Part 27. Trematodes of mammals, II. Jpn J Med Sci (VI. Bacteriol Parasitol) 1, 131–151.

Yamaguti, S. (1958). Part I. The digenetic trematodes of vertebrates. In, Systema Helminthum, Vol. I. New York: Interscience Publishers Inc.

Yamaguti, S. (1975). A Synoptical Review of Life Histories of Digenetic Trematodes of Vertebrates. Kyoto, Japan: Keigaku.

Yang, H.J., Guk, S.M., Han, E.T., and Chai, J.Y. (2000). Molecular differentiation of three species of *Metagonimus* by simple sequence repeat anchored polymerase chain reaction (SSR-PCR) amplification. J Parasitol 86, 1170–1172.

Yokogawa, M., Harinasuta, C., and Charoenlarp, P. (1965a). *Hypoderaeum conoideum* (Block, 1782) Diez, 1909, a common intestinal fluke of man in the north-east Thailand. Jpn J Parasitol 14, 148–153.

Yokogawa, M., Sano, M., Itabashi, T., and Kachi, S. (1965b). Studies on the intestinal flukes II. Epidemiological studies on heterophyid trematodes of man in Chiba Prefecture. Jpn J Parasitol 14, 577–585.

Yu, S.H., and Mott, K.E. (1994). Epidemiology and morbidity of food-borne intestinal trematode infections. Trop Dis Bull 91, R125–R152.

Yu, J.R., Chung, J.S., Huh, S., Lee, S.H., and Chai, J.Y. (1997a). PCR-RFLP patterns of three kinds of *Metagonimus* in Korea. Korean J Parasitol 35, 271–276.

Yu, J.R., Chung, J.S., and Chai, J.Y. (1997b). Different RAPD pattern between *Metagonimus yokogawai* and *Metagonimus* Miyata type. Korean J Parasitol 35, 295–298.

Yu, J.R., Myong, N., and Chai, J.Y. (1997c). Expression patterns of proliferating cell nuclear antigen in the small intestine of mice infected with *Metagonimus yokogawai* and *Metagonimus* Miyata type. Korean J Parasitol 35, 239–244.

Yurlova, N.I., Vodyanitskaya, S.N., Serbina, E.A., Biserkov, V.Y., Georgiev, B.B., and Chipev, N.H. (2006). Temporal variation in prevalence and abundance of metacercariae in the pulmonate snail *Lymnaea stagnalis* in Chany Lake, West Siberia, Russia: long-term patterns and environmental covariates. J Parasitol 92, 249–259.

Zhang, S.L., and Fan, G.H. (1990). Case report: brain abscess caused by heterophyid eggs. Chin J Parasitol Parasitic Dis 8, 178 (in Chinese).

Zhu, D., Lin, J., Cheng, Y., Liang, C., Wang, X., Guo, Z., and Zhuang, H. (1986). Pyquiton treatment of *Echinochasmus japonicus* infection. J Parasitol Parasitic Dis, 4, 1–4 (in Chinese).

3
Paragonimiasis

David Blair, Takeshi Agatsuma and Wenlin Wang

Paragonimiasis is the disease caused by lungflukes of the genus *Paragonimus*. Despite that fact that some 293 million people are at risk of paragonimiasis (Keiser and Utzinger, 2005), and several million are actually infected, the condition is not likely to feature in the curriculum of medical colleges in economically developed countries. Even medical practitioners who are aware of the disease are likely to dismiss it as just another member of the constellation of zoonotic parasitic infections, seen only in the least developed corners of the planet, that are rapidly yielding to advances in primary health care. They are correct—to an extent. However, paragonimiasis is also one of the reemerging parasitic diseases destined to confound the ill-prepared diagnostician in the wealthiest cities and the poorest villages alike. Paragonimiasis can present with an extraordinary array of signs and symptoms. These are a function of the exact species of lung fluke responsible and its interaction with the host. Public health workers are increasingly finding that lung flukes are the agents actually responsible for "nonresponsive tuberculosis" in remote tropical and subtropical areas. Urban clinicians, similarly, are encountering, and failing to recognize, paragonimiasis in their relatively affluent and mobile clientele. Easy and rapid long-distance movement of people and fresh foods harboring infective stages will force attention onto such diseases even in places where they have been disregarded for so long. At least one new journal will focus on "neglected tropical diseases" such as paragonimiasis (see PLOS Neglected Tropical Diseases; www.plosntd.org).

The study of paragonimiasis has a far wider scope than solely clinical diagnosis and treatment. In most places it is essentially a zoonosis, circulating in the natural environment with humans only incidentally involved. Thus the roles of humans and their environmental ecology, and of environmental change, need to be explored. Not surprisingly, some members of the genus *Paragonimus* have become model organisms for the study of trematode interactions with mammalian hosts (invasion biology) and for drug development. The biogeography, evolution, and systematics of *Paragonimus* species are fields of ongoing enquiry. Many species have been placed in the genus *Paragonimus*. A full list of these, and a near-complete catalogue of their hosts, is in Blair (1999).

This chapter focuses on recent developments (especially since the review by Blair et al., 1999) with an emphasis on changing patterns of disease in human populations and the ways in which disease is acquired, detected, and treated. Relevant work on parasite taxonomy and evolution is also reviewed.

Lung flukes are hermaphroditic trematodes that mature in the lungs of mammals. In most cases, two maturing worms meet in the thoracic cavity and move into the lung parenchyma where a fibrous cyst forms around them. The worms, each resembling a small coffee bean, mate and produce numerous eggs that are released into bronchioles and hence to the outside world via sputum or feces. This is classical pulmonary paragonimiasis caused by several species of *Paragonimus*. One variation on this theme concerning the widespread species *P. westermani* needs to be introduced here. In east and northeast China, Japan, Korea, and Taiwan (collectively referred to here as East Asia), sexually reproducing diploid forms of the species are sympatric, or nearly so, with triploid forms that are physically larger and can produce large numbers of eggs parthenogenetically. Triploid individuals can establish themselves in a lung cyst and cause pulmonary paragonimiasis without the need for a mate. Juvenile diploid worms may wander in the pleural spaces and only move into the lungs if they find a mate. Diploid and triploid forms have sometimes been regarded as distinct species (discussed in Blair et al., 1999) and under some circumstances they cause different clinical pictures in humans (see below).

Lung flukes are found in the tropics and subtropics of East and South Asia and in sub-Saharan Africa. In the Americas, they occur from Peru northwards as far as southern Canada. Occasionally they are reported from other regions, but such cases are often imported through the movement of either infected people or contaminated food. Some details of the main species infecting humans are in Table 3.1.

Mammalian hosts of lungflukes, especially in Asia, are often carnivores such as felids and canids (Table 3.1). However, many small mammals, in particular mustelids, viverrids, murids, and (in the Americas) didelphid marsupials, can act as definitive hosts for one or more species. Cercariae of lung flukes occur in freshwater (occasionally brackish water) snails and their metacercariae in freshwater crabs and crayfish. Human infection usually occurs when crustaceans harboring metacercariae are eaten without adequate preparation such as cooking. Some species mature routinely in the lungs of humans; for other species, humans are not very suitable hosts and juvenile worms may wander in the body, often failing to mature but causing pathological effects that can be dramatic. Thus a clinical picture quite different from pulmonary paragonimiasis might be seen in infected people. Some wild mammals, notably wild boars, can act as paratenic hosts. Juvenile worms remain alive for years in the muscles of such hosts and can develop further in humans eating uncooked boar meat. There have been suggestions in the literature that humans can also be infected by drinking untreated water into which metacercariae have been released by decomposition of their crustacean host (Cui et al., 1998). However, there is no clear evidence that this is a feasible route of infection.

TABLE 3.1. The major species of lungflukes responsible for human disease; additional species may occasionally infect humans.

Species	Geographical distribution	Type of paragonimiasis	Natural definitive hosts[1]	Notes
P. westermani	East Asia (China and Taiwan, Korea, Japan, Southeast Siberia); Southeast Asia (Philippines, Malaysia, Thailand, Cambodia, Laos, Vietnam), South Asia (Sri Lanka, India, Nepal, Pakistan); possibly Papua New Guinea.	Usually pulmonary; other forms sometimes	Cercopithecids, canids, felids, herpestids, viverrids, mustelids, murids	A species complex exhibiting considerable geographical genetic variation; infective to humans in East Asia and the Philippines; situation in India not known
P. skrjabini and subspecies such as *P. s. miyazakii*	East Asia (China, Japan), Thailand; may extend westwards into Northeast India.	Pulmonary forms rare (except possibly Japan): pleural and ectopic forms usual	Canids, felids, mustelids, viverrids, murids, hystricids	A species complex, but more restricted in distribution than *P. westermani*; humans not usually suitable definitive host
P. heterotremus	South and Southwest China, Vietnam, Laos, Thailand, Northeast India	Usually pulmonary	Felids, sciurids, murids	
P. africanus	West Africa: Equatorial Guinea, Cameroon, Nigeria, possibly Ivory Coast	Usually pulmonary	Lorids, cercopithecids, herpestids, viverrids	
P. uterobilateralis	West Africa: Gabon, Cameroon, Nigeria, Liberia	Usually pulmonary	Canids, herpestids, mustelids, viverrids	
P. kellicotti	North America: Mississippi Basin and Atlantic coast of the U.S.; Ontario and Quebec in Canada	Usually pulmonary	Didelphids, canids, felids, mustelids, procyonids, suids, bovids, murids	Rare in humans, but some interesting cases reported
P. mexicanus	Central and South America: Mexico, Costa Rica, Panama, Guatemala, Ecuador, Peru; probably other countries in the region	Usually pulmonary	Didelphids, cebids, canids, felids, mustelids, procyonids, suids	Possibly a species complex; taxonomic status not fully resolved

[1]Omits experimental hosts, paratenic hosts, and humans.

For historical accounts of lung flukes and/or the disease they cause, see Blair et al., (1999), Grove (1990), Miyazaki (1991), Yokogawa (1965), and Yokogawa, et al., (1960). Recent publications focusing on particular countries include the following: China (Yang et al., 2000); Japan (Nakamura-Uchiyama et al., 2002; Nishida and Shibahara, 2003; Otsuji, 2003); Korea (Choi, 1990; Cho et al., 1997); Vietnam (De et al., 2003; Doanh 2005); Ecuador (Waikagul et al., 2003); Colombia (Vélez et al., 2000); Thailand: (Waikagul and Yoonuan 2005); Philippines (Cabrera, 1984; de Leon and Piad, 2005).

Invasion Biology and Parasite Proteases

In the journey from the metacercarial cyst within a crustacean to egg-production inside a capsule in the lung of a mammal, juvenile and maturing lung flukes face a number of challenges. The metacercaria must emerge from its cyst in the right place and at the right time, acting on specific cues. The emerged worm, now a juvenile, must penetrate the wall of the digestive tract, facilitate its passage to the pleural spaces by muscular action and digestion of host tissues, find a mate, and establish a home in the lungs (reviewed in Blair et al., 1999). At all times it must gain nourishment from the host (Choi et al., 2006) while evading the host immune system. Central to many of these processes are hydrolytic enzymes, and in particular the cysteine proteases that feature prominently in relevant literature. This class of enzyme (see Sajid and McKerrow, 2002, and Tort et al., 1999, for general reviews) is generally regarded as crucial to the aspects of invasion biology mentioned above, as well as being important in parasite housekeeping and cellular homeostasis. They are present in the tissues of worms and are also major constituents of worm excretory-secretory products (ESPs) that can be harvested for analysis and experimentation from media in which worms are incubated. The ESPs, especially those produced by adult worms, contain a complex mixture of molecules that form the front line of interactions between host and parasite. Lee et al. (2006) detected at least 147 protein spots when ESP from adult worms was subjected to two-dimensional electrophoresis and the gel stained. The mammalian immune system recognizes an increasing number of ESP components, many of them cysteine proteases, during the course of an infection (Lee et al., 2006).

Although cysteine proteases are crucial to metacercarial excystment in the digestive tract of a mammal, they act only after appropriate stimulation of the worm by external cues. In *Paragonimus ohirai*, bile salts (e.g., sodium cholate) present in the mammalian intestine are sufficient to fulfill this role (Ikeda, 2003). Ionophores (molecules that transport ions across cell membranes) also stimulate excystment, whereas ion-channel antagonists as well as serine protease and cysteine protease inhibitors limit excystment (Ikeda, 2001a, 2003, 2006). Ikeda's work suggests that bile salts act by stimulating Ca^{2+} channels within the encysted worm, whereupon the worm begins active movement and releases proteases that lyse the wall of the cyst, permitting its escape. In *P. westermani*, two types of cysteine protease with masses of 27 and 28 kd, respectively, are produced in large quantities at this stage (Chung et al., 1995). Chung et al., (2005b) found the two proteases to be released from the excretory bladder of the metacercaria during excystment. Na et al. (2006) found two isoforms of the 28-kd protease to be most abundant in the anterior tegument of the newly excysted worms, including the oral sucker. Both the 27 and 28-kd enzymes are probably used by the juvenile worm to aid digestion of the fibrillar proteins of the extracellular matrix, thus facilitating movement through host tissues (Na et al., 2006). Serine proteases may also be involved in this process (Na et al., 2006). The 28-kd protease isoforms, at least, continue to be expressed during the few weeks of juvenile life, while the young worms are migrating through host tissues, and one of them is also expressed in the adult (Na et al., 2006). As the juvenile moves actively through the body on its way to the lungs, additional cysteine proteases of 15, 17, and 53 kd are expressed (Chung et al., 1997). Activities of all cysteine proteases decrease as the worm matures (Chung et al., 1997).

Immune evasion is a necessity for large, long-lived parasites such as lung flukes. An important role in immune modulation has been claimed for parasite cysteine proteases, but recent *in vitro* experimental work suggests that these molecules may not be acting alone. Eosinophils, one of the major effector cell types involved in inflammation, are short-lived, terminally differentiated granulocytes undergoing spontaneous apoptosis after a few days in the absence of antigenic stimulation (Shin et al., 2003). They are attracted to the site of injury or inflammation by chemical signals. Once in the presence of foreign antigen, they can be activated via immunoglobulins attached to, for example, parasite surfaces and release a variety of substances that attract more effector cells to the site or that aid in the lysis of invading cellular organisms. Chung et al. (1997) demonstrated that the 27- and 28-kd enzymes excreted/secreted from newly excysted meta cercariae were able to cleave human immunoglobulin G (IgG). This ability apparently diminishes during maturation of worms (Chung et al., 1997). Cleavage of IgG inhibits IgG-induced degranulation and superoxide production by eosinophils (Shin et al., 2001). In addition, work by Shin et al. (2003) and Min et al. (2004) has shown that worm ESP, in relatively high concentrations, greatly increases the rate of eosinophil apoptosis. In addition to abolishing the effector abilities of the eosinophils, the presence of dead eosinophils has the general effect of downregulation of the local inflammatory response (Min et al., 2004). Min et al. (2004) suggested that tissue-invading worms might be able to create a zone of immune privilege in their neighborhood by this means. Apparently paradoxically, low concentrations of ESP can enhance eosinophil survival. A similar effect is seen when ESP is heat-treated to inactivate cysteine proteases, or is treated with cysteine-protease inhibitors (Shin et al., 2003; Min et al., 2004).

Similar phenomena may occur in the brain in cerebral paragonimiasis. Microglia are the representatives of the immune system in the central nervous system. When activated, they release nitric oxide (NO), a useful marker for activation. Low concentrations of ESP stimulate nitric oxide (NO) production and do not injure microglia, while high concentrations lead to the death of microglia and only marginal increases in NO production (Jin et al., 2006). Heat-inactivated protease-free ESP failed to kill microglia and lacked proteolytic activity. This indicates that enzymes in the ESP had been denatured. However, this ESP was still able to induce NO production. The ESP treated with most cysteine protease inhibitors lost much of its proteolytic activity, but not its ability to kill microglia or its ability to induce NO production. Again, molecules other than cysteine proteases might be involved in immune interactions with the host. Proteoglycans, glycocalyx, and other glycoproteins might be involved in this process. Much remains to be learned about worm ESP, its constituent molecules, and their biological functions.

Mature lung flukes induce formation of a cyst that surrounds worm pairs in the lung parenchyma and peribronchiolar spaces. Cysts can be up to about 2 cm in diameter. They are produced by the host in response to the presence of eggs and worm ESP discharged into the lung tissue (Choi, 1990; Marty and Neafie, 2000). There is little recent literature on this topic.

A Spectrum of Clinical Manifestations

All manner of signs and symptoms have been associated with human paragonimiasis, adding to diagnostic difficulties for practitioners unfamiliar with the disease. Infections may yield no symptoms or nonspecific ones, or symptoms may be severe and alarming. Symptoms also vary according to the species of worm involved, stage of infection, numbers of worms invading, and, of course, individual idiosyncrasy (probably of both worms and host). Useful reviews of clinical aspects include those of Marty and Neafie (2000) and Haswell-Elkins and Levri (2003).

Early Migration and Establishing Stages

Lung-fluke metacercariae excyst in the small intestine or stomach and then penetrate the wall of the gastrointestinal tract to enter the abdominal cavity. Depending on the species of worm and of mammalian host, they may linger in the abdominal cavity or its walls for days before migrating into the pleural cavity, in some cases via the liver. If the host is not suitable for maturation, juvenile worms may remain quiescent in the musculature (reviewed in Blair et al., 1999). Symptoms (abdominal pain, fever, and diarrhea) are not often reported during the juvenile migration phase in humans, except in heavy infections (Zhong et al., 1981). Chen et al. (2001) suggested that a distinctive suite of hepatic symptoms seen in 20 patients in China, all but one of them children younger than 10 years, might be due to migration of juvenile diploid *P. westermani*. Liver involvement in children is also mentioned in the report by the World Health Organization (1995). Elevated IgE levels are usual. Eosinophilia is usual, especially in early paragonimiasis.

Established Pulmonary Infection

In a "typical" case of paragonimiasis, two maturing worms meet in the pleural cavity and move into the parenchyma of the lungs where a fibrous cyst is laid down around them in which they reside, mate, and produce eggs. Worm pairs can survive for long periods of time. The record seems to be that cited in Yokogawa et al. (1960) of a man still expectorating eggs at least 20 years after leaving an endemic area. Species of lung flukes producing typical pulmonary symptoms in humans include *P. heterotremus* and some members of the *P. westermani* complex in Asia, *P. mexicanus* in Central and South America, and *P. uterobilateralis* and *P. africanus* in Africa (Table 3.1). In North America, there have been a few human cases of autochthonous pulmonary paragonimiasis due to *P. kellicotti*, which seems to evoke particularly pronounced pulmonary and pleural symptoms (e.g., Procop et al., 2000; DeFrain and Hooker, 2002).

In the early stages of pulmonary infection, symptoms include cough, fatigue, fever, bloody sputum, loss of appetite, chest pain, and headache (Chen et al., 2001). Clinical symptoms of established pulmonary paragonimiasis may be few, or there may be cough and the production of bloody or "rusty" sputum containing parasite eggs, cyst debris, and Charcot-Leyden crystals. Pulmonary symptoms may be more severe if many worms are present. Occasionally, adult worms have been coughed up and expectorated (reviewed in Waikagul and Yoonuan, 2005).

Pleural Manifestations

Pleural lesions may or may not occur in association with pulmonary lesions. Some of the reasons for this are mentioned below. Pleural manifestations include pleural effusion, pneumothorax, and thickening of the pleura. Surgical decortication of thickened pleura has been required in some cases, and drainage of pleural effusion is commonly done (Tomita et al., 1996; DeFrain and Hooker, 2002; Castilla et al., 2003).

Views are changing on the clinical picture produced by *P. skrjabini miyazakii* in Japan. Until recently, it has been supposed that this species rarely matures in humans and causes primarily pleural lesions along with a relatively high incidence of ectopic infection in the brain, skin, peritoneal cavity, and eye (Otsuji, 2003). However, a number of cases are now known of infections producing typical pulmonary symptoms with eggs (Uchiyama et al. 1999).

Similarly, it used to be axiomatic that *P. westermani* in Japan produced only pulmonary and not pleural symptoms. This has also been found to be untrue, particularly in recent Japanese cases (Nawa, 2000; Nakamura-Uchiyama et al., 2002). In light infections, such as are probably now typical in Japan, juvenile diploid worms are likely to wander in the pleural spaces, seeking a mate, and producing pleural symptoms. Single diploid worms can produce eggs, but these are not viable (Miyazaki et al., 1981).

Ectopic Infections: Cerebral Paragonimiasis

Occasionally, adult worms of species causing pulmonary paragonimiasis (Table 3.1) will be found in sites other than the lungs, and there are suggestions that this is more likely in heavy infections (Ashitani et al., 2000; Nakamura-Uchiyama et al., 2002). In such cases, eggs and inflammatory responses may be found in unusual sites (Lee et al., 1997; Jeong et al., 1999). If adult *P. westermani* leave the lungs, they tend to migrate to the brain, where they can cause serious problems. Cerebral cases can exhibit a variety of unpleasant neurological manifestations (headache, convulsions, paralysis, behavioral change, disturbed vision etc.), which might be more pronounced during the early stages when worms are actively moving in brain tissue (Cha et al., 1994). If the patient survives this phase, the worms may become surrounded by inflammatory tissue, which eventually becomes calcified. The striking lesions produced by chronic cerebral paragonimiasis are easily seen using medical imaging methods [computed tomography (CT) scans, magnetic resonance imaging (MRI)] and are variously described as resembling soap bubbles or bunches of grapes. On surgical removal, such lesions very rarely contain worms, but often yield large numbers of eggs (Choo et al., 2003). Cho et al., (1997) estimated about 6000 cerebral paragonimiasis cases occurred in Korea in the 1960s. Nowadays, perhaps two to seven cases in the early stages and 10 to 15 cases of chronic cerebral paragonimiasis are seen annually, the latter usually in old people.

Other Ectopic Infections

Other ectopic sites include, but are not limited to, skin, liver, omentum and elsewhere in abdominal organs, gonads and genitalia, eye, and spinal cord (Choi, 1990; Blair et al., 1999; Marty and Neafie, 2000; Otsuji, 2003).

Several species have been found to produce subcutaneous nodules, usually on the torso, that may be migratory. Juvenile worms or eggs are very rarely found in such nodules on surgical investigation (but see Okamoto et al., 1993; Obara et al., 2004). Chen et al. (2001) found two cases of subcutaneous nodules in 94 patients with paragonimiasis due to diploid *P. westermani* in China. Recent Japanese examples due to *P. westermani* have been reported by Dainichi et al. (2003) and Matsumoto et al. (1998). Miyazaki and Harinasuta (1966) removed two immature specimens of *P. heterotremus* from a subcutaneous swelling in a boy in Thailand. Waikagul et al. (2003) have reviewed relevant cases from Ecuador. Migratory nodules are particularly common in cases due to members of the *P. skrjabini* complex that rarely mature in humans. Jiang et al. (2000) analyzed data for 258 cases of *P. skrjabini* infection recorded between 1986 and 2000 in Hubei Province, China. Diagnosis was based on history, clinical manifestations, and immunological tests. Nearly 56% of patients exhibited subcutaneous nodules.

Some ectopic infections, reported as due to lung flukes, are probably due to very different species of trematodes. In particular, worms of the unrelated genus *Achillurbainia* are found in cavities of the respiratory system, and adjacent tissues, of a range of mammals in Africa, South America, and Asia. Metacercariae are known from freshwater crabs (Waikagul and Yaemput, 1999). A small number of human cases are known, and in most of these, worms or (more commonly) eggs were found in a subcutaneous cyst or abscess close to the ear (Waikagul and Yaemput, 1999). Usually only eggs are found and often assumed to be those of *Paragonimus* species. The most recent report is by Schuster et al. (2007). The cyst in the Nigerian student in this case yielded eggs but no worms. The eggs could also have been those of *Paragonimus uterobilateralis*, but serology for paragonimiasis was negative. The patient admitted to enjoying raw freshwater crabs in his hometown in Nigeria.

Diagnosis and Diagnostic Confusions

Finding of lung-fluke eggs in sputum is definitive. Unfortunately, eggs are not invariably found in every sputum sample, even in active pulmonary paragonimiasis, requiring investigators to take multiple samples over time. Asor et al. (2003), mindful of this problem, determined that peak egg appearance in sputum of patients infected with *P. uterobilateralis* in Nigeria was between 5 a.m. and 9 a.m. Chen et al. (2001), in their review of 94 cases, found eggs in 24-hour sputum of only 42 of 67 patients with pulmonary paragonimiasis caused by diploid *P. westermani*, and in none of the remaining 27 cases presenting with less typical signs and symptoms. Mukae et al. (2001) found eggs in sputum of a minority of pulmonary paragonimiasis cases in Japan. Nakamura-Uchiyama et al. (2001a) found eggs in sputum of only three of 30 patients infected with *P. westermani*. Eggs are also sometimes found in pleural effusion or pleural lesions removed surgically (Castilla et al., 2003). Eggs may be found in stool samples, but this is less certain than in sputum (Cabrera, 1984). However, in the case of very old and very young patients, stool might be a better source of eggs than sputum (Cabrera, 1984).

Given that pulmonary paragonimiasis does not always yield eggs and ectopic infections rarely do, diagnosis often has to be made from history, radiography, and immunological tests. Clinicians need to suspect paragonimiasis in order to seek the appropriate historical details from a patient or to request relevant serology. Given that symptoms and radiological findings of pulmonary paragonimiasis can closely mimic those of pulmonary tuberculosis, frequent misdiagnosis is not surprising. Many recent papers presenting case reports of pulmonary paragonimiasis start by mentioning that the cases being reviewed were initially diagnosed as some other condition, most often tuberculosis (Singh et al., 1986, 2005; Belizario et al., 1997; Nagakura et al., 2002; Narain et al., 2004; Tay et al., 2005). Toscano et al. (1995) have produced the most comprehensive review of this problem to date. Im et al. (1997), Kuroki et al. (2005), and Yoo et al. (2006) have discussed diagnostic imaging of paragonimiasis and tuberculosis.

As a consequence of diagnostic confusion, large numbers of paragonimiasis patients have been needlessly treated for tuberculosis or other conditions. Indeed, the apparent failure to cure certain cases of tuberculosis in rural areas has led to a recent increase of interest in paragonimiasis. This can be illustrated by the situation in northeast India. A major effort is being made throughout India to find and treat tuberculosis patients through the Revised National Tuberculosis Control Program. An alarming rate of apparent failure of treatment for smear-negative pulmonary tuberculosis in remote villages of Arunachal Pradesh was investigated, and most refractory patients were found to have paragonimiasis, not tuberculosis (Narain et al., 2004). As a consequence, research on the prevalence and distribution of paragonimiasis in northeast India is now expanding. The ramifications of misdiagnosis can extend beyond the failure of an individual to receive appropriate treatment. Treatment at the community level can be jeopardized because others see symptomatic friends and relatives fail to benefit from the effort made, and are themselves discouraged from seeking treatment (Narain et al., 2004; Mahajan, 2005). There is also an economic penalty. In Ecuador, up to 13% of people in some tuberculosis treatment centers have been found to have paragonimiasis, the treatment of which costs 1% of that for tuberculosis (World Health Organization, 1995).

Immunological tests, if applied, almost always resolve diagnostic confusion between paragonimiasis and other conditions. Immunological methods have been used for various purposes associated with diagnosis: for population-level screening prior to more detailed examination and treatment of infected individuals; to obtain a definitive diagnosis, especially for ectopic paragonimiasis or other circumstances where parasitological methods fail (e.g., no eggs found); to differentiate between paragonimiasis and conditions that it mimics, especially pulmonary tuberculosis; to identify the actual species of *Paragonimus* infecting an individual; to follow posttreatment changes in antibodies or parasite antigen to demonstrate cure; or as an *a posteriori* confirmation of diagnosis. Material on these topics is reviewed in Blair et al., (1999) and is not repeated here. Laboratory techniques used in recent years have been the "traditional" intradermal test (IDT) and a large family of tests based on the enzyme-linked immunosorbent assay (ELISA) and immunoblotting. Other methods, such as the indirect hemagglutination test (Maleewong et al., 1998), and the indirect immunofluorescent antibody test (Wang et al., 1998), are less commonly reported. Much of this has been reviewed in Maleewong (1997) and Blair et al., (1999).

The simple IDT has been in use for many years (Yokogawa et al., 1960). A small amount of diluted worm antigen is injected into the skin. After some minutes, a positive result is indicated if a wheal significantly larger than a control is raised. The IDT has a few serious drawbacks. It can remain positive for years following cure, and it often gives a false-positive result through cross-reactions if other parasites are present in the same patient (Blair et al., 1999). On the other hand, false negatives are rare, so infected patients are unlikely to be missed. This test is still used extensively in broad-scale population screening as a first step in identifying infected individuals. For example, in a recent survey in Fujian Province, China, 650 of 9197 people tested were positive for the IDT. Of these, 60 were positive for a more definitive serological test (Cheng et al., 2005). In the Three Gorges region of China, 262 of 3325 school students gave a positive reaction, of whom 48 were subsequently diagnosed with paragonimiasis (Pan et al., 2001).

Vélez et al. (2000) found 57% of a surveyed population of Embera Indians in Colombia were positive. Other examples are in Blair et al., (1999).

The ELISA tests can be formatted in many ways. All have in common the eventual use of an antibody that will attach to appropriate antigen-antibody complexes. This antibody is conjugated to a reporter enzyme that produces a color change in a substrate, the intensity of which is proportional to the amount of antigen-antibody complex present. The various approaches tried have been in search of increased stability, sensitivity and specificity, and decreased cost. This search is not over yet. The large number of published studies using different variants of ELISA, the relatively small number of literature citations attracted by many of these, and the fact that each research group tends to use its own home-grown approach all suggest that the field is an active one.

In the ELISA plate format, wells can be variously coated with parasite antigen, with antibody, or with substances that can bind specific antigen or antibody (see Ikeda, 1998, 2001b, for the last two of these options). When antigen is used to coat wells, it may be crude or partially purified parasite antigen (Narain et al., 2005) derived from worm tissues or worm ESPs. The latter usually give better results (Maleewong, 1997). However, cross-reactions due to antibodies against other parasites can occur. This has led to experiments with a range of methods for purifying antigens, such as affinity chromatography using monoclonal antibodies (MAbs) (Maleewong et al., 1997) or expression of cloned recombinant antigens, for example, Kim et al. (2001) for yolk ferritin, and Kim et al. (2000); Ling et al. (2003); and Yun et al. (2000) for cysteine proteases. Antigens from worm ESPs are at the front line of the interaction between the parasites and their hosts. Not surprisingly, therefore, cysteine proteases from ESP are highly immunogenic and have received much attention (Lee et al., 2006).

After antigens are immobilized in wells of ELISA plates, patient body fluids containing antibodies are added. Specific antibodies, if present, bind to the immobilized antigens. Antibodies against *Paragonimus* species can be found in the patient's blood or pleural effusion, the latter, when present, often being a rich source. Antibodies can also be found in cerebrospinal fluid in the early stages of cerebral paragonimiasis, but disappear toward the chronic stage (Nakamura-Uchiyama et al., 2002).

An alternative approach is to coat wells with patient antibodies and use these to trap parasite antigens from the patient's blood (antigen-capture ELISA; Zhang et al., 1996) or fecal extracts (Maleewong et al., 1997). Maleewong et al. (1997) suspected that passage through the digestive system affected parasite antigens, resulting in rather poor results for antigen capture from fecal material.

Typically, enzyme-conjugated anti-human IgG is used as the reporter in ELISA and immunoblot tests. However, particular IgG subclasses or IgE may be better targets, giving more specific results (Guevara et al., 1995; Kong et al., 1998; Wongkham et al., 2005). IgM is generally expressed early in an infection, and then largely superseded by IgG. Anti-IgM antibody may thus be better for detecting early infection (Mukae et al., 2001; Nakamura-Uchiyama et al., 2001b).

Variants on the ELISA theme include a form of immunoblotting (Dekumyoy et al., 1995, 1998) and dot-ELISA (Itoh and Sato, 1990; Maruyama et al., 1996). Both have been promoted as providing a species-specific diagnosis. In the first method, worm antigens are separated electrophoretically, transferred to a nitrocellulose membrane, and probed with antibody from the patient and then with enzyme-conjugated anti-human IgG. Appropriate substrate is then added to permit a color-change reaction. Antibody binding with a particular antigen band has been regarded as diagnostic of infections due to *P. heterotremus*. However, the species-specific nature of this test seems to have been thrown into question by results from South American paragonimiasis cases (Waikagul et al., 2003). In the dot-ELISA method, antigens of a number of species of parasites present in Japan, including *P. westermani* and *P. skrjabini miyazakii*, are dotted onto a membrane strip and dried. Such membranes, which have a long shelf life, have been widely used in Japan. The strip is incubated with diluted serum from a patient. The remaining steps are similar to that for conventional plate-ELISA, but color-scoring is done by eye instead of using an expensive plate-reader. Despite cross-reactions, homologous sera always give stronger signals that heterologous ones. See Nakamura-Uchiyama et al. (2002) for an illustrated example. However, the dot-ELISA method can still deliver ambiguous results and has been supplemented in Japan with two further methods for establishing the exact species of parasite. One of these is the traditional Ouchterlony double-diffusion method in which antigens and antibodies, placed in separate wells, are allowed to diffuse through agarose and form complexes where they meet (Maruyama et al., 1996). See Maruyama et al. (1997) for an illustrated example using parasitic nematodes. The remaining method is inhibition-ELISA (Maruyama et al., 1996, 1997). This method requires batches of sera from a patient to be preincubated with antigens from different parasite species prior to being used in a conventional ELISA. In this case, the inhibition of binding in the conventional ELISA should be greatest when the serum used had been preincubated with antigen from the parasite species present in that patient. Related methods for reducing cross-reactions are reviewed in Blair et al. (1999).

Molecular Methods

Maleewong et al. (1997) have tried a molecular approach to diagnosis. They extracted DNA from feces of experimentally infected cats and probed this with a DNA fragment of about 1500 base pairs that is highly repeated within the *P. heterotremus* genome. This fragment (GenBank accession AZ254640) appears to be a portion of a retrotransposon (unpublished). The lower limit of detection was about 1500 to 2000 eggs of *P. heterotremus* per gram of feces. Intapan et al. (2005) extended this work by designing polymerase chain reaction (PCR) primers to amplify a portion of the DNA probe. This was more sensitive, producing an amplified band when as few as six eggs were added to 0.6 g of feces and DNA extracted. However, bands were also amplified from *P. westermani* and *P. siamensis*. Chang et al. (2000) were able to sequence selected gene regions using DNA from three to five eggs from the sputum of paragonimiasis patients in China and demonstrate unequivocally that the species involved was *P. westermani*. They were also able to sequence from eggs of *P. skrjabini* obtained from laboratory hosts. Le et al. (2006) sequenced DNA from eggs from a human case in Vietnam caused by *P. heterotremus*. DNA sequencing is not likely to become a routine aid for diagnosis in the near future but will remain a valuable research tool.

Chemotherapy

Options for drug treatment of paragonimiasis, and indeed for many other trematode infections, are now excellent (Keiser and Utzinger, 2004). Praziquantel, a spinoff from research to develop new tranquilizers (Cioli and Pica-Mattoccia, 2003) has long been the drug of choice for paragonimiasis. Given at a dose of 25 mg kg^{-1} three times daily for 2 to 3 days, high cure rates are achieved (Harinasuta and Bunnag, 1990; Keiser and Utzinger, 2004). The drug is effective against all forms of ectopic infection as well as pulmonary paragonimiasis. If there is considerable pleural effusion, this should be drained before praziquantel treatment (Nakamura-Uchiyama et al., 2002). These authors mention a patient with pleural thickening and pleural effusion that was compartmentalized by fibrous septa. Chemotherapy had no effect and surgical decortication was required. Similar cases have been reported by Obara et al. (2004), Sumitani et al. (2005), and Tomita et al. (1996). Side effects due to praziquantel are said to be frequent but mild (Cioli and Pica-Mattoccia, 2003) and may include abdominal discomfort, nausea, headache, and dizziness. Treatment during pregnancy is best avoided. Patients with cerebral paragonimiasis should be hospitalized for any drug treatment in case of adverse reactions.

Although praziquantel has long been used for treatment of trematodiases, it is ineffective against fascioliasis. Triclabendazole, however, is active against liver flukes and is now the drug of choice for treatment of fascioliasis. Its efficacy against paragonimiasis is also being actively investigated (Keiser et al., 2005). Dose rates are much lower than for praziquantel: the World Health Organization (2004) recommends two doses each of 10 mg kg^{-1} in a single day. Calvopiña et al. (2003) achieved high cure rates in Ecuadorian paragonimiasis patients with only a single dose of 10 mg kg^{-1} of triclabendazole. Calvopiña et al. (1998) also observed that triclabendazole is better tolerated than praziquantel. Patient compliance in mass treatment or in remote areas is far easier to ensure when only a single dose need be given. Gao et al. (2003) in China successfully used triclabendazole (10 mg kg^{-1} twice a day for 3 days) on five patients infected with *P. skrjabini*. This is useful to know because *P. skrjabini* does not occur in the lungs of humans, but was nonetheless susceptible to triclabendazole.

After chemotherapy, there can be a transient rise in antibody titer (Obara et al., 2004), presumably due to increased release of parasite antigen as worms die (Zhang et al., 1996). Worm death might be expected to lead to an inflammatory flare-up, but these seem rare. In a case reported by Clyti et al. (2006), a girl in

Laos with patent pulmonary paragonimiasis was treated with praziquantel at the standard dose regime. Ten days later, a large inflamed mass had appeared on her right side. Drainage of this cyst was followed by uneventful resolution without further treatment.

There is no clear indication of resistance by *Paragonimus* species to either praziquantel or triclabendazole. However, there is a strong suggestion of incipient resistance by schistosomes to praziquantel (Croft, 1997; Cioli and Pica-Mattoccia, 2003; Alonso et al., 2006). *Fasciola* species are developing resistance against triclabendazole (Fairweather, 2005). Consequently new ways of treating fascioliasis (and by extension, other trematode pathogens such as lung flukes) are being sought (Keiser and Utzinger, 2004). Derivatives of artemisinin were the subject of the paper by Keiser et al. (2006) and these show some promise.

The study of parasite invasion biology has led to a great increase in information about the molecules produced by worms in their interactions with the host. Many of these are components of the ESPs discussed earlier. Since most exert their parasite-protective effects outside the body of the parasite, they are accessible to drugs introduced into the host. Chung et al. (2005a) reviewed the possibilities for development of synthetic inhibitors of ESP cysteine proteases that could be used for chemotherapy. They pointed out that the peptide sequences of crucial portions of the commonest cysteine proteases of many helminth and protist parasites are highly conserved, suggesting that it might be possible to develop broad-spectrum inhibitors.

Rapid development of vaccines against *Paragonimus* species seems unlikely, given the slow progress toward effective vaccination against other, more important trematodes. Nevertheless, any promising results with fasciolids and schistosomes are likely to have applicability for paragonimiasis as well. Dalton et al. (2003) have reviewed the prospects for developing vaccines against parasite ESP proteins. Not surprisingly, proteases, including cysteine and aspartic proteases, remain at the top of the list of potential vaccine candidates.

Changing Epidemiology of Paragonimiasis

Numbers of People Infected

Given the different ways in which estimates of the population at risk of paragonimiasis have been reached, it is hard to gain a clear picture of recent trends. For example, a report by a World Health Organization study group published in 1995 estimated that approximately 20.7 million people are infected with the disease worldwide (20 million of these in China alone) and a further 195 million are at risk. More recently, Keiser and Utzinger (2005) estimated that 292.8 million people are at risk worldwide. They did not estimate the number infected. Taken at face value, these figures suggest that the incidence of paragonimiasis is not dropping. However, the estimates published in 1995 include only figures from China, Ecuador, Laos, Peru, and Republic of South Korea. Keiser and Utzinger (2005) have added estimates for Nigeria, Cameroon, and the Philippines to reach their totals. The 1995 study estimated that about 16% of the population of China was at risk, while Keiser and Utzinger (2005) used the figure of 15%.

In fact, prevalences in China are falling. The World Health Organization report issued in 2004 estimated fewer than 1 million cases there, a substantial downward revision from 1995. With one quarter of the world's population, China is probably the country with the greatest number of people infected and at risk. The Chinese Ministry of Health has conducted nationwide surveys of parasitic disease burden in 1990 and again in 2001–2004. Again, it is rather difficult to summarize figures across these studies. In the more recent survey (China, Coordinating Office of the National Survey on the Important Human Parasitic Diseases, 2005), sero-prevalence of paragonimiasis was 1.71% (1163/68,209 people tested in eight provinces). Taken on a province-by-province basis, paragonimiasis is certainly becoming less common. For example, Cheng et al. (2005) reported considerable declines in positive skin test prevalence in humans and in prevalences of infection in snails and crabs in Fujian Province.

Japan and South Korea are two developed economies that until recently had large numbers of paragonimiasis cases but where prevalences are now very low. Shibahara et al. (2004) noted that during the 1950s, as many as 600,000 people were infected in Japan. Following control and education efforts, there are now at most a few dozen cases annually (Maruyama et al., 1996; Kawanaka et al., 1999; World Health Organization, 2004). A very similar situation exists in Korea. Cho et al., (1997) reported that a survey in 1924 had found 7.9% of 353,729 people examined were positive for eggs in sputa. Similar high infection levels were suggested by surveys in later years up to about the 1970s (Kim, 1969; Kim and Park, 1974; Choi 1990). Recent estimates (World Health Organization, 1995, 2004) are that no more than 1000 people are infected in Korea.

In Taiwan, prevalences of paragonimiasis as high as 24% in some areas were recorded in the past (Cross, 1984). Nowadays, the disease has virtually disappeared, despite the continuing presence of the intermediate hosts (Li and Huang, 2002).

Changing Demographic and Clinical Pictures

Prevalences of paragonimiasis in some foci, where most people are rural poor, are remarkably high. In these places, the great majority of infected people are children (Table 3.2). This pattern is commonly reported in Africa, Asia, and South America, regardless of exact species of lung fluke present. Exceptions, often sex-specific, to this pattern of age-prevalence are known, but in these cases specific cultural habits might lead adults to consume crabs frequently (see below). It remains unclear whether there is any age-immunity to paragonimiasis; probably children are more likely to encounter and catch infected crustaceans while at play or helping their parents in agricultural activities.

Changes in the epidemiology of paragonimiasis in Japan and Korea may give an indication of what could happen in other countries in the future. In Japan in the 1950s and 1960s, paragonimiasis was mainly a disease of children contracted

TABLE 3.2. Summaries of age- and sex-prevalence data on paragonimiasis taken from recent literature.

Place	Overall and sex prevalence	Age distribution	Species	References
Cross R. Basin, Nigeria	12.27% (of 880 randomly selected people) had eggs in sputum samples more females infected	Peak in prevalence (24%) and egg count between ages 17 and 22 years	*P. uterobilateralis*	Arene et al., 1998
Cross R. Basin, Nigeria	9.36% (of 1100 people) had eggs in sputum samples	Prevalence was 19% in 11–16 year age group	*P. uterobilateralis*	Ibanga et al., 2003
Oban, Cross R., Nigeria	8.7% of population had eggs in spurum: more males infected than females	Prevalence was 15.7% in 11–15 year age group	*P. uterobilateralis*	Ibanga and Eyo, 2001
Cameroon	30 egg-positive cases found in village surveys: slightly more females infected than males	No patients over age of 19 years	*P. africanus*	Moyou-Sumou and Tagni-Zukam, 2003
Hubei, China	Review of 258 cases: more common in males	66.3% of the cases were children aged <12 years	*P. skrjabini*	Jiang et al., 2000
Henan, China	16.3% seroprevalence: more females infected than males	Seroprevalence was 35% in children <7 years of age	*P. skrjabini*	Wang et al.,1998
Kyushu, Japan		Anecdotal: formerly seen mostly in children, often with multifocal lesions	*P. westermani*	Nawa 2000; Nakamura-Uchiyama et al., 2002
Korea	~13% of population sputum egg-positive—thought to be an underestimate	No apparent age bias	*P. westermani*	Kim, 1969
Zhejiang, China	Review of 94 cases; more males than females infected	76.6% of patients were <20 years of age	*P. westermani* (diploid)	Chen et al., 2001
Survey in eight provinces in China	Of 68,209 tested, 1163 (1.71%) people were serologically positive. Slightly higher prevalence in females	Highest rates in children <10 years of age	Not specified	China, Coordinating Office of the National Survey on the Important Human Parasitic Diseases, 2005
Pacific coast of Colombia	57% of general population was positive for the intradermal test	75% (of 24 cases) were <14 years of age	Not identified	Vélez et al., 2000
Vietnam		86% of 102 patients were under 15 years of age	probably *P. heterotremus*	Doanh et al., 2005

through eating crabs (Uchiyama et al., 1999; Nawa, 2000). Following education programs there to warn people of the dangers of eating inadequately prepared crabs, the disease all but disappeared. However, by the early 1990s, it appeared to be undergoing resurgence. The reemerging disease differs epidemiologically from that of the 1960s in three striking ways. First, instead of primarily occurring in young people, it is now seen in middle-aged people, usually men (Nawa, 2000). Second, many of the patients denied ever eating freshwater crabs. Third, the clinical picture is rather different from that formerly described. These changes have come about largely due to an emerging mode of human infection: consumption of undercooked meat from wild boars (*Sus scrofa leucomystax*) acting as paratenic hosts for *P. westermani* (and occasionally *P. skrjabini miyazakii*; Nakamura-Uchiyama et al., 2002). Of 104 cases occurring between 1986 and 1998 reviewed by Uchiyama et al. (1999), six were due to *P. miyazakii* and the remainder due to *P. westermani*. For slightly more than half these cases, the patients admitted to having eaten raw meat of wild boars. More than 60,000 wild boars are shot annually in Japan and the meat distributed commercially (Kawanaka et al., 1999). In Kyushu, 44 of 59 (74.6%) serum samples from wild boars were positive for *Paragonimus*-specific antibody and in West Honshu, 16 of 39 sera (41.0%) were positive (Kawanaka et al., 1999).

Very recently, a case of paragonimiasis due to *P. westermani* following ingestion of bear (not boar) meat has been reported. Occasionally, nuisance bears are culled in Japan, and the meat can find its way to local restaurants, as happened in this case (Yoshikawa et al., 2005).

The changing clinical picture of paragonimiasis in Japan has been reviewed by Mukae et al. (2001). Instead of conventional pulmonary paragonimiasis involving cystic lesions in the lungs, pleural involvement is more frequently seen and pleural effusion common. Eggs are rarely found in sputum of these cases, requiring diagnosis by serology, medical imaging, and history. These features suggest infections are due to single, or small numbers of, diploid *P. westermani* that are unable to find mates and cause pleural pathology during their search for one.

In Korea, ingestion of infected crustaceans remains the usual route of infection. Mammalian paratenic hosts are not known. Jeon et al. (2005) have reviewed clinical features of recent cases of paragonimiasis in South Korea, which are in general agreement with the Japanese situation mentioned above. However, in Korean cases, lung lesions were more common than pleural lesions and eggs were more frequently found in sputum. Reasons for these differences are not hard to find. Most *P. westermani* in Korea are triploid, and individuals can induce cyst formation in the lung and start egg production in the absence of a mate.

Role of Zoonotic Hosts

In highly modified agricultural systems now lacking many of the former carnivorous hosts of *P. westermani*, it is easy to imagine humans being the most important mammalian host. In the past, this was almost certainly so in parts of Japan (Yokogawa, 1965), China, and probably Korea (Kim, 1969; Cho et al., 1997) and Taiwan. However, despite the relatively very low current human infection rates in these former hotspots, infected crustaceans can still be found (Shin and Min, 1999 in Korea; Chen et al., 2001, in China; Shibahara et al., 2004 in Japan). Domestic and feral mammals such as dogs and cats are now likely to be principal definitive hosts (Shin and Min, 1999). However, Sohn and Chai (2005) found *P. westermani* in only two of 438 feral cats sourced from many parts of South Korea. In Heilongjiang, northeast China, 7.9% of 178 farm dogs from different parts of the Province were infected with *P. westermani* (Wang et al., 2006). In areas of the Philippines where *P. westermani* in humans is known, Cabrera (1984) noted high prevalences also in field rats and suggested that these might be important zoonotic hosts. In Liberia, West Africa, Sachs and Cumberlidge (1990) found village dogs to be a good reservoir host for *P. uterobilateralis*.

Many species of *Paragonimus* do not apparently infect humans. Of those that do, some (mainly members of the *P. skrjabini* species complex) are unable to mature in humans and thus must be maintained and circulated in the environment entirely by other mammalian definitive hosts. *Paragonimus skrjabini miyazakii* in Japan matures in small mammals such as mustelids and raccoon dogs, but not often in humans. Gyoten (1994) observed no significant change in prevalence and intensity of crab infection with this species between 1979 and 1994 in Ehime Prefecture and considered that infected crabs would be available for unwary humans to eat for the foreseeable future. Conversely, Li et al. (1999) and Cheng et al. (2005) in Fujian, China, noted a substantial reduction over a 20- to 30-year period in infection rates of humans, snails, and crabs with *P. westermani*. Rodent control programs using poison had reduced the numbers of natural definitive and paratenic hosts. Environmental changes were also implicated (see below).

Control Through Education and Cultural Change

Paragonimiasis is closely associated with human cultural activities related to diet and food preparation. Consequently, epidemiologists have been keen to document the kinds of activities leading to human infection, and the ways these are changing in the modern world.

Freshwater crustaceans are regarded as tasty by most cultures around the world, and catching them can be an amusing activity, especially for children. Authors in several countries have told of watching children catching and eating crabs in the field. Wang et al. (1998) said that children in Henan, China, often eat crabs raw out of curiosity and because they think that raw crabs are more nutritious. Doanh et al., (2005) reported that children in endemic areas of Vietnam usually grill crabs for about 2 minutes before eating them. This was not enough to kill the worms; Doanh et al. (2005) found 12 live metacercariae in two grilled crabs. Cabrera (1984) described very similar behavior from the Philippines. But children in developed economies are no longer interested in such things. Cho et al., (1997) commented that rural children in Korea used to catch wild crabs and eat them. They then added, rather wistfully, "This traditional practice of rural children has also disappeared and has been replaced by candy, ice cream, television and joysticks" (Cho et al., 1997, p. 33).

Adults often like to eat crabs marinated in various sauces. In Korea, crabs were soaked in soybean sauce to make the dish called *kejang*. Metacercariae can survive over a week in this mixture (Cho et al., 1997). In parts of China, live crabs are marinated in alcoholic liquors. A bottle of "drunken crabs" (Chinese freshwater crabs steeped in liquor) was the source of *P. westermani* infection in two Chinese residents of Japan (Obara et al., 2004). In parts of South America, raw crabs are eaten with lemon juice and vegetables. See Nakamura-Uchiyama (2002) for further examples.

Many dishes require the use of crab juice, extraction of which might contaminate fingers and utensils with metacercariae (Nakamura-Uchiyama et al., 2002). In the case of cerebral paragonimiasis reported by Choo et al. (2003) in Korea, the elderly woman patient claimed never to have eaten raw crabs or crayfish, but had prepared these for others to eat. Cabrera (1984) described in some detail the preparation of crab-juice dishes in the Philippines. Raw crabs are chopped up and mashed, then juice extracted by filtration through cloth. The juice is added to grated coconut and the mixture wrapped in leaves and boiled until dry. Obviously, only the handling of fresh crab material represents a hazard in this case.

Belief in the medicinal properties of crabs and crayfish provides another avenue of infection. In Henan, China, adults sometimes eat crabs marinated in brine, vinegar, or wines, hoping to relieve back pain and other conditions. Although such marinades coagulate the crab tissues, they do not kill metacercariae in the short term (Wang et al., 1998). Also in Henan, ingestion of raw crabs is thought to aid fertility and lactation in women (Wang et al., 1998). In Korea, juice from crushed crayfish has long been regarded as effective against measles (Cho et al., 1997). Even as recently as the 1960s, many mothers gave this juice to their children (Choi, 1990). However, recent cultural change has ended that. Modern, educated mothers in a nuclear family no longer listen to the advice of their mothers-in-law! (Cho et al., 1997).

Men may place themselves at greater risk of contracting paragonimiasis through social habits. In Taiwan, where the disease is now effectively eradicated, men used to eat salted crabs to induce thirst during drinking parties (Yokogawa et al., 1960). Similarly, in parts of the Philippines, men eat crabs to accompany local wine at parties and festivals (Cabrera and Fevidal, 1974). In Colombia, members of the Embera Indian communities consider that men can become better hunters and more skilled fighters by eating raw crabs (Vélez et al., 2000). The resurgence of paragonimiasis in Japan due to consumption of raw boar meat, mostly by men, has been mentioned earlier. On questioning boar hunters, Kawanaka et al., (1999) learned that many continue in this habit even though they are aware of the risks. Of 44 boar hunters from Kyushu, 29 admitted that they had eaten raw boar meat and three had experienced an episode of lung-fluke infection.

With increasing urban affluence in China, city dwellers can treat themselves to delicacies such as freshwater crabs. This might happen in a city restaurant, such as reported by Liu et al., (2002), with the subsequent pulmonary symptoms of the participants in that feast causing diagnostic confusion for a year. It also happens that city dwellers are more mobile and able to make visits to the countryside for recreation. Cui et al. (1998) reported an outbreak of paragonimiasis in a group of urban dwellers who had collected and eaten raw crabs while on a recreational visit to mountainous areas in Henan Province. In this case, the species responsible was *P. skrjabini*. Infection was accompanied by eosinophilia and symptoms of cough, chest pain, and fever. In one of the four patients, a migratory subcutaneous nodule was present on the abdomen. Eggs and adult worms were not found.

Health education has probably played a major role in the decline of paragonimiasis in East Asia (here defined as Japan, Korea, China, and Taiwan). Unfortunately, few details of education programs have been published. In the early years of the People's Republic of China, education was the only option available for control of paragonimiasis. Some figures on the efficacy of this are in World Health Organization (1995). In one endemic area in Zhejiang Province, egg-positive sputum levels were 31.1% in 1951. This decreased to 0.05% in 1979, largely as a result of education, aided latterly by treatment with the drug bithionol. The infection rates in dogs and cats dropped from 55.5% to 4.3%, and in crabs from levels approaching 100% in some places to 7.8%. In 1975, in some parts of Kuandian County, Liaoning Province, 30% to 50% of schoolchildren gave a positive intradermal test and 10% to 20% had evidence of clinical disease. More recently, however, no active paragonimiasis cases have been found and the prevalence and intensity of infection in crayfish has dropped to low levels. Another health education program in endemic areas in Jiangxi and Anhui Provinces led to the proportion of villagers eating crabs dropping from around 50% to zero over 3 years (World Health Organization, 1995). Similarly, Wang et al. (1998) found that, in a cluster of villages in an area of Henan Province endemic for *P. skrjabini*, prevalence was very low in the one village in which a health education program had been undertaken in 1979, nearly 20 years earlier. The inhabitants of that village remained aware of the consequences of eating raw crabs.

In village schools in Korea, even at the primary level, children are taught about *P. westermani* and the disease it causes (Seo, 1984). This has been the case for decades and was reinforced by the passage of legislation, even as early as the 1920s, prohibiting the collection and transport of crabs and crayfish. The effect of health education was also apparent to Kim (1969). Several years after a large-scale survey and eradication program on Cheju Island, most of the new cases found were among people who had been elsewhere at the time of the original program (Kim, 1969).

Environmental Change

An unwelcome aspect of economic development in many countries has been environmental degradation and pollution. This has greatly impacted on the natural hosts of paragonimiasis and consequently on the incidence of disease. Gross pollution has been implicated in some cases. In Henan Province, China, Wang et al. (1998) found very low prevalences of paragonimiasis in two villages where gold mining had contaminated streams with mercury and cyanide, killing crabs. Widespread use of pesticides in both China (Yan et al., 2004) and Korea (Choi, 1990; Hong et al., 1986) has been blamed for the deaths of many aquatic organisms. In Fujian Province, China, Li et al. (1999) ascribed the extinction of one lungfluke species, *P. fukienensis*, to schistosome control programs that had eliminated its snail host. Other environmental changes, such as deforestation and the direct and indirect effects of programs to poison rodents, greatly reduced the prevalence of *P. westermani* in the province over 20 to 30 years. The snail and crab hosts of this species are found in larger streams that are typical of agricultural landscapes and hence vulnerable to environmental impacts. Interestingly, prevalences of *P. skrjabini* in crabs changed far less over a similar time scale. The aquatic hosts of this species live in tiny mountain streams, often far from population centers, and cycle through local small mammals (Li et al., 1999).

Various authors have cited construction of factories, roads and dams as having negative effects on aquatic animals (Choi, 1990; Li et al., 1999). In Taiwan, construction of a nuclear power plant was followed by the disappearance of crabs from the area (Cross, 1984). However, it is also possible that environmental change may have effects that are positive for intermediate hosts of *Paragonimus* and may enhance transmission. Zhang et al. (2002) surveyed inhabitants in Fengjie County, Chongqing Municipality, an area being impacted by the filling of the Three Gorges Reservoir. Nearly 10% of the population was seropositive for paragonimiasis, and *P. skrjabini* was present in local crabs. They predicted that habitat suitable for crabs will increase, leading to increased prevalence of paragonimiasis. This is especially dangerous given the popularity of the area as a migrant and tourist destination and the low level of local awareness of the dangers of eating crabs (Pan et al., 2001).

Systematics and Evolution

Only one new taxon of lung fluke has been proposed since the taxonomic overview by Blair et al., (1999). This was a subspecies, *P. miyazakii manipurinus*, from northeast India described by Singh et al., (1998). In addition, Vélez et al., (2003) reported the life cycle of a species from Colombia regarded as close to *P. mexicanus* but possibly new. However, in the last few years, there have been some refinements in our understanding of the identities and relationships of species. Adult worms exhibit relatively few morphological characters to justify the ~50 nominal species known (reviewed in Blair et al., 1999). These characters include the shapes and degree of branching of the testes and ovary, the arrangement of spines on the tegument, the relative sizes of the suckers and the body length/width ratio. There is a long-standing assumption that metacercarial cyst morphology is highly conserved. A number of species have been proposed solely on the basis of distinctive cyst morphology. Unfortunately, variation in metacercarial cyst morphology does occur within a species (Blair et al., 1997, 1999, 2005). Given the problem of finding morphological characters for distinguishing between species, molecular data are now being used extensively for lung flukes (reviewed in Nolan and Cribb, 2005). Molecular phylogenies produced to date indicate that *Paragonimus* species infecting humans are scattered on the tree and are not a monophyletic group (Blair et al., 1998).

Molecular studies, usually using DNA sequences, have not always simplified matters but have certainly highlighted taxonomic difficulties and taught us much about the evolution of the genus and the biology of its members. This can be illustrated using studies on two species complexes—*P. westermani* and *P. skrjabini*.

Worms referred to as *P. westermani* occur from the Indian subcontinent and Sri Lanka to the Philippines, Indonesia, and northward to China, Japan, Korea, and into parts of southeast Siberia. We have reconstructed the anatomy of a serially sectioned adult worm from a human case reported in Papua New Guinea in the 1920s (Cilento and Backhouse 1927; Heydon, 1927) that also seems to belong to this complex (unpublished). Molecular data show that the complex contains two nominal species that can be distinguished from each other by their surface spination: *P. westermani* and *P. siamensis*. Across its geographical range, adult morphology of *P. westermani* is rather uniform, but metacercarial morphology is somewhat variable (Iwagami et al., 2000, 2003). Specimens of *P. westermani* from different regions can differ in their DNA sequences as much as do other, well-accepted *Paragonimus* species (Fig. 3.1). There are also biological differences across the range. In East Asia, populations of *P. westermani* are rather uniform genetically, use snail hosts of the family Pleuroceridae and are very infective to humans, causing pulmonary paragonimiasis. Further south, in the Philippines, snail hosts belong to the Thiaridae (see Kohler and Glaubrecht, 2001, for changing views on systematics of the snail families). *Paragonimus westermani* is also very infective to humans there. Snail hosts in most other places are not known with certainty but are likely to be thiarids (pleurocerids are not present). *Paragonimus westermani* does not occur in humans anywhere except East Asia, the Philippines, and (presumably) Papua New Guinea, although it is unclear whether this is due to differing regional human dietary habits (Iwagami et al., 2003) or to biological variation between worm populations. There are also regional differences in host specificity involving mammals other than humans. Rats act as normal definitive hosts in the Philippines, but rarely do so in East Asia. In Malaysia, dogs and cats are poor hosts, whereas they are excellent hosts in East Asia (Blair et al., 1999). Little is yet known about the representatives of the species complex in India.

It remains very unclear how many "species" are included under the name *P. westermani*. There is no simple measure of molecular divergence that can be used to discriminate between species. Furthermore, in the case of *P. westermani*, the full geographical range of the taxon has not yet been systematically surveyed to estimate overall variation—molecular, biological, and morphological. If *P. westermani* is to be split into more than one species, some taxonomic and nomenclatural problems will need to be faced. The type specimens of *P. westermani* came from a Bengal tiger that died in the Amsterdam zoo in 1877 (see Blair et al., 1999). The accepted historical range of the Bengal tiger is the entire area now

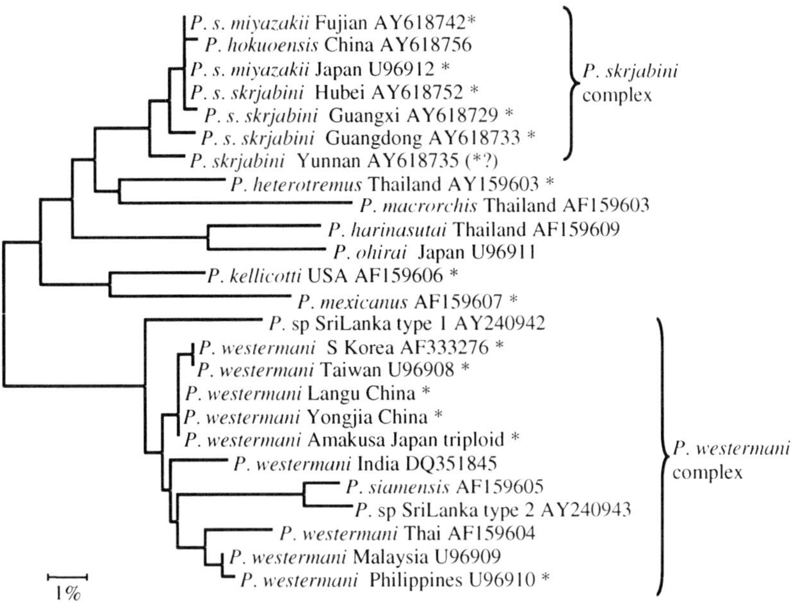

FIGURE 3.1. Phylogenetic tree (constructed using MEGA v3.1; Kumar et al., 2004). Data used were DNA sequences of the ribosomal second internal transcribed spacer (ITS2) from a number of species and populations of *Paragonimus*. GenBank accession numbers are included where available. Note that the exact topology of a phylogenetic tree can vary according to gene regions and program settings used. The purpose of this figure is only to demonstrate the degree of genetic variation within the two species complexes, *P. westermani* and *P. skrjabini*. In particular, some members of the *P. westermani* complex are as distinct from one another in their ITS2 sequences as are clear species pairs such as *P. harinasutai* and *P. ohirai*, or the two species from the Americas (*P. kellicotti* and *P. mexicanus*). Asterisks indicate a form likely to infect humans. Branch lengths on the tree are proportional to the differences between sequences as indicated by the scale bar.

encompassing India and Bangladesh (Luo et al., 2004). If *P. westermani* as currently understood were to be split into two or more species, the rules of zoological nomenclature would dictate that the name "*westermani*" should remain with the Indian form. However, the name *P. westermani* has long been used in the vast literature on lung flukes from East Asia, whereas records of the species in India are scant and human cases not confirmed. One solution would be to petition the International Commission for Zoological Nomenclature for permission to restrict the use of the name *P. westermani* to populations from East Asia, and to propose a different name for the Indian form. Another solution might be to divide the complex into subspecies, thus retaining the name *P. westermani* across the range but recognizing different subspecies in different places. This was the approach taken with the *P. skrjabini* complex (see below).

Molecular phylogenetic studies on *P. westermani* usually place the form from Sri Lanka basal within the complex. Based on this, and the fact that East Asian populations are genetically less diverse than those further south, Blair et al., (2001) proposed an evolutionary scenario, whereby populations of *P. westermani* arose first in Southeast Asia and utilized thiarid snails. Later, by range expansion and the addition of pleurocerid snail hosts, they were able to establish populations in East Asia. When this might have happened is open for speculation, but there is no reason to assume that humans aided the range expansion.

A further complication in the *P. westermani* story is the presence of diploid and triploid forms, often sympatrically, in east and northeast China, Japan, Korea, and Taiwan. A very few tetraploid individuals have also been found in northeast China. Diploid and triploid forms have sometimes been regarded as distinct species (discussed in Blair et al., 1999). Although all polyploid forms are now usually regarded as falling within *P. westermani*, the origins of this condition are still being investigated. Genetic variation between triploid individuals has been detected (van Herwerden et al., 1999; Park et al., 2003). This could indicate multiple origins of triploidy, or variation could have arisen by mutations in parthenogenetic lines descended from a single ancestral triploid. There has been much discussion as to whether polyploids have arisen through hybridization with another taxon of lung flukes (allopolyploidy) or within a single population (autopolyploidy). All genetic studies on triploids have shown them to be very similar to diploids in East Asia, and especially northeast China, but distinctly different from diploid members of the *P. westermani* complex in other parts of Asia (Agatsuma and Hirai, 2005). This implies autopolyploidy. Very recent cytogenetic studies by Agatsuma and Hirai (2005) on worms from northeast China have revealed an extraordinary situation. Diploids, triploids, and a single tetraploid were found in their sample of 144 worms raised to adulthood in experimental hosts. One out of 84 diploids produced eggs and sperm ameiotically; sperm produced were probably nonfunctional, but eggs were diploid. All remaining diploids produced haploid gametes by normal meiosis. A few triploid individuals appeared to produce apparently functional sperm by meiosis. Eggs produced meiotically by triploids had variable numbers of chromosomes; those produced ameiotically were triploid. The tetraploid worm mostly produced tetraploid eggs and probably diploid sperm. Worms of different ploidies frequently shared the same cyst in the lungs of a mammal (Agatsuma and Hirai, 2005), making exchange of sperm between them possible. It is easy to envisage zygotes of various ploidies arising from time to time through fusion of gametes from such parents (Agatsuma and Hirai, 2005).

Molecular studies have also revealed a complex of forms in what has been known as *P. skrjabini*. Confirmed members of this complex occur from western Japan, through east, south, and southwest China (Yang et al., 2000; Blair et al., 2005). The complex probably extends into Thailand and India. Snail hosts are tiny species of pomatiopsids and amnicolids found in small, unpolluted mountain streams. Most members of this complex do not mature in humans but rather in a wide range of small mammals such as mustelids and viverrids (Table 3.1).

The most interesting finding was that nominal *P. skrjabini* from Fujian Province in eastern China was closer genetically (mitochondrial sequences) and morphologically to *P. miyazakii* from Japan than it was to *P. skrjabini* populations from elsewhere in China. Based on this, Blair et al. (2005) proposed that *P. miyazakii* should be reduced to subspecific status within *P. skrjabini*. It remains to be seen whether this proposal will be generally accepted.

Conclusion

Paragonimiasis is probably in decline in many parts of the world. However, it is clearly a disease that will occur in human populations for the foreseeable future. Through movements of people and fresh foods, it will increasingly be seen in nonendemic areas where it will continue to cause diagnostic confusion. Lung flukes, like everything else in the biota, are the products of evolution and therefore complex, variable, and challenging to study. Nevertheless, they offer us model systems suitable for analysis of trematode biochemistry and physiology, as well as host–parasite interactions. Their natural history and taxonomy are still far from being fully understood. Most of the references cited in this chapter have appeared since the review by Blair et al., (1999), and we predict that *Paragonimus* species will continue to feature prominently in the primary parasitology literature.

References

Agatsuma, T., and Hirai, H. (2005). Genetic variation of *Paragonimus* species. In: Arizono, N., Chai, J.Y., Nawa Y., and Takahashi, Y., eds. Asian Parasitology, Vol. 1: Food-Borne Helminthiasis in Asia (pp. 155–165). Japan: FAP Journal.

Alonso, D., Munoz, J., Gascon, J., Valls, M.E., and Corachan, M. (2006). Short report: failure of standard treatment with praziquantel in two returned travelers with *Schistosoma haematobium* infection. Am J Trop Med Hyg 74, 342–344.

Arene, F.O.I., Ibanga, E., and Asor, J.E. (1998). Epidemiology of paragonimiasis in Cross River Basin, Nigeria—prevalence and intensity of infection due to *Paragonimus uterobilateralis* in Yakurr Local Government Area. Public Health 112, 119–122.

Ashitani, J., Kumamoto, K., and Matsukura, S. (2000). Paragonimiasis westermani with multifocal lesions in lungs and skin. Intern Med 39, 433–436.

Asor, J.E., Ibanga, E.S., and Arene, F.O.I. (2003). *Paragonimus uterobilateralis*: peak period of egg output in sputum of infected subjects in Cross River Basin, Nigeria. Mary Slessor J Med 3, 24–27.

Belizario, V., Guan, M., Borja, L., Ortega, A.R., and Tiri, R. (1997). Pulmonary paragonimiasis in non-responding tuberculosis patients in Irosin, Sorsogon. Philippines J Microbiol Infect Dis 26, 13–15.

Blair, D., Agatsuma, T., and Watanobe, T. (1997). Molecular evidence for the synonymy of three species of *Paragonimus, P. ohirai* Miyazaki, 1939, *P. iloktsuenensis* Chen, 1940 and *P. sadoensis* Miyazaki et al., 1968. J Helminthol 71, 305–310.

Blair, D., Waikagul, J., Honzako, Y., and Agatsuma, T. (1998). Phylogenetic relationships among the Thai species of *Paragonimus* inferred from DNA sequences. In: Tada, I.,

Kojima, S., and Tsuji, M., eds. Proceedings of the 9th International Congress of Parasitology, Chiba, Japan, 24–28 August 1998 (pp. 643–647). Bologna: Monduzzi International.

Blair, D., Xu, Z.B., and Agatsuma, T. (1999). Paragonimiasis and the genus *Paragonimus*. Adv Parasitol 42, 113–222.

Blair, D., Davis, G.M., and Wu, B. (2001). Evolutionary relationships between trematodes and snails emphasizing schistosomes and paragonimids. Parasitol 123, S229–S243.

Blair, D., Chang, Z.S., Chen, M.G., Cui, A.L., Wu, B., Agatsuma, T., Iwagami, M., Corlis, D., Fu, C.B., and Zhan, X.M. (2005). *Paragonimus skrjabini* Chen, 1959 (Digenea: Paragonimidae) and related species in eastern Asia: a combined molecular and morphological approach to identification and taxonomy. Syst Parasitol 60, 1–21.

Cabrera, B.D. (1984). Paragonimiasis in the Philippines: current status. Arzneimittelforschung 34, 1188–1192.

Cabrera, B.D., and Fevidal, P.M. (1974). Studies on *Paragonimus* and paragonimiasis in the Philippines. III. Prevalence and treatment of human paragonimiasis with bithionol in Jaro, Leyte, Philippines. Southeast Asian J Trop Med Public Health 5, 39–45.

Calvopiña, M., Guderian, R.H., Paredes, W., Chico, M., and Cooper, P.J. (1998). Treatment of human pulmonary paragonimiasis with triclabendazole—clinical tolerance and drug efficacy. Trans R Soc Trop Med Hyg 92, 566–569.

Calvopiña, M., Guderian, R.H., Paredes, W., and Cooper, P.J. (2003) Comparison of two single-day regimens of triclabendazole for the treatment of human pulmonary paragonimiasis. Trans R Soc Trop Med Hyg 97, 451–454.

Castilla, E.A., Jessen, R., Sheck, D.N., and Procop, G.W. (2003). Cavitary mass lesion and recurrent pneumothoraces due to *Paragonimus kellicotti* infection—North American paragonimiasis. Am J Surg Pathol 27, 1157–1160.

Cha, S.H., Chang, K.H., Cho, S.Y., Han, M.H., Kong, Y., Suh, D.C., Choi, C.G., Kang, H.K., and Kim, M.S. (1994). Cerebral paragonimiasis in early active stage: CT and MR features. AJR 162, 141–145.

Chang, Z.S., Wu, B., Blair, D., Zhang, Y., Hu, L., Chen, S.H., Chen, M., Feng, Z., and Davis, G. (2000). Gene sequences for identification of *Paragonimus* eggs from a human case. Chinese J Parasitol Parasitic Dis 18, 213–215.

Chen, M.G., Chang, Z.S., Shao, X.Y., Liu, M.D., Blair, D., Chen, S.H., Zhang, Y.N., Hong, J.L., Shen, B.G., and Feng, Z. (2001). Paragonimiasis in Yongjia County, Zhejiang Province, China: clinical, parasitological and karyotypic studies on *Paragonimus westermani*. Southeast Asian J Trop Med Public Health 32, 760–769.

Cheng, Y.Z., Xu, L.S., Chen, B.J., Li, L.S., Zhang, R.Y., Lin, C.X., Lin, J.X., Li, Y.S., Li, Y.R., Fang, Y.Y., Lin, K.Q., and Zheng, G.B. (2005). Survey on the current status of important human parasitic infections in Fujian Province. Chin J Parasitol Parasitic Dis 23, 283–287 (in Chinese, English abstract).

China, Coordinating Office of the National Survey on the Important Human Parasitic Diseases (2005). A national survey on current status of the important parasitic diseases in human population. Chin J Parasitol Parasitic Dis 23, 332–340.

Cho, S.Y., Kong, Y., and Kang, S.Y. (1997). Epidemiology of paragonimiasis in Korea. Southeast Asian J Trop Med Public Health 28, 32–36.

Choi, D.W. (1990). *Paragonimus* and paragonimiasis in Korea. Korean J Parasitol 28 (suppl), 79–102.

Choi, J.H., Lee, J.H., Yu, H.S., Jeong, H.J., Kim, J., Hong, Y.C., Kong, H.H., and Chung, D.I. (2006). Molecular and biochemical characterization of hemoglobinase, a cysteine proteinase, in *Paragonimus westermani*. Korean J Parasitol 44, 187–196.

Choo, J.D., Suh, B.S., Lee, H.S., Lee, J.S., Song, C.J., Shin, D.W., and Lee, Y.H. (2003). Chronic cerebral paragonimiasis combined with aneurysmal subarachnoid hemorrhage. Am J Trop Med Hyg 69, 466–469.

Chung, Y.B., Kong, Y., Joo, I.J., Cho, S.Y., and Kang, S.Y. (1995). Excystment of *Paragonimus westermani* metacercariae by endogenous cysteine protease. J Parasitol 81, 137–142.

Chung, Y.B., Kong, Y., Yang, H.J., Kang, S.Y., and Cho, S.Y. (1997). cysteine protease activities during maturation stages of *Paragonimus westermani*. J Parasitol 83, 902–907.

Chung, J.Y., Bae, Y.A., Na, B.K., and Kong, Y. (2005a). Cysteine protease inhibitors as potential antiparasitic agents. Expert Opin Ther Patents 15, 995–1007.

Chung, Y.B., Kim, T.S., and Yang, H.J. (2005b). Early cysteine protease activity in excretory bladder triggers metacercaria excystment of *Paragonimus westermani*. J Parasitol 91, 953–954.

Cilento, R.W., and Backhouse, T.C. (1927). Paragonimiasis. Its first recorded occurrence in the territory of New Guinea. Med J Aus 14, 79–81.

Cioli, D., and Pica-Mattoccia, L. (2003). Praziquantel. Parasitol Res 90, S3–S9.

Clyti, E., Khéosang, P., Huerre, M., Sayasone, S., Martinez-Aussel, B., and Strobel, M. (2006). Cutaneous paragonimiasis with flare up after treatment: a clinical case from Laos. Int J Dermatol 45, 1110–1112.

Croft, S.L. (1997). The current status of antiparasite chemotherapy. Parasitology 114, S3-S15.

Cross, J.H. (1984). Changing patterns of some trematode infections in Asia. Arzneimittelforschung 34, 1224–1226.

Cui, J., Wang, Z.Q., Wu, F., and Jin, X.X. (1998). An outbreak of paragonimiasis in Zhengzhou city, China. Acta Trop 70, 211–216.

Dainichi, T., Nakahara, T., Moroi, Y., Urabe, K., Koga, T., Tanaka, M., Nawa, Y., and Furue, M. (2003). A case of cutaneous paragonimiasis with pleural effusion. Int J Dermatol 42, 699–702.

Dalton, J.P., Brindley, P.J., Knox, D.P., Brady, C.P., Hotez, P.J., Donnelly, S., O'Neill, S.M., Mulcahy, G., and Loukas, A. (2003). Helminth vaccines: from mining genomic information for vaccine targets to systems used for protein expression. Int J Parasitol 33, 621–640.

De, N.V., Murrell, K.D., Cong, L.D., Cam, P.D., Chau, L.V., Toan, N.D., and Dalsgaard, A. (2003). The food-borne trematode zoonoses of Vietnam. Southeast Asian J Trop Med Public Health 34 (suppl 1), 12–34.

DeFrain, M., and Hooker, R. (2002). North American paragonimiasis—case report of a severe clinical infection. Chest 121, 1368–1372.

Dekumyoy, P., Setasuban, P., Waikagul, J., Yaemput, S., and Sanguankiat, S. (1995). Human lung fluke *Paragonimus heterotremus*: differentiation of antigen proteins of adult worms by enzyme-linked immunoeletrotransfer blot technique. Southeast Asian J Trop Med Public Health 26, 434–438.

Dekumyoy, P., Waikagul, J., and Eom, K.S. (1998). Human lung fluke *Paragonimus heterotremus*: differential diagnosis between *Paragonimus heterotremus* and *Paragonimus westermani* infections by EITB. Trop Med Int Health 3, 52–56.

de Leon, W.U., and Piad, J.N.C. (2005). Paragonimiasis in the Philippines. In Arizono, N., Chai, J.Y., Nawa, Y., and Takahashi, Y., eds. Asian Parasitology Vol. 1: Food-Borne Helminthiasis in Asia (pp. 133–137). Japan: FAP Journal.

Doanh, P.N., Le, N.T., and The, D.T. (2005). *Paragonimus* and paragonimiasis in Vietnam. In Arizono, N., Chai, J.Y., Nawa, Y., and Takahashi, Y., eds. Asian Parasitology; Vol 1: Food-Borne Helminthiasis in Asia (pp. 149–153). Japan: FAP Journal.

Fairweather, I. (2005). Triclabendazole: new skills to unravel an old(ish) enigma. J Helminthol 79, 227–234.

Gao, J., Liu, Y., Wang, X., and Hu, P. (2003). Triclabendazole in the treatment of Paragonimiasis skrjabini. Chin J 116, 1683–1686.

Grove, D.I. (1990). A History of Human Helminthology. Wallingford, UK: CAB International.

Guevara, A., Vieira, J.C., Araujo, N., Calvopiña, M., Guderian, R.H., and Carlier, Y. (1995). Antibody isotypes, including igg subclasses, in Ecuadorian patients with pulmonary paragonimiasis. Memórias Inst Oswaldo Cruz 90, 497–502.

Gyoten, J. (1994). Changes of living habits related to the infection in human paragonimiasis in Japan, based on the epidemiological data. Jpn J Parasitol 43, 462–470.

Harinasuta, T., and Bunnag, D. (1990). Liver, lung and intestinal trematodiasis. In: Warren, K.S., and Mahmoud, A.A.F., eds. Tropical and Geographical Medicine 2nd ed (pp. 473–489). New York: McGraw-Hill.

Haswell-Elkins, M.R., and Levri, E. (2003). Food-borne trematodes. In: Cook, G.C., and Zumla, A.I., eds. Manson's Tropical Diseases, 21st ed (pp. 1473–1486). London: Saunders.

Heydon, G.M. (1927). *Paragonimus ringeri* in a New Guinea native. Med J Aus 14, 204–205.

Hong, Y.A., Joo, C.Y., and Pyun, Y.S. (1986). Infestation status of *Paragonimus westermani* metacercariae in the second intermediate host in Ulchin County, Kyungpook Province. Korean J Parasitol 24, 194–200.

Ibanga, E.S., Arene, F.O.I., and Asor, J.E. (2003). Association of pulmonary Paragonimiasis with active pulmonary tuberculosis in rural Yakurr community in Cross River Basin, Nigeria. Mary Slessor J Med 3, 19–23.

Ibanga, E.S., and Eyo, V.M. (2001). Pulmonary paragonimiasis in Oban community in Akamkpa Local Government Area, Cross River State, Nigeria: prevalence and intensity of infection. Trans R Soc Trop Med Hyg 95, 159–160.

Ikeda, T. (1998). Cystatin capture enzyme-linked immunosorbent assay for immunodiagnosis of human paragonimiasis and fascioliasis. Am J Trop Med Hyg 59, 286–290.

Ikeda, T. (2001a). Effect of ionophores on in vitro excystment of *Paragonimus ohirai* metacercariae. Parasitol Res 87, 343–344.

Ikeda, T. (2001b). Protein A immunocapture assay detecting antibodies to fluke cysteine proteinases for immunodiagnosis of human paragonimiasis and fascioliasis. J Helminthol 75, 245–249.

Ikeda, T. (2003). Involvement of cysteine proteinases in excystment of *Paragonimus ohirai* metacercariae induced by sodium cholate and A23187. J Helminthol 77, 21–26.

Ikeda, T. (2006). Effects of L-type Ca2+ channel antagonists in in vitro excystment of *Paragonimus ohirai* metacercariae induced by sodium cholate. Parasitol Res 99, 336–340.

Im, J.G., Chang, K.H., and Reeder, M.M. (1997). Current diagnostic imaging of pulmonary and cerebral paragonimiasis, with pathological correlation. Semin Roentgenol 32, 301–324.

Intapan, P.M., Wongkham, C., Imtawil, K.J., Pumidonming, W., Prasongdee, T.K., Miwa, M., and Maleewong, W. (2005). Detection of *Paragonimus heterotremus* eggs in experimentally infected cats by a polymerase chain reaction-based method. J Parasitol 91, 195–198.

Itoh, M., and Sato, S. (1990). Multi-dot enzyme-linked immunosorbent assay for serodiagnosis of trematodiasis. Southeast Asian J Trop Med Public Health 21, 471–474.

Iwagami, M., Ho, L.Y., Su, K., Lai, P.F., Fukushima, M., Nakano, M., Blair, D., Kawashima, K., and Agatsuma, T. (2000). Molecular phylogeographic studies on *Paragonimus westermani* in Asia. J Helminthol 74, 315–322.

Iwagami, M., Rajapakse, R.P.V.J., Paranagama, W., and Agatsuma, T. (2003). Identities of two *Paragonimus* species from Sri Lanka inferred from molecular sequences. J Helminthol 77, 239–245.

Jeon, K., Koh, W.J., Kim, H., Kwon, O.J., Kim, T.S., Lee, K.S., and Han, J. (2005). Clinical features of recently diagnosed pulmonary paragonimiasis in Korea. Chest 128, 1423–1430.

Jeong, M.G., Yu, J.S., Kim, K.W., Kim, J.K., Kim, S.J., Kim, H.J., and Choi, Y.D. (1999). Retroperitoneal paragonimiasis: a case of ectopic paragonimiasis presenting as periureteral masses. J Comput Assist Tomogr 23, 696–698.

Jiang, X., Fan, Z., Wang, X., and Du, Y. (2000). Clinical analysis of 258 cases with *Pagumogonimus skrjabini* in west of Hubei Province. Endemic Dis Bull 15, 88–90 (in Chinese, English abstract).

Jin, Y.N., Lee, J.C., Choi, I.Y., Kim, E.A., Shin, M.H., and Kim, W.K. (2006). Excretory-secretory products produced by *Paragonimus westermani* differentially regulate the nitric oxide production and viability of microglial cells. Int Arch Allergy Immunol 139, 16–24.

Kawanaka, M., Sugiyama, H., and Kato, K. (1999). Paragonimiasis acquired by eating boar meat: current status in Japan. Jpn J Infect Dis 52, 49.

Keiser, J., and Utzinger, J. (2004). Chemotherapy for major food-borne trematodes: a review. Expert Opin Pharmacother 5, 1711–1726.

Keiser, J., and Utzinger, J. (2005). Emerging foodborne trematodiasis. Emerg Infect Dis 11, 1507–1514.

Keiser, J., Engels, D., Buscher, G., and Utzinger, J. (2005). Triclabendazole for the treatment of fascioliasis and paragonimiasis. Expert Opin Invest Drugs 14, 1513–1526.

Keiser, J., Xiao, S.H., Tanner, M., and Utzinger, J. (2006). Artesunate and artemether are effective fasciolicides in the rat model and in vitro. J Antimicrobial Chemother 57, 1139–1145.

Kim, J.S. (1969). Mass chemotherapy in the control of paragonimiasis. Korean J Parasitol 7, 6–14.

Kim, J.S., and Park, S.S. (1974). Values of intradermal test and multiple sputum examinations for diagnosis of paragonimiasis. Korean J Parasitol 12, 119–125.

Kim, T.Y., Joo, I.J., Kang, S.Y., Cho, S.Y., Kong, Y., Gan, X.X., Sukomtason, K., Sukomtason, K., and Hong, S.J. (2001). Recombinant *Paragonimus westermani* yolk ferritin is a useful serodiagnostic antigen. J Infect Dis 185, 1373–1375.

Kim, T.S., Na, B.K., Park, P.H., Song, K.J., Hontzeas, N., and Song, C.Y. (2000). Cloning and expression of a cysteine proteinase gene from *Paragonimus westermani* adult worms. J Parasitol 86, 333–339.

Kohler, F., and Glaubrecht, M. (2001). Toward a systematic revision of the Southeast Asian freshwater gastropod *Brotia* H. Adams, 1866 (Cerithioidea: Pachychilidae): an account of species from around the South China Sea. J Molluscan Studies 67, 281–318.

Kong, Y., Ito, A., Yang, H.J., Chung, Y.B., Kasuya, S., Kobayashi, M., Liu, Y.H., and Cho, S.Y. (1998). Immunoglobulin G (IgG) subclass and IgE responses in human paragonimiases caused by three different species. Clin Diagn Lab Immunol 5, 474–478.

Kumar, S., Tamura, K., and Nei, M. (2004). MEGA3: integrated software for Molecular Evolutionary Genetics Analysis and sequence alignment. Briefings Bioinformatics 5, 150–163.

Kuroki, M., Hatabu, H., Nakata, H., Hashiguchi, N., Shimizu, T., Uchino, N., and Tamura, S. (2005). High-resolution computed tomography findings of *P. westermani*. J Thoracic Imag 20, 210–213.

Le, T.H., De, N.V., Blair, D., McManus, D.P., Kino, H., and Agatsuma, T. (2006). *Paragonimus heterotremus* Chen and Hsia (1964), in Vietnam: a molecular identification and relationships of isolates from different hosts and geographical origins. Acta Trop 98, 25–33.

Lee, S.C., Jwo, S.C., Hwang, K.P., Lee, N., and Shieh, W.B. (1997). Discovery of encysted *Paragonimus westermani* eggs in the omentum of an asymptomatic elderly woman. Am J Trop Med Hyg 57, 615–618.

Lee, E.G., Na, B.K., Bae, Y.A., Kim, S.H., Je, E.Y., Ju, J.W., Cho, S.H., Kim, T.S., Kang, S.Y., Cho, S.Y., and Kong, Y. (2006). Identification of immunodominant excretory-secretory cysteine proteases of adult *Paragonimus westermani* by proteome analysis. Proteomics 6, 1290–1300.

Li, Y.S., Lin, J.X., Cheng, Y.Z., and Lin, C.X. (1999). Study on reasons and changes of infection rate in 10 epidemic areas of paragonimiasis in Fujian Province. Chin J Parasitic Dis Contr 12, 275–277.

Li, Y., and Huang, Y. (2002). A survey of paragonimiasis in Taiwan. J Trop Med (Guangzhou), 2, 105–107 (in Chinese, English abstract).

Ling, J., Hou, M., Liu, J., Zhang, Z., and Zhang, Y. (2003). Cloning and identification of the gene fragments of *Paragonimus westermani*. Chin J Parasitol Parasitic Dis 21, 144–146 (in Chinese, English abstract).

Liu, S., Wang, Y., and Cai, S. (2002). A collective infection of *Paragonimus westermani* caused by entertaining at a seafood banquet. Chin J Parasitol Parasitic Dis 20, 360 (in Chinese, English abstract).

Luo, S.J., Kim, J.H., Johnson, W.E., van der Welt, J., Martenson, J., Yuhki, N., Miquelle, D.G., Uphyrkina, O., Goodrich, J.M., Quigley, H.B., Tilson, R., Brady, G., Martelli, P., Subramaniam, V., McDougal, C., Hean, S., Huang, S.Q., Pan, W.S., Karanth, U.K., Sunquist, M., Smith, J.L.D., and O'Brien, S.J. (2004). Phylogeography and genetic ancestry of tigers (*Panthera tigris*). PLOS Biol 2, 2275–2293.

Mahajan, R.C. (2005). Paragonimiasis: an emerging public health problem in India. Indian J Med Res 121, 716–718.

Maleewong, W. (1997). Recent advances in diagnosis of paragonimiasis. Southeast Asian J Trop Med Public Health 28 (suppl 1), 134–138.

Maleewong, W., Intapan, P.M., Wongkham, C., Wongratanacheewin, S., Tapchaisri, P., Morakote, N., and Chaicumpa, W. (1997). Detection of *Paragonimus heterotremus* in experimentally infected cat feces by antigen capture-ELISA and by DNA hybridization. J Parasitol 83, 1075–1078.

Maleewong, W., Intapan, P.M., Wongkham, C., Pajongthanasaris, M.P., Morakote, N., and Chaicumpa, W. (1998). Indirect haemagglutination test using monoclonal antibody-affinity purified antigens for diagnosis of human paragonimiasis due to *Paragonimus heterotremus*. Trop Med Int Health 3, 57–59.

Marty, A.M., and Neafie, R.C. (2000). Paragonimiasis. In: Meyers, W.M., Neafie, R.C., Marty, A.M., and Wear, D.J., eds. Pathology of Infectious Diseases: Vol. I—Helminthiases (pp. 49–67). Washington, DC: Armed Forces Institute of Pathology and American Registry of Pathology.

Maruyama, H., Noda, S., and Nawa, Y. (1996). Emerging problems of parasitic diseases in southern Kyushu, Japan. Jpn J Parasitol 45, 192–200.

Maruyama, H., Noda, S., Choi, W.Y., Ohta, N., and Nawa, Y. (1997). Fine binding specificities to *Ascaris suum* and *Ascaris lumbricoides* antigens of the sera from patients of probable visceral larval migrans due to *Ascaris suum*. Parasitol Int 46, 181–188.

Matsumoto, T., Kimura, S., Yamauchi, M., Nawa, Y., and Miike, T. (1998). Soluble CD23 and IL-5 levels in the serum and culture supernatants of peripheral blood mononuclear cells in a girl with cutaneous paragonimiasis: case report. Ann Trop Paediatr 18, 49–53.

Min, D.Y., Lee, Y.A., Ryu, J.S., Ahn, M.H., Chung, Y.B., Sim, S., and Shin, M.H. (2004). Caspase-3-mediated apoptosis of human eosinophils by the tissue-invading helminth *Paragonimus westermani*. Int Arch Allergy Immunol 133, 357–364.

Miyazaki, I. (1991). An Illustrated Book of Helminthic Zoonoses. Tokyo: International Medical Foundation of Japan.

Miyazaki, I., and Harinasuta, T. (1966). The first case of human paragonimiasis caused by *Paragonimus heterotremus* Chen et Hsia, 1964. Ann Trop Med Parasitol 60, 509–514.

Miyazaki, I., Terasaki, K., and Habe, S. (1981). Comparison of single worm infection to cats between *Paragonimus westermani* and *P. pulmonalis*. Med Bull Fukuoka Univ 8, 405–416.

Moyou-Somo, R., and Tagni-Zukam, D. (2003). La paragonimose au Cameroun: tableaux radio-cliniques et evolution sous traitement. Med Trop 63, 163–167 (in French, English abstract).

Mukae, H., Taniguchi, H., Matsumoto, N., Iiboshi, H., Ashitani, J.I., Matsukura, S., and Nawa, Y. (2001). Clinicoradiologic features of pleuropulmonary *Paragonimus westermani* on Kyusyu Island, Japan. Chest 120, 514–520.

Na, B.K., Kim, S.H., Lee, E.G., Kim, T.S., Bae, Y.A., Kang, I., Yu, J.R., Sohn, W.M., Cho, S.Y., and Kong, Y. (2006). Critical roles for excretory-secretory cysteine proteases during tissue invasion of *Paragonimus westermani* newly excysted metacercariae. Cell Microbiol 8, 1034–1046.

Nagakura, K., Oouchi, M., Abe, K., and Araki, K. (2002). Pulmonary paragonimiasis misdiagnosed as tuberculosis: with special references on paragonimiasis. Tokai J Exp Clin Med 27, 97–100.

Nakamura-Uchiyama, F., Onah, D.N., and Nawa, Y. (2001a). Clinical features of paragonimiasis cases recently found in Japan: parasite-specific immunoglobulin M and G antibody classes. Clin Infect Dis 32, e171-e175.

Nakamura-Uchiyama, F., Onah, D.N., and Nawa, Y. (2001b). Clinical features and parasite-specific IgM/IgG antibodies of paragonimiasis patients recently found in Japan. Southeast Asian J Trop Med Public Health 32, 55–58.

Nakamura-Uchiyama, F., Mukae, H., and Nawa, Y. (2002). Paragonimiasis: a Japanese perspective. Clin Chest Med 23, 409–420.

Narain, K., Devi, K.R., and Mahanta, J. (2004). Pulmonary paragonimiasis and smear-negative pulmonary tuberculosis: a diagnostic dilemma. Int J Tuberculosis Lung Dis 8, 621–622.

Narain, K., Devi, K.R., and Mahanta, J. (2005). Development of enzyme-linked immunosorbent assay for serodiagnosis of human paragonimiasis. Indian J Med Res 121, 739–746.

Nawa, Y. (2000). Re-emergence of paragonimiasis. Intern Med 39, 353–354.

Nishida, H., and Shibahara, T. (2003). Epidemiology of paragonimiasis. Prog Med Parasitol Japan 8, 201–217.

Nolan, M.J., and Cribb, T.H. (2005). The use and implications of ribosomal DNA sequencing for the discrimination of digenean species. Adv Parasitol 60, 101–163.

Obara, A., Nakamura-Uchiyama, F., Hiromatsu, K., and Nawa, Y. (2004). Paragonimiasis cases recently found among immigrants in Japan. Intern Med (Tokyo) 43, 388–392.

Okamoto, M., Miyake, Y., Shouji, S.I., Fujikawa, T., Tamaki, A., Takeda, Z., and Nawa, Y. (1993). A case of severe paragonimiasis miyazakii with lung and skin lesions showing massive egg production in sputum and faeces. Jpn J Parasitol 42, 429–433.

Otsuji, Y. (2003). Paragonimiasis. Prog Med Parasitol Japan 8, 183–200.

Pan, H.M., Chen, L.Z., Wang, C.Q., Yi, Z.D., Lu, D.X., Zhang, X.H., Qin, S., and Zhou, X.C. (2001). Case reports and epidemiologic investigation of paragonimiasis in the area of Three Gorges. Chin J Parasitic Dis Control 14, IV-V. (In Chinese)

Park, G.M., Im, K.I., and Yong, T.S. (2003). Phylogenetic relationship of ribosomal ITS2 and mitochondrial COI among diploid and triploid *Paragonimus westermani* isolates. Korean J Parasitol 41, 47–55.

Procop, G.W., Marty, A.M., Scheck, D.N., Mease, D.R., and Maw, G.M. (2000). North American paragonimiasis—a case report. Acta Cytol 44, 75–80.

Sachs, R., and Cumberlidge, N. (1990). The dog as a natural reservoir host for *Paragonimus uterobilateralis* in Liberia, West Africa. Ann Trop Med Parasitol 84, 101–102.

Sajid, M., and McKerrow, J.H. (2002). Cysteine proteases of parasitic organisms. Mol Biochem Parasitol 120, 1–21.

Schuster, H., Agada, F.O., Anderson, A.R., Jackson, R.S., Blair, D., McGann, H., and Kelly, G. (2007). Otitis media and a neck lump—current diagnostic challenges for *Paragonimus*-like trematode infections. J Infect 54, e103–e106.

Seo, B.S. (1984). Socio-economic and cultural aspects of human trematode infections in Korea. Arzneimittel forschung 34, 1116–1118.

Shibahara, T., Nishida, H., Aohagi, Y., Iwaki, T., Tsuboi, T., Torii, M., Kubo, N., Kashiwagi, A., Fukudome, H., Okamoto, M., and Yamamoto, H. (2004). Geographical distribution of the lung fluke, *Paragonimus westermani* -diploid type (Trematoda, Troglotrematidae) in the northern part of Okayama Prefecture, Japan. Biogeography 6, 87–93.

Shin, M., and Min, D. (1999). Infection status of *Paragonimus westermani* metacercariae in crayfish (*Cambaroides similis*) collected from Bogildo (Islet), Wando-gun, Chollanam-do, Korea. Korean J Parasitol 37, 55–57.

Shin, M.H., Kita, H., Park, H.Y., and Seoh, J.Y. (2001). Cysteine protease secreted by *Paragonimus westemani* attenuates effector functions of human eosinophils stimulated with immunoglobulin G. Infect Immun 69, 1599–1604.

Shin, M.H., Seoh, J.Y., Park, H.Y., and Kita, H. (2003). Excretory-secretory products secreted by *Paragonimus westermani* delay the spontaneous cell death of human eosinophils through autocrine production of GM-CSF. Int Arch Allergy Immunol 132, 48–57.

Singh, T.S., Mutum, S.S., and Razaque, M.A. (1986). Pulmonary paragonimiasis: clinical features, diagnosis and treatment of 39 cases in Manipur. Trans R Soc Trop Med Hyg 80, 967–971.

Singh, T.S., Singh, Y.I., and Singh, K.N. (1998) a case of paragonimiasis in a civet cat with a new *Paragonimus* sub species in Manipur, India. Ind J Pathol Microbiol 41, 351–353.

Singh, T.N., Kananbala, S., and Devi, K.S. (2005). Pleuropulmonary paragonimiasis mimicking pulmonary tuberculosis—a report of three cases. Ind J Med Microbiol 23, 131–134.

Sohn, W.M., and Chai, J.Y. (2005). Infection status with helminthes in feral cats purchased from a market in Busan, Republic of Korea. Korean J Parasitol 43, 93–100.

Sumitani, M., Mikawa, T., Miki, Y., Nisida, K., Tochino, Y., Kamimori, T., Fujiwara, H., Fujikawa, T., and Nakamura, F. (2005). [A case of chronic pleuritis by *Paragonimus westermani* infection resistant to standard chemotherapy and cured by three additional cycles of chemotherapy]. Nihon Kokyuki Gakkai Zasshi 43, 427–431 (in Japanese).

Tay, N.S.W.T., Ong, K.C., Tan, S.Y., and Kaw, G.J.L. (2005). Tuberculosis mimicry. Eur Respir J 26, 554–556.

Tomita, M., Ichinari, H., Matsuzaki, Y., Shibata, K., Koga, Y., Yamaguchi, R., Mukai, H., Masumoto, H., Matsukura, S., Maruyama, H., and Nawa, Y. (1996). A case of chronic pleural empyema by *Paragonimus westermani* infection resistant to chemotherapy and cured by surgical decortication. Jpn J Parasitol 45, 242–246.

Tort, J., Brindley, P.J., Knox, D., Wolfe, K.H., and Dalton, J.P. (1999). Proteinases and associated genes of parasitic helminths. Ad Parasitol 43, 161–266.

Toscano, C., Yu, S.H., Nunn, P., and Mott, K.E. (1995). Paragonimiasis and tuberculosis, diagnostic confusion: a review of the literature. Trop Dis Bull 92, R1-R26.

Uchiyama, F., Morimoto, Y., and Nawa, Y. (1999). Re-emergence of paragonimiasis in Kyushu, Japan. Southeast Asian J Trop Med Public Health 30, 686–691.

Van Herwerden, L., Blair, D., and Agatsuma, T. (1999). Genetic diversity in parthenogenetic triploid *Paragonimus westermani*. Int J Parasitol 29, 1477–1482.

Vélez, I., Velásquez, L.E., and Vélez, I.D. (2003). Morphological description and life cycle of *Paragonimus* sp (Trematoda: Troglotrematidae): causal agent of human paragonimiasis in Colombia. J Parasitol 89, 749–755.

Vélez, I.D., Ortega, J., Hurtado, M.I., Salazar, A.L., Robledo, S.M., Jimenez, J.N., and Velásquez, L.E. (2000). Epidemiology of paragonimiasis in Colombia. Trans R Soc Trop Med Hyg 94, 661–663.

Waikagul, J., and Yaemput, S. (1999). Two *Achillurbainia* Metacercariae from Mountain Crabs in Central Thailand. J Trop Med Parasitol 22, 1–6.

Waikagul, J., and Yoonuan, T. (2005). *Paragonimus* and paragonimiasis in Thailand. In Arizono, N., Chai, J.Y., Nawa, Y., and Takahashi, Y., eds. Asian Parasitology; Vol 1: Food-Borne Helminthiasis in Asia (pp. 139–148). Japan: FAP Journal.

Waikagul, J., Lazo, R., Cornejo, E., Lazo, M., Llaguno, M., Barrezueta, J., Llaguno, M., Botta, M., Yoonuan, T., and Akahane, H. (2003). *Paragonimus* infection in Pedernales, Ecuador. Bull Central Res Inst Fukuoka Univ Series E Interdisc Sci 1, 259–273.

Wang, Z.Q., Cui, J., Mao, F.R., Ling, X.M., Jin, X.X., and Wu, F. (1998). Epidemiological study of paragonimosis in central China. Helminthologia 35, 117–121.

Wang, C.R., Qiu, J.H., Zhao, J.P., Xu, L.M., Yu, W.C., and Zhu, X.Q. (2006). Prevalence of helminthes in adult dogs in Heilongjiang Province, the People's Republic of China. Parasitol Res 99, 627–630.

Wongkham, C., Intapan, P.M., Maleewong, W., and Miwa, M. (2005). Evaluation of human IgG subclass antibodies in the serodiagnosis of paragonimiasis heterotremus. Asian Pacific J Allergy Immunol 23, 205–211.

World Health Organization. (1995). Control of Foodborne Trematode Infections. (Technical Report Series No. 849). Geneva: World Health Organization.

World Health Organization. (2004). Report on Joint WHO/FAO Workshop on Foodborne Trematode Infections in Asia. (Held in Hanoi, Viet Nam from 26–28 November 2002.) Report Series Number: RS/2002/GE/40(VTN). Manila: World Health Organization.

Yan, Q., Yan, T., Zhou, X., Li, Y., Zhu, C., Shi, L., Ma, X., and Hu, N. (2004). Epidemiological survey on the infection of *Paragonimus westermani* in Jiangxi Province. Chin J Parasitol Parasitic Dis 22, 250–252 (in Chinese, English abstract).

Yang, J.S., Chen, M.G., Feng, Z., and Blair, D. (2000). *Paragonimus* and paragonimiasis in China: a review of the literature. Chin J Parasitol Parasitic Dis Special Issue, 1–78.

Yokogawa, M. (1965). *Paragonimus* and paragonimiasis. Adv Parasitol 3, 99–158.

Yokogawa, S., Cort, W., and Yokogawa, M. (1960). *Paragonimus* and paragonimiasis. Exp Parasitol 10, 81–137,139–205.

Yoo, I.R., Park, H.J., O, J.H., Chung, Y.A., Sohn, H.S., Chung, S.K., and Kim, S.H. (2006). Two cases of pulmonary paragonimiasis on FDG-PET CT imaging. Ann Nucl Med 20, 311–315.

Yoshikawa, M., Ouji, Y., Moriya, K., Saito, K., Ishizaka, S., Sakurai, M., Morii, T., Fukuoka, A., Hamada, K., Kimura, H., Takamure, M., Tonomura, Y., Yakura, H., Kataoka, H., Ueno, S., Nakamura-Uchiyama, F., Hiromatsu, K., and Nawa, Y. (2005) Immunopositivity for *Trichinella spiralis* presented in a case of *Paragonimiasis westermani*—suspected infection by ingestion of raw bear meat. J Nara Med Assoc 56, 253–259 (in Japanese, English abstract).

Yun, D.H., Chung, J.Y., Chung, Y.B., Bahk, Y.Y., Kang, S.Y., Kong, Y., and Cho, S.Y. (2000). Structural and immunological characteristics of a 28-kilodalton cruzipain-like cysteine protease of *Paragonimus westermani* expressed in the definitive host stage. Clin Diagn Lab Immunol 7, 932–939.

Zhang, X.L., Wang, Y., Huang, Y., Chen, G.G., Huang, Z.G., Huang, Y., Li, J., Qin, Q., Xiao, J., and Chen.C. (2002). Effects of variation in ecological environment of the Three Gorges reservoir areas on the epidemic of paragonimiasis. Chin J Parasitic Dis Control 15, 30–32. (in Chinese, English abstract).

Zhang, Z., Zhang, Y., Liu, L., Dong, C., Zhang, Y., Wu, Z., Piessens, W.F., Zhang, Z.H., Zhang, Y.J., Liu, L.T., Dong, C.X., Zhang, Y.Q., and Wu, Z.Y. (1996). Antigen detection assay to monitor the efficacy of praziquantel for treatment of *Paragonimus westermani* infections. Trans R Soc Trop Med Hyg 90, 43.

Zhong, H.L., He, L.Y., Xu, Z.B., and Cao, W.J. (1981). Recent progress in studies of *Paragonimus* and paragonimiasis control in China. Chin Med J 94, 483–494.

4
Diphyllobothriasis: The *Diphyllobothrium latum* Human Infection Conundrum and Reconciliation with a Worldwide Zoonosis

Terry A. Dick

Tapeworms belonging to the genus *Diphyllobothrium* (Cestoda: Pseudophyllidea) are some of the most conspicuous and best known human parasites since antiquity. One can only imagine what the ancients thought when a meter-long piece of strobila of *Diphyllobothrium* spp. appeared hanging from the anus of a human or a dog (Fig. 4.1A) or heavily parasitized char or whitefish were observed with large cysts along the viscera, too numerous to count (Fig. 4.1B). Not surprising it was named the broad tapeworm of humans. For years most human cases were thought to result from the human form only, that is, *D. latum*, and improvements in sanitation and changes in eating habits would result in its eradication. Even though this view persists to some extent today, the evidence is accumulating that human infections occur from a variety of unrelated sources and that this group of parasites persists as a natural sylvatic cycle in many parts of the world.

A resurgence of interest in the fish tapeworm is tied to the recent increase in cases reported in humans from large urban centers through the consumption of raw fish, the expanding production of aquacultured fish and movement of these fish to distance markets, the implications of climate change on aquatic disease transmission, and observations that circumpolar sylvatic cycles of *Diphyllobothrium* spp. are as robust as ever.

This chapter discusses the historical record, the biology, the species reported to infect humans, problems with taxonomy, biogeography, epidemiology and ecology, world distribution, diagnosis and control, and our gaps in knowledge.

Historical

The fishing-foraging cultures of ancient peoples ensured that larval tapeworms encysted in the flesh of fish would find their way into the intestine of humans. Archaeological evidence indicates that eggs of *Diphyllobothrium* spp. have been preserved in middens for a long time. The first archaeological discovery of

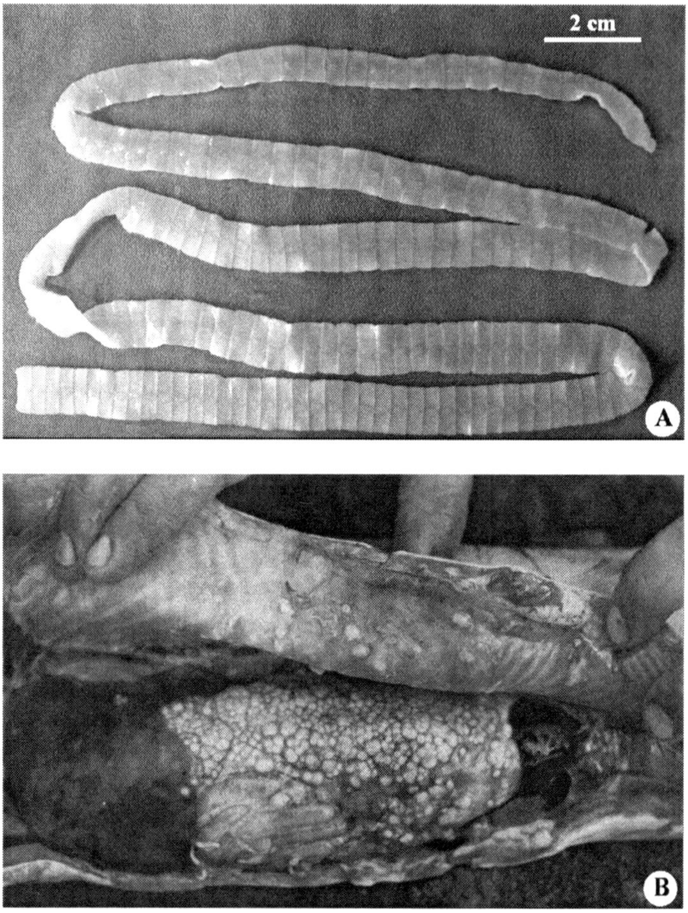

FIGURE 4.1. (A) Segment of strobila from a human experimental infection with *Diphyllobothrium* sp. plerocercoids from a pike from eastern Manitoba, Canada. (B) *D. dendriticum* and *D. ditremum* plerocercoids in a piscivorous Arctic char from a freshwater lake, Baffin Island, Canada. Note plerocercoid in the viscera, liver, attached to the wall of the body cavity and occasionally penetrating the hypaxial muscle.

Diphyllobothrium spp. was in Prussia and was dated to the fifth century A.D. (Szidat, 1944). Since then, eggs of *Diphyllobothrium* spp. have been found in numerous archaeological sites in both the Old and New World pre- and post-Columbian (Goncalves et al., 2003), from the Old World in France and Germany by Jansen and Over (1962) and Hermann (1985), in Switzerland and Israel by Mitchell and Stern (2000) and Goncalves et al. (2003), and from the New World in Chile by Ferriera et al. (1984), Reinhard and Aufderheide (1990), in Peru by Holiday et al. (2003), in the United States by Bouchet et al. (1999, 2001), and in Canada by Bathurst (2005). The records vary, with the oldest reported from Peru around 8000 B.C. by Reinhard and Barnum (1991), followed by the discovery of eggs in 5500-year-old midden sediments along the Pacific coast of British Columbia (Bathurst, 2005) and to Neolithic sites in Switzerland dating from 2900 to 3900 B.C. (Le Bailly et al., 2005). Clearly the genus *Diphyllobothrium* was widespread several thousand years ago and still is today (see Geographic Distribution and Epidemiology, below).

Taxonomy

It is not the intention here to review the taxonomy of the diphyllobothrids but rather to indicate why some key problems arose. Foremost among these problems is the continued use of *D. latum* to identify most human infections when there is strong evidence that this group is a species-complex worldwide zoonosis, with wild mammals, bird, and fishes as the key hosts. The first description of *D. latum* was written by Dunus and Wolpius in 1592 (cited by Guttowa and Moskaw, 2005), based on a specimen without a scolex, and discovery of the microscopic procercoid by Janicki and Rosenin 1917 (cited by Guttowa and Moskaw, 2005) was a key finding in the understanding of transmission. The review of the diphyllobothrids by Stunkard (1965) raised some interesting points: (1) the systematics of the group was in a state of confusion, (2) diphyllobothrids from mammals and birds do not seem to be especially host specific, (3) "the strobila and scolex manifest striking variations," and (4) plerocercoids present in salmon were acquired in freshwater and could be the source of infection to whales and seals (Vik, 1964). The problems with morphological variation led Andersen (1975) and Andersen and Gibson (1989) to show that plerocercoids could be identified using, among other tools, scanning electron microscopy. Based on viewpoints like those of Vik (1964) (item 4 above) and the general opinion that most human infections were *D. latum*, researchers focused on humans as the main definitive host and freshwater as the primary source of infection. Nevertheless, researchers began to note differences in diphyllobothrid biology, morphological differences among adult worms, and the location of plerocercoids in the fish host, and had difficulty conjuring up reasons to explain why humans were responsible for transferring *D. latum* in order to account for every new case of human diphyllobothriasis. Eventually, cases of diphyllobothriasis started to be attributed to other species. *Diphyllobothrium latum* was still the most frequently identified human parasite (Table 4.1) but other species were being diagnosed from human infections. The most commonly reported secondary species was *D. dendriticum* (Freeman and Jamieson, 1976) and other species of importance to humans included *D. ursi*, *D. dalhae*, *D. nihonkaiense*, *D. pacificum*, and *D. klebanovski*. The situation became more confusing as *D. latum* was still being diagnosed in areas where the main source of infection was marine salmonids and where there was no history of freshwater diphyllobothriasis. The application of molecular probes allowed a standard and reproducible approach to identification and was used successfully by Isobe et al. (1998) in Japan, Nicoulaud et al. (2005) in France, and Yera et al. (2006) to identify *D. nihonkaiense* from a human in France infected by eating Pacific salmon from Canada

TABLE 4.1. Species of *Diphyllobothrium* spp. of world wide importance to humans (from Dick et al., 2001).

Species	Distribution	Fish hosts	Definitive host
D. latum	Holarctic	Pike, some percids	Humans, dogs, bears
D. ditremum	Holarctic	Salmonids	Fish-eating birds
D. dendriticum	Holarctic	Salmonids	Fish-eating birds
D. dalliae	Alaska	*Dallia pectoralis*	Gull, dog
D. ursi[*]	Alaska	Salmonines	Bears
? *D. alascense*	Alaska	Salmonines	Dogs (atypical)
D. vogeli	Eurasia	Salmonids	Fish-eating birds
? *D. klebanovski*	Amur R.	Salmonines	Humans (atypical?)
D. nihonkaiense	Japan	Salmonines	Humans (atypical?)
D. pacificum	South America	Sea fish	fur seals and sea lions

[*]Considered a synonym of *D. dendriticum* by some (see Hoffman, 1999) but not by others (Rausch and Adams, 2000, and Ching, 1984).
[?] Likely or possibly a marine species. Modified from Dick et al. (2001).

Biology

Since humans appear to be a good host for most *Diphyllobothrium* spp. research in the past, and even today, has focused on *D. latum*. *Diphyllobothrium* spp. have three important assets for a parasite: (1) high biotic potential, with up to a million eggs produced by a single adult worm; (2) longevity, with the infective plerocercoid stage in fish flesh and the long life of the adult in the intestine of the definitive host; and (3) the ability of the plerocercoid to be laterally transferred from one fish host to another through piscivory. These attributes ensure a complex life cycle involving three trophic levels: an intermediate copepod host; a second intermediate fish host, often a paratenic fish; and finally a mammal or bird definitive host (Fig. 4.2).

Humans become infected when they consume uncooked filets or raw fish dishes, and the parasite varies in size, in the intestine of the definitive host, from 3 to 12 m (Tanowitz, 2001). The fact that this parasite has been reported since antiquity, present in both the Old and New Worlds, and found in fish that frequent both freshwater and marine environments indicates it has a highly successful transmission mechanism. Adult worms found in the ileum and jejunum of humans consist of 3000 to 4000 proglottids. Pieces of strobila and the operculated eggs are passed in faeces. If the eggs enter water, they require 10 to 14 days to develop, and when they hatch, a ciliated hexacanth embryo (coracidium) is released. This free living stage must be ingested within 12 hours by a suitable intermediate crustacean host such as *Cyclops* or *Diaptomus*. The ingested coracidium penetrates the gut, enters the hemocoel, and differentiates into a ~ 0.5-mm procercoid in 10 to 21 days. When the copepod is ingested by a suitable second intermediate fish host, the procercoid penetrates the gut wall and enters the body cavity where it encysts *(D. dendriticum)* or it enters the fish

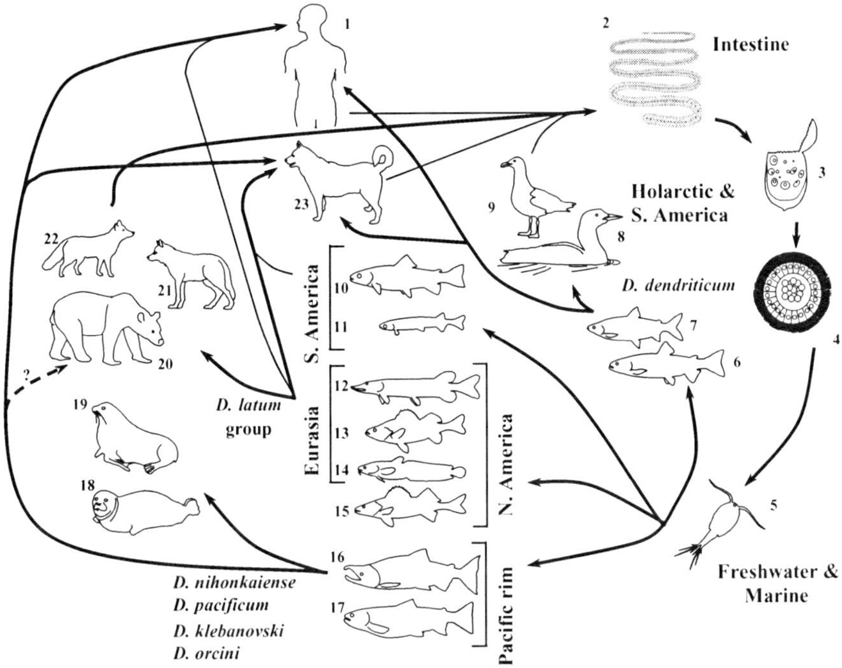

FIGURE 4.2. The life cycles of *Diphyllobothrium* species known to infect humans. 1, human; 2, mature *Diphyllobothrium* spp.; 3, eggs; 4, coracidium; 5, copepod harbors the procercoid. Fish host harboring the plerocercoids; 6, Arctic char (*Salvelinus alpinus*); 7, *Coregonus* spp., 10, Rainbow trout (*Oncorhynchus mykiss*); 11, Inanga (*Galaxius maculatus*); 12, Northern pike (*Esox lucius*); 13, Yellow perch (*Perca flavescens* in North America and *P. fluviatilis* in Eurasia); 14, Burbot (*Lota lota*); 15, Sauger (*Sander canadensis*); 16, Sockeye salmon (*Oncorhynchus nerka*); 17, Cherry salmon (*Oncorhynchus masou masou*). Birds hosts harboring adult worms: 8, Arctic loon (*Gavia arctica*); 9, Herring gull (*Larus argentatus*). Mammal hosts harboring adult worms: 18, seal; 19, sea lion; 20, bear; 21, wolf; 22, fox; 23, dog.

muscle *(D. latum)* and differentiates into a plerocercoid (Fig. 4.2). If a second intermediate host such as a large piscivorous pike *(Esox lucius)* eats a first intermediate host such as yellow perch *(Perca flavescens)*, the plerocercoid reinvades the muscle of the pike. If a bird, wild mammal, or human consumes raw fish, the plerocercoid enters the small intestine and matures in 5 to 6 weeks.

There are two important freshwater life cycles for *Diphyllobothrium* spp.: (1) human to human via fish, and (2) a sylvatic life cycle involving wild animals and fish where humans acquire the parasite through consumption of infected fish (Fig. 4.2). With the exception of the definitive hosts and different intermediate fish hosts, the cycle is quite similar. The source of infection to humans differs between North America and Eurasia. In North America pike *(E.lucius)*, walleye *(Sander vitreus)*, sauger *(S. canadensis)*, and yellow perch *(P flavescens)* and char

(Salvelinus alpinus) are the main hosts that were recognized early on by Wardle (1935) and later *D. dendriticum* in cisco *(Coregonus artedi)* and lake whitefish *(C. clupeaformis)* by Watson and Dick (1979, 1980), Dick and Poole (1985), DeVos and Dick (1989), Szalai et al. (1992), and Dick et al. (2001). McDonald and Margolis (1995) list 17 fish species as hosts of *D. dendriticum* and five as hosts of *D. latum*. Lake trout *(Salvelinus namaycush)* has also been identified as a host for a diphyllobthrid plerocercoids encysted along the viscera and it was suggested that it might be a new species (Freeman and Thompson, 1969). Considering the location of plerocercoids in the viscera and not the flesh, it was most likely *D. dentriticum* or *D. ditremum*, since both species were common in other fish species from the same lakes. Wardle (1935) listed the following fish species as hosts of *Diphyllobothrium* spp. in Europe; pike *(E. lucius)*, perch *(Perca fluviatilis)*, burbot *(Lota lota)*, trout *(Trutta vulgaris)*, lake trout *(Trutta lacustris)*, and grayling *(Thymallus vulgaris)*. Powell and Chubb (1966) discussed the importance of *Diphyllobothrium* plerocercoids in *S. alpinus* and *Salmo trutta* in the United Kingdom, and Dupouy-Camet and Peduzzi (2004) added big whitefish *(Coregonus fera)* and char *(S. alpinus)* to the European list of fish hosts.

The importance of burbot in the transmission of diphyllobothrids in North America is less clear, although, according to Nicholson (1932), Vergeer (1928) infected dogs and cats and Nickolson stated that Vergeer "demonstrated that burbot may contain the larva of *D. latum* and another species of bothriocephalid." The marine plercoercoids in the flesh of salmon are frequently diagnosed as *D. latum* but evidence is accumulating that these are usually distinct from the freshwater *D. latum* (see discussions below on the Pacific rim countries).

Geographic Distribution and Epidemiology

It has been proposed that there are 9 million cases of human diphyllobothriasis worldwide (von Bonsdorff, 1977) largely because many infections are undiagnosed and because in most cases the symptoms of infection are relatively benign and nonspecific. More recently the number of cases worldwide has been estimated to be as high as 20 million (Muller, 2001; Chai et al., 2005)

Diphyllobothrium spp. is a world-class parasite with a wide distribution and well-established endemic populations in North America and Eurasia and new hot spots emerging in South America. It was and continues to be a circumpolar and boreal parasite of wild birds and mammals. New human cases were reported in Europe (Kyronseppa, 1993; Peduzzi and Boucher-Rodoni, 2001; Dupouy-Camet and Peduzzi, 2004; Waloch, 2005), and the increase in cases probably relates to a more thorough screening of patients than new foci of infections, but it does suggest that the parasite has persisted from antiquity in Europe. There is fairly compelling evidence that *D. latum* infections in humans continue to decline in Scadinavian countries such as Finland, but it is being reported more frequently from South America and other parts of the world (Table 4.2).

TABLE 4.2. Records of *D. latum* in humans in North America and South America (from Dick et al., 2001).

Year	Location	Host	Reference
1858	United States	European	Ward (1930)
1879.	North America	Swedish immigrant	von Bonsdorff (1977)
1896	United States	American born	von Bonsdorff (1977), Ward (1930)
1901.	Montreal, Canada	French Canadian	Hamilton in Cushing and Bacal (1934)
1928	Winnipeg, Canada	Residents (15 cases)	Nicholson (1932)
1932	New York City	3 cases	Waters and O'Connor (1932)
1932	Oklahoma	Immigrant Finn	Canavan (1932)
1932	New York City	Jewish residents (21)	Plotz (1932)
1936	Ely, Minnesota	Residents	Thompson (1936)
1937	Syracuse, NY	1 human	Mueller (1937)
1939	Indiana	1 native resident	Headlee et al. (1939)
1943	Florida	3 children (+ family dog)	Summers and Weinstein (1943)
1947	U.S.	11 cases	Sandweiss and Sugarman (1947)
1947	Canada	95 cases to date	Sandweiss and Sugarman (1947)
1947	U.S.	309 cases to date	Sandweiss and Sugarman (1947)
1950–1953	New York City	13 cases	Rosenberg et al. (1955)
1950	Chile	1 case	Neghme et al. (1950a)
1950	Chile	12 cases to date	Neghme et al. (1950b)
1951	Chile	22 cases to date	Neghme and Bertin (1951a,b)
1949–1970	Alaska	Native cases	Rausch and Hilliard (1970)
1961–1971	Chile	0.3% of 51,010	Reyes et al. (1972)
1973	Quebec, Canada	2 of 500 Chinese	Seah (1973)
1974	Ontario, Canada	9 native cases	Turgeon (1974)
1974	Louisiana	4 children	Christian and Perret (1974)
1979	Canada	Human faecal samples	Croll and Gyorkos (1979)
1932	Yellowstone, National Park	Black bears	Rush (1932)
1977–1981	United States	100–200 cases/year	Deardorff and Overstreet (1991)

Human Cases of D. Latum in North America

In the endemic areas of North America, the past few decades have seen very few published reports of *D. latum* in humans in North America (Kingston and Kilbourn, 1989). An analysis of 414,820 fecal samples examined by state and territorial laboratories in the United States found only 25 positive cases of *D. latum* (Ruebush et al., 1976). Some of the earlier reports of *D. latum* in North America were very likely confused *with D. dendriticum*, and this is especially true for higher latitudes and Inuit populations (von Bonsdorff, 1977), where the primary fish eaten is Arctic char (*S. alpinus*) (see Bylund, 2003). The reports of *D. latum* infecting humans from Manitoba, Ontario, and northern Minnesota as well as the reports from Alaska (Rausch and Hilliard, 1970) are probably correct, but the origin of the infection as a human source versus a zoonosis from the boreal region is still controversial. Since human diphyllobothriasis does not appear to be a mandatory reportable disease in Canada or the U.S., it is not clear if there has been a decline in reporting or if in fact human infections have actually declined. There appear to be fewer infections of *D. latum* in humans (Table 4.3), perhaps as result of increased public awareness, public-health monitoring, as well as modern sanitation and hygiene. Research by Wardle and coworkers, (1935) in Manitoba, Michigan, and Minnesota, and by Rausch and Hilliard (1970) in Alaska was followed by a decline in human reports of *D. latum* in the 1960s and 1970s (von Bonsdorff, 1977). There are some interesting reports outside the endemic area. A 1% prevalence of *D. latum* in children in Baton Rouge, Louisiana, is based on 452 stool samples (Christian and Perret, 1974) and from humans (mainly children) in Florida (Summers and Weinstein, 1943). *Diphyllobothrium latum* was acquired by humans in Hawaii from eating raw fish in Alaska (Ho et al., 1979), and from Cuba (Bouza-Suarez et al., 1990).

South America

Diphyollobothriais has an interesting history in South America, with some of the oldest records from archaeological digs in Peru, reports from humans and dogs in Chile as early as 1919 (von Bonsdorff, 1977), and a recent proliferation of human infections from Brazil. Early reports of *D. latum* in Chile were from dogs and humans located in the southern part of the country in several lakes (Colico, Villarria, Panguipulli, Rinihue, and Ranco. The infective plerocercoids were reported from introduced rainbow trout (*Oncorhynchus mykiss*) and brown trout (*Salmo trutta*) (Neghme et al., 1950a,b; Neghme and Bertin,

TABLE 4.3. Reports and publications on human infections of *D. latum* and other diphyllobothrids in chronological order (1975–2006); additions to original table from Dick et al. (2001).

1976: *D. latum* reviewed in Finland. Prevalence dropped in the last 20 years (1956–76) from 20% to 2% over the whole country, and from 100% to 10% in eastern areas.

1977: Eastern Taimyr, Khatangskii region (Russia). Human infections: *D. latum* dominates in upper reaches of the Khatanga (where people eat *Lota lota* and *E. lucius*) while *D. dendriticum* dominates the lower Khatanga region where people eat mainly coregonids.

1975: Perm region, Russia. *D. latum* mainly in housewives (get infected by tasting minced raw fish while preparing fish balls, etc.).

1979: Hawaii. Man infected from eating raw fish while in Alaska.

1979: Okayama Prefecture, Japan; five cases in humans.

1979: Chiba Prefecture, Japan. Human infections.

1979: Kyubishev Reservoir, Russia. Focus of *D. latum*.

1980: Pakistan, Karachi. Human infection.

1982: Novosibirsk region (Russia, Siberia), *D. latum* in humans.

1983: Southern Sukhon River (Russia, SFSR), *D. latum* focus.

1983: Southern Chile. New cases of *D. latum* in humans.

1984: Finland. *D. latum* in 1–4% of the population. Some geographical variability in prevalence.

1984: Experimental infection of Japanese men with *D. latum* of Finnish origin. Pathogenic effects due to the parasitism.

1985: Amur River basin (eastern Siberia). *D. latum*: 4.2% among natives, 0.34% among newcomers (settlers). Fish hosts: *Oncorhynchus keta* and *O. gorbuscha*. This species later became *D. klebanovskii* (see 1988 below).
1985: European part of Russian SFSR. In humans, pike, perch, and ruffe (*Acerina cernua* or *Gymnocephalus cernua*, a perch-like percid).
1986: Kremenchug Reservoir, Russia. Focus of *D. latum*. Fish hosts: *E. lucius* and *P. fluviatilis*. Human hosts not mentioned.
1986: Vanzetur settlement. Lower Ob' river. Siberia. Khanty, Mansi, and Komi aboriginals. *D. latum* in 34.3% of the northern natives, 20.2% of the Russian residents (settlers).
1986: Sao Paulo, Brazil. Eggs identified as *D. latum* in sand on beaches.
1986: Korea, Seoul. *D. latum* eggs in human stool samples.
1987: Japan. 3 cases of *D. latum* treated between 1970–1974 at the Department of Medical Zoology, Prefectural University of Medicine.
1987: Iraq. First record of *D. latum* in fishes in Iraq.
1988: Amur River, Siberia. Lower reaches of the river. *D. latum* described as a new species, *D. klebanovskii*, with *Oncorhynchus* as the fish intermediate host.
1988: Okhostsk Sea, coastal areas. *D. latum* in humans (1%).
1988: Sverdlovsk region, Russia. *D. latum*: human infections.
1989: Korea: seven cases of *D. latum* in humans, from eating marine fish.
1990: Cuba. First case of *D. latum* in human (proglottides recovered).
1991: Jordan. *D. latum* in dog feces.
1992: Czech Republic. Czech man returned from 6-month stay in Finland infected with *D. latum*.
1993: Japan. *D. pacificum* one human case.
1993: Finland. Low levels of *D. latum* still detected in humans.
1994: Korea. Claimed to be "first record" in humans in Korea (but see 1986).
1995: Switzerland, *D. latum*, 73 human cases.
1996: Pakistan. Faislabad. *D. latum* in dogs.
1996: Upper Pur River basin, Russia. (Yamal-Neto Autonomous district). *D. latum* in humans.
1997: Korea. 5 cases of *D. latum* in humans from eating raw *Liza haematocheilus*.
1999: Japan. D. nihonkaiense. One human case from eating raw filet of *Oncorhynchus keta*
1999: Poland. *D. latum*, two human cases.
2000: Japan. *D. nihkonkaiense*, five cases in humans eating *Onchorhynchus masou ishikawae*.
2000: Japan. *D. orinci* (marine diphyllobothrid); one case, human from eating raw fish
2001: Korea. *D. latum*? Two human cases from eating raw cherry salmon, *Oncorhynchus masou*.
2001: Chile. *D. pacificum*. New human cases between 1975–2000 correlated with El Nino.
2001: Switzerland, 1982 to 1999, *D. latum*, 31 human cases.
2001: Argentina. *D. latum*, four human cases speculated to be from salmonids?
2002: Malaysia. Diphyllobothrium sp. Human from eating fish roe, liver, or flesh.
2002: Japan. *D. latum*, 86-year-old man.
2002: Spain. Three-year-old child.
2003: Japan. *D. nihkonkaiense*. Human case from eating *Onchorhynchus masou*.
2003: Poland. *D. latum*, one human case.
1996–2003: Japan. Diphyllobothriidae (seven cases), *D. latum* (three cases), *D. orcini* (one case).
2004: France, probably *D. latum* in humans, probably from eating pike, perch, and burbot.
2004: Switzerland, probably *D. latum* in humans, probably from eating pike perch and burbot.
2005: Austria. *Diphyllobothrium* sp. human case from eating fish while on a trip to Alaska
2005: Brazil, first confirmed case in a woman who ate sushi.
2005: Brazil. Eggs of *Diphyllobothrium* in stool of five humans eating sushi and sashimi.
2005: Brazil, *D. latum*, in 18 human cases from eating raw fish in sushi and sashimi.
2005: Brazil. *D. latum*, in 25 human cases from eating in sushi and sashimi.
2006: France, *D. nihonkaiense*, one human case from eating *O. nerka* from Canada.
2006: Brazil *D. latum*, one case from eating fish.
2006: Japan. *D. latum/nihonkaiense*, one case from eating raw salmon.
2006: Taiwan. *D. latum*, one case from eating uncooked fish.

1951a,b; Neghme, 1953). Neghme and Bertin (1951b) concluded that rainbow trout was the main fish host in most of rivers and lakes in the Lake Colico region. Humans, dogs, and cats harbored the adult worm (Neghme and Bertin, 1951b, 1953), and Faust et al. (1951) reported infections in humans and dogs along the lakes at approximately 40° S. Dogs were experimentally infected with plerocercoids from rainbow trout. Human infections in other parts of Chile were traced back to the foci in the Lake Colico area (Faigenbaum and Donckaster, 1955). Examination of 51,010 people in Santiago, Chile, reported a 0.3% infection level (Reyes et al., 1972) but an examination of 60 people from the Lake Colico found no eggs in stool samples (Ramirez et al., 1977). Further studies by Torres et al. (1983) from five lakes in the endemic region reported a prevalence of ~87% for *D. dendriticum* and 13% for *D. latum* in rainbow trout. The diphyllobothriid complex became more complicated, as in addition to humans infected with *D. latum,* dogs were found infected with *D. pacificum*. Clearly, diphyllobothrids are well established in the area as subsequent studies found *D. latum* in rainbow trout and brown trout in lakes and rivers of the Valdivia River basin and in humans (1.2%) and dogs (5.3% and 9.8%) (Torres et al. 1989a,b). Fifteen new cases were reported from humans between 1981 and 1991, although the incidence of infections has remained relatively constant over this period (Torres et al., 1993). Initially transmission was thought to be limited to rainbow trout, humans, and dogs (Torres et al., 1998), but recent research (Torres et al., 2000, 2004) found *Diphyllobothrium* sp. plerocercoids in salmon (*Oncorhynchus kisutch*) and *D. latum* and *D. dendriticum* in native fishes such as perch (*Percichthys trucha)*, puye (*Galaxias maculatus*), silverside (*Basilichthys australis*), and *Odonthestes (Cauque) mauleanum*. Torres et al. (2004) also reported *D. latum* infections of 2.8% humans at Choshuengo, an area where 12.5% of the population consumed smoked fish. An interesting paper by Sagua et al. (2001) reports an increase in human infections of *Diphylobothrium* spp. with El Nino in the Chilean Pacific coast during 1975–2000, and suggests it is related to the migration of sylvatic mammals, fish, and birds.

The situation in Argentina appears to mirror that in Chile, with similar host–parasite associations involving mainly rainbow trout, humans, and dogs. Four new cases of *D. latum* were reported in 2001, bringing the total cases in Argentina to 18 (Semenas et al., 2001). Semenas et al. (2001) considered sushi, sashami, and smoked itic meat as the main route of infection to humans. Szidat and Soria (1952) reported *Diphyllobothrium* spp. plerocercoids in introduced rainbow trout from Lake Nahuel Huapi, near the Argentina–Chile border, close to the endemic focus in Chile. Based on morphology, the plerocercoids were identified as *D. dendriticum* and *D. latum*, and this was confirmed with adult worms recovered from experimental infections in golden hamsters (Revenga and Seminas, 1991; Revenga, 1993). The introduced rainbow trout, brown trout, and brook trout (*Salvelinus fontinalis*) were considered hosts of *Diphyllobothrium* spp., and a native species of fish (*Percichthys* sp.) was also found infected with *D. latum* (Sagua et al., 2001)

Prior to 2005 in Brazil there were no reports of diphyllobothriasis, but in the past two years it has become a public health concern with a total of 49 human cases reported (Sampaio et al., 2005; Santos and de Faro, 2005; Tavares et al., 2005, Emmel et al., 2006). The most likely source of infection is thought to be fresh aquacultured Atlantic salmon but a native fish species (*Cetropomus undecimalis*) is also consumed raw and has been suggested as a possible source of infection (Sampaio et al., 2005). Since most of the imported salmon came from Chile, the species was most likely *D. pacificum*, but since native species in South America are part of the transmission in Argentina and Chile, an intensive survey of native freshwater and marine fish species for the presence of *Diphyllobothrium* spp. in Brazil is needed. Also a survey of consumption patterns of raw fish from local areas in Brazil needs to be done to determine if there are local foci of infection.

Eurasia

Europe

A thorough review of the European situation has been provided by von Bonsdorff (1977), and more recently Dupouy-Camet and Peduzzi (2004) described the current situation in Europe. According to Dupouy-Camet and Peduzzi, since 1980 the countries with >10 cases/year are Finland, Sweden, Switzerland, and Lithuania, those with 2 to 4 cases/year are Poland, Italy, and France, and those with <1 case /year are Spain, Norway, the Czech Republic, and Romania. Human cases of diphyllobothriasis are reported infrequently from Slovkia. Finland has at least 20 cases/year, Sweden 10 to 50 cases/year, and Estonia had 715 cases in 1990 and 440 cases in 1997 (Dupouy-Camet and Peduzzi, 2004). Recent cases of diphyllobothriasis have been reported from Spain (Colomina et al., 2002) and from Austria by Stadlbauer et al., (2005). There are no recent reports of human infections from Denmark, Croatia, Belgium, the United Kingdom, Netherlands, Yugoslavia, Macedonia, Hungary, and Germany. The intermediate fish hosts has not changed much from the report by Wardle (1935), as Dupouy-Camet and Peduzzi (2004) list perch (*P. fluviatilis*), pike (*E. lucius*), big whitefish (*C. fera*), char (*S. alpinus*), lake trout (*Salmo trutta lacustris*), and rainbow trout (*O. mykiss*). It is assumed that it is *D. dendriticum* that is reported in rainbow trout from Europe, which is similar to North America (McDonald and Margolis, 1995). Lake Leman on the Swiss/French border is of interest because there has been a recent increase in human cases. According to Dupouy-Camet and Peduzzi (2004), 58% of perch and burbot had plercercoids in 1909 compared to 12.5% of perch in 1963 and 8% to 12% of perch fillets in 2003. The plerocercoids were identified as *D. latum* (Nicoulaud et al., 2005). Dupouy-Camet and Peduzzi (2004) suggest that the parasite is imported by infected humans or infected fish, and this helps to maintain or reintroduce the parasite to areas where the parasite had disappeared. Further, it was also suggested that poor local sanitation systems and fecal pollution by the many yachts that use the lake contribute to the maintenance of the life cycle (Dupouy-Camet and Peduzzi, 2004). Dupouy-Camet and Peduzzi (2004) point out that since local foxes and dogs are infected at very low rates, they are not likely very important in maintaining the life cycle of the parasite, and Peduzzi and Boucher-Rodoni (2001) report that the parasite was maintained by domestic animals (presumably dogs) when there were no human cases. Recent surveys of wild animals and dogs from Europe found *D. latum* in 5% of lynx (*Lynx lynx*) in Estonia (Valdmann et al., 2004), in 0.4% of dogs in Finland (Pullola et al., 2006), in 0.1% of 8438 dogs in Germany (Barutzki and Schaper, 2003), and in 31.6% of wolves (*Canis lupus*) and 2.6% of otter (*Lutra lutra*) in a primeval forest in Poland (Gorski et al., 2006). With the recent comeback of populations of brown bear (*Ursus arctos*), wolf, lynx, and wolverine (*Gulo gulo*) into western Europe (Enserink and Vogel, 2006), it is very likely that the *Diphyllobothrium* will also be reintroduced.

Siberia

Dick et al. (2001) reported that most of the historical published information on the distributions on *D. latum* implied that it was a predominantly European parasite, with occasional introductions into distant locations (as in North America). Numerous studies showed that the tapeworm is common east of the Urals in Asian Siberia, and foci of infection have been reported in humans and dogs from the Ob and the Yenisei River drainages. *Diphyllobothrium latum, D. ditremum*, and *D. dendriticum* were identified from the Lena, Kolyma, and Indigirka drainages (Suvorina and Simonova, 1993). Human infections of *D. latum* were also prevalent in eastern Taimyr (Khatangskii region), particularly along the Khatanga and Khata Rivers, where the favorite food fishes were pike and burbot (Klebanovskii et al., 1977). *Diphyllobrothrium latum* was reported from humans in the Amur River basin (Muratov, 1985), and the infective source was considered to be plerocerocids from *Oncorhynchus* spp. Further study revealed these infections were likely due to *D. klebanovskii* (Muratov and Semenova, 1986; Muratov and Posokhov, 1988), which is thought to have marine mammals as a definitive host. *D. latum* and *D. dendriticum* have been reported from the far eastern (Okhotsk) regions of Russia based on plerocercoid morphology and experimental human infections (Dovgalev et al., 1991). Interestingly, the plerocercoid from salmon from the Okhotsk region was unencapsulated and was identified as *D. latum* (Dovgalev, 1988; Dovgalev et al., 1991). Since both *D. latum* and *D. klebanovskii* have unencysted plercerocoids, it is likely that *D. latum* in salmon is actually *D. klebanovskii*. In eastern Siberia *D. dendriticum* is known to infect humans, and the reported observations are similar to those of other circumpolar regions in the northern hemisphere. Clearly, the human infections acquired from freshwater fish route appears similar to the infection route in Europe and North America, but infections from salmon are more similar to the reports in human infections from South America and Japan.

Japan

Due to the consumption of raw fish, it is not surprisingly that diphyllobothiriais has been known from Japan for some time. What is surprising is that *D. latum* has been continually identified as one of the main species infecting humans in Japan.

One of the earliest reports of a human infection of *D. latum* was Yamaguti (1935), who diagnosed *D. latum* based on a piece of tapeworm from his feces in May 1925 while visiting Hamburg, and he thought it likely it was acquired in Japan. Recently, Okino et al. (2005) reported three humans infected with *D. latum* and one patient infected with *D. orcini.* Eguchi (1973) was the first researcher to question the distribution of the typical European *D. latum* in Japan, since most of the fish consumed were Pacific salmonids and other marine fishes. Human diphyllobothriasis has been reported from several locations in Japan by Tomita et al. (1979) and Yokogawa et al. (1979). Even after Yamane et al. (1988) described *D. nihonkaiense* and revised the taxonomy of Japanese *Diphyllobothrium* spp. the identification of *D. latum* persisted. Additional studies by Fukumoto et al. (1988) using immunoelectrophoresis, by Fukumoto et al. (1990) using isozyme patterns and soluble protein profiles, and by Matsuura et al. (1992) using restriction fragment length polymorphisms of rDNA found that *D. nihonkaiense* was different from *D. latum* in Europe. Experimental infections of two Japanese men with *D. latum* of Finnish origin (Yazaki et al., 1984) reported symptoms of general fatigue, epigastric pain, fever and diarrhea, increased eosinophilia, as well as hypochromatic anemia in one man. While these symptoms and pathology were not generally associated with "*D. latum*" from Japan, these symptoms were not usually associated with *D. latum* human infections, acquired from freshwater fishes, reported from Europe and North America.

In spite of the proposed use of "diphyllobothriasis *latum*" as a synonym of "diphyllobothriasis *nihonkaiense*" in Japan (Nishiyama, 1994), and recognition that *D. nihonkaiense* has usually been misidentified as *D. latum* (Yamane et al. 1988, 1989; Hatsushika et al. 1995), the identification of *D. latum* from humans continues. For example, *D. latum* has been reported in humans in Japan by Nishiyama (1994), Hatsushika et al. (1997), Yamaguchi et al. (1997), and Okino et al. (2005). The presence of other diphyllobothrids in the region, particularly marine species such as *D. yonagoensis, D. cordatum, Diplogonoporus* spp., and possibly others (*D. dendriticum* and *D. ursi*) complicates species identification. The Japanese *D. latum* reported by Eguchi (1973) from brown bears that fed on migratory salmon is very likely either *D. ursi* or *D. nihonkaiense*, making it a zoonosis rather than the European human *D. latum*. Recently, human infections of *D. orcini* (Kifune et al., 2000), *D. latum* and *D. orcini* (Okino et al., 2005), *D. nihonkaiense* (Yoshida et al., 1999, Ando et al., 2001; Fuchizaki et al., 2003) and *D. latum/nihonkaiense* (Hirata et al., 2006) have been reported from Japan.

Korea

The earliest reported cases of human diphyllobothriasis appears to be by Cho et al. (1971), followed by a report of a human case of *D. latum* infection in Kangwon Do (Cho et al., 1974). Moon (1976) and Min (1990) reported on human infections in Korea due to the consumption of raw salmonid (sashimi). In a study in the Seoul area, between 1985 and 1986, a prevalence of 0.2% *D. latum* from 5251 human fecal samples was reported by Min et al. (1987). Lee et al. (1989) reported seven additional cases, bringing the total number of cases in Korea to 28. A second larger study of 52,552 human fecal samples from the Seoul Park Hospital between 1984 and 1992 revealed a low prevalence by 0.004% (Lee et al., 1994a). The "dwarf-type" of *D. latum*, namely *D. latum parvum*, was recovered from a human case (Lee et al., 1994b). More recently, five cases of human *D. latum* infections have been reported in Korea from the consumption of the redlip mullet, *Liza haematocheila* (Chung et al., 1997). Given the fish host involved, it is unlikely that this involves *D. latum* and more likely to be a marine diphyllobothriid. Rim (1998) reviewed the fish-borne parasites in Korea and lists the following species *D. latum, D. yonagoense, D. pacifium, D. cameroni, D. scoticum, D. hians*, and *D. nihonkaiense* from humans from the consumption of raw fish *(O. masou, O. gorbuscha, O. keta*, and *O. nerka)* from many countries, presumably from Pacific rim. Interestingly, Rim (1998) mentions the increasing popularity of eating raw salmon, trout, and perch as a cause for the increase in diphyllobothriasis in Korea, but the origin of infection, that is, freshwater or marine, is not given. Lee et al. (2001) reported a young girl and mother were infected with *D. latum* from eating raw salmon flesh, and stated that 37 cases had been reported in Korea since 1921.

Subtropical and Tropical Asia

Diphyllobothriasis appears to be rare in China but two human cases of "*D. latum*" have been reported (Fan et al., 1995; Zhang et al., 1996). The presence of diphyllobothrids in the Amur River basin indicates that Chinese citizens inhabiting that region could acquire the parasite. Mar et al. (1999) reported *D. latum* in 1% each of 96 stray dogs and 95 stray cats in Taipei, Taiwan, and recently Chou et al. (2006) reported the first case of *D. latum* in a child from Taiwan who acquired the parasite from eating raw fish, probably salmon. Chou et al. (2006) mentioned two cases of diphyllobothriasis in aborigines more than 40 years ago. The first case of diphyllobothriasis was reported from Malaysia by Rohela et al. (2002) and identified as *D. latum*, but since the patient had a history of eating sashami it is probably *D. nihonkaiense*.

A human case of *D. latum* was reported from southern India (Pancharatnam et al., 1998) and from Karachi, Pakistan (Bilqees et al., 1984). *Diphyllobothrium latum* has been reported from 1.2% of 756 dogs examined in Faislabad (Maqbool et al., 1998), from dogs in Kerala (Jacob and Pillai, 1991), from a clouded leopard in Alipore Zoo, Calcutta (Sen Gupta, 1974), and from a tiger in Nehru Zoological Park, Hyderabad (Rao and Singh, 1998). *Diphyllobothrium latum* has also been reported from 1.5% of 756 dog faecal samples from Jordan (Abo-Shehada and Ziyadeh, 1991). *Diphyllobothrium latum* has also been reported from 1.5% of 756 dog fecal deposits collected from five governorates in Jordan (Abo-Shehada and Ziyadeh, 1991). All these reports are likely misidentifications. The identification of *D. latum* from a cyprinid, *Acanthobrama centisquama*, in Iraq (Ali et al., 1987) and in *Harpodon nehereus* in Bangladesh (Uddin et al., 1980) are also likely incorrect. However, the identification of *D. latum* from Sefid Rud river drainage (Caspian basin) in Iran in the proximity of Eurasia freshwater systems known to harbor diphyllobothrids warrants a closer look (Mokhayer, 1981).

The New World North American *D. Latum*: Native or Introduced?

North America:

The discovery of human infections in New York as result of consuming pike and percids from Manitoba generated intensive research efforts in the region since New York was a major export market for fish. Lubinsky and Loch (1979) reported that researchers investigating the biology of the parasite at the time were either based in Manitoba (Bajkov, 1933; Wardle 1932, 1933, 1935; Wardle and McLeod, 1952) or worked with material from Manitoba (Magath and Essex, 1931; Magath, 1933, 1937). Intense research efforts during the 1930s found 85% of 500 sled dogs in the area of Lake Winnipeg were infected (Wardle, 1933). *Diphyllobothrium latum* was reported from 85% of pike (*E. lucius*), 7% of sauger (*S. canadense*), and 28% of yellow perch (*P. flavescens*), walleye (*S. vitreus*), and burbot (*L. lota*). Wardle (1935) listed all known hosts at the time known to be infected with adult worms of *D. latum*. The list included dogs (*Canis familiaris, C. azarae, C. cinereoargentatus, C. occidentalis*), cats (*Felis domesticus, F. concolor, F. mellivora, F. hernandesii, F. macroura, F. pardus, F. milis*), mongoose (*Herpestes leucurus*), walrus (*Odobaenus rosmarus*), seals and sea-lions (*Leptonyx monachus, Phoca barbata, P. hispida, P. vitulina, Phocaena phocaena*), bears (*Thalarctos maritimus, Ursus americanus, Urus horribilis, foxes (Vulpes fulva)*, and mink (*Mustelus vison*). Even if some of the identifications were incorrect for *D. latum*, it is difficult to misidentify a diphyllobothrid and clearly illustrates an extensive host list. Although not often discussed in the literature, Wardle noted the absence of pathology around the plerocercoid in the fish flesh, which was also reported by Dick and Poole (1985), and both groups of researchers found that the plerocercoids extend into adjacent myotomes. The absence of pathology due to *D. latum* from Manitoba can be contrasted with reports by Davydov (1978) in which pathology was noted around the worm in pike and burbot with capsules being located in the liver and striated muscles. Wardle (1933) noted that distribution of *D. latum* did not coincide precisely with the areas frequented by European immigrants and that it was common in Eskimos and Indians prior to immigration by Europeans. Humans, bear, mink, cats, dogs, particularly the husky dog of the Eskimo, and Indian and immigrant fishers were considered the main hosts (Wardle, 1933).

These early researchers from central Canada had advantages over European researchers as intermediate fish hosts (pike, yellow perch, walleye, sauger, and possibly burbot) were identified early on and were either identical or closely related to fish hosts eventually determined to be fish intermediate hosts in Europe and Siberia. Unlike Europe Manitoba does not have native salmon, and while coregonines are common they are not a host of *D. latum*. Experimental infections using plerocercoids from pike and perch fish from Manitoba to infect black bears (*Ursus americanus*) by Vergeer was further proof that the source of infection could be a sylvatic host, in addition to humans. Numerous studies on fish parasites in Manitoba over the years (see Lubinsky and Loch, 1979; Watson and Dick, 1979, 1980; McDonald and Margolis, 1995; and Dick and Poole, 1985) clearly demonstrated that plerocercoids of *D. latum* did not occur in any other species of fish.

There are two opposing views on the origins *D. latum* in North America. Some believe that *D. latum* in North America was introduced by northern Europeans (Vergeer, 1928, 1929a; Magath and Essex, 1931), especially the Finns, and this was supported by von Bonsdorff (1977). Others (Bajkov, 1930; Wardle, 1932; Lubinsky and Loch, 1979) believe that the parasite was present prior to European immigration. As early as the late 1920s, Vergeer (1929b) stated: "Fish in the smaller lakes far distant from towns were moderately infested with broad tapeworm. This immediately suggested the possibility of wild carnivores as a source of infection and again raised the question whether the white man or the tapeworm was first in North America." But despite the evidence, Vergeer concluded that *D. latum* was introduced by European immigrants. By contrast, Bajkov (1933) was convinced that "the American *Diphyllobothrium* was not introduced from Europe, but is a native form." He also stated that "Canadian Indians along the shores of Lake Winnipeg knew and observed *D. latum* in connection with their dogs a long time before it was discovered by white man" and concluded that the tapeworm was "very abundant in all eastern and western tributaries of Lake Winnipeg, in the Nelson River, also practically all lakes of northern Manitoba" (Bajkov, 1933).

Another hurdle to the idea that *D. latum* was introduced by European immigrants is the presence of this parasite in more western regions of the continent, such as Alaska (Rausch and Hilliard, 1970), where immigrants were much more sparse than in central North America in Minnesota and Manitoba. It is likely that the *D. latum* of Rausch and Hilliard (1970) is similar to the *D. latum* of more interior North American boreal regions. A similar view has been raised by Dick et al. (2001). It is very likely that there was an older existing population of *D. latum*, based on its presence today in very isolated boreal lakes, where no humans resided, prior to European immigration. Furthermore, since the parasite is still present in fish flesh it has to be maintained by natural definitive hosts such as wolves, foxes, bears, and otter. As with the fish hosts, these mammalian hosts also have close relatives (dogs, foxes) in Eurasia capable of being hosts of this parasite. The unresolved question, even today is the possibility that this parasite is *D. ursi* or another closely related species to the original European *D. latum*.

A study by Dick et al. (2001), including experimental human infections, found that a *Diphyllobothrium* species encysted in pike flesh is widely distributed in lakes of various sizes in the forested Whiteshell Area of Manitoba. Indeed, *D. latum* is widely distributed in central Canada today in a variety of locations and fish hosts, and *D. dendriticum* is found throughout the Canadian Arctic (Fig. 4.3). The three common fish hosts of *D. latum*—pike, walleye, and perch—had prevalances ranging from 28% to 70%, 2% to 51% and 0.5% to 50%, respectively, in the boreal region of Manitoba and present in all but one of 16 lakes (Fig. 4.4, Tables 4.4 and 4.5). The lake without *D. latum* is a large

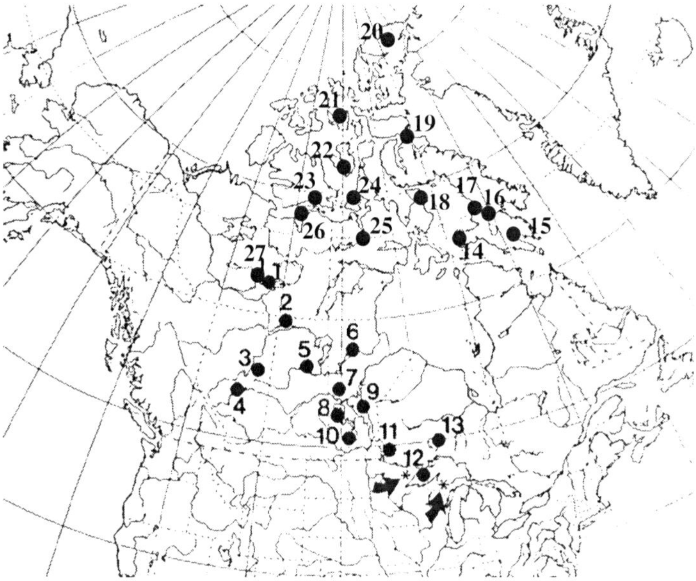

FIGURE 4.3. Map of North America showing major drainages where *Diphyllobothrium* spp. from freshwater have been reported in fish and is based on data from surveys by the author and coworkers. Note that Alaska has been excluded because it includes both freshwater and marine diphyllobothrids but the reader is referred to the excellent work of Rausch and coworkers (1970, 2000) for additional details. 1, Great Slave Lake; 2, Lake Athabaska (Mackenzie /Churchill River drainage); 3, Lac LaBiche (Athabaska River drainage); 4, Lac Ste. Anne (North Saskatchewan River drainage); 5, Lac La Ronge (Churchill River drainage); 6, Southern Indian Lake (Churchill River drainage); 7, lakes of the Pas area; 8, Lake Winnipegosis; 9, Lake Winnipeg; 10, Lake Manitoba; 11, Whiteshell area and Lake of the Woods (Winnipeg River drainage); 12, Lake Superior (Laurentian Great lakes drainage); 13, Lake Nipigon; 14, Scarp Lake; 15, Iqalugaajuruluit Lake; 16, Nettilling Lake; 17, Koukdjuak River; 18, Igloolik; 19, Salmon River; 20, Hazen Lake; 21, unnamed lakes; 22, Crooked Lake; 23, Freshwater Creek; 24, unnamed lake; 25, Hayes River; 26, Little Nauyuk Lake; 27, Chitty Lake. Heavy arrows point to locations (asterisks) in northern Minnesota and Michigan where *D. latum* were reported from pike, and/or percids (perch and walleye). The records are from northern pike, perch, walleye sauger and Arctic char. Records from 14 to 26 are for *D. dendriticum* in Arctic char, and 27 is *Diphyllobothrium* from the flesh of pike, walleye, and liver of nine spine sticklebacks (*Pungitius pungitius*).

riverine lake. One might argue in the Whiteshell area (Fig. 4.4C; Table 4.5) that poor sanitation in "cottage country" allows humans to contribute to levels in the lakes, but there is no report of human infections other than the three experimental human infections reported by Dick et al. (2001). By contrast, the northern lakes are even more isolated (Fig. 4.4A,B), but prevalences are high in the fish hosts (Table 4.4). While pike seem to be the most frequently infected fish host, it is also common in walleye and perch. Other lakes where pike were found to be infected with *D. latum* include Hawnck, Sheep, and Rae

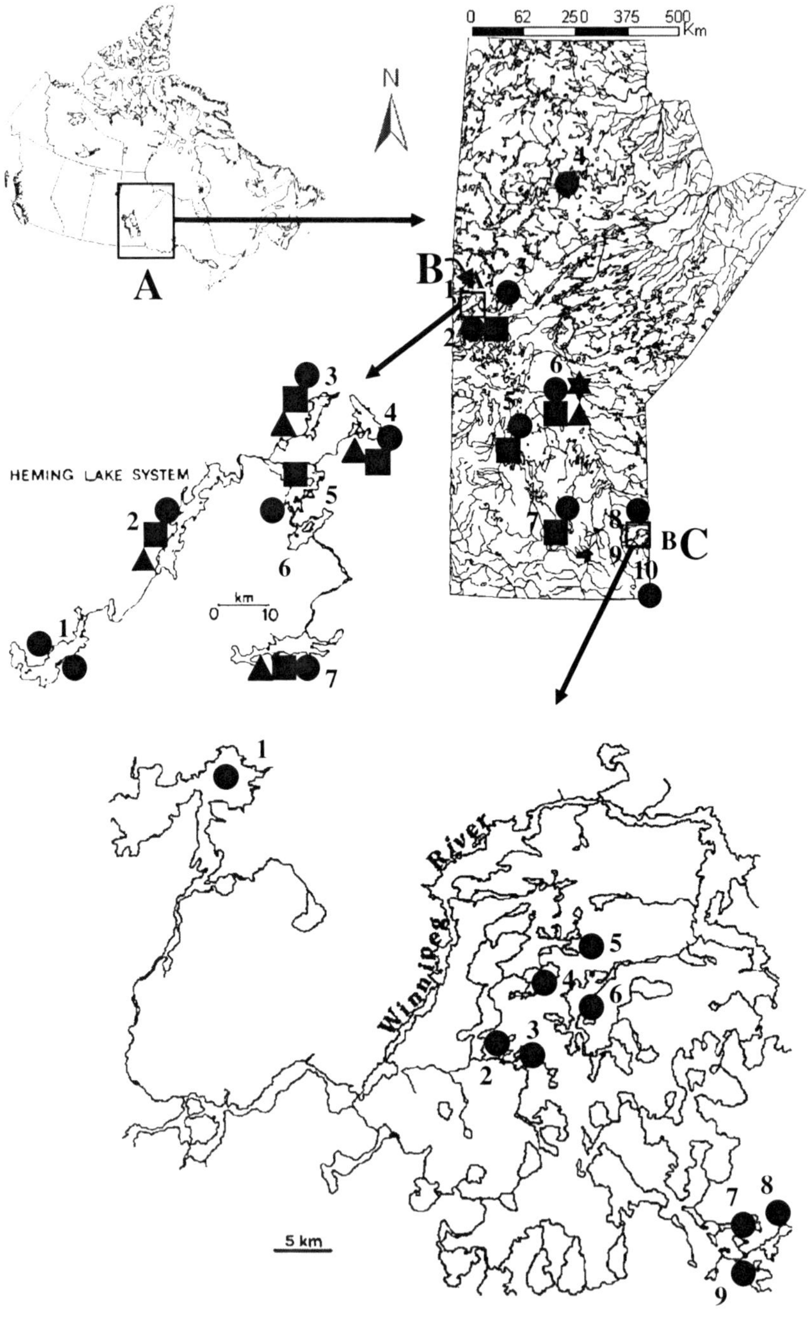

FIGURE 4.4. Map of study areas in Manitoba, Canada, where detail surveys have been conducted in the fish host and from which infective plerocercoids of *Diphyllobothrium* were used for experimental infections. A = study areas where *Diphylobthrium* plerocercoids were recovered from fish flesh 1) Heming Lake, 2) Other Lakes of The Pas area,

TABLE 4.4. *Diphyllobothrium* spp. recovered from fish flesh in fishes from Manitoba lakes (from Dick et al., 2001).

Lake	*Esox lucius*	*Sander vitreum*	*Perca flavescens*
Heming	P (187)	P (48)	0.5 (201)
Home	P (75)	P (75)	1.3 (78)
Demarch	P (75)	2.5 (76)	0 (40)
Quigly	46.9 (33)	51 (78)	0 (11)
South Indian	54.8 (444)	NS	NS
North Sailing	70 (30)	NS	50 (42)
Echo	28.5 (32)	50 (28)	6.25 (16)

Values refer to prevalence, that is, percent infected in sample, and value inside parentheses is the sample size. P, present; NS, no sample.

The relatively high prevalences in fish in certain areas and the wide distribution of this parasite throughout central Canada (Figs. 4.3 and 4.4) indicate that the life cycle can be maintained without the participation of human hosts. It seems more likely that *D. latum* is not originally a human parasite, but humans are accidental participants in what is really a sylvatic life cycle involving wild fish-eating carnivores.

It is often stated that infected humans are responsible for local foci of infections by contaminating the local water systems, and that may happen occasionally. By contrast, human behavior can also play a significant role in contributing to increased levels of *D. latum* in fish, locally. In central Canada, McLaren and Reindeer lakes are commercially fished for walleye, and the prevalences in walleye suddenly increased to >50%. These fish from these lakes were rejected for human consumption, and the

TABLE 4.5. *Diphyllobothrium* in the flesh of pike: Whiteshell area of Manitoba (see Fig. 4.4) (from Dick et al., 2001).

Lake	*n*	Prevalence
Lac De Bonnet	18	0
Quesnel/Manigatogan	30	10
Beck	7	55
Boatfield	3	66.7
Boon	5	60
Johnstone	10	20
Falcon	55	100
Horse Shoe	38	50

FIGURE 4.4. (*Continued*) 3) Snow Lake, 4) Southern Indian Lake, 5) Lake Winnipegosis; 6) Lake Winnipeg, Lake Manitoba. 7) Lake Manigotagan, 9) Whiteshell, 10) Lake of the Woods, 11). Boxes A = Heming Lake area (1) Wapun Lake, 2) Heming Lake, 3) Unger Lake, 4) Martin Lake, 5) Home Lake, 6) deMarch Lake) and B = Whiteshell area (1) Lac DuBonnet, 2) Beck Lake, 3) North Sailing Lake, 4) Boon Lake, 5) Boatfield Lake, 6) Echo Lake, 7) Sheep Lake, 8) Jadel Lake, 9) Mantario Lake; some of these lakes are accessible by road and some by canoe). Symbols: solid circles = pike (*E. lucius*), solid squares = walleye (*S. vitreus*). Solid triangles = yellow perch (*P. flavescens*), star = sauger (*S. canadense*),

author was asked to determine the cause. McLaren Lake is a relatively small lake of a few square kilometers, while Reindeer Lake is a very large lake in northern Manitoba. It is evident from Table 4.4 that walleye are usually not this heavily infected, so the reason for the higher prevalences was investigated. McLaren Lake is isolated with no permanent human settlements and only an occasional itinerant fisher or sport fisherman on the lake. Upon questioning, the fishers revealed that at McLaren Lake, during the commercial fishing season, fish remains were left on shore and they noted extensive feeding activity by wolves and foxes, which likely contributed to the increased levels in fish hosts. In the case of Reindeer Lake there are several sections on the lake where the fishing effort is concentrated. One of these areas is isolated but close to a large island where dogs were held during the summer and these dogs are fed whole fish and fish remains. These two examples indicate a sylvatic cycle involving wild carnivorous mammals where local transmission, in these two examples, were amplified by human activities.

The situation in South America is quite complicated. Diphyllobothrids are thought to have been introduced by fish stocking and human immigrants, but are now reported from native species and Pacific salmonids. One source of infection is considered to be rainbow trout from North America, where its original geographic range was the streams and lakes of western North America and in some river systems it was anadromous. Pacific salmon (*Oncorhynchus* spp.) obviously harbors a number of species of *Diphyllobothrium,* but *D. latum* is likely not one of them. Brook trout were also introduced to South America from North America but has not been identified as a natural host of *D. latum* either. Speculation that rainbow trout, which is not a natural host of *D. latum* in North America, is a key host in South America is problematic. It is very doubtful if there are any bona fide records of *D. latum* in salmon species in North America (Margolis and Arthur, 1979; Hoffman, 1999). Studies for over a century in the holoarctic region have consistently reported the main fish hosts of *D. latum* as pike, the North American and Eurasian perches, the North American and Eurasian pike-perches, the ruffe (*Acerina cernua*), which has no counterpart in North America, and the burbot in Europe and Siberia. Neither the amphi-Atlantic trouts (*Salmo* spp.) nor the amphi-Pacific trouts (*Oncorhynchus* spp.) or the chars (*Salvelinus* spp.) are considered significant natural hosts of *D. latum*. Recently, the species thought to be *D. latum* in Pacific trouts (e.g., *Oncorhynchus masou*) causing diphyllobothriasis in Japan is a distinct species, *D. nihonkaiense*. A species infecting humans in the Amur basin with plerocercoids in *Oncorhynchus* spp. is now considered a distinct species, *D. klebanovskii*, and *D. pacificum*, a parasite of fur seals and sea lions, has been reported from humans in Peru, Chile, Ecuador, and Japan. Clearly there are a number of species of *Diphyllobothrium* in South America from both the freshwater and marine environments. The flesh dwelling plercocercoids from salmon in South America appear to follow the pattern of other diphyllobothrids from salmon species around the Pacific rim in that they readily infect humans, have low pathogenicity, and are not *D. latum*. The freshwater diphyllobothrid are still an enigma and given the location in the mesenteries, gonads and viscera of fish fit the pattern observed for *D. dendriticum* from chars and coregonids in the Northern hemisphere. In extremely heavy infections (Fig. 4.4), *D. dendriticum* may be found along the wall of the body cavity where it occasionally penetrates the hypaxial muscle, but this is very much a secondary site. The observation that some plercercoids are found in the fish flesh of native South American freshwater species is reminiscent of the "*D. latum*" from the Northern Hemisphere. A large body of knowledge from human infections shows that the diphyllobothrids species most likely to infect humans are those with plerocercoids located in fish flesh and not the viscera. Perhaps the original source of human infections from freshwater systems in South America was *D. latum* introduced by European immigrants. However, a similar interpretation for the source of *D. latum* in North America is now questioned, based on the wide distribution of a diphyllobothrid infective to humans in the northern boreal regions today and its presence in aboriginal peoples and dogs prior to European immigration. The presence of a *Diphyllobothrium* in native fish species in South America suggests it may have been in south temperate South America long before European immigrants arrived. Perhaps *D. latum* or a *Diphyllobothrium* spp. of bears and/or canids, which also infects humans, could have been carried south along the Pacific coast with the first human migrants and their dogs from Beringia. This parasite became established in native fish species and mammals associated with freshwater systems and infected rainbow trout when they were introduced. The bird transmitted *D. dendriticum*, which is also present in southern South America, could easily have been transported to the region through bird migrations in the past, unless there is evidence of stocking with infected char or coregonids from the northern hemisphere. *D. dendriticum* could also have been introduced by infected rainbow trout as it has been reported from rainbow trout in Europe (Dupouy-Camet and Peduzzi, 2004) and North America (McDonald and Margolis, 1995).

Control

There are a number of controls measures that can be used for *Diphyllobothrium spp.*: (1) treatment of fecal matter and prevention of contamination of drinking and natural waterways, (2) adequate inspection of all sources of raw fish for food and restaurant markets, (3) public health education on the potential for infections from a variety of raw fish products, (4) temperature or brine treatment of fish fillets to eliminate infectivity, (5) knowledge of the local biology of *Diphyllobothrium*, and (6) drug therapy.

Control includes adequate cooking of fish, freezing at −10°C for 24 hours and if smoke fish is brined before smoking there is less risk of infection (Tanowitz et al., 2001). Peduzzi and Boucher-Rodoni (2001) recommended freezing and heating to kill the plerocercoids of *D. latum* (−10 °C for 1 to 7 days, +40°C for 72 hours, and +50°C for 10 minutes). The consumption of raw fish is the major problem and requires either careful inspection of raw fillets or changing eating habits.

Drug therapy is widely employed after an infection has been diagnosed. The most complete history on the application of drugs to treat diphyllobothriasis is presented by Harder (2002) for the Bayer drug company; as early as 1940 Acranil (Sostol) was used to treat *D. latum* and in 1975 the drugs Droncit and Biltricide (praziqantel Embay 8440) were was used to treat *D. dendriticum* in cat and dogs.

Today the drug of choice is prazinquantel, but there still are a variety of therapies used. Rohela et al. (2002) treated patient with praziquantel (750 mg) in a single dose, and Lee et al. (2001) treated patients with 600 mg praziquantel. Yoshida et al. (1999) treated patients with oral administration of 200 mL of Gastrografin (meglumine sodium amidotrizoate; Schering A.G., Germany), in combination with intramuscular injection of Vagostigmin (including 0.5 mg of neostigmine; Shionogi Co. Ltd., Japan). Fujita et al. (2002) and Fuchizaki et al. (2003) treated patients with 300 mL of amidotrizoic acid through an intraduodenal tube. Yera et al. (2006) treated patients with praziquantel (Biltricide R) at 10 mg/kg. Treatment with praziqantel followed by ingestion of a cathartic (30 g magnesium sulfate and 300 mL of water 2 hours later) was reported by Hirata et al. (2006). A single dose of praziquantel (600 mg) was used by Santos and de Faro (2005). Chou et al. (2006) treated patients with a single dose of praziquantel at 8.5 mg/kg with a second dose one week later.

Symptoms

The classic symptoms of diphyllobothriais have been reported by numerous researchers, but the compilation of symptoms by Marty and Neafie (2000) is quite thorough. Briefly, a patient may present with diarrhea, abdominal pain, or anemia. Obstruction of the bowel may occur, and occasionally pieces of the worm are vomited. Slight leukocytosis with eosinophilia may also occur. According to Marty and Neafie (2000), "2% of patients suffer from bothriocephalus anemia, a non-lethal form of pernicious anemia." These patients often show fatigue, weakness, sore tongue, paresthesias of hands and feet, diarrhea, and low levels of vitamin B_{12}. According to Bylund (2003), if the parasite resides high up in the intestine, in the jejunum, it utilizes and absorbs large quantities of vitamin B_{12}. A survey in 1978–79 from 350 worm carriers in Finland did not find a significant reduction in B_{12} nor evidence of anemia, and Bylund (2003) attributed these findings to better diets. The adult worm rarely causes biliary obstruction. An interesting case reported by Norman (1937) shows how infections with *Diphyllobothrium* sp. can be confused with other disorders. The patient was a male who had lived in Massachusetts for 8 years and visited Mexico 3 years prior to presenting with "severe epigastric pains that radiated to the back and were only partially relieved by milk and Sippy powders." The patient also complained of considerable heart burn, weakness, and headaches. Initially the patient was thought to have peptic ulcers, was hospitalized, and at one point passed segments of a tapeworm. The patient after taking capsules of oleoresinae aspidi (half a dram, in divided doses), passed an 18-foot tapeworm, and the "patient's symptoms vanished immediately."

The following symptoms have been reported in the recent literature by several authors: watery stools and abdominal discomfort by Rohela et al. (2002); abdominal discomfort and intermittent diarrhea by Sampaio et al. (2005); sudden nausea, epigastria pain, and abundant diarrhea by Yera et al. (2006); and mild abdominal cramping (Chou et al., 2006). The author and two volunteers were experimentally infected with *D. latum*, but the symptoms were nonspecific and mild (Dick et al. 2001). It appears that generally the symptoms of infection are quite mild, across species of *Diphyllobothrium* and ethnic groups. The one exception may be people of Scandinavian ethnicity, in whom more cases of anemia have been reported in the past (Bylund, 2003).

Gaps in Knowledge

Clarification of species infective to humans is a priority, not because there is strong evidence for species-related pathology, but to establish clearly the source of infection so that public health measures can be taken. Either the original *D. latum* (the so-named European form) has moved relatively freely around the world via humans and become established as new endemic populations or it is has been misdiagnosed. The general consensus is that *Diphyllobothrium* spp. in the Northern hemisphere is *D. latum* and it does not infect salmon species. *Diphyllobothrium dendriticum* does occasionally infect humans, but its intermediate hosts are primarily char and coregonid fishes.

The source *D. latum* in the North America is problematic, as it has been speculated to have arrived with the early settlers, but prior to settlers arriving in North American, First Nation communities reported infections in humans and dogs. Further, a study by Dick et al. (2001) on its distribution in fish and human experimental infections clearly showed that this parasite is still widely distributed in the boreal region of North America. It acts and looks like *D. latum* in Europe (and is found in similar fish species), and readily infects humans (Dick et al., 2001). There is a need, therefore, for a thorough study of the molecular biology of the *Diphyllobothrium latum* forms from the boreal region of North America by comparing to the "true human *D. latum* form" from Europe and plerocercoids recovered from fish-eating mammals from isolated areas of North America, Europe, and Russia.

South America and some Pacific rim countries have a mixture of marine and freshwater diphyllobothrids with several species being reported from humans. Added to this species complex is the popularity of raw fish cuisine around the world, in the form sushi and sashimi, plus the continued consumption of raw fish from either freshwater or marine sources by locals. There is a need to establish the source of infection, that is, whether it is imported or endemic in regions such as Brazil, where there has been a dramatic increase in reporting the disease. Interestingly, Rienhard and Urban (2003) identified *D. pacificum* from Chinchorro mummies 4000 to 5000 years ago and suggested that trade in infected fish reached inland or that inland peoples migrated to the coast to fish. Not much has changed in 5000 year, except that trade in fresh fish is now worldwide, and infections of *Dipyllobothrium* spp. can occur in large urban areas anywhere in the world.

More recently, Skerikova et al. (2006), using molecular techniques, indicated that *D. pacificum* is a valid species. Interestingly in the same paper it is reported that diphyllobothrids recovered from a human infected with plerocercoids from salmon in Canada and plerocercoids recovered from the mesenteries of burbot in Russia are both *D. latum*. The picture gets even more confusing where *D. latum* is reported from mesenteries, muscles, stomach, and intestine from rainbow trout from Chile, and *D. dendriticum* is reported from mesenteries and muscle of rainbow trout (an atypical site based on all the information from North America and Europe). Further, in native fish hosts in Chile, *D. latum* is recovered from muscle, liver, gonads, and mesenteries, and *D. dendriticum* was found in mesenteries and liver of native fish species. Vik (1964) suggested that all marine salmon species were infected with the freshwater *D. latum* during the freshwater stage of their life history, and Margolis et al. (1979) also suggested that some of the salmons may be infected with freshwater diphyllobothrids prior to going to sea. Along the west coast of North America, *D. ursi* is a likely candidate since brown bear and salmon definitely interact, but the plerocercoids of *D. ursi* were restricted to the viscera according to Margolis et al. (1979). By contrast, Ching (1984) reports *D. ursi* from the viscera and flesh of salmon and from the liver and flesh of rainbow trout.

Clearly the origin of plerocercoids as the source of infection to humans needs more work. If the results from Skerivkova et al. (2006) and the observations reported by researchers in Chile are correct, then the location of plerocercoids in the flesh of a fish host may not be a good distinguishing characteristic to define species infecting humans. Plereocercoids in fish flesh is the best route to ensure infections to humans, but plerocercoids located in the viscera is certainly the best route to infect birds and likely the smaller fish eating mammal.

Since some species of marine and freshwater fishes overlap during their life histories, there is likely some transfer of the freshwater diphyllobothrids to anadromous fish in the Northern hemisphere where the original named *D. latum* is endemic in freshwater systems. Superimposed on this transmission is the presence of marine forms of *Diphyllobothrium* in fish returning to spawn in freshwater, that is, anadromous salmon infected with diphyllobothrids with a marine life cycle. The role of rainbow trout in the transmission of *Diphyllobothrium* species is still an enigma, but many populations of this species migrate to estuaries; some populations move several hundred kilometers offshore, and some are exclusively freshwater. The trout species of the salmonids need to be reevaluated.

Molecular probes will help identify the sources of infections of human cases, whether of marine or freshwater fish origin. However, it is equally important to appreciate that this group of parasites is a sylvatic zoonosis, and begin to interpret patterns of transmission from that perspective. Surveys of intestinal helminths from marine and terrestrial carnivorous mammals and fish surveys are needed as well as detailed records of the location of plercercoids in a fish for the same species of *Diphyllobothrium*. Unquestionably, the diphyllobothrids are a well-established worldwide zoonosis, and in all likelihood, as gourmet interactions continue to increase between humans and fish, and predator–prey interactions between fish and fish-eating mammals and birds expand, so will the number of humans cases, diagnosed and undiagnosed.

Acknowledgments. I thank graduate students Reg Watson, Brent Poole, Alex Szalai, Theo deVos, Patrick Nelson, and Anindo Choudhury for help over the years. I also thank Jim Beyette, Manitoba Natural Resources, for the collection of some of the pike fish from the Whiteshell area of Manitoba, and Lionel Bernier for data from some of the Arctic char surveys. I also thank M. Lu and D. Block for technical help, and especially Colin Gallagher for technical help with the life cycle and the maps. The Natural Sciences and Engineering Council of Canada (NSERC) is gratefully acknowledged for financial support through operating grants over the years and with current support from an NSERC Northern Research Chair

References

Abo-Shehada, M.N., and Ziyadeh, Y. (1991). Prevalence of endoparasites in dog faecal deposits in Jordan. J Helminthol 65(4), 313–314.

Ali, N.M., Salih, N.E., and Abdul-Ameer, K.N. (1987). Parasitic fauna of some freshwater fishes from Tigris river, Baghdad, Iraq. III. Cestoda. J Biol Sci Res 18(3), 25–33.

Ando, K., Ishikura, K., Nakakugi, T., Shimono, Y., Tamai, T., Sugawa, M., Limviroj, W., and Chinzel, Y. (2001). Five cases of *Dhyllobothrium nihonkaiense* infection with discovery of plerocercoids from an infective source, *Oncorhynchus masou ishikawae*. J Parasitol 87, 96–100.

Andersen, K. (1975). Comparison of surface topography of three species of *Diphyllobothrium* Cestoda, Pseudophyllidea) by scanning electron microscopy. Int J Parasitol 5, 293–300.

Andersen, K.I., and Gibson, D.I. (1989). A key to three species of larval *Diphyllobothrium* Cobbold, 1858 (Cestoda: Pseudophyllidae) occurring in European and North American freshwater fishes. Syst Parasitol 13, 3–10.

Archer, J., Taft, S.J., and Thiel, R.P. (1986). Parasites of wolves, *Canis lupus*, in Wisconsin, as determined from fecal examinations. Proc Helminthol Soc Wash 53(2), 290–291.

Bajkov, A.D. (1933). Report on fisheries investigation. Investigations in connection with the broad tapeworm (*Diphyllobothrium latum*) in Manitoba waters. Rep Man Prov Fish Br, 19p.

Barutzki, D., and Schaper, R. (2003). Endoparasites in dogs and cats in Germany 1999–2002. Parasitol Res 90, 148–150.

Bathurst, R.R. (2005). Archeological evidence of intestinal parasites from coastal shell middens. J Archeol Sci 32, 115–123.

Bilqees, F.M., Khan, A., and Ali, N. (1984). *Diphyllobothrium latum* infestation in Karachi—a case report. Pakistan J Med Res 23, 55–57.

Bouza-Suarez, M., Hormilla-Manson, G., Dumenigo-Ripoll, B., Quintana-Olmos, R., and Cordovi-Prado, R. (1990). First certain case of *Diphyllobothrium latum* in Cuba (West Indies). Rev Cubana Med Trop 42(1), 9–11.

Bouchet, C., Lefevre, C., West, D., and Corbett, D. (1999). First paleoparasitological analysis of a midden in the Aleutian Islands (Alaska): results and limits. J Parasitol 85, 369–372.

Bouchet, C., West, D., Lefevre, C., and Corbett, D. (2001). Identification of parasitoses in child burial from Adak Island (Central Aleutian Island, Alaska). Compte-Rendu Acad Sci Paris (series III) 324, 123–127

Bylund, G. (2003). *Diphyllobothrium latum*. In: Akuffo, H., Ewert, L., Lungstrom, I., and Wahlgren, M., eds. Parasites of the colder climates. New York: Taylor and Frances, 169–176.

Canavan, P.N. (1932). Spread of broad fish tapeworm of man. Science 75, 382.

Chai, J-Y., Murrell, J.D., and Lymbery, A.J. (2005). Fish-borne parasitic zoonoses: status and issues. Int J Parasitol 35, 1233–1254.

Ching, H. (1984). Fish tapeworm infections (diphyllobothriasis) in Canada, particularly British Columbia. Can Med Assoc J 130, 1125–1127.

Cho, S.Y., Cho, S.J., Ahn, J.H., and Seo, B.S. (1971). One case report of *Diphyllobothrium latum* infection in Korea. Seoul J Med 12(3), 157–163.

Cho, S.Y., Lee, S.H., Seo, B.S., and Lee, S.K. (1974). Cases of *Diphyllobothrium latum* infection in Kangwon Do. Korean J Parasitol 12(1), 83.

Chou, H-S., Yen, C.-M., Liang, W.-C., and Jong, Y.-J. (2006). *Diphyllobothrium latum*: The first child case report in Taiwan. Kaohsiung J Med Sci 22, 346–350.

Christian, F.A., and Perret, J.T. (1974). Incidence of intestinal parasites in children from Scotlandville area of Baton Rouge in Louisiana. Proc Helminthol Soc Wash 41(2), 249–250.

Chung, P.R., Sohn, W.M., Jung, Y., Pai, S.H., and Nam, M.S. (1997). Five human cases of *Diphyllobothrium latum* infection acquired through eating raw flesh of redlip mullet, *Liza haematocheila*. Korean J Parasitol 35(4), 283–289.

Colomina, J., Villar, J., and Estaban, G. (2002). Asymptomatic infection by *Diphyllobthrium latum* in a Spanish three year old child. Med Clin 118, 279.

Croll, N.A., and Gyorkos, T.W. (1979). Parasitic disease in humans: the extent in Canada. Can Med Assoc J 120(3), 310–312.

Cushing, H.B., and Bacal, H.L. (1934). *Diphyllobothrium latum*: with particular reference to its increasing prevalence. Can Med Assoc J 30, 377–384.

Davydov, V.G. (1978). Differences in the tissue reactions of fish in response to infection with *Diphyllobothrium latum* (L., 1758) plerocercoids. Biol Vnutr Vod Infor Byul 39, 68–71.

Deardorff, T.L., and Overstreet, R.M. (1991). Seafood transmitted zoonoses in the United States: the fishes, the dishes and the worms. In: Microbiology of marine food products (Ward, D.R., and Hickey, C.), eds. New York: Van Nostrand Reinhold, 211–265.

DeVos, T., and Dick, T.A. (1989). Differentiation between *Diphyllobothrium dendriticum* and *D. latum* using isozymes, restriction profiles and ribosomal probes. Syst Parasitol 13, 161–166.

Dick, T.A., and Poole, B.C. (1985). Identification of *Diphyllobothrium dendriticum* and *Diphyllobothrium latum* from some freshwater fishes of central Canada. Can J Zool 63, 1, 196–201.

Dick, T.A., Nelson, P., and Choudhury, A. (2001). Diphyllobothriasis: update on human cases, foci, patterns and sources of human infections and future considerations. Southeast Asian J Trop Med Public Health 32, 59–76.

Dovgalev, A.S. (1988). Diphyllobothriasis in the western coastal area of the Sea of Okhotsk (USSR). Med Parazit Parazit Bolezni 4, 67–71.

Dovgalev, A.S., Valovaya, M.A., Piskunova, IuA., Yu, A., Romanenko, N.A., Khodakova, V.I., and Artamoshin, A.S. (1991). Title: The morphology of human diphyllobothriasis agent in the Far East. Med Parazit Parazit Bolezni 6, 42–46.

Dupouy-Camet, J., and Peduzzi, R. (2004). Current situation of human Diphyllobothriaisis in Europe. Eur Surveill 9, 31–35.

Eguchi, S. (1973). *Diphyllobothrium latum* (Linnaeus, 1758). In: Morishita, K., Komiya, Y., and Matsubayashi, H., eds. Prog Med Parasitol Jpn 5, 125–144.

Emmel, V.E., Inamine, E., and Secchii, C. (2006). *Diphyllobothrium latum*: case report in Brazil. Rev Soc Bras Med Trop 39, 82–84.

Enserink, M., and Vogel, G. (2006). The carnivore comeback. Science 314, 746–749.

Faigenbaum, J., and Donckaster, R. (1955). Consideraciones clinicas y epidemilogicas en relacion con dos nuevos casos de difilobotriasis humanas. Bol Chil Parasitol 10(1), 15–17.

Fan, S.Q., Wan, G.Y., Sun, M.F., Fan, S.Q. (1995). A case of *Diphyllobothrium latum* infection in Heilongjiang Province. Chinese J Parasitol Parasitic Dis 13(3), 240.

Faust, E.C., Neghme, R.A., and Tagle, V.I. (1951). *Diphyllobothrium latum* indigenous in the Lake District of Chile J Parasitol 37, 24.

Ferreira, L.F.A., Araujo, A., Confalonieri, U.E.C., and Nunez, I. (1984). The finding of eggs of *Diphyllobothjium latum* L., 1758 in human coprolites (4,100–1,950 BC) from northern Chile Mem Inst Oswaldo Cruz 79, 175–180.

Freeman, R.S., and Thompson, B.H. (1969). Observations on transmission of *Diphyllobothrium* sp. (Cestoda) to lake trout in Algonquin park. Canada J Fish Res Bd Canada 26, 871–878.

Freeman, R.S., and Jamieson, J. (1976). Parasites of Eskimos and Igloolik and Hall Beach. *In*: Sheperd, R.J., and Itoh, S. eds. Third International Symposium Circumpolar Health. Yellowknife, NWT, Canada: University of Toronto Press, 306–315.

Fuchizaki, U., Hideki, O., and Sugimoto, T. (2003). Diphyllobothriais. Lancet 3, 32.

Fujita, M., Koga, H., Iida, M., Hirakawa, K., Hoshika, K., Haruma, K., and Okino, T. (2002). The diagnostic yield of colonoscopy and the therapeutic value of intraduodenal amidotrizoic acid injection in intestinal *Diphyllobothrium latum* infection: report of a case. Am J Gastro enterol 97, 2468–2469.

Fukumoto, S., Yazaki, S., Kamo, H., Yamane, Y., and Tsuji, M. (1988). Distinction between *Diphyllobothrium nihonkaiense* and *Diphyllobothrium latum* by immunoelectrophoresis. Jpn J Parasitol 37, 91–95.

Fukumoto, S., Yazaki, S., Maejima, J., Tsuboi, T., and Hirai, K. (1990). Soluble protein profiles and isozyme patterns of *Diphyllobothrium pacificum* by isoelectric focusing: comparison with those of related diphyllobothriid species. Yonago Acta Med 33, 61–70.

Goncalves, M.L.C, Araujo, A., and Ferreira, I.F. (2003). Human intestinal parasites in the past; new findings and review. Mem Inst Oswaldo Cruz 98, 103–118.

Gorski, P., Zalewski, A., and Lakomy, M. (2006). Parasites of carnivorous mammals in Bialowieza primeval forest. Wiad Parazytol 52, 49–53.

Guttawa, A., and Moskwa, B. (2005). Histroia odkrycia i opisania cyklu rozwojowego *Diphyllobothrium latum*. L Wiad Parasitol 51, 359–364.

Harder, A. (2002). Milestones of helminthic research at Bayer. Parasitol Res 88, 477–480.

Hatsushika, R., Okino, T., and Okazawa, T. (1995). A case study of human infection with broad tapeworm, *Diphyllobothrium latum* Linnaeus, 1758) Luhe, 1910 found in Okayama Prefecture, Japan. Jpn J Parasitol 44, 311–320.

Hatsushika, R., Okino, T., and Tsutsui, Y. (1997). A case study of human infection with small-size strobilae of diphyllobothriid tapeworm discharged from a man in Okayama Prefecture, Japan. Kawasaki Med J 23, 143–150.

Headlee, W.H., Kmezca, J.M., and Cable, R.M. (1939). Report of a native case of infection by the fish tapeworm, *Diphyllobothrium latum*. J Indiana State Med Assoc 32, 188–189.

Hermann, B. (1985). Parasitologische-Epidemiologische Auswertungen Mittlaterlicher Kloaken. Z Archaol Mittelalters 13, 131–161.

Hirata, M., Yamaguchi, Y., Ikei, Y., and Koyama, G. (2006). A case of *Diphyllobothrium latum/nihonkaiense* infection identified by capsule endoscopy in small intestine. Gastrointest Endosc 64, 129–130.

Ho, P.W.L, Pien, F.D., and Guerrero, R.C. (1979). *Diphyllobothrium latum* infection in a Hawaiian male. Hawaiian Med J 38, 401–402.

Holiday, D.M., Guillen, S., and Richardson, D.J. (2003). Diphyllobothriasis of the Chiribaya culture (700–1476 AD) of Southern Peru Comp Parasitol 70, 167–171.

Hoffman, G.L. (1999). Parasites of North American Freshwater Fishes. Ithaca, NY: Cornell University Press.

Isobe, A., Shiwaku, K., Fukushima, T., Gao, T., Gibson, D.I., Bray, R., Margolis, L., and Yamane, Y. (1998). Molecular phylogeny of *Diphyllobothrium nihonkaiense* Yamane 1986) and other diphyllobothriid tapeworms based on mitochondrial cytochrome c oxidase subunit 1 gene sequence. Parasitol Int 476, 138.

Jacob, L., and Pillai, K.M. (1991). Incidence of parasitic infection in dogs in Thrissur, Kerala. J Veter Anim Sci 22(1), 149–150.

Jansen, J., and Over, H.J. (1962). The occurrence of parasites in prehistorical material from north-west Germany. Tijdschr Diergeneesk 87, 1377–1379.

Kifune, T., Hatsushika, R., Ushirogawa, H., Takeda, S., Kono, K., and Shimzu, T. (2000). A case study of human infection with diphyllobothriid tapeworm (Cestoda: Pseudophyllidea) found in a man in Fukuoka Prefecture, Japan. Med Bull Fukuoka Univ 27, 93–100.

Kingston, S., and Kilbourn, J.P. (1989). Diagnosis of a fourth reported case of intestinal anisakiasis in the United States? Am J Clin Pathol 92, 256.

Klebanovskii, V.A., Smirnov, P.L., Klebanovskaya, I.A., and Obgol'ts, A.A. (1977). Human helminthiases in eastern Taimyr (Katangskii region). Problemy epidemiologii I profilaktiki prirodnoochagovykh boleznei v Zapolyar'e. Sbornik nauchnykh rabot. Omsk Med Inst; Omsk; USSR. 144–164.

Kyronseppa, H. (1993). The occurrence of human intestinal parasites in Finland. Scand. J Infect Dis 25, 671–673.

Lee, S.H., Chai, J.Y., Hong, S.T., Sohn, W.M., Huh, S., Cheong, E.H., and Kang, S.B. (1989). Seven cases of *Diphyllobothrium latum* infection. Kor J Parasitol 27, 213–216.

Lee, S.K., Shin, B.M., Chung, N.S., Chai, J.Y., and Lee, S.H. (1994a). Second report on intestinal parasites among the patients of Seoul Paik Hospital (1984–1992). Kor J Parasitol 32(1), 27–33.

Lee, S.H., Chai, J.Y., Seo, M., Kook, J., Huh, S., Ryang, Y.S., and Ahn, Y.K. (1994b). Two rare cases of *Diphyllobothrium latum parvum* type infection in Korea. Kor J Parasitol 32, 117–120.

Lee, K.W., Suhk, H.-C., Pai, K.-S., Shin, H.-J., Jung, S.-Y., Han, E.-Y., and Chai, J.-Y. (2001). *Diphyllobothrium latum* infection after eating domestic salmon flesh. Kor J Parasitol 39, 319–321.

Le Bailly, M., Leuzinger, U., Schlichtherle, H., and Bouchet, F. (2005). *Diphyllobothrium*: Neolithic parasite? J Parasitol 9, 957–959.

Lubinsky, G.A., and Loch, J.S. (1979). Ichthyoparasites of Manitoba: literature review and bibliography. Can Fish Mar Serv Manus Rep, 513, 29p.

Magath, T.B. (1933). The relationship of *Diphyllobothrium latum* infestation to the public health. JAMA 101, 337–341.

Magath, T.B. (1937). Factors influencing the geographic distribution of *Diphyllobothrium latum*. Papers in Helminthology published in commemoration of the 30 year jubileum of K.I. Skrjabin and of the 15th anniversary of the All-Union Institute of Helminthology, Moscow, 366–380.

Magath, T.B., and Essex, H.E. (1931). Concerning the distribution of *Diphyllobothrium latum* in North America J Prev Med 5, 227–242.

Maqbool, A., Saleem, A., Awan, M.A., and Amin, M.K. (1998). Prevalence and chemotherapy of zoonotic parasitic diseases in dogs in Faisalabad (Punjab), Pakistan. Vet Arhiv 68, 121–125.

Mar, P.H., Su, Y.C., Fei, A.C.Y., and Bowman, D.D. (1999). A survey of endoparasitic zoonoses of stray dogs and cats in Taipei city. An Asia Seasonly Report of Environmental Microbiology in the Republic of China. 8, 77–86.

Margolis, L., Rausch, R.L., and Robertson, E. (1979). Diphyllobothrium ursi from man in British Columbia—first report of this tapeworm in Canada. Can J Public Health 64, 588–589.

Margolis, L., and Arthur, J.R. (1979). Synopsis of the parasites of fishes of Canada. Bull Fish Res Bd Can 199, Ottawa, Ontario, Canada.

Marty, A.M., and Neafie, R.C. (2000). Diphyllobothriasis and sparaganosis. In: Meyers, W.M., Neafie, R.C., Marty, A.M., Wear, D.J., eds. Pathology of Infectious Diseases Vol. 1. Helminthiases. Armed Forces Inst Pathol 165–183.

Matsuura, T., Bylund, G., and Sugane, K. (1992). Comparison of restriction fragment length polymorphisms of ribosomal DNA between *Diphyllobothrium nihonkaiense* and *D. latum*. J Helminthol 66, 261–266.

McDonald, T.E., and Margolis, L. (1995). Synopsis of the parasites of fishes of Canada: Supplement (1978–1993). Can Spec Publ Fish Aqua Sci 122, Ottawa, Ontario, 265p.

Min, D.Y. (1990). Cestode infections in Korea. Kor J Parasitol 28(suppl), 123–144.

Min, D.Y., Ahn, M.H., Kim, K.M., and Kim, C.W. (1987). Intestinal parasite survey in Seoul by stool examination at Hanyang University Hospital. Kor J Parasitol 24, 209–212.

Mitchell, P.D., and Stern, E. (2000). Parasitic intestinal helminth ova from the latrines of the 13th century crusader hospital of St. John in Ancre, Israel. Paleopathology Association 13th Biennal European Members meeting, Cheiti, 21, 22.

Mokhayer, B. (1981). Parasites of fish in the Sefid Rud Basin. J Vet Fac Univ Tehran 36, 61–75.

Moon, J.R. (1976). Public health significance of zoonotic tapeworms in Korea. Intern J Zoonon 3, 1–18.

Mueller, J.F. (1937). A repartition of the genus *Diphyllobothrium*. J Parasitol 23, 308–310.

Muller, R. (2001). Worms and human diseases. Second edition. Wallingford. CABI publishing, 320 p.

Muratov, I.V. (1985). Additional hosts of broad tapeworms and factors of transmission of diphyllobothriasis in the Lower Amur River area (Russian SFSR, USSR). Med Parazit Parazit Bolezni 0(2), 24–28.

Muratov, I.V., and Poshokov, P.S. (1988). *Diphyllobothrium klebanovskii*, new species, a parasite of man. Parazitologiya (Lenin.). 22, 165–170.

Muratov, I.V., and Semenova, T.A. (1986). Characteristics of the early stages in the development of *Diphyllobothrium latum* and *Diphyllobothrium* sp. from the Amur region (Russian SFSR, USSR). Med Parazit Parazit Bolezni 0(3), 48–50.

Neghme, A. (1953). An autochthonous focus of *Diphyllobothrium latum* in the southern hemisphere. In: Dayal, J., and Singh, K.S., eds. Thapar Commemoration Volume. A collection of Articles Presented to Prof. G.S. Thapar on his 60th Birthday. 223–226.

Neghme, A., and Bertin, V. (1951a). *Diphyllobothrium latum* en Chile. IV. Estado actual de las investigaciones epidemiologicas. Rev Chil Hig Med Prev 13, 8–11.

Neghme, A., and Bertin, V. (1951b). Estado actual de las investigaciones sobre *Diphyllobothrium* latum en Chile. Rev Med Chile 79, 637–640.

Neghme, A., Bertin, V., Tagle, I., Silva, R., and Artigas, J. (1950a). *Diphyllobothrium latum* en Chile. II.- Primera encuseta en el Lago Colico. Bol Infor Parasit Chilenas 5, 16–17.

Neghme, A., Bertin, V., Donckaster, R., Silva, R., Artigas, J., Bunster, A., Munoz, T., Faundes, F., and Basso, B. (1950b). *Diphyllobothrium latum* en Chile. III. Nuevas encusetas coprologicas en el sur del pais y comprobacion de un caso autoctono en Santiago. Bol Infor Parasit Chilenas 5, 42.

Nicholson, D. (1932). *Diphyllobothrium* infection in *Esox lucius*. Can J Res 6, 166–170.

Nicoulaud, J., Yera, H., and Dupouy-Camet, J. (2005). Prevalance de l'infestation par *Dipyllobothrium latum* L 1758 chez les perches (*Perca fluviatilis*) du Lac Leman. Parasite 12, 362–364.

Nishiyama, T. (1994). Environmental changes and tapeworm diseases in Japan: Special reference to diphyllobothriasis nihonkaiense (diphyllobothriasis latum). Jpn J Parasitol 43, 471–476.

Norman, J.P. (1937). *Bothriocepahalus latus* infection simulating peptic ulcer. Med Rec 146, 533–534.

Okino, T., Shimzu, M., Ohyama, F., Ushirogawa, H., Furukawa, N., Matoba, K., and Hatsushika, R. (2005). Survey of parasitological examinations performed by the Department of Parasitology, Kawasaki Medical School 3. Results: (1996–2003). Kawasaki Igakkaishi 31, 31–37.

Pancharatnam, S., Jacob, E., and Kang, G. (1998). Human diphyllobothriasis: first report from India. Trans R Soc Trop Med Hyg 92, 179–180.

Peduzzi, R., and Boucher-Rodoni, R. (2001). Resurgence of human bothriocephalosis (*Diphyllobothrium latum*) in subalpine lake region. J Limnol 60, 41–44.

Plotz, M. (1932). *Diphyllobothrium latum*: Infestation on the eastern seabord. JAMA 98, 312–314.

Powell, A.M., and Chubb, J.C. (1966). A decline in the occurrence of *Diphyllobothrium* plerocercoids in the trout *Salmo trutta* L. of Llyn Padarn, Caernarvonshire. Nature 211, 439.

Pullola, T., Vierimaa, J., Saari, S., Virtala, A.-M., Nikander, S., and Sukura, A. (2006). Canine intestinal helminths in Finland: prevalence, risk factors and endoparasites control practices. Vet Parasitol 140, 321–326.

Ramirez, L.S., Standen, I.D., Lizana, M.E., and Biolley, H.M.A. (1977). Preliminary study of diphyllobothriasis in Lake Colico (Cautin Province, IX Region, Chile). Acta Zool. Lillo., 35, 1, 21–28. [Actas, Septimo Congr. Latinoamericano de Zool 2, 15–21 Mayo, 1977, Tucuman, Argentina.].

Rao, P.B., and Singh, K.R. (1998). Diphyllobothriasis in a tiger (*Panthera tigris*). J Vet Parasitol 12, 148.

Rausch, R.L., and Hilliard, D.K. (1970). Studies on the helminth fauna of Alaska. XLIX. The occurrence of *Diphyllobothrium latum* (Linnaeus, 1758) (Cestoda: Diphyllobothriidae) in Alaska, with notes on other species. Can J Zool 48, 131–140.

Rausch, R.L., and Adams, A.M. (2000). Natural transfer of helminths of marine origin to freshwater fishes, with observations on the development of *Diphyllobothrium alascense*. J Parasitol 86, 319–327.

Reinhard, K.J., and Aufderheide, A.C. (1990). Diphyllobothriasis in pre-Columbian Chile and Peru: Adaptative radiations of a helminth species to native American populations. Papers in Paleopathology, 8th European Members Meeting, Cambridge.

Reinhard, K.J., and Barnum, S.V. (1991). Parasitology as an interpretative tool in archaeology. Am Antiq 57, 231–245.

Reinhard, K., and Urban, O. (2003). Diagnosing ancient Diphyllobothriasis form Chinchorro mummies. Mem Inst Oswaldo Cruz 98, 191–193.

Revenga, J., and Semenas, L. (1991). Diphyllobothriasis of introduced salmonids in the Nahuel Huapi National Park and Reserve, Argentina: morphology of plerocercoids. Arch Med Vet 23, 157–164.

Revenga, J.E. (1993). *Diphyllobothrium dendriticum* and *Diphyllobothrium latum* in fishes from southern Argentina: association, abundance, distribution, pathological effects, and risk of human infection. J Parasitol 79, 379–383.

Reyes, H., Doren, G., and Inzunza, E. (1972). Human taeniasis. Current frequency of infection by different species in Santiago, Chile. Bol Chil Parasitol 27, 23–29.

Rim, H-J. (1998). Field investigations on epidemiology and control of fish-borne parasites in Korea. Int J Food Sci Tech 33, 1547–168.

Rohela, M., Jamaiah, I., Chan, K.W., and Wan Yusoff, W.S. (2002). Diphyllobothriasis: the first case report from Malyasia. Southeast Asian J Trop Med Public Health 33, 229–230.

Rosenberg, J., Heumann, E., and Matzner, M.J. (1955). The regulation and treatment of endemic fish tapeworm infestation (diphyllobothriasis). Am J Gastroenterol 24, 121–136.

Ruebush, T.K. II, Juranek, D.D., and Brodsky, R.E. (1978). Diagnoses of intestinal parasites by state and territorial public health laboratories, 1976. J Infect Dis 138, 114–117.

Rush, W.M. (1932). *Diphyllobothrium latum* in bear. J Mammol 13, 274–275.

Sagua, H., Neira, I., Araya, J., and Gonzalez, J. (2001). New cases of *Diphyllobothrium pacificum* (Nyelin, 1931) Margolis, 1956 human infections in North of Chile, probably related with El Nino phenomenon, 1975–2000. Bol Chil Parasitol 56, 22–25.

Sampaio, J.L.M., de Andrade, V.P., da Conceicao Lucas, M., Fung, L., Gagliardi, M.B., Santos, S.R.P., Mendes, C.M.F., de Paula Eduardo, M.B., and Dick, T. (2005). Diphyllobothriasis, Brazil. Emerg Infect Dis 10, 1598–1600.

Sandweiss, D.J., and Sugarmann, M.H. (1947). Fish tapeworm infestation due to sampling of gefülte fish or its soup before adequate cooking. Report of eleven cases with a discussion on the epidemiology of the disease. J Mich Med Soc 46, 1156–1164.

Santos, F., and de Faro, L.B. (2005). The first confirmed case of *Diphyllobothrium latum* in Brazil. Memo Inst Oswaldo Cruz 6, 685–686.

Seah, S.K.K. (1973). Intestinal parasites in Chinese immigrants in a Canadian city. J Trop Med Hyg 76, 291–293.

Semenas, L., Krieter, A., and Urbnaski, J. (2001). New cases of human diphyllobothriasis in Patagonia, Argentine. Rev Saude Publica, 35, 214–216.

Sen Gupta, M.R. (1974). A preliminary report on diseases and parasites of zoo animals, birds and reptiles. Ind J Anim Health 13, 15–24.

Skerikova, A., Brabec, J., Kuchta, R., Jimenez, J.A., Garcia, H.H., and Scholz, T. (2006). Is the human-infecting *Diphyllobothrium pacificum* a valid species or just a South American population of the Holarctic fish broad tapeworm *D. latum*? Am J Trop Med Hyg 75, 307–310

Stadlbauer, V., Harberl, R., Langner, C., Krejs, G.J., and Eherer, A. (2005). Annoying vacation souvenir: fish tapeworm (Diphyllobothrium sp.) infestation in an Austrian fisherman. Wien Klin Wochenschr 117, 776–779.

Stunkard, H.R. (1965). Variation and criteria for generic and specific determination of diphyllobothriid cestodes. J Helminthol 39, 281–296.

Summers, W.A., and Wienstein, P.P. (1943). *Diphyllobothrium latum* in Florida. Am J Trop Med 23, 363–367.

Suvorina, V.I., and Simonova, N.F. (1993). Epidemiological aspects of diphyllobothrioses in Yakutiya. Med Parazit Parazit Bolezni 4, 23–26.

Szalai, A.J., Craig, J.F., and Dick, T.A. (1992). Parasites of fishes from Dauphin Lake, Manitoba 1985–1987. Can Tech Rep Fish Aquat Sci 1735, 36p.

Szidat, L. (1944). Uber die Erhaltungsfahigkeit von Helmintheneiern in Vor- und Fruhgeschichtlichen Moorleichen. Z Parasitenkd 13, 265–274.

Szidat, L., and Soria, M. (1952). Difilobotriasis en nuestro pais. Nota preliminar. Pren Med Argentina 39, 77–78.

Tanowitz, H.B., Weiss, L.M., and Wittner, M. (2001). Tapeworms. Curr Infect Dis Rep 3, 77–84.

Tavares, L.E.R., Luque, J.L., and do Bomfim, T.C.B. (2005). Human diphyllobothriasis; reports from Rio de Janeiro, Brazil. Re Bras Parasitol Vet 14, 85–87.

Thompson, J.E. (1936). Some observations on the European broad fish tapeworm *Diphyllobothrium latum*. J Am Vet Med Assoc 89, 77–86.

Tomita, S., Tongu, Y., Sakumoto, D., Suguri, S., Itano, K., Inatomi, S., and Kawaguchi, K. (1979). Epidemiological survey for *Diphyllobothrium latum* in Okayama prefecture. Jpn J Parasitol 28, 317–32.

Torres, P., Figueroa, L., and Franjola, R. (1983). Pseudophyllidea in the south of Chile. IX. Types of plerocercoids in trouts from five lakes and new cases of *Diphyllobothrium latum* in man and *D. pacificum* in a dog. Int J Zoonoses 10, 15–21.

Torres, P., Franjola, R., Perez, J., Auad, S., Uherek, F., Miranda, J.C., Flores, L., Riquelme, J., Salazar, S., Hermosilla, C., and Rojo, R. (1989a). Epidemiology of *Diphyllobothrium* spp. in the Valdiva river basin, Chile. (Epidemiologia de la difilobotriasis en la cuenca del Rio Valdivia, Chile). Rev Saude Publica 23, 45–57.

Torres, P., Torres, J., Garrido, O., and Thibaut, J. (1989b). Research on Pseudophyllidea (Carus, 1813) in southern Chile. X. Experimental observations on the coexistence of *Diphyllobothrium latum* (L.) and *D. dendriticum* (Nitzch) plerocercoids in salmonids of the river Valdivia basin. Arch Med Vet 21, 51–57.

Torres, P., Franjola, R., Weitz, J.C., Pena, G., and Morales, E. (1993). New records of human diphyllobothriasis in Chile (1981–1992), including a case of multiple *Diphyllobothrium latum* infection. Bol Chil Parasitol 48, 39–43.

Torres, P., Gesche, W., Montefusco, A., Miranda, J.C., Dietz, P., and Huijse, R. (1998). Diphyllobothriasis in man and fishes from Lake Rinihue, Chile: effect of health education, seasonal distribution and relationship to sex, size and diet of the fish. Arch Med Vet 30, 31–45.

Torres, P., Aedo, E., Figueroa, L., Siegmund, I., Silva, R., Navarrete, N., Puga, S., Marin, F., and Aedo, E. (2000). Infection by parasitic helminths in coho salmon, *Oncorhynchus kitutch*, during their return to Simpson Tiver, Chile Bol Chil Parasitol 55, 1–7.

Torres, P., Cuevas, C., Tang, M., Barra, M., Franjola, R., Navarrete, N., Montefusco, A., Otth, L., Wilson, G., Puga, S., Figueroa, L., and Cerda, O. (2004). Introduced and native fishes as infection foci of *Diphyllobothrium* spp. in humans and dogs from two localities at Lake Panguipulli in southern Chile. Comp Parasitol 71, 111–117.

Turgeon, E.W.T. (1974). *Diphyllobothrium latum* (fish tapeworm) in the Sioux Lookout zone. Can Med Assoc J 111, 507.

Uddin, M., Dewan, M.L., Hossain, M.I., and Huq, M.M. (1980). Occurrence of *Diphyllobothrium latum* larvae (plerocercoid) in lotiya (*Harpodon nehereus*) fish. Bangladesh Vet J 14, 33–35.

Valdmann, H., Moks, E., and Talvik, H. (2004). Helminth fauna of Eurasian Lynx (*Lynx lynx*) in Estonia. J Wildl Dis 40, 356–360.

Vergeer, T. (1928). *Diphyllobothrium latum* (Linne 1758) the broad tapeworm of man. JAMA 90, 673–678.

Vergeer, T. (1929a). The broad tapeworm in America, with suggestions for its control. J Infect Dis 44, 1–12.

Vergeer, T. (1929b). The dog, a reservoir of the broad tapeworm. JAMA 92, 607–608.

Vik, R. (1964). Studies of the helminth fauna of Norway. V. Pleroceroids of *Diphyllobothrium* spp. From the Rossaga water system, Nordland County. Nytt Magas Zool 12, 1–9.

Von Bonsdorff, B. (1977). Diphyllobothriasis in Man. London: Academic Press.

Waloch, M. (2005). Cestode infections in Poland in 2003. Przegl Epidem 59, 331–335.

Ward, H.B. (1930). The introduction and spread of the fish tapeworm (*Diphyllobothrium latum*) in the United States. De Lamar Lect 1929–1030, 1–36.

Wardle, R.A. (1932). The cestoda of Canadian fishes. II. The Hudson bay drainage system. Contri Can Biol Fish 7, 377–403.

Wardle, R.A. (1933). Significant factors in the plerocercoid environment of *Diphyllobothrium latum* (Linn.). J Helminthol 1, 25–44

Wardle, R.A. (1935). Fish tapeworm. Bull Biol Bd Can 45, 25 p.

Wardle, R.A., and McLeod, J.A. (1952). The Zoology of Tapeworms. Minneapolis: University of Minnesota Press.

Waters, H.S., and O'Connor, F.W. (1932). *Diphyllobothrium latum* sources of native infestations in New York City. JAMA 99, 1941.

Watson, R.A., and Dick, T.A. (1979). Metazoan parasites of whitefish *Coregonus clupeaformis* (Mitchill) and cisco *C. artedii* Lesueur from Southern Indian Lake, Manitoba. J Fish Biol 15(5), 579–587.

Watson, R.A., and Dick, T.A. (1980). Metazoan parasites of pike, Esox lucius Linnaeus, from Southern Indian Lake, Manitoba, Canada. J Fish Biol 17(3), 255–261.

Yamaguchi, T., Kamoshida, S., Shimizu, T., Morita, T., Okada, Y., Matsuda, M., Kano, M., Hosokawa, Y., Sawada, S., and Arakawa, Y. (1997). Two cases of *Diphyllobothrium latum* infection. Nichidai Igaku Zasshi 1997; 56, 264–268.

Yamaguti, S. (1935). Studies on the helminth fauna of Japan. Part 7: Cestodes of Mammals and snakes. Jpn J Zool 6, 233–246.

Yamane, Y., Kamo, H., Bylund, G., and Wikgren, B.J. (1988). *Diphyllobothrium nihonkaiense* sp. nov. (Cestoda: Diphyllobothriidae) – Revised identification of Japanese broad tapeworm. Shimane J Med Sci 10, 29–48.

Yamane, Y., Shiwaku, K., Abe, K., Osaki, Y., and Okamoto, T. (1989). The taxonomic differences of embryonic hooks in *Diphyllobothrium nihonkaiense, D. latum* and *D. dendriticum*. Parasitol Res 75, 549–553.

Yazaki, S., Kamo, H., Kawasaki, H., and Yamane, Y. (1984). Experimental infection of Japanese men with *Diphyllobothrium latum* from Finland. Yonago Acta Med 27, 19–30.

Yera, H., Estran, C., Delaunay, P., Gari-Toussaint, M., Dupouy-Camet, H., and Marty, P. (2006). Putative *Diphyllobothrium nihonkaiense* acquired from a Pacific salmon (*Oncorhynchus nerka*) eaten in France; genomic identification and case report. Parasitol Int 55, 45–49.

Yokogawa, M., Niimura, M., Kobayashi, M., Hata, H., Tazaki, T., Tokita, K., and Kojima, S. (1979). Epidemiological survey for diphyllobothriasis latum in Chiba prefecture and treatment with bithionol. Jpn J Parasitol 28: 12. (Japanese Society of Parasitology: Proceedings of the 38th East Japan Regional Meeting in Mitaka City, 14 October 1978. Abstracts.)

Yoshida, M., Hasegawa, H., Takaoka, H., and Miyata, A. (1999). A case of *Diphyllobothrium nihonkaiense* infection successfully treated with oral administration of Gastrografin. Parasitol Int 48, 151–155.

Zhang, Y.R.A. (1996). A case report of *Diphyllobothrium latum* infection. Chin. J Parasit Dis Contrl 9, 211.

5
Anisakid Nematodes and Anisakiasis

A.J. Lymbery and F.Y. Cheah

Anisakiasis (anisakidosis) refers to infection of people with larval stages of ascaridoid nematodes belonging to the family Anisakidae (and possibly also Raphidascarididae). These worms, commonly called anisakids, utilize aquatic mammals, piscivorous birds, aquatic reptiles, or fish as definitive hosts, and aquatic invertebrates and fish as intermediate or paratenic hosts. Adult and larval anisakids often have major pathological effects in the alimentary tract and associated organs of their natural host species (reviewed by Smith, 1999).

Humans become infected by consuming fish or cephalopod mollusks with larval anisakids in their flesh, viscera, or body cavity. Although mammalian hosts have been experimentally infected with worms from a number of species within the families Anisakidae and Raphidascarididae, human infections almost always involve *Anisakis simplex* and *Pseudoterranova decipiens,* both of which have recently been found to constitute a complex of morphologically similar sibling species. Humans are accidental hosts in the life cycle of anisakid nematodes, and, although the parasites almost never develop further within the human alimentary tract, they may penetrate the tract and associated organs, with severe pathological consequences. In addition, there is growing evidence that the parasites may produce a strong allergic reaction, often culminating in anaphylactic shock. Anisakiasis is therefore a serious zoonotic disease, and there has been a dramatic increase in its reported prevalence throughout the world in the last two decades.

This chapter briefly describes the taxonomy, geographic distribution, and biology of anisakids, concentrating principally on the *Anisakis simplex* and *Pseudoterranova decipiens* species complexes, before discussing the pathology, diagnosis, treatment, and control of anisakiasis. We end with a description of what we believe are the major areas where more research effort is required to further understand the diseases caused by this important group of parasites.

Anisakid Nematodes

Taxonomy and Geographic Distribution

Species within the superfamily Ascaridoidea are among the most thoroughly studied nematode parasites of vertebrates. Ascaridoids have been used extensively for studies of respiratory biochemistry, immunology, molecular genetics, and population genetics (Nadler and Hudspeth, 2000). Paradoxically, however, the evolutionary taxonomy of the superfamily is very uncertain, largely because of the great variation of external features and life cycle patterns among different species (Fagerholm, 1991; Anderson, 1992).

Before the widespread use of cladistic analysis, most hypotheses of ascaridoid phylogeny were based on a few key morphological structures or life history features, such as the presence or absence of the ventriculus (Hsu, 1933), the structure of the secretory-excretory system (Hartwich, 1974), or male caudal morphology (Fagerholm, 1991). Differences in features used for phylogenetic reconstruction led to an array of contrasting interpretations and hypotheses of relationships, in turn leading to instability of ascaridoid classification, although the classification schemes of Hartwich (1974) and Fagerholm (1991) have been most commonly used.

The anisakids, broadly defined, constitute those ascaridoids with an aquatic definitive host (fish, reptile, piscivorous bird, or mammal), whose transmission is dependent on water and usually involves aquatic invertebrate and fish intermediate or paratenic hosts (Anderson, 1992). At least 20 different genera of anisakids have been described. Hartwich (1974) recognized only one family within the group (Anisakidae), with three subfamilies: Anisakinae; Geoziinae, and Raphidascaridinae (Table 5.1). Fagerholm (1991), however, split the anisakids into two families: Anisakidae (containing the subfamilies Anisakinae and Contracaecinae) and Raphidascarididae (Table 5.1).

Molecular data have been used to investigate phylogenetic relationships in the superfamily Ascaridoidea (e.g. Nadler, 1992; Nadler and Hudspeth, 1998, 2000; Zhu et al., 1998). Although these studies have not fully resolved taxonomic uncertainty within the superfamily, they tend to support the classification proposed by Fagerholm (1991). In particular, combined analysis of both mitochondrial DNA (mtDNA) and nuclear ribosomal DNA (rDNA) sequences provide provisional support for a monophyletic origin of the family Anisakidae, including the genera *Anisakis, Pseudoterranova, Terranova*, and *Contracaecum*, distinct from the monophyletic family Raphidascarididae, including the genera *Geozia* and *Hysterothylacium* (Nadler and Hudspeth, 2000).

Species identification in the Anisakidae has traditionally been complicated by a lack of distinguishing morphological characteristics, particularly in larval worms. Historically, therefore, only two major zoonotic species were recognized: the herring worm or whale worm *Anisakis simplex*, and the codworm or seal worm *Pseudoterranova decipiens*, both with an apparently cosmopolitan distribution (Smith and Wootten, 1978; Oshima, 1987). Recent molecular genetic studies,

TABLE 5.1. Contrasting classification schemes of anisakid nematodes by Hartwich (1974) and Fagerholm (1991).

Authority	Family	Subfamily	Genera
Hartwich (1974)	Anisakidae	Anisakinae	*Anisakis*, *Phocanema* (= *Pseudoterranova*), *Terranova*, *Sulcascaris*, *Duplicaecum*, *Galeiceps*, *Contracaecum*, *Phosascaris*
		Geoziinae	*Goezia*
		Raphidascaridinae	*Raphidascaris*, *Raphidascaroides*, *Thynnascaris* (= *Hysterothylacium*), *Lappetascaris*, *Aliascaris*, *Heterotyphlum*, *Paranisakis*, *Paranisakiopsis*
Fagerholm (1991)	Anisakidae	Anisakinae	*Anisakis*, *Pseudoterranova*, *Terranova*, *Sulcascaris*, *Peritrachelius*, *Pulchrascaris*, *Paranisakiopsis*
		Contracaecinae	*Contracaecum*, *Galeiceps*, *Phosascaris*
	Raphidascarididae		*Raphidascaris*, *Raphidascaroides*, *Hysterothylacium*, *Lappetascaris*, *Heterotyphlum*, *Paranisakis*, *Goezia*, *Sprentascaris*, *Paraheterotyphlum*

however, have shown that both of these morphospecies actually comprise a number of sibling species, genetically differentiated and often with distinct geographic ranges.

Three different species have been described within the *Anisakis simplex* complex (Mattiucci et al., 1997). *Anisakis simplex* (sensu stricto) is found in the north Atlantic Ocean between 30°N and the Arctic polar circle, *A. pegreffi* is distributed in southern oceans from 35°S to 55°S as well as in the Mediterranean Sea, and *A. simplex* C is found in the northern Pacific and southern oceans below 30°N (Mattiucci et al., 1997). Within each sibling species, there is very little genetic differentiation between populations located thousands of kilometers apart. This is thought to be caused by the homogenizing effects of gene flow, enhanced by the high mobility of fish hosts (Mattiucci et al., 1997). In addition to these three sibling species, four other species of *Anisakis* have been confirmed using genetic markers: *A. typica*, from the Atlantic Ocean, Indian Ocean, and Mediterranean Sea; *A. physeteris*, from the Atlantic and Mediterranean; *A. brevispiculata*, from the south east Atlantic; and *A. zhiphidarum*, from the southeast Atlantic and Mediterranean (Mattiucci et al., 2005).

Six different species have been described within the *Pseudoterranova decipiens* species complex (Paggi et al., 1991, 2000; Mattiucci et al., 1998; George-Nascimento and Urrutia, 2000; McClelland, 2002). *Pseudoterranova decipiens* (sensu stricto) is found in the northeast Atlantic, in waters off northern Europe and Iceland, and in the northwest Atlantic, off eastern Canada. *Pseudoterranova krabbei* is found only in the northeast Atlantic, where it is sympatric with *P. decipiens* (sensu stricto), while *P. bulbosa* is confined to the Barents Sea in the northeast Atlantic, and is also found in the north Pacific, off Japan, where it is sympatric with *P. azarasi* (McClelland, 2002). Two species have been described from southern oceans; *P. decipiens* E in the Antarctic and *P. cattani* in the south Pacific, off Chile. In addition to these six sibling species, two other species of *Pseudoterranova* have been described, based on morphological criteria: *P. kogiae* and *P. ceticola* (Anderson, 1992).

Neither the species lists for *Anisakis* and *Pseudoterranova* nor the described geographic ranges of these species can be regarded in any sense as definitive. Further genetic studies will undoubtedly uncover more species of anisakid nematodes and extend the geographic ranges of those species that have already been described.

Biology

Life Cycle

Anisakids utilize aquatic mammals, piscivorous birds, aquatic reptiles, or fish as definitive hosts (Anderson, 1992). Larval anisakids are found in aquatic invertebrates and fish, although for most species the precise details of the life cycle are uncertain and it is not clear whether the invertebrate and fish hosts are obligatory or whether larval development occurs within them. For this reason, they have been referred to by different authors as both intermediate and paratenic hosts; for the purposes of this review we will refer to both invertebrates and fish as intermediate hosts, on the (untested) assumption that they are both required for successful completion of the parasite life cycle.

A generalized life cycle for species of *Anisakis* and *Pseudoterranova* is shown in Figure 5.1. Adults of *Anisakis* spp. are found in the alimentary tract (particularly the stomach), mainly of cetaceans (dolphins, porpoises, and whales), and adults of *Pseudoterranova* spp. in pinnipeds (seals, sea lions, and walrus). Eggs are shed in the feces, and require an incubation period before hatching into free living larvae. There is uncertainty over whether one or two molts occur within the egg and therefore whether the free-living larvae are second stage (L2) or third stage (L3) (Køie and Fagerholm, 1993; Køie et al., 1995; Measures and Hong, 1995). Free-living L2 or L3 larvae are ingested by invertebrates, particularly crustaceans, where they grow within the hemocoel, perhaps undergoing one molt (Acha and Szyfres, 1987; Klimpel et al., 2004). Fish and cephalopod mollusks become infected by eating crustaceans containing L3 larvae, which penetrate the intestine and invade the tissues, where they may continue to grow or become encapsulated (Anderson, 1992). Fish and cephalopods are referred to as primary hosts when they obtain their larval infection from crustaceans, and as secondary hosts when they become infected by eating other infected fish or cephalopods. Definitive hosts are usually infected by eating fish or cephalopods containing L3 larvae, which then develop to adults in the alimentary tract.

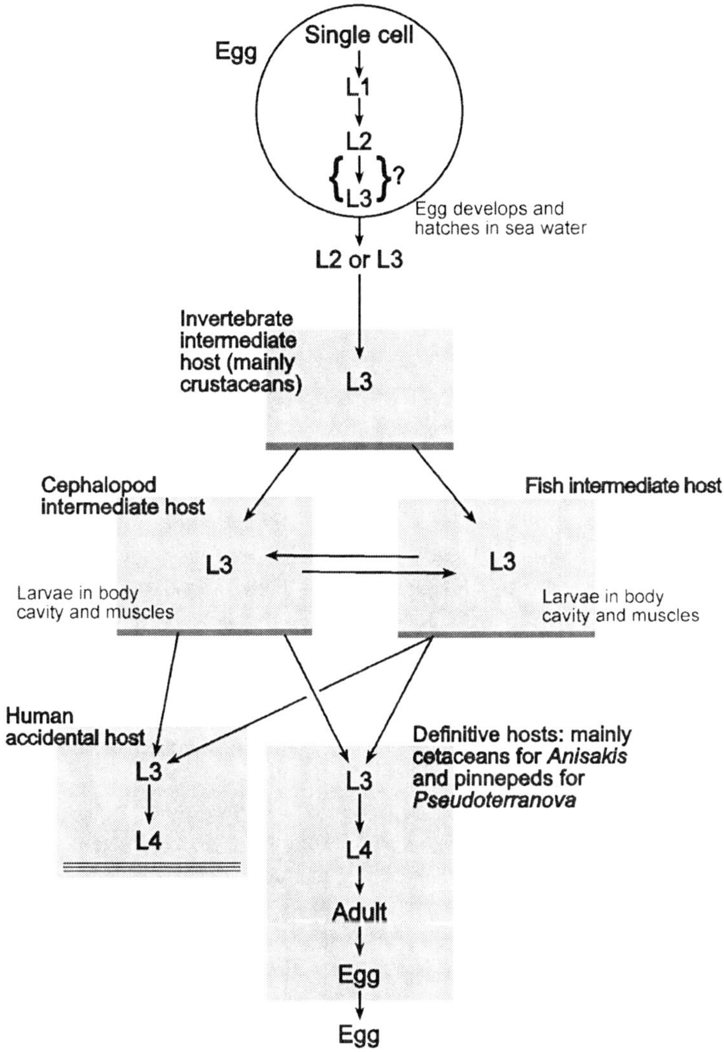

FIGURE 5.1. Life cycle of anisakid nematodes belonging to the genera *Anisakis* and *Pseudoterranova*.

Host Range

The definitive and intermediate host ranges have not been completely described for any anisakid species. This is partly due to the uncertain species-level taxonomy and confusion over identification of different morphospecies, but also to low host specificity of the group.

Definitive Hosts

Adult worms in the *Anisakis simplex* complex appear to be associated principally with oceanic dolphins and porpoises, such as harbour porpoises (*Phocoena*

phocoena), common dolphins (*Delphinus delphis*), bottle-nosed dolphins (*Tursiops truncatus*), and white-beaked dolphins (*Lagenorhynchus albirostrus*), although they are also frequently found in beluga whales (*Delphinapterus leucas*), and have been recorded from a wide range of other cetaceans, including minke whales (*Balaenoptera acutorostrata*), false killer whales (*Pseudorca crassidens*), and long-finned pilot whales (*Globicephala melaena*) (Greenwood and Taylor, 1978; Smith and Wootten, 1978; Pascoe, 1986; Wazura et al., 1986; Kuramochi et al., 1996; Ugland et al., 2004; Herreras et al., 2004; Mattiucci et al., 2005). They have also been recorded, although less frequently, from pinnipeds, such as harbor seals (*Phoca vitulina*), gray seals (*Halichoerus grypus*), subantarctic fur seals (*Arctocephalus tropicalis*), and Stellar's sea lions (*Eumatopius jubatus*) (Smith and Wootten, 1978; Stroud and Dailey, 1978; Bester, 1989).

To our knowledge, adult worms in the *Pseudoterranova decipiens* complex have been found naturally only in pinnipeds. They seem to be associated most commonly with true seals (in the family Phocidae), particularly gray seals (*H. grypus*) and harbor seals (*P. vitulina*), although they have also been recorded from many other species, such as bearded seals (*Erignathus barbatus*), ribbon seals (*Phoca fasciata*), Weddell seals (*Leptonychotes weddelli*), and harp seals (*Phoca groenlandica*) (Stroud and Roffe, 1979; Paggi et al., 1991, 2000; Brattey and Stenson, 1993; Mattiucci et al., 1998; McClelland, 2002). Eared seals (Otariidae) appear to be infected less frequently, although adult worms have been reported in northern fur seals (*Callorhinus ursinus*), Stellar's sea lions (*E. jubatus*), Californian sea lions (*Zalophus californianus*), and South American sea lions (*Otario byronia*) (Keyes, 1965; Stroud and Roffe, 1979; George-Nascimento and Urrutia, 2000; McClelland, 2002).

Although there appear to be differences in host preference between different species in both the *Anisakis simplex* and *Pseudoterranova decipiens* complexes, different parasite species may be found in the same definitive host species, and sometimes in the same individual definitive host (McClelland, 2002; Stobo et al., 2002; Mattiucci et al., 2005). The population structure of anisakid nematodes has been studied more frequently in intermediate fish hosts than in definitive hosts. Prevalences and burdens of anisakids in definitive hosts vary widely with host species, geographic location, and season (McClelland, 2002; Herraras et al., 2004). As with all parasitic infections, adult worm abundances are positively skewed, with infection intensities of more than 20,000 worms per host being recorded (Ólafsdóttir, 2001). Intensity of infection is generally positively related to host age and size (McClelland, 2002).

Crustacean Intermediate Hosts

Larvae (L2 or L3) of *Anisakis* and *Pseudoterranova* have been reported from a wide range of crustaceans, including copepods, amphipods, isopods, euphausiids, and decapods, and occasionally from polychaete worms and mollusks (Smith and Wootten, 1978; McClelland, 2002). Euphausiids (krill) appear to be the most important hosts in the life cycles of *Anisakis* species, and copepods in the life cycles of *Pseudoterranova* species (McClelland, 2002; Smith and Snyder, 2005).

Prevalences and intensities of infection are usually low, with <1% of hosts being infected and rarely more than one worm per infected host (McClelland, 2002; Smith and Snyder, 2005).

Fish Intermediate Hosts.

A very large number of fish and cephalopod species act as hosts for species of *Anisakis* and *Pseudoterranova*. For example, *Anisakis* larvae have been found in 200 fish species and 25 cephalopod species throughout the world (Abollo et al., 2001; Klimpel et al., 2004), while *Pseudoterranova* larvae have been reported from more than 75 fish species in the north Atlantic alone (McClelland et al., 1990; Desportes and McClelland, 2001). Primary fish hosts are planktivores, or predominantly planktivores, such as herring (*Clupea harengus*), haddock (*Melanogrammus aeglefinus*), blue whiting (*Micromesistius poutassou*), and juvenile plaice (*Hippoglossoides platessoides*), mackerel (*Scomber scombrus*), and cod (*Gadus morhua*), which acquire the parasite directly from crustacean invertebrate hosts (Abollo et al., 2001; McClelland and Martell, 2001). Secondary fish hosts are piscivores, such as blue shark (*Prionace glauca*), barracuda (*Sphyraena barracuda*), monkfish (*Lophius americanus*), and conger eel (*Conger conger*), which usually acquire the parasite from infected planktivorous fish (Laffon-Leal et al., 2000; Abollo et al., 2001; McClelland and Martell, 2001).

Both prevalences and parasite burdens can be very large in fish hosts. They tend to increase with host age and size, and are usually greater in secondary than in primary hosts. For example, Costa et al. (2003) found *Anisakis* larvae in 97% of black-scabbard fish (*Aphanopus carbo*) in waters off Portugal, with a mean intensity of 70 worms per fish, while over 80% of sculpins (*Myoxocephalus scorpius*) from Vega, Norway, were infected with *Pseudoterranova decipiens*, with a maximum intensity of 300 worms per fish (Jensen and Andersen, 1992). Prevalences and intensities of infection vary widely between fish hosts, both within and between anisakid species (e.g., Wharton et al., 1999; Abollo et al., 2001; Álvarez et al., 2002; Costa et al., 2003). These differences appear to be related more to geographic distribution, feeding habit, and growth rate of hosts than to behavioral or physiological host preferences of the parasites (Konishi and Sakurai, 2002; McClelland, 2002).

Accidental Hosts

Accidental hosts become infected by eating intermediate hosts (typically fish or cephalopods) that contain larval anisakids. The anisakid larvae do not complete development in the accidental host, but may penetrate the alimentary tract and invade associated organs, causing a range of pathological effects. Humans are, of course, the accidental hosts of most interest to us, but invasive anisakid larvae have also been reported from other fish-eating mammals, such as sea otters (*Enhydra lutris*) and brown bear (*Ursus arctos*) (Rausch, 1953; Davey, 1971; Jefferies et al., 1990), and experimental infections have been established in a wide range of laboratory mammals, including rats, mice, guinea pigs, rabbits, dogs, and cats (Smith, 1999). Fish-eating birds act as natural definitive hosts for a range of anisakid species in the genus *Contracaecum*, but a number of species, including fulmars (*Fulmarus glacialis*), have also been reported as accidental hosts of *Anisakis* and *Pseudoterranova* (Riley, 1972; Smith, 1999).

Site of Infection in Fish Hosts

As people usually become infected with anisakids by eating larvae contained within fish hosts, the distribution of larval nematodes within the tissues of fish is epidemiologically important. After L3 larvae, contained within infected invertebrate or primary fish hosts, are ingested by a fish, they penetrate the intestinal wall. They may then remain within the body cavity, or migrate into the musculature or internal organs. Differences in relative abundance among these microhabitats may be affected by the species of parasite, the species and age of fish infected, and the environmental conditions to which the fish are subjected after capture.

The effects of parasite species, host species, and host age on microhabitat distribution are often difficult to disentangle. Smith (1984), for example, found that the distribution of *Anisakis simplex* (sensu lato) larvae changed with age (size) in Northern Hemisphere cod (*G. morhua*) and whiting (*Merlangius merlangus*), but not in herring (*C. harengus*), blue whiting (*Micromesistius poutassou*), walleye pollock (*Theragra chalcogramma*), or mackerel (*S. scombrus*). In younger cod (<30 cm), more than 40% of the worm burden was found in the flesh, whereas in older cod (>30 cm) less than 12% of worms occurred in the flesh. In whiting, the worm burden in the flesh was higher in older fish (>40%) and lower (3%) in younger, smaller fish. Proper interpretation of these data is complicated by the fact that different morphospecies of the *A. simplex* complex were not differentiated. A study of the larvae of *Pseudoterranova decipiens* (sensu stricto) in a number of fish species in eastern Canadian waters found that worms were almost totally confined to the flesh of young, demersal fish, but became increasingly prevalent in the body cavity and surrounding musculature of older, benthic fish (McClelland et al., 1990). On the other hand, the larvae of *P. bulbosa* are usually confined to the surface of the liver in plaice (*H. platessoides*) from the Barents Sea (Bristow and Berland, 1992). Smith (1984) suggested that an understanding of microhabitat preference in anisakid larvae required the sites of L3 penetration from the lumen of the alimentary tract into the body cavity to be studied in terms of distribution, arrangement, and connection of organs of different fish species at different ages. Such studies are hampered by the fact that the mechanisms of larval penetration within hosts are still unknown.

Some studies have found that larval nematodes migrate from the visceral organs to the muscle after the death of the fish host, and that this migration may be enhanced by the cold storage or processing of ungutted fish (Van Thiel et al., 1960; Smith and Wootten, 1975; Hauck, 1977; Smith, 1984; Abollo et al., 2001). Other studies, however, have not been able to demonstrate postmortem migration of larvae (Cattan and Carvajal, 1984; Roepstorff et al., 1993). Smith and Wooten (1975) suggested that differences observed in larval migration in different studies could be due to different temperatures at which fish were kept, and to different techniques for detecting larvae, with candling much less efficient than digestion for detecting larvae in the flesh. Differences between fish species may also be important. Smith (1984) reported that storage of ungutted herring (*C. harengus*) and mackerel (*S. scombrus*) on ice (3–5°C) resulted in postmortem migration of *Anisakis simplex* (sensu lato) larvae into the flesh, but no significant migration was seen in blue whiting (*Micromesistius poutassou*), whiting (*Merlangius merlangus*), and walleye pollock (*T. chalcogramma*). On the basis of these results, Smith (1984) suggested that larval migration was related to the location of lipid deposits, with mackerel and herring being fatty species with higher lipid storage in the flesh. Roepstorff et al. (1993), however, found no migration of *Anisakis* larvae into the flesh of herring kept over a range of temperatures and examined with pepsin-HCl digestion; there were no increases in larval numbers in the flesh even though after 5 days the viscera had disintegrated completely and many larvae had migrated out of the fish via anal or gill openings. It is probable that the postmortem migration behavior of anisakids is affected by a complex of parasite, host, and external environmental variables.

Anisakiasis

Pathology

Of the 14,000 to 15,000 cases of anisakiasis that have been reported throughout the world, the vast majority (>90%) are due to infection with *Anisakis simplex* (sensu lato), with most of the remainder due to infection with *Pseudoterranova decipiens* (sensu lato) (Oshima, 1987; Bouree et al., 1995; Rosales et al., 1999; Smith, 1999; Audicana et al., 2002). The only other anisakid species reliably implicated in anisakiasis have been a few cases of infection with *Anisakis physeteris* and one case of infection with *Contracaecun osculatum* (Rosales et al., 1999; Smith, 1999). Almost all reported cases of anisakiasis have involved L3 anisakid larvae, although L4 larvae have been identified in a small number of cases (Smith, 1999).

Clinically, human anisakiasis can take a number of forms, depending on the location and histopathological lesions caused by the larvae (Table 5.2). In noninvasive anisakiasis, often (but not exclusively) associated with infections with *Pseudoterranova* (Smith, 1999; Amin et al., 2000), larvae remain in the alimentary tract, without penetrating the mucosa wall. This often causes an asymptomatic infection, which may only be discovered when the worms are expelled by coughing, vomiting, or defecating (Acha and Szyfres, 1987; Smith, 1999). Occasionally, noninvasive infections give rise to a "tingling throat" syndrome, which happens when worms migrate back up the esophagus into the oropharynx (Sakanari and McKerrow, 1989).

TABLE 5.2. Classification of human anisakiasis.

Form of anisakiasis	Description	Pathology
Noninvasive	Larvae do not penetrate mucosa	Usually asymptomatic
Invasive		
Oropharyngeal	Larvae penetrate tissues of oropharyngeal cavity	Slight
Gastrointestinal	Larvae penetrate gastric or intestinal mucosa	Slight-severe
Extra-alimentary	Larvae enter body cavity	Severe
Allergic	Allergic response to larval antigens	Slight-severe

Note: In the table and associated text, we use *alimentary* to refer to the whole tract by which food passes through the body, and *gastrointestinal* in a more restricted sense to refer only to the stomach and intestine.

In invasive anisakiasis, larvae penetrate the alimentary tract and associated organs. Penetration of the buccal mucosa or pharyngeal mucosa occurs only rarely, but has been reported for the larvae of both *Anisakis* and *Pseudoterranova* (Smith, 1999; Amin et al., 2000). These cases of oropharyngeal anisakiasis have usually been associated with slight pain, feelings of discomfort, and difficulty in swallowing (Smith, 1999).

Penetration of the gastric or intestinal mucosa is the most common form of anisakiasis. Gastric anisakiasis is more common than intestinal anisakiasis. Infections with *Anisakis* are associated with both gastric and intestinal invasions, while infections by *Pseudoterranova* usually lead only to gastric invasions (Bouree et al., 1995; Smith, 1999). Symptoms of acute gastric anisakiasis appear 1 to 12 hours after consumption of fish, and include sudden stomach pains, nausea, and vomiting. Occult blood is often found in gastric juices and stools (Sakanari and McKerrow, 1989; Smith, 1999). Acute cases are often misdiagnosed and gastric anisakiasis then becomes a chronic disease, with clinical features very similar to peptic ulcer, gastric tumor, acute gastritis, and cholecystitis (Acha and Szyfres, 1987). Intestinal anisakiasis usually manifests as an acute disease, occurring 5 to 7 days after fish consumption. Clinical symptoms include nausea, vomiting, fever, diarrhea with occult blood, and severe lower abdominal pain, similar to acute abdominal syndromes such as intestinal obstruction, appendicitis, or peritonitis (Acha and Szyfres, 1987; Bouree et al., 1995; Noh et al., 2003). Histopathological examination of invasive gastrointestinal anisakiasis usually reveals the worm embedded in a dense eosinophilic granuloma in the mucosa, often with localized or diffuse tumors in the stomach or intestinal wall (Beaver et al., 1984; Acha and Szyfres, 1987).

Occasionally, anisakid larvae have been found to completely penetrate the wall of the alimentary tract and enter the body cavity. Larvae then usually lodge in the peritoneum or subcutaneous tissues, forming tumour-like, eosinophilic granulomas or abscesses (Acha and Szyfres, 1987). Associated symptoms include abdominal pain, vomiting, and bloody stools (Smith, 1999). In all but one (unconfirmed) case, penetrating larvae have been identified as *Anisakis* (Smith, 1999).

In recent years, it has become clear that anisakiasis is often associated with a strong allergic response, with clinical symptoms ranging from isolated swellings to urticaria and life-threatening anaphylactic shock (Alonso et al., 1999; Audicana et al., 2002). The first signs of an allergic reaction usually occur within 2 hours after eating infected fish, although they may take up to 6 hours to appear (Audicana et al., 2002). Most cases of allergy reported to date have originated in Spain, and have involved an elevated immunoglobulin E (IgE) response to *Anisakis simplex* (sensu lato); the strength of allergic reactions to other anisakid species is not yet known. The *Anisakis* allergens that invoke a hypersensitivity reaction appear to be highly resistant to heat and freezing (Audicana et al., 2002; Caballero and Ignacio, 2004), raising the prospect of an allergic response to parasitized fish products that have been prepared in a way that would normally kill nematode larvae. It has been suggested that handling or even inhaling *Anisakis* allergens from contaminated fish might cause an allergic response (Purello-D'Ambrosio et al., 2000), although there is currently no evidence to support this (Alonso-Gomez et al., 2004). In any case, a priming infection with live parasites may be required to induce sensitisation (Alonso et al., 1999; Valiñas et al., 2001; Baeza et al., 2004).

Diagnosis

The symptoms of gastrointestinal anisakiasis, by far the most common form of the disease, are nonspecific, and clinical diagnosis requires careful examination of clinical symptoms and patient history. From clinical symptoms, gastric anisakiasis is often misdiagnosed as peptic ulcer, stomach tumor, or stomach polyps, and intestinal anisakiasis as appendicitis or peritonitis (Acha and Szyfres, 1987; Sakanari and McKerrow, 1989).

Definitive diagnosis of anisakiasis requires identification of the causative agent, which is much easier for gastric anisakiasis than for intestinal anisakiasis. With gastric disease, endoscopy and radiologic films can be used, but in the case of intestinal infection, verifying the presence of the organism is problematic (Ido et al., 1998). Endoscopy, first used to diagnose and treat gastric anisakiasis in Japan, utilizes a fiberoptic endoscope that is equipped with a miniature camera with biopsy forceps so that the physician can remove the larva (Sakanari and McKerrow, 1989; see Fig. 5.2). This method allows for the identification of the worm morphologically or genetically. Radiologic films of the upper gastrointestinal tract have also been used to diagnose gastric anisakiasis. Radiology can detect morphological changes in the host, such as thickened, narrowed, and obstructed areas due to host response, and sometimes the larval worm itself (Sugimachi et al., 1985; Sakanari and McKerrow, 1989). A possible alternative to radiology is sonography, but the efficacy of sonagraphic identification has not been properly tested (Ido et al., 1998).

A variety of immunological assays, including the in vivo skin prick test, complement fixation, immunofluorescent-antibody test, enzyme-linked immunosorbent assay (ELISA), Western blotting, and the radioallergosorbent test

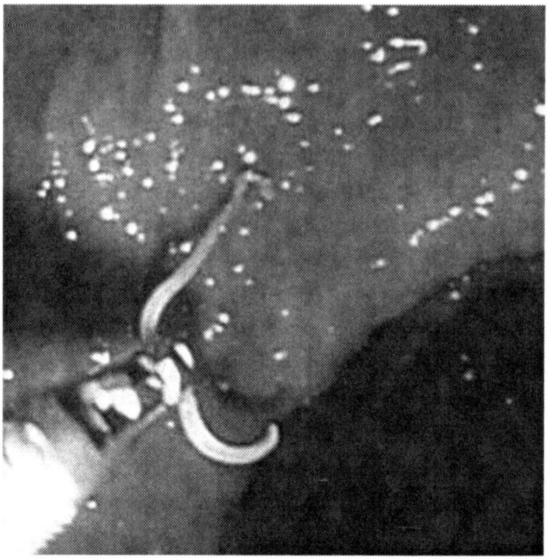

FIGURE 5.2. Anisakid nematode in gastric mucosa of Japanese patient who developed acute gastric pain after consuming a meal of raw fish. (From Hokama et al., 2005, with permission from Blackwell Publishing.) See also color insert.

(RAST) have been developed for the diagnosis of anisakiasis (Sakanari and McKerrow, 1989; Lorenzo et al., 2000). Although immunological diagnosis of anisakid infection shows potential, the interpretation of most immunological assays used to date has been complicated by antigenic cross-reactivity with other ascarids, which is partly a function of the use of multiple parasite antigens in the tests (Inglesias et al., 1996; Lorenzo et al., 2000; Lozano et al., 2004). Sensitivity and specificity could be improved by using specific antigens (Lorenzo et al., 2000). Recent studies have found an IgE-binding protein, Ani s 1, which is secreted from the excretory gland and appears to be specific for *Anisakis simplex* (sensu lato), although specificity has not been compared among different species in the *A. simplex* complex (Moneo et al., 2000; Caballero and Moneo, 2002). The detection of IgE directed at Ani s 1 has provided accurate diagnosis in clinical tests (86% sensitivity and 90% specificity) and provides much promise for the future diagnosis of *Anisakis* infections (Caballero and Moneo, 2002; Toro et al., 2004).

Treatment

Preferred treatment options for anisakiasis are endoscopy and surgical intervention. Removal of worms using a fiberoptic endoscope is recommended for acute gastric anisakiasis, leading to immediate improvement of symptoms (Noh et al., 2003). In the case of intestinal anisakiasis, if preoperative diagnosis is possible, conservative treatments with antibiotics and isotonic glucose solution have been recommended (Smith and Wootten, 1978). If these treatments are not effective, however, removal of the affected tissue by surgery is usually required (Moschella et al., 2005). There are no prescribed drugs for treatment of anisakiasis at present, although albendazole and ivermectin have shown high efficacy both in vitro and in animal trials (Dziekonska-Rynko et al., 2002). Moore et al. (2002) reported the clinical use of albendazole, with a dramatic improvement of symptoms, in a patient who was subsequently found to have positive serological results for *Anisakis*.

Epidemiology and Control

Prevalence

Anisakiasis occurs through the world, with foci in North Asia and Western Europe. Of the total (approximately 20,000) cases of anisakiasis reported to date, over 90% are from Japan (where approximately 2000 cases are diagnosed annually), with most of the rest from the Netherlands, France and Spain (Bouree et al., 1995; Smith, 1999; Audicana et al., 2002). Cases of anisakiasis, however, have also been reported from many other areas of the world, including the United States (Amin et al., 2000), Mexico (Laffon-Leal et al., 2000), Canada (Couture et al., 2003), the United Kingdom (Lewis and Shore, 1985), Belgium (Vercammen et al., 1997), Egypt (Cocheton et al., 1991), Korea (Im et al., 1995), Philippines (Petersen et al., 1993), Chile (Mercado et al., 2001), and New Zealand (Paltridge, 1984).

In the last 30 years, there has been a marked increase in the prevalence of anisakiasis throughout the world. This increase in reported cases of anisakiasis is probably due in large part to the use of new diagnostic techniques, particularly endoscopy. Prior to the development of the gastrofiberscope, many cases of gastric anisakiasis were misdiagnosed and therefore went unreported (Oshima, 1987). The development of serological tests with higher specificity for anisakid larvae (Moneo et al., 2000; Caballero and Moneo, 2002) will undoubtedly lead to an even greater rate of diagnosis of anisakiasis in the future, because it will not be limited to cases with severe symptoms, as endoscopy is at present. For example, in Spain, Toro et al. (2004) found that 13.8% of 174 patients with generalized dyspeptic symptoms, and Del Rey Moreno et al. (2006) found that 22.1% of randomly selected blood donors, were seropositive to the Ani s 1 antigen.

It is likely, however, that the recent increased prevalence of anisakiasis is not solely due to improved diagnostic methods, but that it also reflects a greater risk of contracting parasitic infections. As with other fish-borne parasitic diseases, the increasing global demand for seafood and a growing preference for raw or lightly cooked food, especially in many Western countries, increase the risk of parasite exposure (McCarthy and Moore, 2000; Chai et al., 2005). There has also been speculation that the risk of exposure to anisakids has increased because greater regulatory controls over the exploitation of marine mammals has led to increasing population sizes of potential definitive hosts (Oshima, 1987; Bouree et al., 1995; McCarthy and Moore, 2000). There is some evidence to support this view, with positive correlations found between seal population sizes and abundance of larval *Pseudoterranova* in fish in eastern Canadian waters (McClelland et al., 2000; McClelland and Martell, 2001). Other studies in Canada and Norway, however, have found little empirical evidence that changes in seal population sizes have led to changes in larval nematode abundances in fish (McClelland, 2002). The relationship between definitive host population size and parasite population size is not straightforward for parasites such as anisakid nematodes, which have a complex, multihost life cycle. Population dynamic modeling by Lunneryd et al. (2001), for example, showed that heavy larval sealworm infections in fish may be maintained by a relatively small number of seals; increases in seal population numbers above this threshold had very little effect on parasite abundance in fish.

Risk Factors

The prevalence of anisakiasis and other fish-borne parasitic zoonoses is clearly related to traditions of consuming raw, lightly cooked or marinated fish, such as Japanese sushi and sashimi, Dutch salted or smoked herring, Scandinavian gravlax (dry, cured salmon), Spanish boquerones en vinagre (pickled anchovies), Hawaiian lomi-lomi (raw salmon), Filipino kinilaw (chopped, marinated fish), and Latin American ceviche (raw fish seasoned with lemon juice) (Chai et al., 2005). The risk of anisakid larvae in these dishes depends on the species of fish being used and may be enhanced if the fish are eaten whole (because worms are often found in the viscera rather than the flesh of fish) or if the fish have been kept whole for some time after capture, rather than gutted immediately (because worms may migrate from the viscera to the flesh after death of the fish). In Japan, for example, commonly infected fish such as Pacific cod (*Gadus morhua macrocephala*), halibut (*Hippoglossus stenolepsis*), greenling (*Hexagramos otakii*), and mackerel (*Scomber japonicus*) can be obtained cheaply and are mainly consumed in home prepared sushi dishes (Oshima, 1987). Laffon-Leal et al. (2000) studied five different fish species that are commonly used in preparing ceviche in Mexico and found larvae of *Pseudoterranova* in barracuda (*Sphyraena barracuda*) at a prevalence of 33%, with a mean intensity of 10.2 worms per fish, and red grouper (*Epinephelus morio*) at a prevalence of 83% and a mean intensity of 6.5 worms per fish, while larvae of *Contracaecum* were found in yellowfin mojarra (*Gerres cinereus*) with 57% prevalence and a mean intensity of 7.6 worms per fish. In Galicia (northwestern Spain), anchovies (*Engraulis encrasicholus*) are traditionally prepared without freezing and eaten raw with vinegar sauce, while sardines (*Sardina pilchardus*) are eaten ungutted and charcoal-grilled; both species may be heavily infected with anisakid larvae (Moreno-Ancillo et al., 1997; Alonso-Gómez et al., 2004).

Control Measures

Preventive or control measures for anisakiasis focus on postharvest handling, storage, and cooking procedures for fish. Most control measures for anisakiasis emphasize the importance of immediate evisceration of captured fish, and cooking or freezing fish product prior to eating (Acha and Szyfres, 1987; Abollo et al., 2001). Consumption of live larvae in raw or undercooked fish, however, is not necessarily the only way in which the parasite can cause disease; in some cases there is evidence that occupational exposure to fish products may be sufficient to trigger an allergic response to anisakid allergens (Purello-D'Ambrosio et al., 2000).

Selective fishing for lightly infected populations or size classes of fish has been suggested as a method of reducing the risk of anisakiasis, but such concentration of fishing effort is not likely to be either economically feasible or sustainable for the stocks in question (McClelland, 2002). Preventive measures for anisakiasis, therefore, focus on postharvest handling, storage, and preparation procedures for fish.

Postharvest Handling

Immediate evisceration of fish may reduce the zoonotic potential of the parasite by preventing migration of worms into the flesh of host fish, but this practice may also cause heavier parasite infections for fish that feed on the discarded viscera (Acha and Szyfres, 1987; McClelland et al., 1990). In many countries with high prevalences of anisakid larvae in fish, such as Canada, fish are examined for infection at processing, with heavily infected fillets trimmed or discarded (McClelland, 2002). The usual procedure is to examine fillets by candling on a light table. The candling procedure, however, is very inefficient, often detecting as few as 33% of heavily infected fish (i.e., those with more than three worms per kilogram) (McClelland, 2002). Candling efficiency can be improved by slicing fillets longitudinally or by using different wavelengths of light and illumination sources to increase the contrast between the worm and flesh of the fish (Hafsteinsson and Rizvi, 1987). Other approaches, such as laser candling, radiography, scanning laser acoustic microscope, pulse-echo technology, and electromagnetic detection have also been tested, but none of these is able to improve efficiency of detection to an extent that justifies increased processing costs or reduction in marketable fillet yield (McClelland, 2002). Recent research has concentrated on molecular genetic approaches, with the development of primers for polymerase chain reaction (PCR) amplification of anisakid gene sequences. Such techniques have demonstrated good specificity and sensitivity in experimental tests, but have not yet been developed for commercial application (Santos et al., 2006).

Storage

The U.S. Food and Drug Administration (1992) recommends that fish intended for raw or semiraw (marinated or partly cooked) consumption, be frozen to − 35°C or below for 15 hours, or be frozen at -20°C or below for a minimum of 7 days. Wharton and Aalders (2002) pointed out that the mass of fish in the container for freezing should also be taken into account as 20 kg containers of fish did not reach -35°C even after 28 hours of exposure. Marine fish used for aquaculture or farm feed should also be frozen adequately before use. For example, turtles that were farmed at Torres Strait, between Australia and Papua New Guinea, acquired heavy infections of *Anisakis* sp. when fed with raw sardines (Stevenson and Hughes, 1980).

Preparation

Anisakid larvae are resistant to salting, smoke-curing, and marinading, and also do not appear to be killed by microwaving (Sakanari and McKerrow, 1989; Bouree et al., 1995). For home consumption, therefore, fish should be cooked until the core temperature reaches 60°C or higher, for at least 10 minutes (Sakanari and McKerrow, 1989).

Research Needs

Parasite Biology

Further basic research is necessary to better understand the life cycle and transmission dynamics of anisakid nematodes. Of primary importance is a thorough systematic study of the family using molecular genetic techniques. Allozyme electrophoresis and sequencing of ribosomal and mitochondrial DNA have already produced unexpected findings at a variety of taxonomic levels. These studies have identified non-monophyletic groupings within the superfamily Ascaridoidea (Nadler and Hudspeth, 2000) and found previously described morphospecies to consist of several genetically differentiated biological species (Paggi et al., 1991; Mattiucci et al., 1997). Molecular genetic analyses will also provide an essential tool for basic ecological studies of anisakids, providing us with a clearer understanding of species diversity within the Anisakidae, and the extent of their geographic distribution, host range, and prevalence rates in definitive and intermediate hosts.

Epidemiology of Anisakiasis

A major issue that has not been addressed is the reason why anisakiasis is most often associated with certain species of anisakid larvae. Almost all recorded cases involve infection with *Anisakis simplex* (sensu lato) or *Pseudoterranova decipiens* (sensu lato). It is not clear, however, to what extent this is a function of the geographic distribution of these parasite species, their fish host range, their microhabitat distribution within fish, their propensity for postmortem migration, or their invasive ability when ingested by humans. It is possible that all of these factors are involved, but determining their relative importance is essential, because it will impact upon the advice we can provide about risk factors for anisakiasis and critical control points for lessening the risk of anisakid infection from the products of commercial and recreational fishing.

Hypersensitivity Reactions

One of the most important findings about anisakiasis to emerge in recent years has been the discovery of hypersensitivity reactions to anisakid allergens (Audicana et al., 2002). This not only has led to the recognition of a different suite of pathological reactions to anisakid infections, which may be manifested in people handling, as well as ingesting fish products, but also the thermostability of the allergens involved means that standard precautions of freezing or cooking fish may not provide protection against an allergic response. An important issue, which remains unresolved, is whether a priming infection with a living parasite is needed to induce sensitization.

Summary

Anisakiasis is a disease caused by infection of people with larval (usually L3) nematodes belonging to the family Anisakidae. The two species most often associated with anisakiasis are *Anisakis simplex* and *Pseudoterranova decipiens*, both of which have recently been found to constitute a complex of morphologically similar sibling species. Molecular genetic studies have identified three different species within the *Anisakis simplex* complex, and six different species within the *Pseudoterranova decipiens* complex. Adults of all these species parasitize the alimentary tract of marine mammals. Eggs are shed in the feces and ingested by invertebrate intermediate hosts, which are in turn ingested by fish or cephalopod mollusk intermediate (or paratenic) hosts. Humans are accidental hosts, who usually become infected by eating raw or undercooked fish or cephalopods.

Anisakiasis can take a number of different clinical forms, depending on the location and histopathological lesions caused by the larvae. The most common and important of these clinical forms is invasive gastrointestinal anisakiasis, where the live larva penetrates the gastric or intestinal mucosa, producing either acute or chronic symptoms. Recently, it has been found that anisakid larvae can also produce a strong allergic response (allergic anisakiasis), with symptoms ranging from isolated swellings to anaphylactic shock.

Anisakiasis occurs throughout the world, with foci in North Asia and Western Europe. In the last 30 years, there has been a marked increase in prevalence of the disease. This is due in part to improved diagnostic methods, but may also be a consequence of increased risk of parasite exposure due to increasing global demand for seafood and a growing preference for raw or lightly cooked food. Control measures for anisakiasis focus on postharvest handling, storage, and cooking procedures for fish.

Acknowledgments. Thanks to Darwin Murrell for inviting us to write this chapter. Darwin Murrell, Una Ryan, and Jennifer Walters commented on different drafts of the manuscript. Russ Hobbs kindly produced the life cycle diagram shown in Figure 5.1.

References

Abollo, E., Gestal, C., and Pascual, S. (2001). *Anisakis* infestation in marine fish and cephalopods from Galician waters: an updated perspective. Parasitol Res 87, 492–499.

Acha, P., and Szyfres, B. (1987). Zoonoses and Communicable Diseases Common to Man and Animals, 2nd ed. Washington, DC: Pan American Health Organisation.

Alonso, A., Moreno-Ancillo, A., Daschner, A., and López-Serrano, M.C. (1999). Dietary assessment in five cases of allergic reactions due to gastroallergic anisakiasis. Allergy 54, 517–520.

Alonso-Gómez, A., Moreno-Ancillo, A., López-Serrano, M.C., Suarez-de-Parga, J.M., Daschner, A., Caballero, M.T., Barranco, P., and Cabañas, R. (2004). *Anisakis simplex* only provokes allergic symptoms when the worm parasitises the gastrointestinal tract. Parasitol Res 93, 378–384.

Álvarez, F., Iglesias, R., Paramá, A.I., Leiro, J., and Sanmartín, M. (2002). Abdominal macroparasites of commercially important flatfishes (Teleostei: Scophthalmidae, Pleuronectidae, Soleidae) in northwest Spain (ICES IXa). Aquaculture 213, 31–53.

Amin, O.M., Eidelman, W.S., Domke, W., Bailey, J., and Pfeifer, G. (2000). An unusual case of anisakiasis in California, USA. Comp Parasitol 67, 71–75.

Anderson, R.C. (1992). Nematode Parasites of Vertebrates: Their Development and Transmission. Cambridge: CAB International.

Audicana, M.T., Ansotegui, I.J., de Corres, L.F., and Kennedy, M.W. (2002). *Anisakis simplex*: dangerous—dead and alive? Trends Parasitol 18, 20–25.

Baeza, M.L., Rodríguez, A., Matheu, V., Rubio, M., Tornero, P., de Barrio, M., Herrero, T., Santaolalla, M., and Zubeldia, J.M. (2004). Characterization of allergens secreted by *Anisakis simplex* parasite: clinical relevance in comparison with somatic allergens. Clin Exp Allergy 34, 296–302.

Beaver, P.C., Jung, R.C., and Cupp, E.W. (1984). Clinical Parasitology, 9th ed. Philadelphia: Lea & Febiger.

Bester, M.N. (1989). Endoparasites of the subantarctic fur seal *Arctocephalus tropicalis* from Gough Island. S Afr J Zool 24, 363–365.

Bouree, P., Paugam, A., and Petithory, J.-C. (1995). Anisakidosis: report of 25 cases and review of the literature. Comp Immunol Microbiol Infect Dis 18, 75–84.

Brattey, J., and Stenson, G.B. (1993). Host specificity and abundance of parasitic nematodes (Ascaridoidea) from the stomachs of five phocid species from Newfoundland and Labrador. Can J Zool 71, 2156–2166.

Bristow, G.A., and Berland, B. (1992). On the ecology and distribution of *Pseudoterranova decipiens* (Nematoda: Anisakidae) in an intermediate host, *Hippoglossoides platessoides*, in northern Norwegian waters. Int J Parasitol 22, 203–208.

Caballero, M.L., and Ignacio, M. (2004). Several allergens from *Anisakis simplex* are highly resistant to heat and pepsin treatments. Parasitol Res 93, 248–251.

Caballero, M.L., and Moneo, I. (2002). Specific IgE determination to Ani s 1, a major allergen from *Anisakis simplex*, is a useful tool for diagnosis. Ann Allergy Asthma Immunol 89, 74–77.

Cattan, P.E., and Carvajal, J. (1984). A study of the migration of larval *Anisakis simplex* (Nematoda: Ascaridida) in the Chilean hake, *Merluccius gayi* (Guichenot). J Fish Biol 24, 649–654.

Chai, J.Y., Murrell, K.D., and Lymbery, A.J. (2005). Fish-borne zoonoses: status and issues. Int J Parasitol 35, 1233–1254.

Cocheton, J.J., Cabou, I., and Lecomte, I. (1991). Anisakiase et infections par les Anisakidés. Ann Med Int 142, 121–130.

Costa, G., Pontes, T., Mattiucci, S., and D'Amélio, S. (2003). The occurrence and infection dynamics of *Anisakis* larvae in the black-scabbard fish, *Aphanopus carbo*, chub mackerel, *Scomber japonicus*, and oceanic horse mackerel, *Trachurus picturatus* from Madeira, Portugal. J Helminthol 77, 163–166.

Couture, C., Measures, L., Gagnon, J., and Desbiens, C. (2003). Human intestinal anisakiosis due to consumption of raw salmon. Am J Surg Pathol 27, 1167–1172.

Davey, J.T. (1971). A case of *Anisakis* sp. infection in a brown bear from Alaska. Trans R Soc Trop Med Hyg 65, 433–434.

Del Rey Moreno, A., Valero, A., Mayorga, C., Gómez, B., Torres, M.J., Hernández, J., Ortiz, M., and Lozano Maldonado, J. (2006). Sensitization to *Anisakis simplex* s.l. in a healthy population. Acta Trop 97, 265–269.

Desportes, G., and McClelland, G. (2001). Sealworms in the North Atlantic: Ecology and Population Dynamics. Tromso, Norway: North Atlantic Marine Mammal Commission.

Dziekonska-Rynko, J., Rokicki, J., and Jablonowski Z (2002). Effects of ivermectin and albendazole against *Anisakis simplex* in vitro and in guinea pigs. J Parasitol 88, 395–398.

Fagerholm, H.P. (1991). Systematic implications of male caudal morphology in ascaridoid nematode parasites. Syst Parasitol 19, 215–228.

George-Nascimento, M., and Urrutia, X. (2000). *Pseudoterranova cattani* sp. nov. (Ascaridoidea: Anisakidae), a parasite of the South American sea lion *Otaria byronia* De Blainville from Chile. Rev Chil Hist Natural 73, 93–98.

Greenwood, A.G., and Taylor, D.C. (1978). Clinical and pathological findings in dolphins in 1977. Aquatic Mammals 6, 33–38.

Hafsteinsson, H., and Rizvi, S.S.H. (1987). A review of the sealworm problem: biology, implications and solutions. J Food Protect 50, 70–84.

Hartwich, G. (1974). Keys to genera of the Ascaridoidea. In: Anderson, R.C., Chabaud, A.G., and Willmott, S., eds., CIH keys to the Nematode Parasites of Vertebrates (pp. 1–15). Farnham Royal: Commonwealth Agricultural Bureaux.

Hauck, A.K. (1977). Occurrence and survival of the larval nematode *Anisakis* sp. in the flesh of fresh, frozen, brined, and smoked Pacific herring, *Clupea harengus pallasi*. J Parasitol 63, 515–519.

Herraras, M.V., Balbuena, J.A., Aznar, F.J., Kaarstad, S.E., Fernández, M., and Raga, J.A. (2004). Population structure of *Anisakis simplex* (Nematoda) in harbor porpoises *Phocena phocena* off Denmark. J Parasitol 90, 933–938.

Hokama, A., Gakiya, I., Miyagi, T., Fukuchi, J., Kinjo, F., and Saito, A. (2005). Gastrointestinal: acute gastric anisakiasis. J Gastroenterol Hepatol 20, 1121.

Hsu, H.F. (1933). Study on the oesophagael glands of parasitic nematode superfamily Ascaroidea. Chin Med J 47, 1247–1288.

Ido, K., Yuasa, H., Ide, M., Kimura, K., Toshimitsu, K., and Suzuki, T. (1998). Sonographic diagnosis of small intestinal anisakiasis. J Clin Ultrasound 26, 125–130.

Im, K.I., Shin, H.J., Kim, B.H., and Moon, S.I. (1995). Gastric anisakiasis cases in Cheju-do, Korea. Korean J Parasitol 33, 179–186.

Inglesias, R., Leiro, J., Ubeira, F.M., Santamarina, M.T., Navarrete, I., and Sanmartín, M.L. (1996). Antigenic cross-reactivity in mice between third stage larvae of *Anisakis simplex* and other nematodes. Parasitol Res 82, 378–381.

Jefferies, D.J., Hanson, H.M., and Harris, E.A. (1990). The prevalence of *Pseudoterranova decipiens* (Nematoda) and *Corynosoma strumosum* (Acanthocephala) in otters *Lutra lutra* from coastal sites in Britain. J Zool 221, 316–321.

Jensen, T., and Andersen, K. (1992). The importance of sculpin *(Myoxocephalus scorpius)* as intermediate host and transmitter of the sealworm *Pseudoterranova decipiens*. Int J Parasitol 22, 665–668.

Keyes, M.C. (1965). Pathology of the northern fur seal. J Am Vet Med Assoc 147, 1090–1095.

Klimpel, S., Palm, H.W., Rückert, S., and Piatkowski, U. (2004). The life cycle of *Anisakis simplex* in the Norwegian Deep (northern North Sea). Parasitol Res 94, 1–9.

Køie, M., and Fagerholm, H.-P. (1993). Third-stage larvae emerge from eggs of *Contracaecum osculatum*. (Nematoda, Anisakidae). J Parasitol 79, 777–780.

Køie, M., Berland, B., and Burt, M.D.B. (1995). Development to third-stage larvae occurs in the eggs of *Anisakis simplex* and *Pseudoterranova decipiens* (Nematoda, Ascaridoidea, Anisakidae). Can J Fisheries Aquatic Sci 52, 134–139.

Konishi, K., and Sakurai, Y. (2002). Geographical variations in infection by larval *Anisakis simplex* and *Contracaecum osculatum* (Nematoda, Anisakidae) in walleye pollock *Theragra chalcogramma* stocks off Hokkaido, Japan. Fisheries Sci 68, 534–542.

Kuramochi, T., Machida, M., Araki, J., Uchida, A., Kishiro, T., and Nagasawa, K. (1996). Minke whales *(Balaenoptera acutorostrata)* are one of the major final hosts of *Anisakis simplex* (Nematoda:Anisakidae) in the northwestern North Pacific Ocean. Rep Int Whaling Commun 46, 415–419.

Laffon-Leal, S.M., Vidal-Martinez, V.M., and Arjona-Torres, G. (2000). "Cebiche"—a potential source of human anisakiasis in Mexico? J Helminthol 74, 151–154.

Lewis, R., and Shore, J.H. (1985). Anisakiasis in the United Kingdom. Lancet 2, 1019.

Lorenzo, S., Iglesias, R., Leiro, J., Ubeira, F.M., Ansotegui, I., García, M., and Fernández de Corres, L. (2000). Usefulness of currently available methods for the diagnosis of *Anisakis simplex* allergy. Allergy 55, 627–633.

Lozano, M.J., Mañas, A.I., Valero, L.A., Martín, H.L., Díaz, S.V., and Campos, B.M. (2004). Cross-reactivity between antigens of *Anisakis simplex* s.l. and other ascarid nematodes. Parasite 11, 219–223.

Lunneryd, S.-G., Ugland, K.I., and Aspholm, P.E. (2001). Sealworm *(Pseudoterranova decipiens)* infection in the benthic cottid *(Taurulus bubalis)* in relation to population increase of harbour seal *(Phoca vitulina)* in Skagerrak, Sweden. In: Desportes, G., and McClelland, G., eds. Sealworms in the North Atlantic: Ecology and Population Dynamics (pp. 47–55). Tromsø, Norway: NAMCCO Scientific Publications.

Mattiucci, S., Nascetti, G., Cianchi, R., Paggi, L., Arduino, P., Margolis, L., Brattey, J., Webb, S., D'Amelio, S., Orecchia, P., and Bullini, L. (1997). Genetic and ecological data on the *Anisakis simplex* complex, with evidence for a new species (Nematoda, Ascaridoedea, Anisakidae). J Parasitol 83, 401–416.

Mattiucci, S., Nascetti, G., Dailey, M., Webb, S.C., Barros, N.B., Cianchi, R., and Bullini, L. (2005). Evidence for a new species of *Anisakis* Dujardin, 1845: morphological description and genetic relationships between congeners (Nematoda: Anisakidae). Syst Parasitol 61, 157–171.

Mattiucci, S., Paggi, L., Nascetti, G., Ishikura, H., Kikuchi, K., Sato, N., Cianchi, R., and Bullini, L. (1998). Allozyme and morphological identification of *Anisakis, Contracaecum* and *Pseudoterranova* (Nematoda: Ascaridoidea) from Japanese waters. Syst Parasitol 40, 81–92.

McCarthy, J., and Moore, T.A. (2000). Emerging helminth zoonoses. Int J Parasitol 30, 1351–1360.

McClelland, G. (2002). The trouble with sealworms (*Pseudoterranova decipiens* species complex, Nematoda): a review. Parasitology 124, S183-S203.

McClelland, G., and Martell, D.J. (2001). Surveys of larval sealworm *(Pseudoterranova decipiens)* infection in various fish species sampled from Nova Scotian waters between 1988 and 1996, with an assessment of examination procedures. In: Desportes, G., and McClelland, G., eds. Sealworms in the North Atlantic: Ecology and Population Dynamics (pp. 57–76). Tromsø, Norway: North Atlantic Marine Mammal Commission.

McClelland, G., Misra, R.K., and Martell, D.J. (1990). Larval anisakine nematodes in various fish species from Sable Island Bank and vicinity. Can Bull Fisheries Aquatic Sci 222, 83–118.

McClelland, G., Misra, R.K., and Martell, D.J. (2000). Spatial and temporal distributions of larval sealworm (*Pseudoterranova decipiens*, Nematoda: Anisakinae) in *Hippoglossoides platessoides* (Pleuronectidae) in eastern Canada from 1980 to 1990. ICES J Marine Sci 57, 69–88.

Measures, L.N., and Hong, H. (1995). The number of moults in the egg of sealworm, *Pseudoterranova decipiens* (Nematoda: Ascaridoidea): an ultrastructural study. Can J Fisheries Aquatic Sci 52 (suppl 1), 156–160.

Mercado, R., Torres, P., Muñoz, V., and Apt, W. (2001). Human infection by *Pseudoterranova decipiens* (Nematoda, Anisakidae) in Chile: report of seven cases. Memórias Inst Oswaldo Cruz 96, 653–655.

Moneo, I., Caballero, M.L., Gómez, F., Ortega, E., and Alonso, M.J. (2000). Isolation and characterization of a major allergen from the fish parasite *Anisakis simplex*. J Allergy Clin Immunol 106, 177–182.

Moore, D.A.J., Girdwood, R.W.A., and Chiodini, P.L. (2002). Treatment of anisakiasis with albendazole. Lancet 360, 54.

Moreno-Ancillo, A., Caballero, M.T., Cabañas, R., Contreras, J., Martin-Barroso, J.A., Barranco, P., and López-Serrano, M.C. (1997). Allergic reactions to *Anisakis simplex* parasitizing seafood. Ann Allergy Asthma Immunol 79, 246–250.

Moschella, C.M., Mattiucci, S., Mingazzini, P., Mongardini, M., Chein, A., Miccolis, D., and Modini, C. (2005). Intestinal anisakiasis in Italy: a case treated by emergency surgery. Giornale Chir 26, 201–205.

Nadler, S.A. (1992). Phylogeny of some ascaridoid nematodes, inferred from comparison of 18S and 28S rRNA sequences. Mol Biol Evolution 9, 932–944.

Nadler, S.A., and Hudspeth, D.S.S. (1998). Ribosomal DNA and phylogeny of the Ascaridoidea (Nemata: Secernentea): implications for morphological evolution and classification. Mol Phylogenet Evol 10, 221–236.

Nadler, S.A., and Hudspeth, D.S.S. (2000). Phylogeny of the Ascaridoidea (Nematoda: Ascaridida) based on three genes and morphology: hypothesis of structural and sequence variation. J Parasitol 86, 380–393.

Noh, J.H., Kim, B.-J., Kim, S.M., Ock, M.-S., Park, M.I., and Goo, J.Y. (2003). A case of acute gastric anisakiasis provoking severe clinical problems by multiple infection. Korean J Parasitol 41, 97–100.

Ólafsdóttir, D. (2001). A review of the ecology of sealworm, *Pseudoterranova decipiens* (Nematoda: Ascaridoidea) in Icelandic waters. In: Desportes, G., and McClelland, G., eds. Sealworms in the North Atlantic: Ecology and Population Dynamics (pp. 95–111). Tromsø, Norway: North Atlantic Marine Mammal Commission.

Oshima, T. (1987). Anisakiasis—is the sushi bar guilty? Parasitol Today 3, 44–48.

Paggi, L., Mattiucci, S., Gibson, D.I., Berland, B., Nascetti, G., Cianchi, R., and Bullini, L. (2000). *Pseudoterranova decipiens* species A and B (Nematoda: Ascaridoidea): nomenclatural designation, morphological diagnostic characters and genetic markers. Syst Parasitol 45, 185–197.

Paggi, L., Nascetti, G., Cianchi, R., Orecchia, P., Mattiucci, S., D'Amelio, S., Berland, B., Brattey, J., Smith, J.W., and Bullini, L. (1991). Genetic evidence for three species within *Pseudoterranova decipiens* (Nematoda, Ascaridida, Ascaridoidea) in the North Atlantic and Norwegian and Barents Seas. Int J Parasitol 21, 195–212.

Paltridge, G.P. (1984). Anisakiasis: a new New Zealand disease. N Z Med J 97, 558–559.

Pascoe, P.L. (1986). Size data and stomach contents of common dolphins, *Delphinus delphis*, near Plymouth. J Marine Biol Assoc UK 66, 319–322.

Petersen, F., Palm, H., Möller, H., and Cuzi, M.A. (1993). Flesh parasites of fish from central Philippine waters. Dis Aquatic Organisms 15, 81–86.

Purello-D'Ambrosio, F., Pastorello, E., Gangemi, S., Lombardo, G., Ricciardi, L., Fogliani, O., and Merendino, R.A. (2000). Incidence of sensitivity to *Anisakis simplex* in a risk population of fishermen/fishmongers. Ann Allergy Asthma Immunol 84, 439–444.

Rausch, R. (1953). Studies on the helminth fauna of Alaska. XIII. Disease in the sea otter, with special reference to helminth parasites. Ecology 34, 584–604.

Riley, J. (1972). The pathology of *Anisakis* nematode infections of the fulmar *Fulmarus glacialis*. Ibis 114, 102–104.

Roepstorff, A., Karl, H., Bloemsma, B., and Huss, H.H. (1993). Catch handling and the possible migration of *Anisakis* larvae in herring, *Clupea harengus*. J Food Protection 56, 783–787.

Rosales, M.J., Mascaró, C., Fernandez, C., Luque, F., Moreno, M.S., Parras, L., Cosano, A., and Muñoz, J.R. (1999). Acute intestinal anisakiasis in Spain: a fourth-stage *Anisakis simplex* larva. Memórias Inst Oswaldo Cruz 94, 823–826.

Sakanari, J.A., and McKerrow, J.H. (1989). Anisakiasis. Clin Microbiol Rev 2, 278–284.

Santos, A.T., P., S., Verneau, O., and Lenfant, P. (2006). A method to detect the parasitic nematodes from the family Anisakidae, in *Sardina pilchardus*, using specific primers of 18S DNA gene. Eur Food Res Technol 222, 71–77.

Smith, J.W. (1984). The abundance of *Anisakis simplex* L3 in the body cavity and flesh of marine teleosts. Int J Parasitol 14, 491–495.

Smith, J.W. (1999). Ascaridoid nematodes and pathology of the alimentary tract and its associated organs in vertebrates, including man: a literature review. Helminthol Abstr 68, 49–96.

Smith, J.W., and Snyder, J.M. (2005). New locality records for third-stage larvae of *Anisakis simplex* (sensu lato) (Nematoda: Ascaridoidea) in euphausiids *Euphausia pacifica* and *Thysanoessa raschii* from Prince William Sound, Alaska. Parasitol Res 97, 539–542.

Smith, J.W., and Wootten, R. (1975). Experimental studies on the migration of *Anisakis* sp larvae (Nematoda: Ascaridida) into the flesh of herring, *Clupea harengus* L. J Parasitol 5, 133–136.

Smith, J.W., and Wootten, R. (1978). *Anisakis* and anisakiasis. Adv Parasitol 16, 93–163.

Stevenson, W.J., and Hughes, K.L. (1980). *Anisakis* visceral larva migrans. In: Synopsis of Zoonoses in Australia (pp. 230–231). Canberra: Commonwealth Department of Health.

Stobo, W.T., Fanning, P.L., Beck, B., and Fowler, M.G. (2002). Abundance and maturity of three species of parasitic nematodes *(Pseudoterranova decipiens, Contracaecum osculatum, Anisakis simplex)* occurring in Sable Island harbour seals *(Phoca vitulina)*. Can J Zool 80, 442–449.

Stroud, R.K., and Dailey, M.D. (1978). Parasites and associated pathology observed in pinnipeds stranded along the Oregon coast. J Wildlife Dis 14, 292–298.

Stroud, R.K., and Roffe, T.J. (1979). Causes of death in marine mammals stranded along the Oregon coast. J Wildlife Dis 15, 91–97.

Sugimachi, K., Inokuchi, K., Ooiwa, T., Fujino, T., and Ishii, Y. (1985). Acute gastric anisakiasis. JAMA 253, 1012–1013.

Toro, C., Caballero, M.L., Baquero, M., García-Samaniego, J., Casado, I., Rubio, M., and Moneo, I. (2004). High prevalence of seropositivity to a major allergen of *Anisakis simplex* Ani s 1 in dyspeptic patients. Clin Diagn Lab Immunol 11, 115–118.

Ugland, K.L., Strømnes, E., Berland, B., and Aspholm, P.E. (2004). Growth, fecundity and sex ratio of adult whaleworm (*Anisakis simplex*; Nematoda, Ascaridoidea, Anisakidae) in three whale species from the North-East Atlantic. Parasitol Res 92, 484–489.

U.S. Food and Drug Administration (1992). Rockville, MD (June 14, 2006); http://www.cfsan.fda.gov/~mow/chap25.html.

Valiñas, B., Lorenzo, S., Eiras, A., Figueiras, A., Sanmartín, M.L., and Ubeira, F.M. (2001). Prevalence of and risk factors for IgE sensitization to *Anisakis simplex* in a Spanish population. Allergy 56, 667–671.

Van Thiel, P.H., Kuipers, F.C., and Roskam, R.T. (1960). A nematode parasitic to herring, causing acute abdominal syndromes in man. Trop Geographical Med 2, 97–113.

Vercammen, F., Kumar, V., Bollen, J., Lievens, C., Van den Bergh, L., and Vervoort, T. (1997). Gastric involvement with *Anisakis* sp. larva in a Belgian patient after consumption of cod. Acta Gastroenterol Belg 60, 302–303.

Wazura, K.W., Strong, J.T., Glenn, C.L., and Bush, A.O. (1986). Helminths of the beluga whale (*Delphinapterus leucas*) from the Mackenzie River delta, Northwest Territories. J Wildlife Dis 22, 440–442.

Wharton, D.A., and Aalders, O. (2002). The response of *Anisakis* larvae to freezing. J Helminthol 76, 363–368.

Wharton, D.A., Hassall, M.L., and Aalders, O. (1999). *Anisakis* (Nematoda) in some New Zealand inshore fish. N Z J Marine Freshwater Res 33, 643–648.

Zhu, X., Gasser, R.B., Podolska, M., and Chilton, N.B. (1998). Characterisation of anisakid nematodes with zoonotic potential by nuclear ribosomal DNA sequences. Int J Parasitol 28, 1911–1921.

6 Capillariasis

J. H. Cross and V. Belizario

There are many newly emerging and reemerging food-borne parasitic diseases that are gaining international significance such as fascioliasis, anisakiasis, gnathostomiasis, trichinellosis, echinostomiasis, cysticercosis, angiostrongyliasis, and capillariasis. *Capillaria philippinensis* is one of the more recently discovered food-borne zoonotic parasites and has become one of the most biologically interesting, not only because of its high risk of death in infected people, but also because of its unusual life cycle and reproductive behavior. The parasite was unknown until the latter part of the 20th century when it was found to be the cause of an epidemic that often resulted in death to a number of those infected with the nematode.

History

In 1966 a parish priest in the municipality of Tagudin, Ilocus Sur, in Northern Luzon in the Philippines Islands became concerned when a number of people in Pudoc West, a barrio in the municipality, were reported to be suffering from a severe gastroenteritis and many were dying at an unusual high rate. The priest told a member of the Philippine Congress about the outbreak, who subsequently informed the Philippine Department of Health in Manila. Upon investigation, government epidemiologists were told of 24 people, mostly adult males, who suffered a profuse diarrhea, gurgling stomach, abdominal pain, and wasting for an average of 2 to 3 months before dying. The etiology of the disease was not known and was initially considered to be a chronic poisoning. Toxicological tests on food and drink specimens, however, were negative. Viruses, bacteria, and fungi were also considered as the causative agent, and testing results were also negative. Routine stool examinations were made and while eggs of *Capillaria* were found they were considered to be unusual eggs of human whipworm (*Trichuris trichuria*). Eventually, a simple autopsy of a fatal case was carried out, and many tiny nematodes were recovered in the intestines that were later found to be similar to those from a patient who died in a hospital in Manila in 1963.

This first patient (1963) with the parasitosis was a 29-years-old man, an alcoholic schoolteacher, from Bacarra, Ilocos Norte, Northern Luzon in the Philippines. He was admitted to the Philippine General Hospital in May 1963. He had an intractable diarrhea for 3 weeks, recurrent ascites, emaciation, and cachexia, and died a few days after being hospitalized. At autopsy, a large number of tiny worms were recovered from the intestines that were identified as belonging to the genus *Capillaria* and later described as a new species *Capillaria philippinensis* (Chitwood et al., 1964, 1968).

The epidemic that occurred in Pudoc West probably began in 1965, and over the next 2½ years, 32% of the barrio population of 600 acquired intestinal capillariasis (Detels et al., 1969). The people of Pudoc West suffered a great deal during the epidemic. The disease was found to occur in all age groups, although most deaths occurred in middle-aged men. Families were disrupted and many fatherless families were adopted by another family, with many males acquiring a second wife. Many families abandoned their homes (Fig. 6.1) and moved to live with relatives living in other barrios.

The people of Pudoc West developed a superstition that the barrio was cursed by the god present in the nearby Amburan River. They believed that the river god was punishing them for killing two of the god's children when a branch from a mango tree that was chopped down fell and killed the children. A second superstition was that punishment was due to the people eating a water buffalo belonging to the river god that floated into the barrio lagoon. The villagers subsequently hired two spiritualists or witch doctors *(herbolario)* to exorcise the village. These individuals recommended that an altar be built under the mango tree (Fig. 6.2) and instructed the people to put tribute on the altar each evening. During the night, the tribute, composed of items such as chickens, eggs, shrimps, snails, fish, alcoholic spirits, and money that were placed on the altar, disappeared. The witch doctors claimed that the river god came by and took the items. This practice continued for a number of months until one of the witch doctors died of the disease and the other one absconded. In the meantime, although the government doctors told the people that they were being swindled, the superstitious beliefs continued

FIGURE 6.1. Home in Pudoc West, Tagudin, Ilocus Sur. Abandoned because of deaths in the family due to intestinal capillariasis.

FIGURE 6.2. To ward off the curse of the God River, a shrine was built so the Pudoc West villagers could place daily tribute on it.

among the villagers that stayed in Pudoc, and they hung talismans such as rice-hull–stuffed dolls and painted crosses on the houses (Fig. 6.3). It took many months before the people accepted the fact that the disease was due to a parasitic worm.

FIGURE 6.3. Tallisman hanging from a house in Pudoc West to guard against the curse of intestinal capillariasis.

The Parasite

Capillaria philippinensis is a trichurid nematode related morphologically to the group that includes *Trichuris* (whipworm) and *Trichinella* (trichina). *Capillaria philippinensis* could be considered a bridge between these two genera: *Trichuris* is oviparous, *Trichinella* is larviparous, and *C. philippinensis* is both oviparous and larviparous.

The species was described initially by Chitwood et al., (1968); however, Moravec (2001) redescribed it placing it in the genus *Paracapillaria*. This taxonomic revision, however, has not found widespread acceptance.

The nematode is one of the smallest worms infecting humans; the males (Fig. 6.4) are 1.5 to 3.9 mm long with width of 3 to 5 μm at the head, 23 to 28 μm at the stichosome, and 18 μm at the cloaca. The esophagus measures 1.18 to 1.73 mm, posterior body 0.79 to 1.07 mm, spicule 230 to 300 by 2.6 to 2.8 μm, and spicular sheath about 440 μm, and it is without spines (Fig. 6.5). The tail has ventrolateral expansions with two pairs of papillae, and the anus is subterminal. The ratio of the esophageal region to the posterior body is about 1.24:0.90. Females (Fig. 6.6) measure 2.3 to 5.3 mm in length with a width of 5 to 8 μm at the head, and a width of 25 μm at the widest portion of the stichosome, 28 to 36 μm at the vulva and 29 to 47 μm postvulva. The ratio of esophageal region to the posterior body is 1.6:0.8 μm. The vulva is salient, without a flap and located immediately behind the end of the esophagus. The anus is subterminal. The ovary is anteriorly directed at onset, 75 μm from the tail joining the oviduct 70 μm from onset, the oviduct reflexing twice near the middle of the posterior body. Females are unique and may contain typical bioperculated thick-shelled eggs (Fig. 6.7), thin-shelled eggs (Fig. 6.8), and uteri containing embryos (Fig. 6.9). Females may be both oviparous and larviparous (Fig. 6.10). The ratio of oviparous to larviparous females recovered from experimentally infected monkeys was approximately

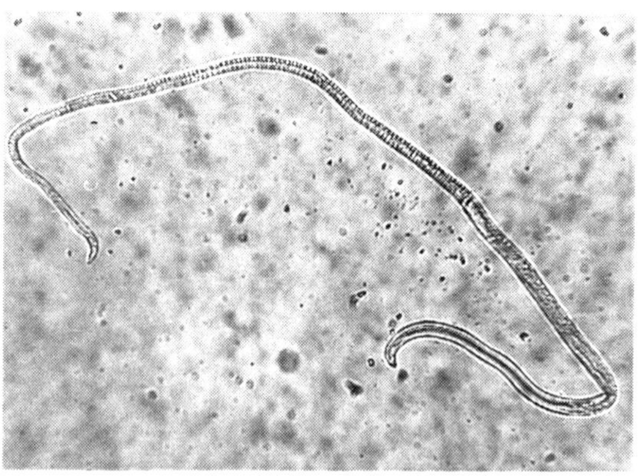

FIGURE 6.4. Male *Capillaria philippinensis*, note the row of stichocytes in the anterior two thirds of the nematode and the spicule in the posterior end (arrow).

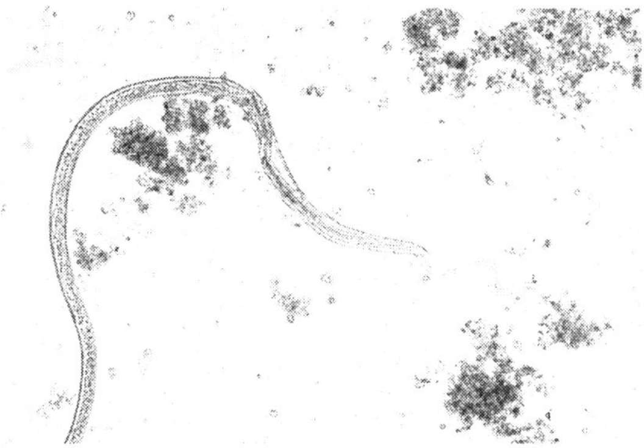

FIGURE 6.5. Posterior end of male *Capillaria philippinensis* showing the external spicule sheath (arrow).

200:1. Eggs recovered from feces are peanut-shaped with flattened bioperculated plugs and measure 36 to 45 by 20 μm (Fig. 6.11). All stages of larvae may be found in the intestines and passed in the feces (Chitwood et al., 1968). The first-stage larvae (Fig. 6.12) may have stichocytes in a double row. The esophagus is

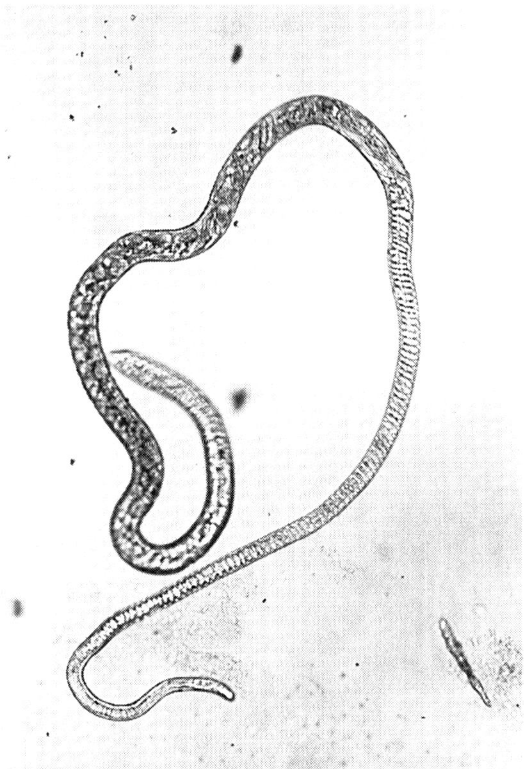

FIGURE 6.6. Female *Capillaria philippinensis*, note of stichocytes in the anterior part of the worm and the uterus filled with eggs in the posterior half of the body.

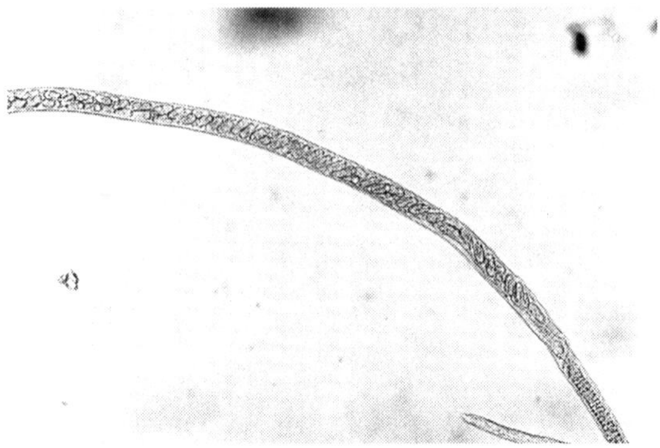

FIGURE 6.7. Female *Capillaria philippinensis* showing thick-shelled eggs in the uterus.

approximately four-fifths of the body and is bluntly rounded on both ends. The genital primordium is well developed. Second-stage larvae (Fig. 6.13) measure 415 to 550 μm long and have stichocytes in a double row. The esophagus is about two thirds of the body length and the genital column about two thirds the distance from the rectal region and a developing vagina is evident. Third-stage larvae (Fig. 6.14) measure about 1.25 mm with an interior genital column almost complete and the posterior one beginning to develop. The vagina is differentiated and most of the stichocytes are in a single row. An oral spear may be seen. In the third-stage male, the cuticle begins to separate at the posterior end and the spicule may be faintly visible. Fourth-stage larvae (Fig. 6.15) differ from adults in size, body proportions, and sclerotization of the spicule and a sheath may be visible on the posterior end. In the females a loop appears in the ovary at about the middle of the posterior body.

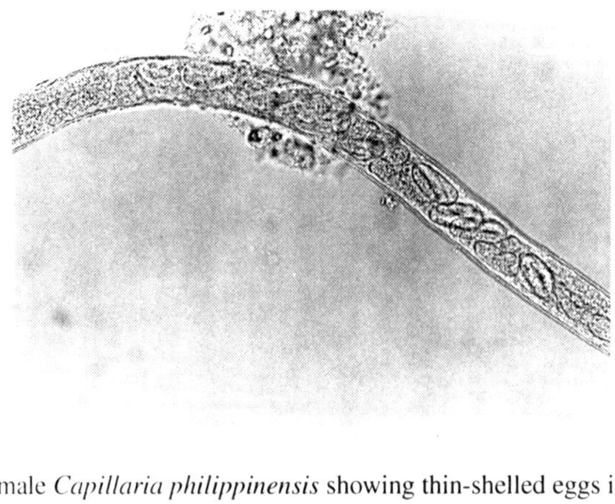

FIGURE 6.8. Female *Capillaria philippinensis* showing thin-shelled eggs in the uterus.

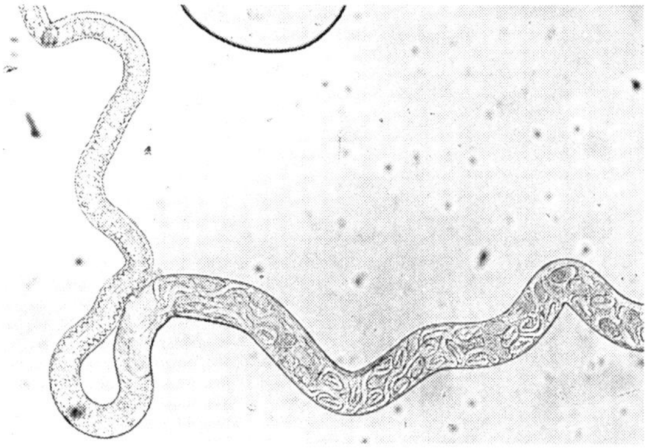

FIGURE 6.9. Larviparous *Capillaria philippinensis* recovered from the intestines of an infected monkey. (From Cross et al., 1972, Trans Roy Soc Trop Med Hyg with permission)

Life Cycle

In the early studies on intestinal capillariasis, research was directed toward determining the life cycle and means of transmission of the nematode. At that time about 250 *Capillaria* species had been reported in the literature, but the life cycles of only a few members of the genus had been described. Some of the known species had a direct life cycle like the rat *Capillaria hepatica,* while others used an intermediate host such as earthworms in the case of *Capillaria annulta* found in birds. In attempts to identify the means of transmission of the nematode and the possible natural hosts, over 150,000 specimens of animal life from the Ilocano endemic areas of the Philippines were examined. The specimens included rodents, bats, cats, dogs, pigs, birds, cows, lizards, snakes, frogs, mollusks, insects, earthworms, shrimp, crabs, copepods, fish as well as plants, soil, and water, but evidence of the parasite was not found. Eggs, however, were found in soil samples collected from around houses of infected person.

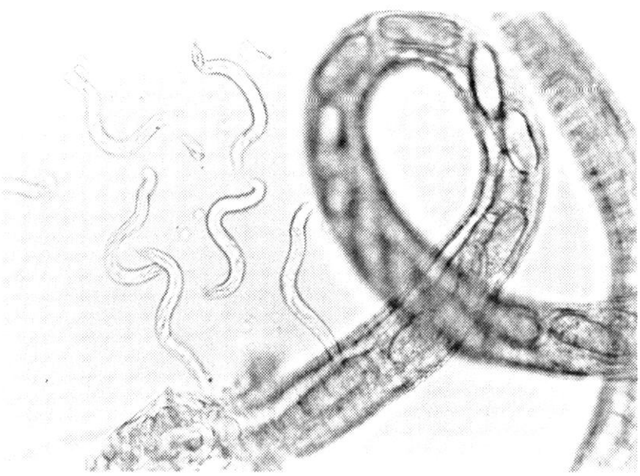

FIGURE 6.10. Larviparous-oviparous. *Capillaria philippinensis.*

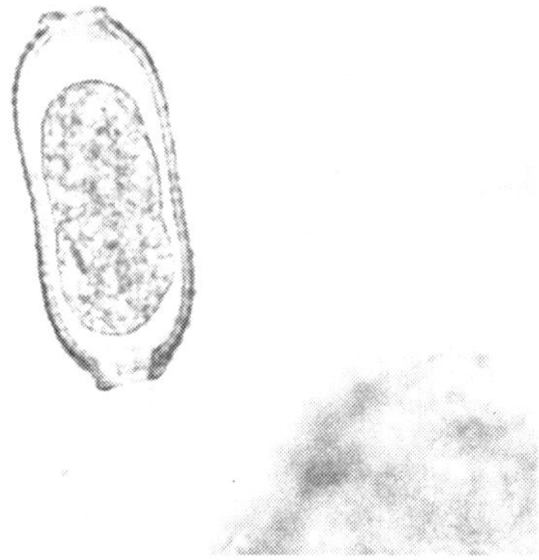

FIGURE 6.11. Egg of *Capillaria philippinensis* with a striated shell and flattened bioperculated plugs measuring 36 to 45 μm by 20 to 22 μm.

In early experimental studies, eggs from human feces were cultured at room temperature in distilled water, and the eggs were found to embryonate in 5 to 10 days. A variety of laboratory animals and wild caught vertebrate and invertebrate animals were fed embryonated eggs by stomach tube or allowed to swallow eggs in water. The eggs usually passed in the feces unchanged in a few days in most experimental animals. Embryonated eggs were also swallowed by human volunteers and these eggs also passed unchanged in the feces within 3 days. Eggs were finally found to hatch in fish intestines and the hatched larvae increased in size. The larvae from eggs measured 130 to 150 μm (Fig. 6.16), but after 3 weeks in the fish intestine, they measure 250 to 300 μm (Fig. 6.17). There was no further growth after 3 weeks. Larvae were found to remain viable in the fish intestines for as long as 7 months. The larvae were found only in the fish intestines and no other

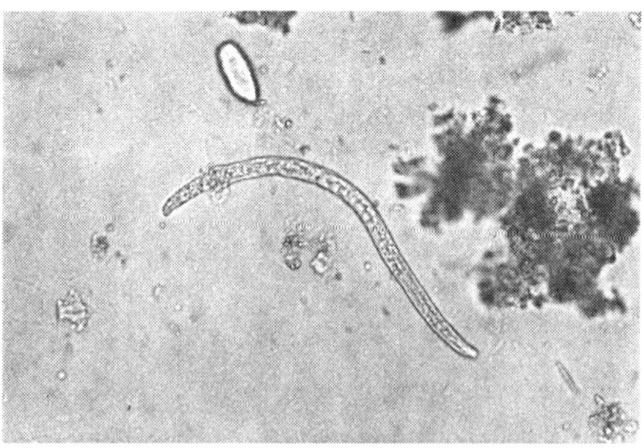

FIGURE 6.12. First-stage *Capillaria philippinensis* larva recovered from human feces.

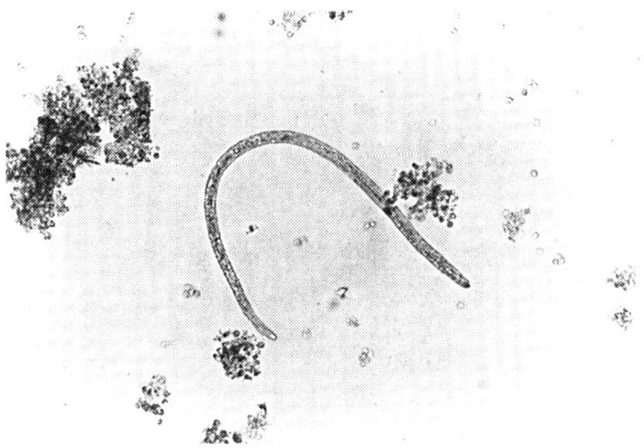

FIGURE 6.13. Second-stage *Capillaria philippinensis* larvae from human intestine.

organ. Although there was growth, the larvae did not appear to molt and were considered to be still in the first stage. Experimentally eggs were found to hatch in a number of indigenous fish from fresh and brackish waters in the Philippines and Thailand (Cross and Bhaibulaya, 1983).

Experimental attempts to infect monkeys were attempted with larvae dissected from fish and administered either by stomach tube, by feeding the intestinal tract of fish, or by feeding whole infected fish to fasting monkeys. Patent infections developed in three species of monkeys (*Macaca cyclopsis, M. fascicularis*, and *M. mulata*), with eggs first appearing in the feces at 22 to 96 days (average of 46 days), with some infections persisting for over 1 year. The monkeys never developed diarrhea or other symptoms during the infection, and the parasite was not found in any organ other than the intestines at necropsy. Analysis of fecal egg counts were from eight monkeys indicated peak egg outputs occurred at 22 ± 6 days. Although the peak egg outputs were not constant, a cyclical pattern was noticeable. The reason for this was probably due to the die-off of senile worms

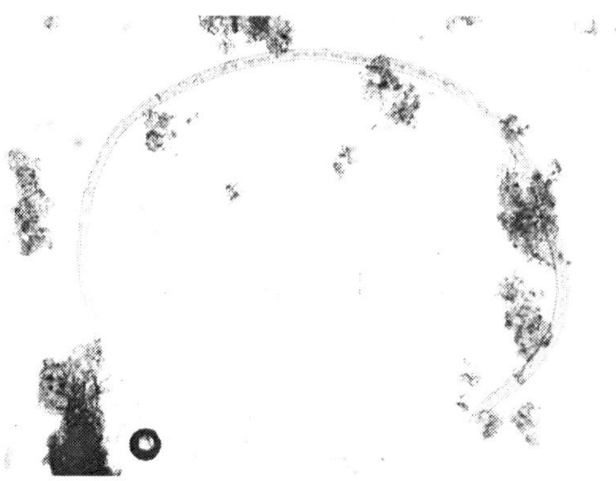

FIGURE 6.14. Third-stage *Capillaria philippinensis* larvae from human intestine.

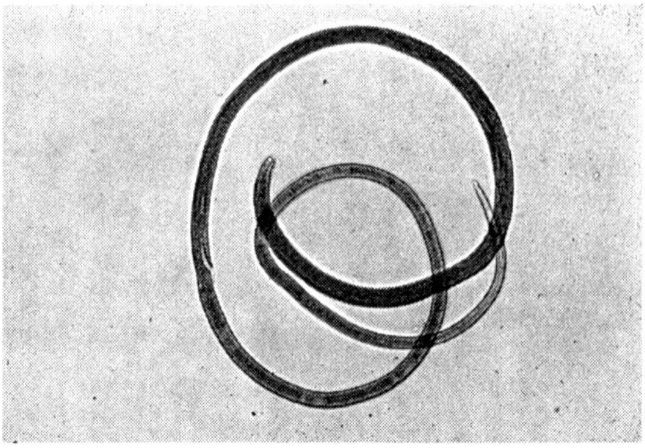

FIGURE 6.15. Fourth-stage, pre-adult *Capillaria philippinensis* from human intestine.

and the production of new ones by larviparous females. Necropsy of some monkeys given 30 to 50 larvae from fish yielded 10,000 to 30,000 worms of all stages were recovered 3 to 4 months after infection (Cross et al., 1972). These results confirmed human autopsy findings that autoinfection was part of the life cycle of the parasite.

Many other laboratory animals as well as wild and domestic animals were also experimentally given infective-stage larvae from fish or fed whole infected fish. Only transient infections resulted in a few wild rats (*Rattus* spp.) and multimammate rats (*Mastomies natalensis*) (Cross and Bhaibulaya, 1983), but patent infections occurred in Mongolian gerbils (*Meriones unguiculatus*), developing in an average 27 days, with death occurring on average at 46 days. In further studies with the gerbils, larvae from fish were found to develop into adult worms in the gerbil intestines in 10 to 11 days with the female worms producing larvae in 13 to 14 days. These larvae were retained and subsequently developed into adult parasites. The adults of this second generation of worms began producing eggs at 24 to 25 days postinfection. Some of the first-generation worms continued to

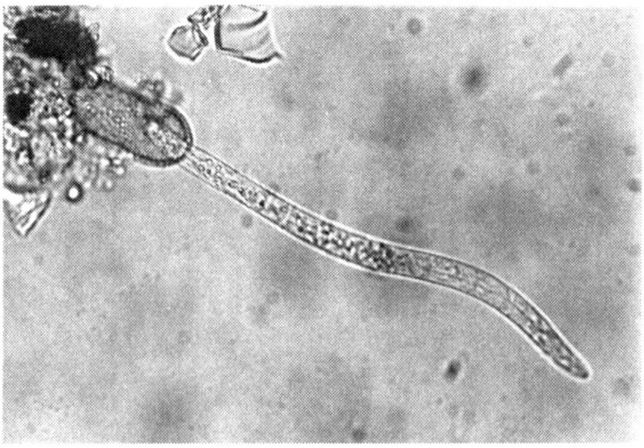

FIGURE 6.16. Larva of *Capillaria philippinensis* emerging from an egg in fish intestines.

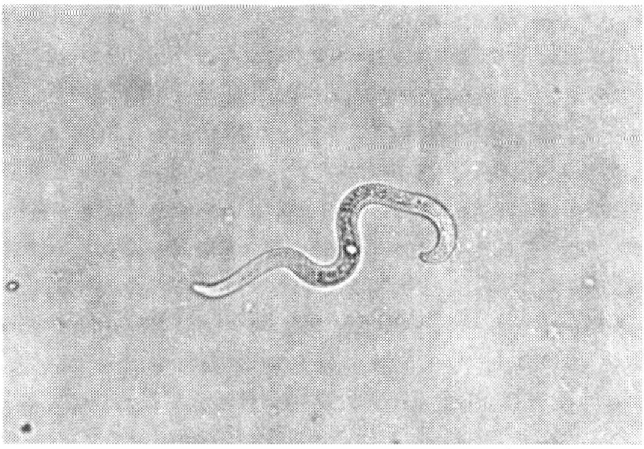

FIGURE 6.17. Larva of *Capillaria philippinensis* after 3 weeks in a fish intestine; considered an infective stage for the definitive host.

produce larvae until they died. Most of the second-generation worms, however, produced eggs but a few continued to produce larvae, which maintained a high population of worms. The reproductive potential in these hosts was high, and thousands of worms were produced. The infected gerbils lived for an average of 46 days and the highest number of worms was recovered between 36 and 46 days. All stages of the parasite were recovered from the small intestines at necropsy. In another experimental study, of gerbils given two larvae from fish and two animals at necropsy at 29 and 31 days postinfection, 2520 and 5353 worms in all stages of development were recovered, respectively (Cross et al., 1978). In another experiment, it was found that infections could be maintained in gerbils by serial passage of worms from a necropsied animal to another gerbil. In other studies, animals given larviparous adult female worms developed massive infections. Infections were also able to develop in animals given only oviparous female; in this instance the female worms switched from egg production to larval production, and the parasite population continued to increase until the death of the animals.

Disease

The clinical aspects of capillariasis philippinensis have been well documented by Whalen et al. (1969), Watten et al. (1972), and more recently by Bair et al. (2004). It appears that everyone who becomes infected with *C. philippinensis* eventually develops symptoms. The incubation period is unknown, but early symptoms in Filipinos appeared at about 3 weeks after eating raw fish; this corresponds well with the prepatent period observed in experimentally infected monkeys. The major symptoms are gurgling stomach (borborygami), abdominal pain, and tenderness. As the disease progresses with subsequent multiplication of the worms, the pain increases and an intractable diarrhea develops with 8 to 10 voluminous stools per day. The stools often resemble that seen in cholera especially during the terminal stage of the disease when the diarrhea is continuous. There is also weight loss, weakness, malaise, edema, hyporeflexia, anorexia, and cachexia. There may be distant heart sounds, hypotension, gallop rhythm, and pulsus alterans. There is malabsorption of fats, sugars, and a severe protein-losing enteropathy. Some patients experience decreased excretion of xylose and low levels of potassium, sodium, calcium, carotene, and total protein in the serum. Immunoglobin E (IgE) increases and levels of IgG, IgM, and IgA are low; however, several months after treatment all immunoglobulins return to normal (Rosenberg et al., 1970). Early in the disease, the patients are ambulatory but become bedridden during the severe stages of the disease. When the disease is undiagnosed, the patients become critical and may die as a result of the irreversible effect of electrolyte loss, resulting in heart failure and effects on the other organs. Intercurrent bacterial septicemia also contributes to death (Whalen et al., 1969). The terminal stage of the disease may be reached in 2 to 5 months.

Pathology

Canlas et al. (1967) and Fresh et al. (1972) published necropsy finding in a small number of Filipino patients. Their bodies were emaciated, dehydrated, and pale. Serous fluid was found in the abdominal, pleural, and pericardial cavities. The weights of the visceral organs for the most part were reduced. Bacteria (*Streptococcus pyrogenes, Klebsiella pneumoniae*) were isolated from the lungs and blood. Most pathology occurred in the small intestines. Bowel fluid was watery and contained flecks of mucus and thousands of worms. In one patient over 200,000 adult and larvae *C. philippinensis* were estimated in 1 L of bowel fluid, and in two others 10,000 and 40,000 were estimated. The parasite was found primarily in the intestines; however, in one autopsy, Capillaria-like worm sections were found in the liver (Fresh et al., 1972). Many worms were found in the lumen of the jejunum (Fig. 6.18), and in the histological sections

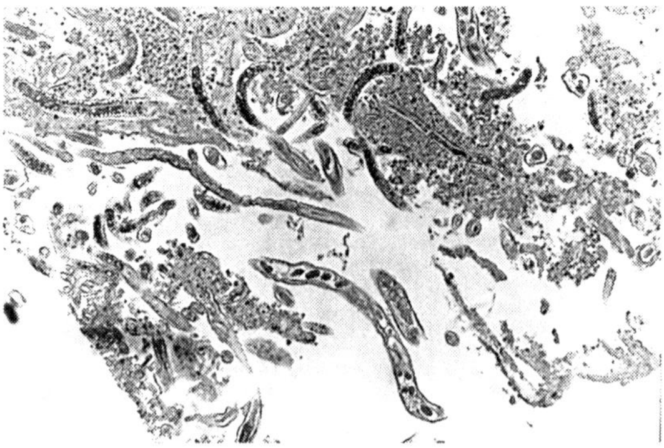

FIGURE 6.18. Histological preparation of human intestines tissue showing sections of *Capillaria philippinensis* in the lumen.

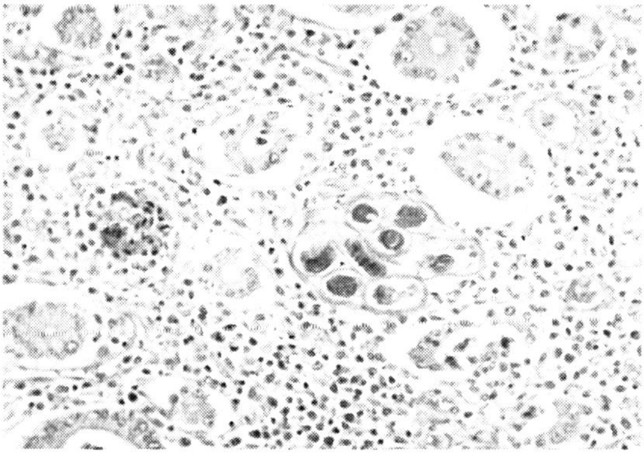

FIGURE 6.19. Histological preparation of human intestinal tissue showing sections of worms in the glands.

large numbers of worms were found in the glands (Fig. 6.19), with tortuous tracts in the lamina propria and the crypts of Liebekuhn (Fig. 6.20). There was infiltration in the lamina propria with plasma cells, lymphocytes, macrophages, neutrophils, and eosinophils. Many glands were filled with eosinophilic debris and dilated (Fig. 6.21). The villi were blunted, flattened, or completely destroyed. In ultrastructural studies, the jejunum (Figs. 6.22 to 6.24) showed a loss of adhesion specialization, degenerative changes of epithelial cells, and widened intracellular spaces. In experimental gerbil infections, the intestine showed erosion of cells and the formation of microulcers in the epithelium. These changes were probably caused by the loss in protein, fluid, and electrolyte from tissue spaces (Sun et al., 1974).

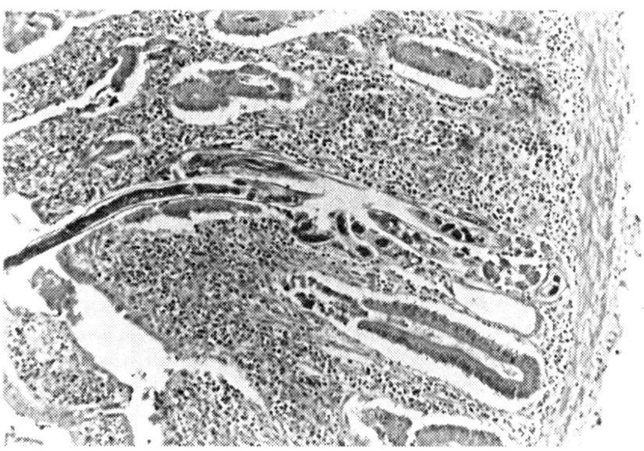

FIGURE 6.20. Sections of human intestinal tissue showing *C. philippinensis* in the Crypt of Lieberkuhn.

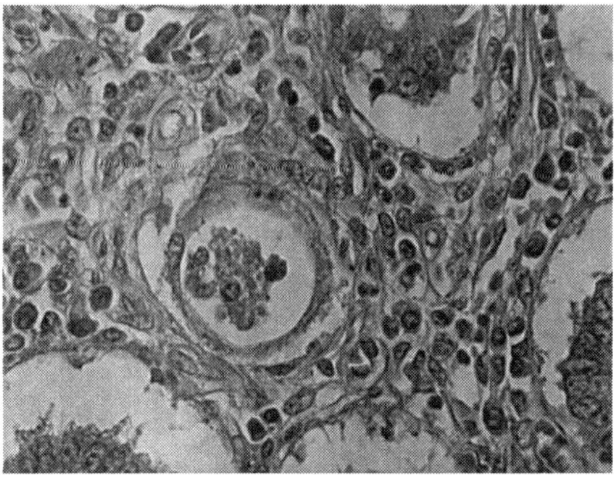

FIGURE 6.21. Histological section of the human small intestine showing debris in atrophied dialated glands.

Diagnosis

During the epidemic and subsequent cases in the Philippines, a clinical diagnosis was made based on the symptoms of diarrhea, abdominal pain, borborygmus, and weight loss. The diagnosis was confirmed, however, by detecting characteristic parasite eggs in the feces. The eggs have a striated shell, are peanut-shaped with flattened bipolar plugs, and measure 36 to 46 μm by 20 μm (see Fig. 6.11). The stools may also contain other stages of the parasites in addition to eggs: larvae and adults. Direct and concentration stool examination methods can be used in diagnosis; however, when the parasite is not found in the stool, aspirates from the small

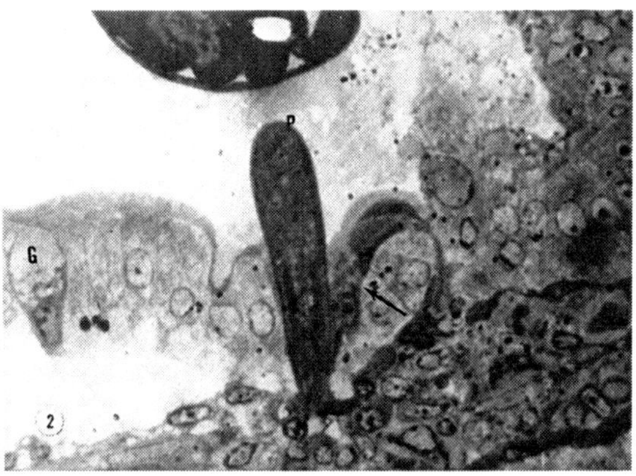

FIGURE 6.22. Electromicroscopic view of a parasite penetrating through the intestinal villus epithelium to the basal lamina of gerbil intestines showing compression of the epithelial cells. (From Sun et al., 1974, with permission of the SouthEast Asian Journal Trop Med Pud Hith)

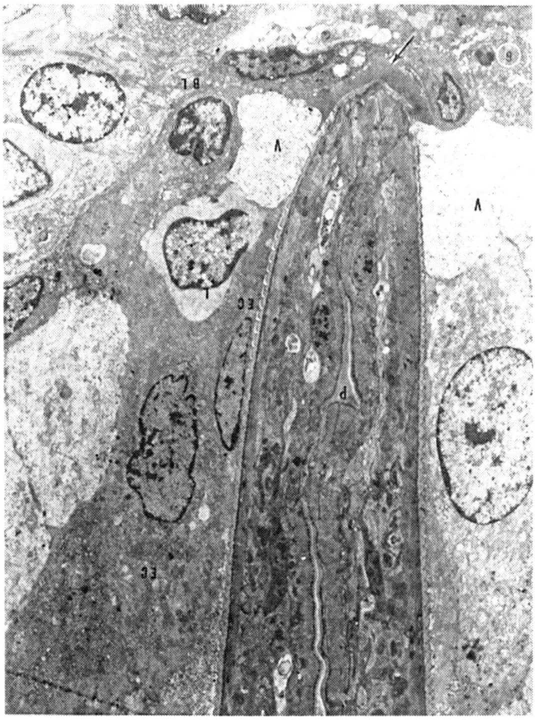

FIGURE 6.23. Higher magnification of an electron micrograph showing penetration site of gerbil intestine. Note thick layer of an electron dense homogeneous material around the oval tip of the parasite. (From Sun et al., 1974, with permission of the southeast Asian J. Trop Med Pub Hith)

intestines by Crosby capsule (Watten et al., 1972), gastroendoscopy (Wongsawadi et al., 2002), or possibly string capsule may reveal the parasite. Multiple intestinal parasitic infections were common among patients (Cross and Bhaibulaya, 1983).

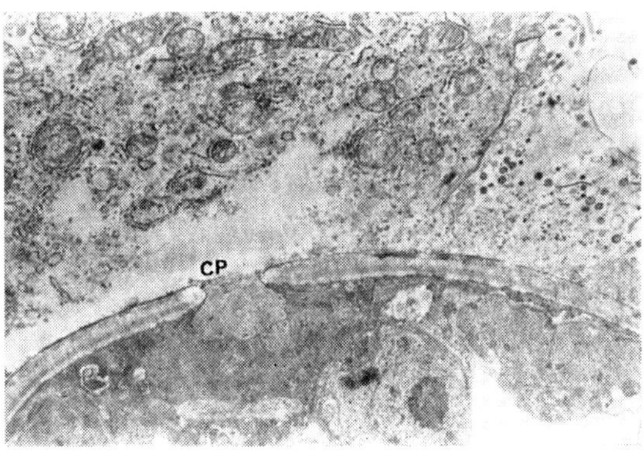

FIGURE 6.24. Cross section of *C. philippinensis* in gerbil intestines showing in the cuticula pore of the parasite and dissolution of the plasma membrane. (From Sun et al., 1974, with permission of the southeast Asian J. Trop Med Pub Hith)

The immunological diagnosis has not been satisfactory and is considered inconclusive (Banzon et al., 1975; Cross and Chi, 1978). However, recently an assay using *Trichinella spiralis* antigen by immunoblot analysis has demonstrated potential (Intapan et al., 2006).

Treatment

Patients with intestinal capillariasis are treated by replacing electrolytes, given an antidiarrheal, an anthelminthic, and a high-protein diet (Whalen et al., 1971). During the early days of the epidemic, thiabendazole, 25 mg/kg/d was given to patients for 3 days. Although the drug was effective initially, causing a disappearance of *C. philippinensis* eggs and worms and improvement in symptoms, the patients relapsed after several days after they stopped taking the drug. It was apparent that this treatment was incomplete, and consequently the drug was continued for several more days. However, because of the continuous relapses in some patients, long-term therapy became necessary (Whalen et al., 1969), and the drug was given in doses of 1 g a day for 10 days followed by the same dosage every other day for 16 to 20 weeks (Singson, 1969, 1974). In spite of side effects with thiabendazole and a high relapse rate, the drug remained the choice for many years. Singson (1974) reported that 68% of over 1200 patients had recurrence of the disease at least once during the second and third years of the epidemic; some of the affected persons relapsed as many as 16 times (Alcantara et al., 1985). Many anthelminthics were tested without effect and the problem of relapse continued. These were considered relapses rather than reinfection because at that time there were more relapse cases than new cases. The relapses were attributed to the inability of the drugs to affect the larval stages of the parasite. With extended treatment, the drugs were effective against the adult worms, and when the larvae matured, the drugs became effective. On many occasions, patients had to be hospitalized for several weeks even after symptoms disappeared because even if the patients were released early and given thiabendazole on an outpatient basis, they often relapsed (once home, they felt good and stopped taking the drug or they gave the drug to others who had intestinal complaints). Retention in the hospital reduced the relapse rate.

Thiabendazole continued to be used until mebendazole became available (Singson et al., 1975). After a series of trials, this drug in dosages of 200 mg twice a day for 20 days was found to be the most effective for first-time infections: however, treatment was given 30 days because of the risk of relapse (Basaca-Sevilla and Cross, 1985). Subsequently, mebendazole was replaced by albendazole when the latter became available (Bhaibulaya and Kobwanthalokun, 1984). Albendazole has been found to be very effective when given at 200 mg twice a day for 10 days (Cross and Basaca-Sevilla, 1987); with this course relapses are no longer are a problem. Few side effects are evident with either mebendazole and albendazole, with symptoms and fecal eggs disappearing as early as 4 days. Early treatment is recommended since prolonged untreated infections often result in death. Deaths are rare when the disease is recognized early and treated appropriately.

In an attempt to identify people with capillariasis before symptoms developed, stool surveys were carried out in a number of barrios in the endemic area (Cross and Bhaibulaya, 1983a). A number of cases were found, and the infected person was advised to obtain treatment at the hospital. Most did not follow the advice since they did not have symptoms at that time. Once symptoms developed, however, the patients went to the hospital for treatment. In an interesting event, in one of the surveys an entire family was found to be egg-positive. But when the entire family of seven was advised to receive treatment, it developed that the parents had submitted stool from one of the children for all members of the family. The only infected child was subsequently identified and treated.

Epidemiology

Although intestinal capillariasis is reported from several countries, most infections have been in Asians. Information on the parasite, disease, and epidemiology has come primarily from studies carried out in the Northern Luzon, Philippines, where there were nearly 2000 human cases and over 100 deaths occurred. The peak of the epidemic occurred in 1967–1969, and once effective treatment and knowledge of the disease was established, the incidence and deaths decreased dramatically. Figure 6.25 shows the incidence and deaths attributed to *C. philippinensis* from 1967 to 2005. The disease is now considered endemic with a few reports recorded by public health agency. The parasite and diseases are documented from several

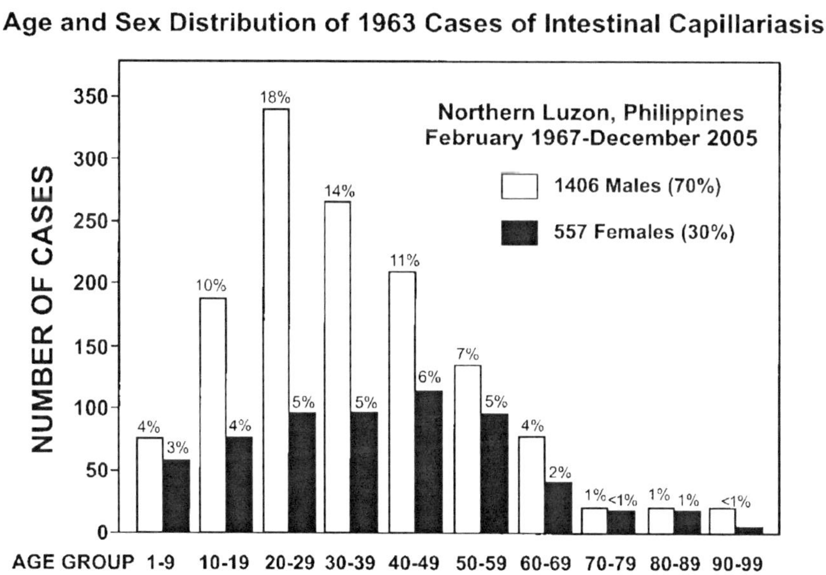

FIGURE 6.25. Documented cases and death due to intestinal capillariasis in Northern Luzon in the Philippines, 1967–2005.

areas of the Philippines in Luzon, Leyte, and Mindanao (Fig. 6.26). Belizario and coworkers (2002) found 22% of a population in the Compostela Valley province in Mindanao infected.

The infection has been seen more consistently in males (70%) than females (30%) and in the 20- to 40-year age group (Fig. 6.27). The reason for the age and sex differences are not known, but it is believed that middle-aged men are more often exposed to infection than are women. They work in the fields all day, and in the late afternoon they go to lagoons and bathe, and being hungry, they eat raw freshly caught fish and other aquatic wildlife caught in fish traps (Fig. 6.28). Some of the catch then may be taken home, where the fishes are more often cooked before being served at meals, although some people may consume fish raw.

The people of Northern Luzon are known as Ilocanos and their eating habits are unlike eating habits of most people living elsewhere in the Philippines. Dietary histories were taken from the villagers of the affected Tagudin, Ilocus Sur, area, and because of economic hardships, the people reported eating everything that they could raise or catch. The basic foods are rice, fish, vegetables, and occasionally meat from chicken, pork, goats, water buffalo, and, on special occasions, dog. Marine and aquatic plant and animal life (fish, snails, crabs, etc.) are eaten when available. A popular dish is "jumping salad," consisting of live shrimp seasoned

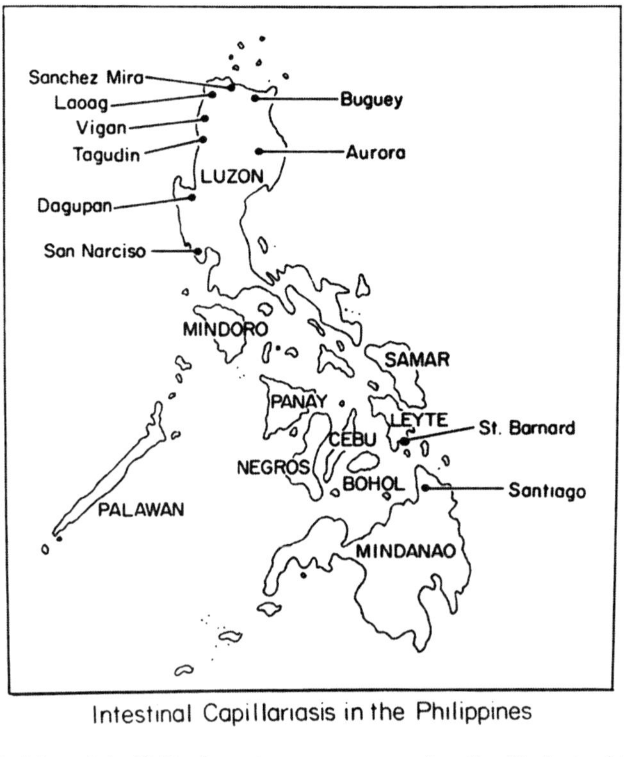

FIGURE 6.26. Map of the Philippines shown areas reporting *Capillariasis philippinensis*.

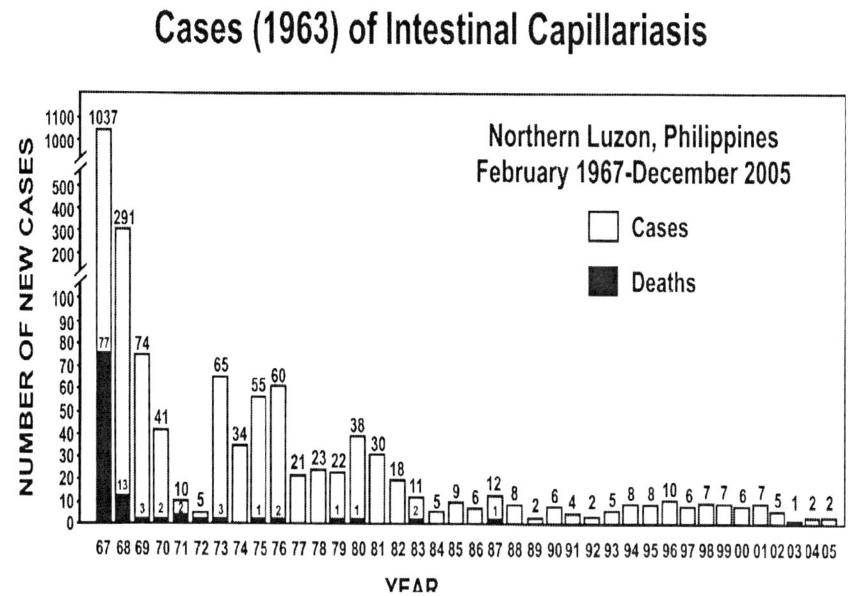

FIGURE 6.27. Age and sex distribution of intestinal capillariasis cases documented from the Philippines, 1967–2005.

with vinegar, garlic, and chili. Organs of animals are often eaten raw (kinilaw) especially when the men drink (usually a local sugar cane wine called basi) while socializing with their neighbors at night. Intestinal juices such as bile are used to season rice. People of Thailand, where cases have been reported, also enjoy eating raw fish (Bhaibulaya et al., 1979), and in Taiwan, most capillariasis cases occur in aboriginal groups who eat fish and other animal life uncooked (Lu et al., 2006).

FIGURE 6.28. Lagoon in West Pudoc, Tagudin, Ilocus Sur, in the Philippines, with several fish traps made from tree branches. Villagers regularly ate fish and other animal life from this lagoon uncooked.

As described above, experimental studies show that many species of fresh and brackish water fish can serve as intermediate host for this parasite (Cross and Bhaibulaya, 1983; Cross and Basaca-Sevilla, 1991), and many of these species of fish are eaten raw, dried, smoked, or cooked. The fish that were incriminated in the Pudoc outbreak were usually caught in fish traps in the lagoons. The people were especially fond of eating gravid female fish called "bagsit" (*Hypeselotris bipartita*) (Fig. 6.29). Natural transmission of the parasite to gerbils was obtained by feeding fish caught in the lagoons or fish sold in the local markets. In most countries reporting the disease, especially in Asia, raw fish is frequent part of the diet.

Fish-eating birds are now considered the natural host of *C. philippinensis*. Bhaibulaya and Indra-Ngarm (1979) experimentally transmitted the parasite to fish-eating birds in Thailand, and in other studies in Taiwan, bird susceptibility was demonstrated with fish-eating migratory birds (Cross and Basaca-Sevilla, 1983b). Birds probably acquire and transmit the infection to fish in water bodies along migratory flyways. Birds migrating from all parts of the world may acquire and transmit *C. philippinensis* while nesting in the Northern Hemisphere.

As mentioned above, during the epidemic in the Philippines in 1967–1968, fecal contamination of the environment was considered the major source of infection. Indiscriminate defecation in the fields is common in the endemic area, and after torrential rains the feces are pulverized and carried by drainage to water bodies. Barrio Pudoc West had over 200 cases, and the lagoon water was contaminated directly with feces from patients. Bedpans as well as fecal soiled bed linen were washed in the lagoon waters, and fish from the lagoon was a major food source for the barrio population. Clothes washing and bathing were done in the nearest waters, especially in the brackish water lagoons. Drinking water from wells and natural sources was not boiled or chemically treated.

There is little doubt that the parasitosis has been endemic in the Luzon area for generations, but it was undiagnosed. People have been dying from gastroenteritis

FIGURE 6.29. Bagsit (*Hypseleotris bipartita*) from the lagoon in Pudoc West, Tagudin, Ilocus Sur, in the Philippines, often eaten raw, gravid female fish are the most favored.

in this region for years, but the causes were not determined because autopsies are rarely carried out, as is the case throughout most of the rural Philippines. If this spectacular epidemic had not occurred, the disease and etiology might still be unknown.

The following is an epidemiological scenario that at present best fits the epidemiological and biological data: The Ilocano people, a very industrious and hard-working people, strive to educate their children. To do this, male Ilocanos frequently travel far and wide to find employment, especially when home-farming chores slacken. It is now believed that a male from Pudoc West may have spent time in Bacarra, Ilocos Norte, the *barrio* where the index case of intestinal capillariasis originated (1963), and may have acquired the parasite while working there. A retrospective investigation was carried out in Bacarra, and the physician caring for the index case revealed that as many as 13 other people had symptoms of intestinal capillariasis and died between 1959 and 1965. It is possible that the person from Pudoc West, while in Bacarra, acquired the parasite, returned to Pudoc West with the infection, developed the disease, and consequently contaminated the environment and lagoon. It is notable that the people found infected in other endemic areas in the Philippines were originally, for the most part, from the Ilocano area of Northern Luzon (and carried their eating behavior and other habits with them).

Capillaria philippinensis is now recognized worldwide, with numerous cases being reported from Korea, Japan, and Egypt. Taiwan recently reported 30 cases from 1983 to 2003 (Lu et al., 2006). In addition, single cases have been reported from Iran, India, Italy, and Spain. The latter two were imported cases from Indonesia and Colombia, and cases among indigenous people have not been reported from these countries (Cross, 1992).

Impact

During the epidemic years, intestinal capillariasis had a significant impact in the Philippines, especially in the Ilocano populations of Northern Luzon. Life in some of the villages was disrupted. As described above, parents, especially the fathers, were dying and the children often orphaned. It was not unusual to find fatherless families joining another family.

People in endemic *barrios* were essentially frightened and believed that they were all going to die of the mysterious disease. It was only after the health department convinced the people that it was a parasitic disease and treatment regimens were developed that the population began to cooperate. Their eating habits did not change until it was shown that the disease came from eating uncooked fresh and brackish water fish. Even then the habit has not completely disappeared. It is difficult to change traditional habits that had been practiced for many generations. The widespread use of anthelminthics and improvements in sanitary conditions have impacted the zoonosis, and the risk of infection has declined.

The economy of the endemic area is essentially agricultural, with rice and tobacco crops exported from the area. Most other agricultural products are for

FIGURE 6.30. Market in Tagudin, Ilocus Sur, Philippines, selling fresh fish and other aquatic animal life from the lagoons in the area.

local consumption. Fishing is also important, and the catch is usually for local consumption. Some products, such as fish, shrimp, snails, and meats, are sold in markets (Fig. 6.30) in the surrounding towns. Any events that have a negative influence on the wholesomeness of such foods have serious consequences for farming. For example, during the epidemic, one scientist reported in a newspaper that he had found the source of the infection in marine fish. It was eventually determined that the worms he found were larval stages of *Anisakis* sp., a parasite of marine fish, a fish-borne parasite unrelated to *Capillaria*. This report impacted the fishing industry because people were afraid to buy and eat fish. It was only after then President Ferdinand Marcos of the Philippines went on television and ate fish that the people went back to eating one of their favorite and economically important foods.

Unresolved Problems

There remains much to learn about the life cycle of *C. philippinensis*. Most of the information available has been obtained experimentally. Additional details on the life cycle and natural transmission need to be documented. The greatest challenge would be to establish the factors involved in autoinfection. The parasite multiplies by autoinfection in the intestines of monkeys, gerbils, birds, and humans, and in gerbils, female worms can switch from egg production to larvae production. Larvae were found in the receiving gerbils at necropsy a few days after exposure, and after 7 to 10 days the number of worms increased substantially. It would be interesting to determine the mechanism involved that enables these worms to switch from egg production to larval production. It is a unique means of perpetuating the species.

Although birds are experimental hosts of the parasite (Bhaibulaya and Indra-Ngarm, 1979; Cross and Basaca-Sevilla, 1983b) and migrating fish-eating birds may pass the egg into water bodies along the flyways, it is not known how much risk this presents to other birds that frequent these habitats and consume fish.

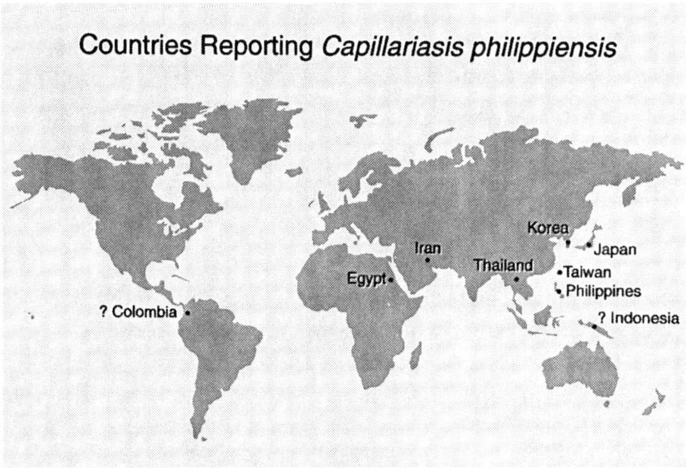

FIGURE 6.31. Countries reporting cases of capillariasis philippinensis.

Some of the experimentally infected birds died of the infections but other survived. Although one male worm was recovered from one *Ixobrychus* species in the Philippines and one of the species was experimentally infected, this and other species of fish-eating birds should be surveyed for natural infections to learn more of the host range of the parasite, and the relationship to the geographic distribution of the parasite in birds and fish, worldwide.

Human infections are widespread, with cases from the Philippines, Thailand, Taiwan, Korea, Japan, and also from Indonesia and Colombia in South America (one case each) (Fig. 6.31). Is it possible that infections are occurring elsewhere but are not recognized? Many cases of gastroenteritis occur in rural areas with an etiology that is not established. This was found to be important in the Philippines and could also be occurring in other countries with populations that eat raw freshwater fish.

The current treatment is effective if initiated soon enough; however, the administration of mebendazole or albendazole over 10 to 20 days, respectively, may be too long. Newly developed anthelminthics should be evaluated that would provide a more rapid death of the adult and larval stages.

Control

As mentioned earlier, cases of capillariasis decreased with the use of anthelminthics and education. During the campaign against the disease, emphasis was placed on education, on improving sanitation, and on changing eating habits. There was an emphasis on cooking animal parts and products before eating. Indiscriminate defecation is also a common practice, so people were encouraged to install and use low-cost water-sealed toilets. Although this can be an effective measure, these practices can also be difficult to alter. In this case, people reported a dislike of defecating in a closet; a one farmer stated that his bowels would not move until he saw the sky and a breeze was passing over his head. More research is needed to find practical and acceptable means for people to dispose of their waste to avoid environmental contamination.

Summary

Capillaria philippinensis is somewhat unique among helminthic diseases in that infection has a high probability, if untreated, of ending in death. This zoonotic infection was first discovered as a cause of epidemic diarrheal disease in 1966 in the Philippines, where nearly 2000 human infections and nearly 100 deaths were documented. Fresh and brackish water fish were found to serve as intermediate hosts, and fish-eating birds were the definitive hosts. Humans acquire the infection by eating fish raw. Most female worms are oviparous, but a few produce larvae, which lead to autoinfection and hyperinfection. Eggs reaching water infect fish, which are then eaten raw by humans and birds. The life cycle was determined experimentally in monkeys and Mongolian gerbils; disease did not develop in monkeys, but gerbils died. Human symptoms were diarrhea, abdominal pain, and gurgling stomach. If untreated, there is weight loss, weakness, anorexia, and cachexia, and death may occur in 4 or 5 months. Mebendazole and albendazole are the drugs of choice. Further information regarding the life cycle and factors associated with autoinfection need to be determined.

References

Alcantara, A.K., Uylangco, C.V., and Cross, J.H. (1985). An obstinate case of intestinal capillariasis. Southeast Asian J Trop Med Public Health 16, 410–413.

Bair, M.J., Hwang, K.P. Wang, T.E. Liou, T.C., Lin, S.C., and Kao, C.R. (2004). Clinical features of human intestinal capillariasis on Taiwan. World J Gastroenterol 16, 2391–2393.

Banzon, T.C., Lewert, R.M., and Yogore, M.G. (1975). Serology of *Capillaria philippinensis* infection: reactivity of human sera to antigens prepared from *Capillaria obsignata* and other helminths. Am J Trop Med Hyg 24, 256–263.

Basaca-Sevilla, V., and Cross, J.H. (1985). Attempts to modify treatment of intestinal capillariasis. Southwest Asian J Trop Med Public Health 16, 546–548.

Belizario, V.Y., de Leon, W.U., Esparar, D.G., Galang, J.M., Fantone, J., and Verdadero, C. (2002). Compostela Valley: 2002. A new endemic focus for capillariasis philippinensis. Southeast Asian J Trop Med Pub Hlth 31, 478–481.

Bhaibulaya, M., and Indra-Ngarm, S. (1979). *Amaurornis phoenicurus* and *Ardeola bacchus* as experimental definitive hosts for Capillaria philippinensis in Thailand. Int J Parasitol 9, 321–322.

Bhaibulaya, M., Indra-Ngarm, S., and Anathapruit, M. (1979). Freshwater fishes of Thailand as experimental intermediate host for *Capillaria philippinensis*. Int J Parasitol 9, 105–108.

Bhaibulaya, M., and Kobwanthalokun, S. (1984). Nonfisheating intestinal capillariasis and the efficacy of albendazole. J Parasitol Trop Med Assoc Thailand 71, 47–50.

Canlas, B.C., Cabrera, B.D., and Davis, U. (1967). Human intestinal capillariasis. II. Pathological features. Acta Med Philipp 4, 84–91.

Chitwood, M.B., Velasquez, C., and Salazar, N.G. (1964). Physiological changes in a species of *Capillaria* (*Trichuroidea*) causing a fatal case of human intestinal capillariasis. Proc Int Cong Parasitol 1(2), 797.

Chitwood, M.B., Valasquez, C., and Salazar, N.G. (1968). *Capillaria philippinensis* sp. n. (Nematoda: Trichinellida) from intestine of man in the Philippines. J Parasitol 54, 368–371.

Cross, J.H. (1992). Intestinal capillariasis. Clin Microbiol Rev 5, 120–129.

Cross, J.H., Banzon, T.C., Clarke, M.D., Basaca-Sevilla, V., Watten, R.H., and Dizon, J.J. (1972). Studies on the experimental transmission of *Capillaria philippinensis* in monkeys. Trans R Soc Trop Med Hyg 66, 819–827.

Cross, J.H., Banzon, T.C., and Singson, T.C. (1978). Further studies on *Capillaria philippinensis*: development of the parasite in the Mongolian gerbil. J Parasitol 64, 208–213.

Cross, J.H., and Basaca-Sevilla, V. (1983a). Experimental infection of *Capillaria philippinensis* in multimammate rats (*Mastomys natalensis*). Southeast Asian J Trop Med Public Health 14:264.

Cross, J.H., and Basaca-Sevilla, V. (1983b). Experimental transmission of *Capillaria philippinensis* to birds. Trans R Soc Trop Med Hyg 77, 511–514.

Cross, J.H., and Basaca-Sevilla, V. (1987). Albendazole in the treatment of intestinal capillariasis. Southeast Asian J Trop Med Public Health 18, 507–510.

Cross, J.H., and Basaca-Sevilla, V. (1991). Capillariasis philippinensis: a fish-borne parasitic zoonosis. Southeast Asian J Trop Med Public Health 22(suppl):153–157.

Cross, J.H., and Bhaibulaya, M. (1983). Intestinal capillariasis in the Philippinensis and Thailand. In: Croll, N., and Cross, J.H., eds. Human Ecology and Infectious Diseases. New York: Academic Press, 103–136.

Cross, J.H., and Chi, J.C.H. (1978). The ELISA test in the detection of antibodies to some parasitic diseases in Asia. In: Current Concepts in the Diagnosis and Treatment of Parasitic and Other Tropical Diseases in Southeast Asia. Proc. 18th SEAMEO TROPMED Semin., Kuala Lumpur, 2 to 5 August, 1977: 178–182.

Detels, R., L. Gutman, J. Jaramillo, E. Zerrudo, T. Banzon, J. Valera, Murrell, K.D., Cross, J.H., and Dizon, J.J. (1969). An epidemic of human intestinal capillariasis: a study in a barrio in Northern Luzon. Am J Trop Med Hyg 21, 169–173.

Fresh, J.W., Cross, J.H., Reyes, V., Whalen, G.E., Uylangco, C.V., and Dizon, J.J. (1972). Necropsy findings in intestinal capillariasis. Am J Trop Med Hyg 21, 169–173.

Intapan, P.M., Maleewong, W., Sukeepaisarnjaroen, W., and Morakote, N. (2006). Potential use of Trichinella spiralis antigen for the serodiagnosis of human capillariasis philippinensis by imminoblot analysis. Parasitol Res 3, 227–231.

Lu, L.A., Lin, M.R., Choi, W.M., Hwang, K.P., Hsu, Y.H., Bair, M.J., Liu, J.D., Wang, T.E., Liu, T.P., and Chung, W.C. (2006). Human intestinal capillariasis (*Capillaria philippinensis*) in Taiwan. Am J Trop Med Hyg 74, 810–813.

Moravec, F. (2001). Redescription and systematic status of *Capillaria philippinensis*, an intestinal parasite of human beings. J Parasitol 87, 161–164.

Rosenberg, E.B., Whalen, G.E., Bennich, H., and Johansson, S.G.O. (1970). Increased circulating IgE in a new parasitic disease—human intestinal capillariasis. N Engl J Med 283, 1148–1149.

Singson, C.N. (1969). Human intestinal capillariasis. Philippine J Int Med 7, 189–200.

Singson, C.N. (1974). Recurrences in human intestinal capillariasis. Philipp J Microbiol Infect Dis 3, 7–13.

Singson, C.N., Banzon, T.C., and Cross, J.H. (1975). Mebendazole in the treatment of intestinal capillariasis. Am J Trop Med Hyg 24, 932–934.

Sun, S.C., Cross, J.H., Berg, H.S., Kau, S.L., Singson, C., Banzon, T., and Watten, R.H. (1974). Ultrastructural studies of intestinal capillariasis *Capillaria philippinensis* in human and gerbil hosts. Southeast Asian J Trop Med Public Health 5, 524–533.

Watten, R.H., Beckner, W.M., Cross, J.H., Gunning, J.J., and Jaramillo, J. (1972). Clinical studies on Capillariasis philippinensis. Trans R Soc Trop Med Hyg 66, 828–834.

Whalen, G.E., Rosenberg, E.B., Strickland, G.T., Gutman, R.A., Cross, J.H., Watten, C. Uylangco, R.N., and Dizon, J.J. (1969). Intestinal capillariasis—a new disease in man. Lancet 1, 13–19a.

Whalen, G.E., Rosenberg, E.B., Gutman, R.A., Cross, J.H., Fresh, J.W., Strickland, T., and Uylangco, C. (1971). Treatment of intestinal capillariasis with Thiabendazole, Bithionol and Bephenium. Am J Trop Med Hyg 20, 95–100.

Wongsawadi, L., Ukarapol, N., and Lertprasertsuk, N. (2002). The endoscopic diagnosis of intestinal capillariasis in a child: a case report. Southeast Asian J Trop Med Public Health 33, 730–732.

7 Gnathostomiasis

J. Waikagul and S. P. Diaz Chamacho

Gnathostomiasis is a major food-borne parasitic zoonosis and an important public health problem in regions where raw freshwater fish is a favorite dish, such as koipla in Thailand, koi ga in Vietnam, sashimi and sushi in Japan, and cebiche and callos in Central and South America. Although these dishes are considered delicious and safe in nonendemic areas, they are a real disease culprit in endemic localities. Recently, gnathostomiasis has been considered an emerging parasitic infection for tourists who travel in Central and South America, and Southeast Asia, who, after returning to their respective countries, develop migratory swelling in skin of limbs and other organs

Gnathostomiasis is caused by a zoonotic nematode in the genus *Gnathostoma*; 12 species in this genus have been considered valid species (Table 7.1), but only five are causative agents of human infection. Four species *G. spinigerum, G. doloresi, G. hispidum*, and *G. malaysiae* are distributed in Asia, and one species *G. binucleatum* occurs in Central and South America (Daengsvang, 1980; Almeyda-Artigas, 1991; Almeyda-Artigas *et al.*, 2000). The countries with high gnathostomiasis prevalence are Mexico, Japan, Thailand, and Vietnam. More than 9000 people were infected in Mexico and Thailand, and over 4000 cases reported in Japan. In Vietnam, though the numbers of cases were only 600, they reflect an increase (Table 7.2).

Biology

Taxonomy

Gnathostoma spp. belong to the phylum Nematoda or *Nemathelminthes* Rudolphi, 1808, class *Secernentea (Phasmidia)* (Linstow, 1905) Dougherthy, 1958, order *Spirurida* Chitwood, 1933, superfamily Gnathostomatoidea (Railliet, 1895) Nicoll, 1927, family Gnathostomatidae Railliet, 1985, subfamily Gnathostomatinae (Railliet, 1895) Baylis and Lane, 1920, and genus *Gnathostoma* Owen, 1836.

The genus *Gnathostoma* was described by Owen in 1836 when he found adult worms in the stomach walls of a tiger (Owen, 1836) and the genotype was called

TABLE 7.1. Valid *Gnathostoma* species (modified from Daengsvang, 1980).

Gnathostoma species	Locality	Definitive host
G. americanum Travassos, 1925	South America	Cat
G. binucleatum Almeyda-Artigas, 1991	North America	Cat
G. brasiliense Ruiz, 1952	Brazil	Otter
G. doloresi Tubangui, 1925	Asia, Oceania	Wild boar, pig
G. hispidum Fedchenco, 1872	Asia, Europe	Pig, wild pig
G. lamothei Bertoni-Ruiz, 2005	North America	Raccoon
G. malaysiae Miyazaki and Dunn, 1965	Malaysia, Thailand	Rat
G. miyasakii Anderson, 1964	North America	Otter
G. nipponicum Yamaguti, 1941	Japan	Weasel
G. procyonis Chandler, 1942	North America	Raccoon
G. spinigerum Owen, 1836	Asia, Oceania	Cat, dog
G. turgidum Stossich, 1902	North, Central and South America	Opossum
G. vietnamicum Le-Van-Hoa, 1965	Vietnam, Thailand	Otter

Gnathostoma spinigerum. Thereafter, 18 more species were registered in Asia and America. In 1960, Miyazaki reviewed the existing literature and concluded that only seven of the 18 species described could be considered different from each other (Table 7.1): *G. spinigerum* Owen, 1836, *G. hispidum* Fedchenco, 1872, *G. turgidum* Stossich, 1902, *G. americanum* Travasos, 1925, *G. doloresi* Tubangui, 1925, *G. nipponicum* Yamaguti, 1941, and *G. procyonis* Chandler, 1942 (Miyazaki, 1960). Recently, five more species have been added to this list: *G. miyasakii* Anderson, 1964, *G. vietnamicum* Le Van Hoa *et al.*, 1965, *G. malaysiae* Miyazaki and Dunn, 1965, *G. binucleatum* Almeyda-Artigas, 1991, and *Gnathostoma lamothei* Bertoni-Ruiz, 2005, which was discovered in Mexico as adult worms in the stomach wall of *Procyon lotor* (Miyazaki and Dunn, 1965; Bertoni *et al.*, 2005). In America, mammals carry six of these *Gnathostoma*

TABLE 7.2. Summary of gnathostomiasis cases in high prevalence countries.

Country	Duration	Number of infected people[a]
Mexico	–1985	–
	1986–1995	3118
	1996–2005	6548
Japan	–1985	3107 (Ando, 2005)
	1986–1995	79
	1996–2005	39
Thailand	–1985	>3228
	1986–1995	>3550
	1996–2005	>3173 (Serodiagnosis records)
Vietnam	–1985	1
	1986–1995	4
	1996–2005	600 (Xuan et al., 2004)

[a] Summarized from hospital records and published documents where not otherwise indicated.

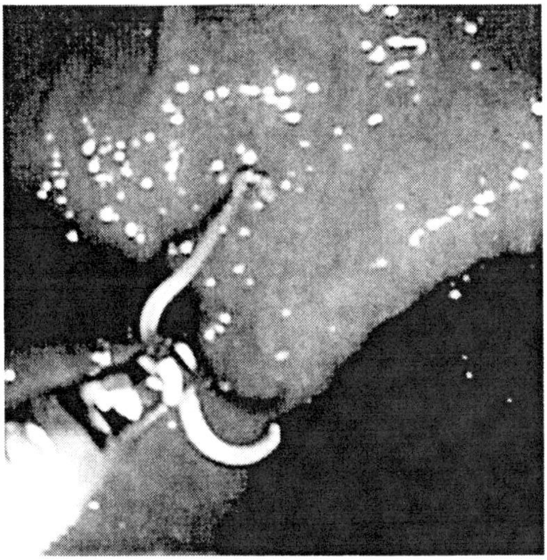

FIGURE 5.2. Anisakid nematode in gastric mucosa of Japanese patient who developed acute gastric pain after consuming a meal of raw fish. (From Hokama et al., 2005, with permission from Blackwell Publishing.)

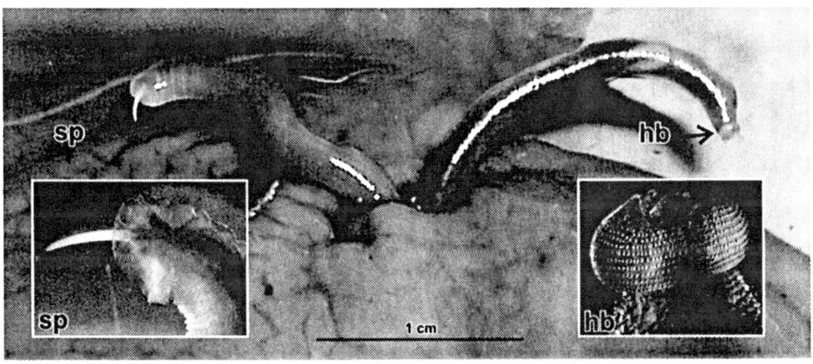

FIGURE 7.1. Adult worms of *Gnathostoma* in stomach wall of opossum *(Didelphis virginiana)* with enlarged spicule (sp) (left) and head-bulb (hb) (right).

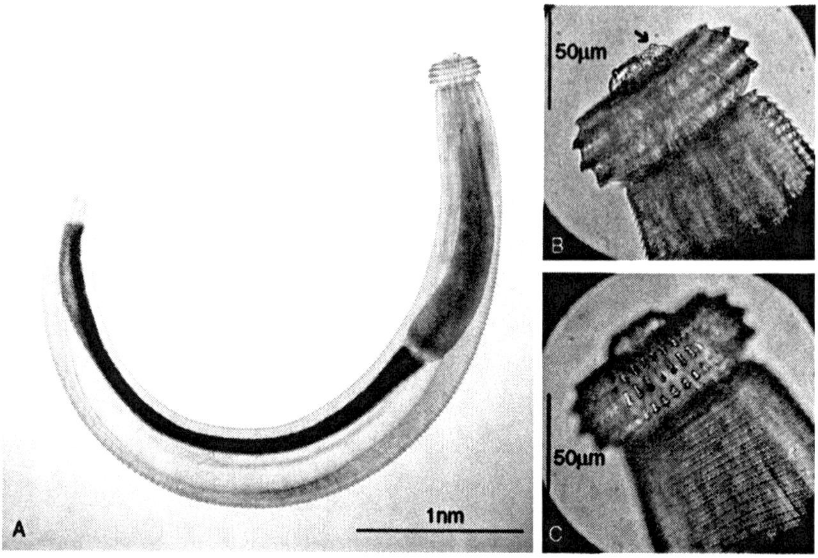

FIGURE 7.2. Advanced third-stage larva of *Gnathostoma spinigerum*. (A) Whole worm. (B) Anterior part showing lips and papillae (arrow). (C) Anterior part showing hooklets and body spines.

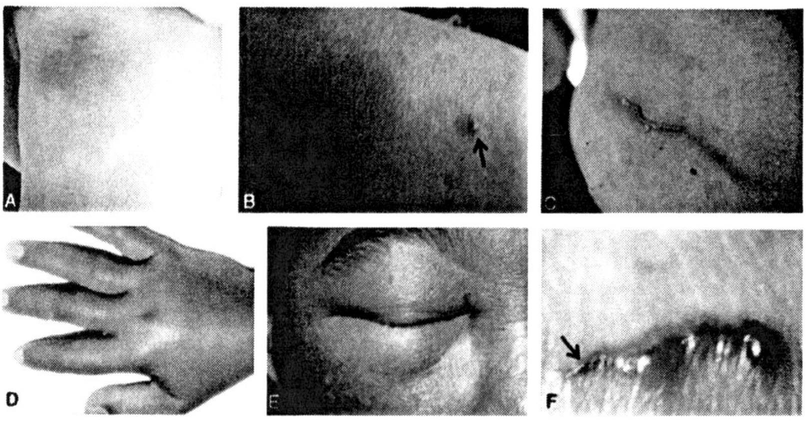

FIGURE 7.6. Cutaneous manifestations of gnathostomiasis caused by *Gnathostoma binucleatum* (A–C) and *Gnathostoma spinigerum* (D–F) courtesy of Dr. Ponganant Nontasut). Intermittent swelling (D,E) with indurated erythematous plaques (A,B) and creeping lesions (C,F). Larva (left arrows in B,F) and nodule (right arrow in B). D, E and F courtesy of Dr. Ponganant Nontasut.

species: *G. turgidum, G. americanum, G. procyonis, G. miyazakii, G. binucleatum,* and *G. lamothei*. The other six species are distributed in Asia.

Morphology

The adult parasite, located either in the stomach or the esophagus of the final host, has a cylindrical body with a variable length according to its age (Fig. 7.1). The worm varies from 2.0 to 2.6 cm for the male and 1.4 to 2.3 cm for the female, which is always thinner than the male in *G. binucleatum*. The *G. spinigerum* male is 1.2 to 4.0 cm in length, and the female 1.3 to 5.5 cm. The anterior end of the body has a round head-bulb with 8 to 10 concentric rows of hooks and a pair of lips surrounding an elongated mouth in the center (Fig. 7.2). The surface of the body is covered by rows of cuticular spines varying in distribution, size, and form according to species. Male genitals are located in the caudal area and are conformed by papillae in different in number, size, and distribution; this area is curved toward the ventral side and shows two spicules of different length (0.9–2.1 mm and 0.4–0.5 mm). Female genitalia comprise a true vagina linked to a double uterus, in which there are eggs in different developmental stages.

Eggs are expelled to the exterior through a vulva, which is located in the midsection of the body. Fertilized eggs are eliminated with the stool of the definitive host and are oval, usually colored yellow or brown due to their direct contact with bile juice. Fertilized egg size is approximately 70×40 μm; the eggshell has either some small pits or a smooth surface, with one or two polar bulges (Koga *et al.*, 1991; Miyazaki, 1991).

The advanced third-stage larva (AL3) (Fig. 7.2) is the infective form for intermediate hosts such as the human, who is an accidental host in that this transmission path is a blind alley for further worm development (see below). It usually encysts within the muscles or liver of intermediate and paratenic hosts; it has a cylindrical body with round extremes, and it is 2.8 to 5.2 mm in length. At the

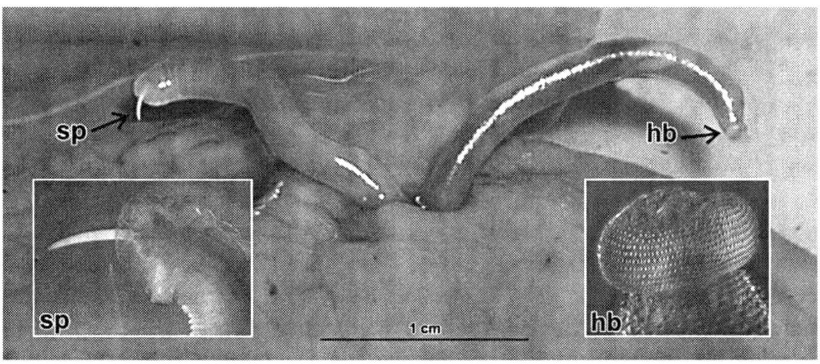

FIGURE 7.1. Adult worms of *Gnathostoma* in stomach wall of opossum (*Didelphis virginiana*) with enlarged spicule (sp) (left) and head-bulb (hb) (right). See also color insert.

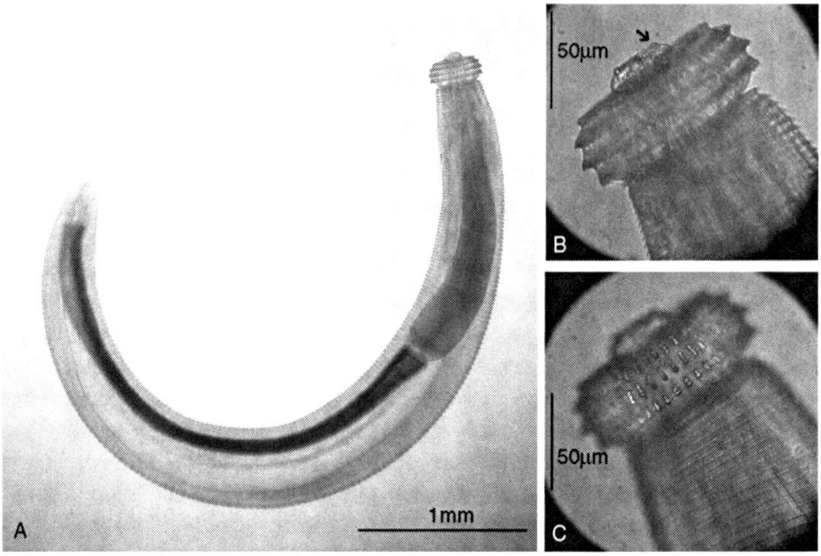

FIGURE 7.2. Advanced third-stage larva of *Gnathostoma spinigerum*. (A) Whole worm. (B) Anterior part showing lips and papillae (arrow). (C) Anterior part showing hooklets and body spines. See also color insert.

anterior it possess a head-bulb with hooks aligned in three or four rows; the number and arrangement of these hooks are of diagnostic importance. In the middle of the head-bulb there are two voluminous lips with a pair of papillae each, surrounding an elongated mouth. Within the head-bulb, larva and adult stages have four empty chambers, called ballonets, which are independently linked with four cervical sacs in a mass form located in the pseudocoel. The ballonet-cervical sacs system constitutes the locomotive apparatus used by larvae and adult parasites to move through the host muscle mass. The esophagus continues from the mouth and has a well-developed gland that ousts some substances to ease migration, and for external digestion. The intestine ends in an anal pore toward the posterior of the body and is formed by a one-cell monolayer with two to six nuclei and numerous brown granules. The larval surface has rows of cuticular spines diminishing in size and number to the posterior end. Usually, a pair of cervical papillae is located on rows 12 and 14; farther down, in rows 30 to 32, is an excretory pore. The morphological characteristics of larvae and adult parasites (hook number on the bulb, shape and distribution of cuticular spines, and nucleus number of intestinal cells) have traditionally constituted patterns of importance in identifying *Gnathostoma* species (Akahane *et al.*, 1986; Miyazaki, 1991). The body lengths and average numbers of hooklets on the head-bulb of gnathostome larvae are summarized in Table 7.2. These morphological features serve as an important complement to the molecular genetic information, the DNA sequences, which are very sensitive and specific in identifying many species (Almeyda-Artigas *et al.*, 2000).

TABLE 7.3. Diagnostic characters for the six *Gnathostoma* species reported from humans: body length and average number of hooklets in the four rows of the head-bulb of advanced third-stage larvae.

Species	Body length (mm)	No. of hooklets 1st	2nd	3rd	4th	References
G. binucleatum	4.0	38	41	45	47	Akahane *et al.*, 1994
G. doloresi	3.2	39	39	36	38	Koga and Ishii, 1987
G. hispidum	3.1	40	41	47	48	Koga *et al.*, 1988
G. malaysiae	5.2	44	45	49	55	Akahane *et al.*, 1995
G. nipponicum	1.9	37	37	41	–	Ando, 2005
G. spinigerum	4.0	43	44	45	49	Anantapruti *et al.*, 1982

Among the six species of gnathostomes reported from humans (Table 7.3), the AL3 of *G. nipponicum* is the smallest (body <2 mm in length), while *G. malaysiae* is the longest (>5 mm). The AL3 of *G. doloresi* and *G. hispidum* have similar body lengths of about 3 mm, but *G. hispidum* has more hooklets, particularly in rows 3 and 4. The AL3 of *G. spinigerum* and *G. binucleatum* are similar in size and number of hooklets, but differ in the number of nuclei of the intestinal epithelial cell, with two to five in *G. binucleatum* and three to seven in *G. spinigerum.*

Life Cycle

There are numerous hosts in the life cycle of *Gnathostoma*—final hosts (domestic or wild mammals), intermediate hosts (crustacea and fresh/saltwater fish), and paratenic hosts (fish-eating birds, reptiles, and small mammals) (Fig. 7.3). Adult worms inhabit the stomach or esophagus of felines (ocelot), canines (dogs), marsupials (opossum), procyons (raccoon), and suides (e.g., domestic pigs and wild boars), among others. *Gnathostoma* spp. adults usually inhabit a cavity formed through a proliferation of connective tissue, which is sometimes followed by calcification of the area surrounding the parasite. Depending on the species, it is possible to find one or more cavities (tumors) in a host, in which there are one or more parasites of both genders. Females expel fertilized eggs into the stomach, and the eggs are excreted via the stool. Fertilized eggs begin to embyonate and develop, from a first-stage larva (L1) to a rhabditoid larva (L2) when deposited in freshwater bodies, such as rivers, lakes, dams, or canals. Optimal development occurs within a temperature range of 24° to 28°C. L2 larvae emerge through an egg operculum and swim in fresh water until devoured by small crustacea (copepods) such as *Cyclops, Eucyclops, Mesocyclops, Tropocyclops,* and *Acantocyclops,* among others. In the hemocoel of this first intermediate host, L2 larvae evolve into early third-stage larvae (EL3) within 7 to 10 days. Infected copepods, when eaten by fresh- or salt-water fish (second or intermediate host) release the EL3 larvae, which migrate from the fish stomach through perforations and move into the skeletal muscle tissue or liver, where they encyst and develop into an advanced third-stage larvae (AL3). Subsequently, the life cycle of the parasite

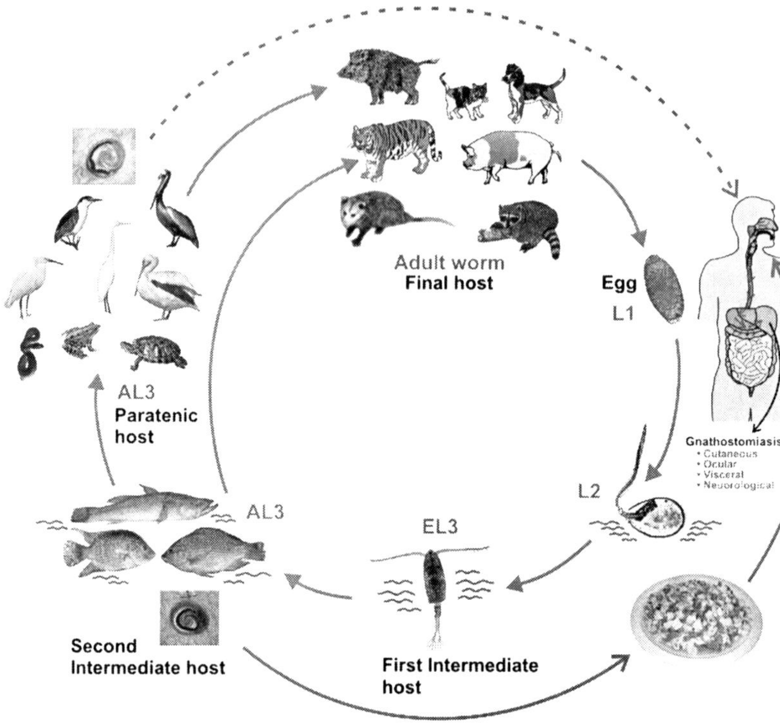

FIGURE 7.3. Life cycle of *Gnathostoma*.

may follow one of several paths: if the fish is eaten by a bird predator, or by ectothermic vertebrates such as frogs and snakes, or by small mammals, they become "paratenic hosts," in which further larval development does not occur even though the larvae remain alive and infective. But if eaten by dogs, cats, pigs, raccoons, or marsupials, which are suitable as final hosts, the larvae develop to the final adult stage to complete the life cycle. Further, if one of these potential final hosts eats a paratenic host harboring AL3 larvae, the larvae can then develop into the adult stage. This process takes approximately 100 days (Fig. 7.3). Gnathostomiasis is not only transmitted by first, second intermediate, and paratenic hosts harboring AL3, but can also gain entry into a vertebrate host by direct penetration of the skin of the palm (Daengsvang, 1980) while handling the meat of infected animals (fish, chicken, pig) during the cooking process. The fetus of an infected pregnant mother can receive larvae migrating from the mother via the placenta; a midwife found AL3 on the abdominal skin of a newborn baby in central Thailand while laboring (Waikagul, unpublished data).

Humans have been considered an accidental host of gnathostomes. But there is a question as to whether gnathostomes can develop to sexually mature, egg-laying adults in humans. Gnathostome eggs had never been confirmed in human feces, but Chandler and Reed (1961) reported that worm eggs, "presumably" those of a gnathostome, were found in human feces on two occasions in Burma. Immature males and females have been reported many times in human hosts. One worm was found in the greater omentum of the stomach of one patient (Kiriratana, 1989). Recently an adult male was removed by endoscopy from the stomach wall of a man living in northeastern Thailand (Lertanekwattana *et al.*, 2004). A tumor with a worm on the stomach wall was reported in one case in Vietnam (Vinh, 2001). Most case reports of gnathostomiasis have reported single worm infections, with only a few reports of multiple worm infections, including one case with five to six worms (Daengsvang, 1980). Although it is possible that a pair could be united in the human stomach wall and become a definitive host of gnathostomiasis, this is probably very rare. Because of this, the ability of AL3 to develop to adults in humans cannot be definitely concluded.

Epidemiology

Many species of fish, birds, reptiles, amphibians, and mammals have been reported as hosts of gnathostomes in endemic areas of Central and South America, and Asia. Fresh meat from these animals can serve as food for humans, some of which are often served raw or only slightly cooked. Recently, several cases of gnathostomiasis have been reported in travelers returning home after visiting Southeast Asia or Central and South America. Gnathostomiasis is considered an emerging imported disease in Europe and other Western countries. The following reviews are status reports from different geographical regions:

The Americas

The *Gnathostoma* species identified in America are *G. procyonis, G. turgidum, G. binucleatum*, and *G. lamothei*. Although only *G. binucleatum* is a proven cause of human gnathostomosis in America, all species probably are infective to humans.

Mexico

Peláez and Pérez-Reyes (1970) documented the first autochthonous cases of human gnathostomiasis in Mexico and is considered the first cases on this continent. To date, there have been >9000 documented cases of cutaneous gnathostomiasis, including 11 ocular and five visceral cases in Mexico. However, the actual frequency is unknown because registration of this disease has not been made obligatory by the Public Health Secretariat of Mexico. The endemic areas are located in six states, four of them along the Pacific coast (Sinaloa, Nayarit, Guerrero, and Oaxaca), and two in the Gulf of México (Veracruz and Tamaulipas) (Martínez-Cruz *et al.*, 1989; Díaz-Camacho *et al.*, 1998; Ogata *et al.*, 1998; Rojas-Molina *et al.*, 1999; Baquera-Heredia *et al.*, 2002; Magaña *et al.*, 2004). It has also been detected in animal and humans hosts in central Mexico, where the human diet includes fish transported from Mexican endemic areas (Fig. 7.4).

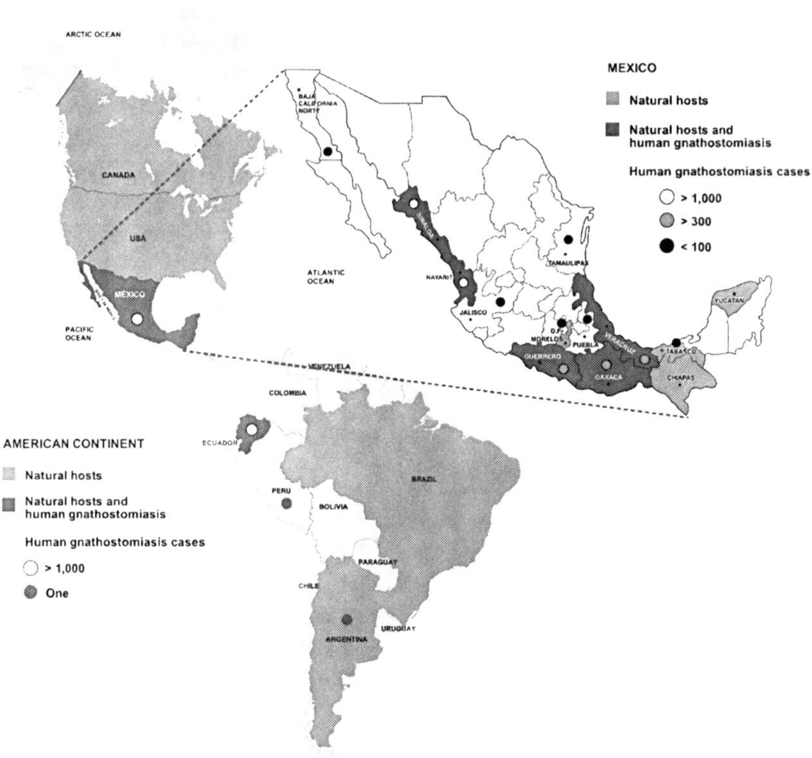

FIGURE 7.4. Distribution of gnathostomiasis in the Americas.

Epidemiological studies of gnathostomiasis in Mexico are scarce. In the state of Sinaloa, Mexico, four populations classified as either high or low risk for gnathostomiasis were compared for their consumption of raw freshwater fish. In a high-risk community (high fish consumption), a seroepidemiological survey and study of living conditions in a random sample of 309 individuals in 74 households revealed five adult patients with acute gnathostomiasis symptoms after eating cebiche, which is a regional salad dish containing pieces of raw fish marinated with lemon juice, in this case prepared from a spotted sleeper perch (*Eleotris picta*). Twelve individuals had a clinical history of migrating skin lesions, and the seroprevalence of the sample was 34.95%, as determined by enzyme-linked immunosorbent assay (ELISA). Morphological and molecular studies showed that *G. binucleatum* was the causal agent of gnathostomiasis in freshwater fish species (Díaz-Camacho *et al.*, 2002, 2003).

Regarding natural hosts in Mexico, the genus *Gnathostoma* is recorded from a variety of animals. Eight species of mammals were found as final hosts: *Canis familiaris, Didelphis marsupialis, D. virginiana, Procyon opossum, P. lotor, Felis catus, F. pardalis*, and *Sus scrofa*. These were identified in Oaxaca, Veracruz,

Sinaloa, Nayarit, Guerrero, Tabasco, Jalisco, Morelos, Chiapas, and Colima. Twenty-five species of fish have been found infected in Veracruz, Oaxaca, Sinaloa, Nayarit, Guerrero, Tabasco, Michoacán, and Tamaulipas. At least 13 species of birds are recorded as hosts in Sinaloa, Nayarit, Oaxaca, Veracruz, and Guerrero, and six species of reptiles in Veracruz and Oaxaca. In Sinaloa, Oaxaca, Veracruz, and Guerrero, four species of amphibians are recorded as hosts (Fig. 7.4).

Fish host species include the Eleotridae, Cichlidae, and Aridae families; in some endemic areas, *Eleotris picta* is the most frequently infected species and is considered the major source of infection in rural areas. However, it is not a commercial species; therefore, human gnathostomiasis in urban areas is probably more related to consumption of introduced and cultured fish species such as mojarra (tilapia) of the genus *Oreochromis* (mojarra tilapia), *Petenia* (mojarra tenhuayaca), or *Cichlasoma* (mojarra criolla, a native species).

Canada

Gnathostoma miyazakii has been described from an adult worm found in otter (*Lutra canadiensis)* kidney tissue (Anderson, 1964).

United States

Alligators *(Alligator missisippiensis)*, mink *(Mustela vision)*, opossum *(Didelphis virginiana)*, and raccoons (*Procyon lotor)* have been found with adult worms in the stomach lumen, stomach wall, liver, and stomach wall, respectively; one *Gnathostoma* species was described as *G. procyonis* (Chandler, 1932, 1942).

Ecuador

In Ecuador, human gnathostomiasis has been described as eosinophilic migratory nodular panniculitis, and was detected in 1979 in individuals living near the source of the Guayas River, which runs from Daule-Babahoyo to northern Guayaquil City (Ollage *et al.*, 1981; Feinstein and Rodriguez-Valdez, 1984). The same authors reported on 15 patients with migratory swellings: 12 cases presented migratory skin swelling, creeping lesions in one, and pneumonitis in two (Ollage *et al.*, 1984). Later, they found *Gnathostoma* adult worms in the stomach walls of domestic dogs and cats captured near the Daule River. Surveys of fish (*Hoplias microlepis)* revealed a high prevalence in specimens obtained from rice fields in areas with reported cases of gnathostomiasis (Ollage, 1985).

Brazil

Two mammals, puma (*Felis concolor)* and little water opossum (*Lutreolina crassicaudata*), and one fish (*Arapaima gigas*) were identified as natural hosts for undetermined *Gnathostoma* species (Travassos, 1925; Ruiz, 1952).

Argentina

Opossum (*Didelphis albiventris*) harboring *Gnathostoma turgidum* adult worms in the stomach wall were reported (Miyazaki, 1960). Kaminsky *et al.* (1989) documented a case with thoracic abdominal and left-arm migratory erythema-edematous plaques, and histopathological studies revealed eosinophilic paniculitis due to probable gnathostomiasis. This case appears to be the first for human gnathostomosis in Argentina.

Peru

A 21-year-old woman from Switzerland visited her family in Peru, and 1 week before her return, presented with epigastric pain, diarrhea, nausea, and anorexia; 5 days later, a migratory skin swelling appeared on the left lower limb. The clinical diagnosis was confirmed by use of indirect ELISA that detected specific antibodies to a purified 24-kD antigen obtained from AL3 larvae from *G. spinigerum* (Chappuis *et al.*, 2001). Three other cases of eosinophilic migratory nodular paniculitis have been reported (one woman and two men). In one case, the disease was acquired in Ecuador, but in the other two this was uncertain (one patient had stayed on the North Coast of Peru in the previous year and 8 years prior to that had lived in Asia, while the third patient had visited cities in the Caribbean over a previous 3.5-month period) (Villar de Cipriani, 2003).

Asia

Gnathostomiasis is widespread in Asia (Fig. 7.5). The first human case was reported in 1889, when Levinsen found an immature female worm, initially described as *Cheiracanthus siamensis*, but now known as *Gnathostoma spinigerum*, in the breast skin of a young Thai woman in Bangkok (Daengsvang, 1983). Since then, human gnathostomiasis cases and animal infections have been reported in many Asian countries.

Bangladesh

Dogs and cats in Bangladesh have been reported infected with *G. spinigerum* (Ahmad, 1962; Shaikh *et al.*, 1968; Bashirullah, 1972). A case involving a woman from Bangladesh reported a 3-year history of intermittent swelling of the right forearm and upper arm to the mid-biceps area associated with pruritus, myalgia, and arthralgia, and the symptoms resolved after treatment with two 21-day courses of albendazole (Moore *et al.*, 2003). In another case, a woman from Bangladesh who had lived in Germany for more than 2 years presented with migratory, painful swellings on her left hand and arm for 5 months. Treatment with albendazole resulted in outward migration of a larva and complete resolution of clinical symptoms (Grobusch *et al.*, 2000). Three cases of ocular infections with *Gnathostoma* have been reported, two cases from Rangpur and one from Nilphamari, in the northern districts (Rahman and Moula, 2006).

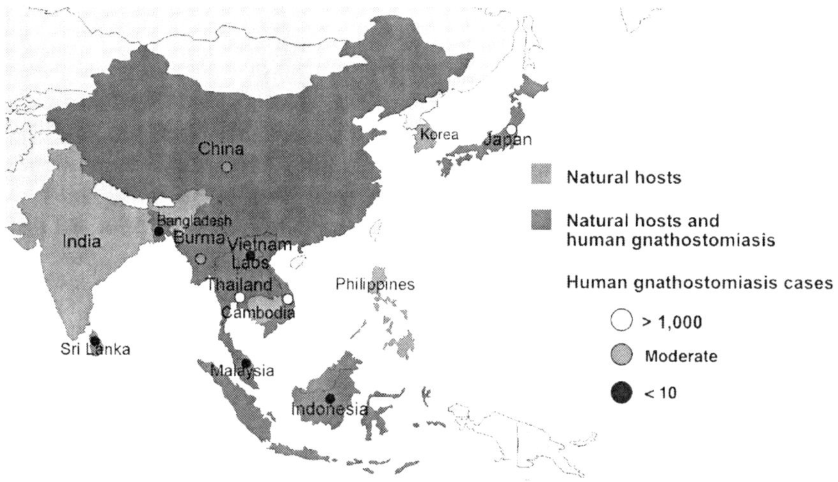

FIGURE 7.5. Distribution of gnathostomiasis in Asia.

Cambodia

Gnathostoma hispidum was reported from pigs in Cambodia in 1963 to 1976 (Daengsvang, 1981).

China

Animal and human gnathostomiasis has been reported in many cities—Hankow, Beijing, Shanghai, and Hong Kong. One of 85 dogs, two of 58 cats, and two human cases have been reported from the Hankow area (Morishita and Faust, 1925). In Shanghai, 0.7% of dogs and 3.8% of cats were found in a survey to be infected with gnathostomes (Andrew, 1937). In a case reported from Hong Kong, a woman resident who complained of subcutaneous migratory swelling admitted to frequently eating raw fish, which possibly contained infective larva of gnathostome (Moore *et al.*, 2003). Sohn *et al.* (1993) detected larvae of *G. doloresi* from the muscle of loaches imported from China to Korea. It had been reported that *G. spinigerum* and *G. hispidum* has been introduced to Japan via fish imported from China.

India

Gnathostoma spinigerum was found in domestic cats, a dog, a leopard, and several species of wild cats by Mitter (1910), in which the infection rate was 12.0% for larvae and 31.4% for adults (Chandler, 1925). Moore *et al.* (2003) mentioned that India was the country most frequently visited by their gnathostomiasis patients.

Indonesia

Three human cases had been reported from Indonesia. The first case was described from Java, where a male *G. spinigerum* was recovered from a subcutaneous swelling of the left abdominal wall of a Chinese man who had never left Java (Lie, 1949). The second case was reported from North Sumatra, in which a male worm was recovered from a subcutaneous swelling mass on the right abdominal wall of a woman who liked to eat naniura, a dish made from raw carp (Margono *et al.*, 1978). In the third case, a male worm was recovered from the cervix uteri of a 29-year-old woman from West Java (Hadidjaja *et al.*, 1979).

Japan

Japan is a hyperendemic area for gnathostomiasis. Four species of *Gnathostoma* are endemic: *G. nipponicum, G. doloresi, G. spinigerum,* and *G. hispidum.* The first two species were indigenous to Japan and the last two were introduced from China through imported fish. Thirty-two species of animals are reported as natural hosts of *G. spinigerum* in Japan (Miyazaki, 1960). The total number of treated patients for the period 1911–2002 was 3225, of whom 86 were originally infected in China and 34 in other Asian countries. Over 3000 cases were caused by *G. spinigerum*, 119 by *G. hispidum*, 26 by *G. nipponicum*, 45 by *G. doloresi*, and one by *G. malaysiae.* Japanese patient ages range from 1 year and 11 months to 76 years, with most patients in the 30- to 40-year age range. Among 751 patients, the males (549) numbered more than double the females (202) (Ando, 2005).

A total of 120 worms were removed from these patients, which included 20 adult *G. spinigerum.* The highest prevelance was reported in 1950, when 1264 patients were detected among 39,000 residents of Saga and Fukuoka prefectures; they were examined because of their high consumption of raw snakehead fish. However, the incidence decreased sharply after the community received a health-education campaign aimed at stopping the eating of snakehead fish raw (Ando, 2005).

Korea

Gnathostome larvae were reported in freshwater fish collected in Gyensang Nam do (Kim, 1973), and larvae of *G. hispidum* and *G. nipponicum* were recorded for the first time in snakes in Korea (Sohn and Lee, 1998, Han *et al.*, 2002).

Laos

A total of 55 cats and one Bengal cat from Vientiane Province, Central Laos were examined for helminthes and two were found to be infected with *Gnathostoma spinigerum* (Scholz *et al.*, 2003). Recently, a 37-year-old Laotian woman, currently a resident in Germany but who had regularly visited her homeland, developed a high-grade fever and sore throat 1 week after returning from a recent trip. One week later, a pruritic swelling appeared on the left cheek and persisted for 14 days. During the following 6 months, three other episodes of swelling occurred, each time lasting 10 to 14 days. Eventually, gnathostomiasis was diagnosed and she was treated with albendazole 400 mg b.i.d. for 3 weeks. Although the swelling disappeared, a relapse occurred 4 months later (Hennies *et al.*, 2006).

Malaysia

Adult *G. spinigerum* has been recorded in domestic cats (Adams, 1933), one tiger, and one leopard (Miyazaki, 1960). Gnathostomiasis was diagnosed in a male resident of Kuala Lumpur who had never traveled out of the country. This 32-year-old taxi driver presented with complaints of headache, nausea, vomiting, and blurred vision of the left eye for 2 days. He was found to have anterior uveitis, glaucoma, neuroretinitis, and phlebitis. A worm was removed from the anterior chamber of the eye, and was identified as *G. spinigerum* (Kamala *et al.*, 1997).

Myanmar

The presence of *G. spinigerum* adults in cats and dogs has occasionally been observed, and two human cases of intraocular gnathostomiasis have been reported (Gyi, 1960; Khin, 1968). A gnathostome larva, later identified as *G. malaysiae,* was found in the subcutaneous tissue of one of two Japanese men who developed symptoms typical of gnathostomiasis after eating freshwater shrimp in Myanmar; this is the first record of this species causing disease in humans, although this needs confirmation (Nomura *et al.*, 2000). In 2003, an outbreak of gnathostomiasis among Korean emigrants in Myanmar was reported (Chai *et al.*, 2003). Thirty-eight of 60 Korean emigrants who consumed raw freshwater fish in a Korean restaurant in Yangon developed symptoms diagnosed as gnathostomiasis by serology (ELISA) using *G. doloresi* adult-worm antigen. The patients were treated with a combination of albendazole and ivermectin for 7 to 14 days.

Philippines

Larval and adult *G. spinigerum* have been reported from four of 125 dogs two of six cats, and a civet in the Philippines (Wharton, 1917; Africa *et al.*, 1936; Tubangui, 1947).

Sri Lanka

Two cases of gnathostomiasis were reported by the Medical Research Institute (Samarasinghe *et al.*, 2002). Sri Lanka was also identified as one of the countries visited by two gnathostomiasis cases diagnosed at the walk-in emergency clinic of the Hospital for Tropical Diseases, London, between April 2000 and March 2001 (Moore *et al.*, 2003).

Thailand

Five species of the genus *Gnathostoma* have been reported from Thailand: *G. spinigerum, G. hispidum, G. doloresi, G. vietnamicum*, and *G. malaysiae*. Forty-eight species of vertebrates have been identified as natural hosts for *G. spinigerum* AL3 (Daengsvang, 1980; Rojekittikhun, 2005). Many human cases have been reported, but there is no record of total gnathostomiasis cases for Thailand. For example, during 1997 to 2002, 100 to 400 new suspected cases visited the Gnathostomiasis Clinic of the Hospital for Tropical Diseases, Bangkok (Rojekittikhun, 2005). The Department of Helminthology, Faculty of Tropical Medicine, Mahidol University, which offers gnathostomiasis serodiagnostic services received (from 1995 to 2005) 7136 serum samples of Thai patients for gnathostomiasis diagnosis, of which 3173 (44.5%) were positive, with an estimated 288 Thai patients per year being diagnosed (Paron Dekumyoy, personal communication). The Department of Helminthology receives about 2000 serum samples yearly for testing for helminthic infections, in which requests for gnathostomiasis testing are the most frequent; this is also true for samples sent from 25 countries around the world. Although some of these countries are not endemic areas for gnathostomiasis, they are the home countries of travelers who visited endemic areas and became infected.

The first author (J.W.) reviewed records of 74 Thai patients, and cases published in official documents, and found that the host ages ranged from 3 to 80 years, that most patients were in the 21 to 40-year age range, and that there were more infected females (46) than males (28).

Vietnam

Vietnam is the third country in Asia in which large numbers of gnathostomiasis patients have been reported from among both local residents and overseas travelers. Four species of *Gnathostoma* have been reported from Vietnam: *G. spinigerum, G. hispidum, G. doloresi*, and *G. vietnamicum*. Since the first case in 1963, there does not appear to be an attempt to track the total number of cases in Vietnam, so the true incidence is difficult to ascertain. However, between 1999 and 2003 about 600 cases were diagnosed using a combination of clinical symptoms and serodiagnosis (ELISA). An average of 125 to 150 patients per year were distributed among several cities in South Vietnam (Xuan *et al.*, 2004). For example, from June to September 1999, 15 cases were diagnosed, with 11 showing migratory cutaneous swelling, two with pleural effusion and eosinophilia, and one with a tumor in the stomach wall (Vinh, 2001). Four cases of cerebromyelitis caused by larvae of *G. spinigerum* were reported from Choray Hospital (Hoan *et al.*, 2001). A case of intraocular gnathostomiasis was diagnosed with *G. spinigerum* larva embedded in the vitreous cavity of the right eye and uveitis (Xuan *et al.*, 2002).

Other Localities

Australia

Nineteen male and 30 female *G. spinigerum* were found in a tumor of a cat out of 40 examined in Townsville (Heydon, 1929), and *G. spinigerum* was reported in five cats examined in North Queensland (Olds, 1952). From 1995 to 2005, 35 of 92 human sera sent from Australia to the Department of Helminthology, Faculty of Tropical Medicine, Mahidol University, for testing by the immunoblot technique were found positive for gnathostomiasis. However, all these positive cases occurred among people who had toured gnathostomiasis-endemic areas outside Australia.

Palestinian Territory

Two male and two female *G. spinigerum* were found in the stomach of a lynx caught in the Emerk Valley (Witenberg *et al.*, 1960).

Rhodesia

Gnathostoma spinigerum was found in a stomach tumor of a lion in North Rhodesia (LeRouk, 1958).

Spain

Gnathostoma hispidum is considered relatively common in pigs and wild boar in Europe (Anderson, 1992). Two human cases were reported in Granada in 1998. The first case occurred in a 76-year-old woman with no history of international travel, whose clinical signs presented as acute epigastric pain and vomiting. A tumor was found near the ileocecal valve. The second case involved a 43-year-old woman who also had no history of international travel. She was hospitalized for severe pain in the right iliac fossa, and diagnosed with acute appendicitis. During intervention, a tumor was located in the cecal wall. Microscopical examination of tissue sections from these two cases suggested that both were due to *G. hispidum* (Montero *et al.*, 2001).

Clinical Aspects

Gnathostomiasis presents a wide spectrum of clinical manifestations that are directly related, among other factors, to the number of larvae ingested, the *Gnathostoma* species, and the organs and tissues affected by parasite migration.

Daengsvang (1980) described the clinical signs and symptoms of *G. spinigerum* infection in detail. After entering the human body, AL3 migrate throughout the body; therefore, the signs and symptoms vary according to the organs affected. However, they can be summarized as consisting of two formsexternal and internal. In the external form, worms migrate in the cutaneous and subcutaneous tissues, with symptoms characterized by intermittent migratory swellings of different sizes, sometimes appearing as a larval migratory track, most commonly on the torso rather than the limbs or face. Migration has no fixed direction, with speeds varying from 1.8 mm to 3.0 cm per hour. Each attack lasts for 10 days, with about six attacks per year (Daengsvang, 1980). The duration of infection varies; the longest case on the record was 17 years (Komiya, 1965).

The internal (visceral) form presents various signs and symptoms according to the organ affected. Vision impairment caused by larva invading the eye has been reported. Patients with worms recovered from the eye had the following symptoms: swelling of the eyelid, pain and itching of the eyelid, photophobia, vision impairment, headache, nausea, and vomiting. Infection of the brain and spinal cord resulted in limb weakness and/or paralysis. Symptoms of brain infection included pain; fever; meningitis; nausea and vomiting; impairment of sensation of pain, touch, and temperature; weakness or numbness of limbs; paralysis; unconsciousness; and death (Daengsvang, 1980). Punyagupta (1969) recognized two groups of clinical manifestations due to *G. spinigerum* infection: abdominopulmonary hypereosinophilic syndrome and eosinophilic myeloencephalitis. The first syndrome consisted of fever, abdominal symptoms, including liver tenderness, chest pain, hydropneumothorax, malaise, and hypereosinophilia. The second syndrome included severe pain in the trunk or limbs for a few days, followed by paralysis of the extremities.

In Mexico and Ecuador, cutaneous manifestations are more frequent and are similar to those caused by *G. spinigerum*. Intermittent migratory swellings with indurated erythematous plaques, itching, generally with little pain, are the most common manifestations; initial edema disappears in a 4 to 6-day period and almost always reappears near the last-affected area. When inflammation disappears in the same area, it is possible to observe a hemorrhagic zone or a pigmented plaque that tends to vanish in 2 to 5 weeks. In these cases, the distribution of skin lesions is more frequent in the trunk and extremities. In addition, an important number of cases present swelling without erythema, which appears mainly in the face and higher extremities. Creeping lesions are less frequent and are erythematose and indurated, wide and short in length; these characteristics are different from those produced by other *Gnathostoma* species, such as *G. hispidum, G. nipponicum,* or *G. doloresi*, which produce longer, more sinuous, lesions that tend to disappear spontaneously after 2 or 3 months, even without medical treatment. They are observed mainly in the stomach area and the back of the body. Gnathostomiasis creeping lesions may be mistakenly attributed to other nematodes that cause cutaneous larva migrans, such as *Ancylostoma caninum*, human hookworm, and *Strongyloides stercoralis,* among others (Ollage *et al.*, 1984; Diaz Camacho *et al.*, 1998; Nawa, 1991).

Another type of skin lesion that appears in a few cases is a nodular form showing an indurated red papula, which sometimes contains larvae that may emerge from it spontaneously. This cutaneous lesion appears during the natural evolutionary pattern of disease or it can be induced by antiparasitic drug administration, such as albendazole or ivermectin, which in turn induce larval migration toward more superficial layers of the skin. Most patients present one or more of the symptoms just described, which may occur in different stages of the development of the disease (Fig. 7.6).

It is important to note that when one or two larvae are ingested, sometimes in the initial stage of infection, they produce nausea, vomiting, and epigastric pain. These symptoms can be attributed to gastrointestinal infections. However, when more parasites are present, symptoms differ significantly. In Sinaloa, Mexico, five individuals who ate cebiche made from highly infected spotted sleeper fish, developed within a few minutes both the described symptoms, and exhibited acute throat pain, chest and joint pain, headache, fever, and general discomfort. One of the patients, a 55-year-old man, was hospitalized with suspected pancreatitis and pneumonia. All five developed cutaneous gnathostomiasis 8 to 9 days later (Diaz Camacho *et al.*, 2003). In addition, in Mexico, AL3 larvae have been identified in mucosae and eyes (Baquera-Heredia *et al.*, 2002), but there are no records of neurological gnathostomiasis.

Diagnosis

Although detecting *Gnathostoma* larvae in skin lesions is the only way to reach a definitive diagnosis of disease, this is often difficult because of the migratory nature of the parasite. For that reason, diagnosis should be based on four criteria (Daengsvang, 1980):

1. Clinical signs and symptoms: The specific symptoms and signs of gnathostomiasis are intermittent migratory circumscribed swellings of various sizes in various places all over the body, followed by itching and pain. Internal infection shows various symptoms according to the organs affected (see above).

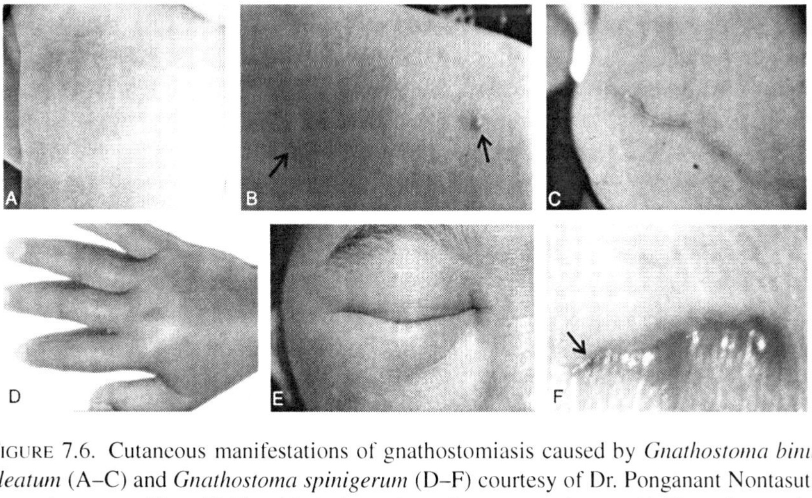

FIGURE 7.6. Cutaneous manifestations of gnathostomiasis caused by *Gnathostoma binucleatum* (A–C) and *Gnathostoma spinigerum* (D–F) courtesy of Dr. Ponganant Nontasut). Intermittent swelling (D,E) with indurated erythematous plaques (A,B) and creeping lesions (C,F). Larva (left arrows in B,F) and nodule (right arrow in B). D, E and F courtesy of Dr. Ponganant Nontasut. See also color insert.

2. History of consuming raw meat dishes made from fish, amphibians, reptiles, birds, or mammals.
3. Presence of an eosinophilia of 10% to 96%, with or without leukocytosis.
4. A positive skin test using a crude somatic extract of AL3 *G. spinigerum* (50 μL of 50 μg/mL antigen) injected intracutaneously for 15 minutes (Kraivichian *et al.*, 2004). Appearance of a wheal surrounded by erythema 9 mm in diameter or larger is considered positive. The sensitivity of the test is high, but specificity is not, since it can cross-react with paragonimiasis, schistosomiasis, and other helminthic infections (Daengsvang, 1980). A positive seological test for antibody, by indirect ELISA, using either crude or purified gnathostome larval antigen has a sensitivity of more than 90%, but again specificity is also variable. At present, the immunoblot technique is accepted as the test of choice; a 24-kD band is considered specific to gnathostomiasis (Tapchaisri *et al.*, 1991).

Treatment and Prevention

In former times, treatment of gnathostomiasis was not successful, and there was no evidence of the efficacy of antiparasitic drugs such as thiabendazole and praziquantel, which were found ineffective (Waikagul *et al.*, 1994). Surgery (removing migrating AL3 from cutaneous sites) was considered the only cure for gnathostomiasis. More recently, trials using other drugs effective against other helminth parasites, such as mebendazole, showed high worm-reduction rates (82.8–96.4%) (Waikagul *et al.*, 1997).

Present treatment of gnathostomiasis requires multiple dosages of antinematode drugs, particularly the benzimidazole derivatives, such as albendazole, and ivermectin. Albendazole given at a dosage of 90 mg/kg twice daily for 21 consecutive days induced a complete larvicidal effect on gnathostome larvae in mice (Maleewong *et al.*, 1992). At a dose of 400 mg twice daily for 10 to 14 consecutive days, no migratory swellings occurred in patients during 24 months' follow-up, and eosinophil counts returned to normal levels (Chitchang, 1987). A dose of 400 mg once or twice daily for 21 consecutive days showed 93.9% to 94.1% cure rates, and reduced immunoglobulin G (IgG) and eosinophil levels (Kraivichian *et al.*, 1992; Nontasut *et al.*, 2000). A single dose of ivermectin was found less effective than albendazole (400 mg/kg for 21 days) for treatment of cutaneous gnathostomiais, but no significant difference (Kraivichian *et al.*, 2004). Stimulation of outward migration of gnathostome larvae to the human dermis was recognized (Suntharasamai *et al.*, 1992). Albendazole inhibits larval glucose intake, resulting in restlessness of the worms, followed by weakness and death.

In patients, oral single-dose ivermectin at 0.2 mg/kg gave 95.2% cure (Nontasut *et al.*, 2000). A combination of albendazole 400 mg, three times daily for 3 weeks and a single dose of ivermectin 0.2 mg/kg were highly satisfactory, with few relapses (Chai *et al.*, 2003). Research is needed for the better treatment of gnathostomiais.

Prevention

Though there are two effective drugs for treatment of gnathostomiasis at present, the cure rate is not 100% and symptoms and signs can persist and infection can be fatal if the vital organs become infected. Because the source of infection is now clearly understood, prevention of infection is better and more effective than treatment. However, changing the eating habits of people is not an easy task. Many studies have been carried out on the effects of temperature, chemicals, and radiation on gnathostome larval viability. The larvae were killed within a few minutes in boiling water, but survive 9 to 12 days in the freezer ($-9°$ to $-4°C$) (Setasuban *et al.*, 1981; Rojekittikhun and Buchachart, 2002). The larvae can survive for 1 month at 4°C (Daengsvang, 1980). Encysted larvae survive 5 to 7 days in lime juice, 5.5 hours in 4% citric acid, 18 to 20 hours in saline solution (23–30% salinity) and 8 to 9 days in 28% to 35% ethyl alcohol (Setasuban *et al.*, 1981). AL3 are highly resistant to radiation, even with exposure to 0.3 to 10 kGy of gamma rays motility persists; however, larval infectivity to mice was gradually reduced to 0% after being irradiated at 8 to 10 kGy. A very high level of irradiation (8 kGy) is considered necessary to inhibit the infectivity of AL3 (Setasuban *et al.*, 1992).

Impact

Gnathostomiasis has an impact on the quality of a patient's life. Migration of the worm causes irritation, itching, and pain. The patient is uncomfortable and cannot sleep well. The parasite's migration may also cause mental problems. For example, after several episodes of migratory swelling, one female patient, in frustration, placed a hot iron over her swelling arm; there was no recurrent of swelling but the burned mark remained (Waikagul, unpublished data). Although disability-adjusted life years (DALYs) have not been estimated for gnathostomiasis, it seems likely that for many victims of infection there is a loss of activity and work ability and interest.

The spread of gnathostomiasis is promoted by the export and importation of culinary traits, international travel, and export of fish. To prevent transmitting gnathostomes to nonendemic regions, scientific knowledge and experience related to the prevention and control of this disease must be widely distributed. For both endemic and nonendemic countries, there is a need to establish a national surveillance system to ensure prompt detection, diagnosis, and treatment of gnathostomiasis.

There remain many areas for scientific advancement in the study of gnathostomiasis around the world. The start to ensuring more adequate world attention is acknowledging that gnathostomiasis is a parasitic zoonosis that is a serious public health issue, not only for some countries in Asia and America, but for the whole world.

Research Needs

To better understand this disease, it is imperative to reevaluate the way it is expanding its range, to develop standardized diagnostic criteria, and to promote research into alternate diagnostic/treatment techniques. It is also imperative to enhance molecular studies on the pathogenicity and pathology of gnathostomiasis. There is a need to establish a consensus on the methodologies for taxonomic determinations, and agreements on the systematics of this genus. There is also a need to create a parasitological communications network among researchers and hygiene and aquaculture specialists to promote collaboration and information sharing.

Other knowledge gaps needing attention are as follows:

Parasite Biology

Gnathostoma spinigerum advanced third-stage larvae present several fixed types of hooklets morphological abnormality, and COI sequences show several variations, although not related to morphological differences (Ngarmamonpirat *et al.*, 2005). However, this information raises the possibility that *G. spinigerum* is a species complex and entails the presence of cryptic species.

The migratory habit of the gnathostome nematode in host tissues is perplexing; therefore, it is of importance to understand the nature of the physiological stimulus responsible for larval migration. Studies in animals show that after being ingested by the host, gnathostome larvae migrated to the liver, then to the muscles, and finally to the stomach, where they mature. Molecular pathology studies may reveal some of the clues and mediators of this migrating habit, which will be very useful for improving the infection treatment.

Diagnosis

Gnathostomiasis (external form) has a very specific characteristic of cutaneous migratory swellings, which is easily recognizable and enables determining the specific diagnosis. Internal gnathostomiasis, however, is more severe than the external type of infection, but it is easily diagnosed. Research on highly specific antigens for immunodiagnosis is needed. The method to evaluate the outcomes of the treatment is another area of research that is needed, as the subsiding of symptoms cannot be interpreted as a cure because of the frequency of relapses.

Summary

This chapter reviewed the biology, epidemiology, public health impact, and control of gnathostomiasis. This zoonotic infection is caused by nematodes in the genus *Gnathostoma*; species reported from humans are *G. spinigerum*, *G. doloresi*, *G. hispidum*, *G. nipponicum*, *G. malaysiae*, and *G. binucleatum*. In

Asia, *G. spinigerum* is widely distributed in Japan, Thailand, and Vietnam, and to a lesser extent in Bangladesh, Sri Lanka, Indonesia, Laos, Myanmar, and China. Recently in Japan the number of human cases decreased significantly, but the number of cases remained high in Thailand and even increased in Vietnam. In the Americas, the main causative agent of gnathosomiasis is *G. binucleatum*, which is distributed in Mexico and Ecuador. In Mexico an increasing number of human cases have been reported since 1989, most of whom had a custom of eating raw fish meat in dishes such as cebiche. Because the infective stage for humans, the advanced third-stage gnathostome larvae, can utilize more than 40 species of fish, reptiles, amphibians, birds, and small mammals as intermediate and paratenic hosts, the opportunity for transmisson to humans in food is considerable, if eaten raw or improperly cooked. The larvae, after ingestion, wander around the human body, causing the larval migratory swelling syndrome. If the affected organ is vital, death or serious impairment may occur. Direct penetration while handling contaminated meat, or transplacental infection of the uterine embryo of an infected pregnant mother is also possible. Diagnosis is dependent on the morphological analysis of larvae recovered from patients during surgery, on clinical manifestations, and on immunological tests. Treatment with albendazole or ivermectin is considered effective. Prevention of the infection is effectively possible through consumption of only well-cooked food, but it is difficult in practice. Food preparation is a cultural art resistant to change and favored by visitors to the area. Recent reports on gnathostomiasis were increased, particularly in nonendemic countries as a result of more extensive international travel and migration to the endemic areas in Southeast Asia and Central and South America. Physician awareness in nonendemic areas of the clinical features of the infection would increase the rate of early and prompt diagnosis. Research on finding better therapy, performing faster laboratory diagnostic tests, and evaluating drug efficacy methods is needed.

Acknowledgments We are grateful to Dr. Le Thi Xuan for kindly providing English abstracts for reports of gnathostomiasis in Vietnam, and to Tippayarat Yoonuan, Phaibul Vasanakomut, Jesus Espinosa, and Maria del Carmen de la Cruz, for their technical support.

References

Adams, A.R.D. (1933). Report on a collection of nematodes from the Federation Malay States. Ann Trop Med Parasitol 27, 1.

Africa, C.M., Refuerzo, P.G., and Garzia, E.Y. (1936). Observation on the life cycle of *Gnathostoma spinigerum*. Phillip J Sci 59, 513–521.

Ahmad, R. (1962). Some aspects of parasites of east Pakistan. Agric Pakist 13, 109.

Akahane, H., Lamothe-Argumedo, R., Martinez-Cruz, J.M., Osorio-Sarabia, D., and Garcia-Priento, L. (1994). A morphological observation of the advanced third-stage larvae of Mexican *Gnathostoma*. Jpn J Prasitol 43(1), 18–22.

Akahane, H., Sano, T., and Mako, T. (1986). Morphological differences of the advanced third-stage larvae of *G. spinigerum*, *G. hispidum* and *G. doloresi*. Jpn J Parasitol 35, 465–467.

Akahane, H., Setasuban, P., Nuamtanong, S., Horiuchi, S., Koga, M., and Kojima, S. (1995). A new type of advanced third-stage larvae of the genus *Gnathostoma* in freshwater eels, *Fluta alba*, from Nakhon Nayok, Central Thailand. Southeast Asian J Trop Med Public Health 26(4), 743–747.

Almeyda-Artigas, R.J. (1991). Hallazgo de *Gnathostoma binucleatum* n. sp. (Nematoda: Spirurida) en felinos silvestres y el papel de peces dulceacuícolas y oligohalinos como vectores de la gnatostomiasis humana en la cuenca baja del Río Papaloapan, Oaxaca, Veracruz, México. An Inst Cienc Del Mar Limnol UNAM 18, 137–155.

Almeyda-Artigas, R.J., Bargues, M.D., and Mas–Coma, S. (2000). ITS-2 rDNA sequencing of *Gnathostoma* species (Nematoda) and elucidation of the species causing gnathostomiasis in the Americas. J Parasitol 86, 537–544.

Anantapruti, M., Setasuban, P., Daengsvang, S., and Vajrasthira S. (1982). Electron microscopy of the advanced third-stage larvae of *Gnathostoma spinigerum*. Southeast Asian J Trop Med Public Health 13(4), 531–540.

Anderson, R.C. (1964). *Gnathostoma miyazaki* n. sp. from the Otter *(Lutra c. canadiensis)* with comments on *G. sociale* (Leidy 1858) of mink *(Mustela vison)*. Can J Zool 42, 249–254.

Anderson, R.C. (1992). Nematode parasites of Vertebrates: Their Development and Transmission. Wallingford: C.A.B. International, 339–342.

Ando, K. (2005). Gnathostomiasis in Japan. In: Asian Parasitology, vol 1: Food-Borne Helminthiasis in Asia. Arizono, N., Chai, J.Y., Nawa, Y., and Takahashi, Y., eds. Japan: FAP Journal, 231–239.

Andrew, M.N. (1937). The helminth parasites of dogs and cats in Shanghai, China. J Helminthol 15, 145–152.

Baquera-Heredia, J., Cruz-Reyes, A., Flores-Gaxiola, A., Lopez-Pulido, G., Diaz-Simental, E., and Valderrama-Valenzuela L. (2002). Case report: ocular gnathostomiasis in northwestern Mexico. Am J Trop Med Hyg 66(5), 572–574.

Bashirullah, A.K.M. (1972). Occurrence of *Gnathostoma spinigerum* Owen, 1836, in Dacca, Bangladesh. J Parasitol 58(1), 187–188.

Bertoni-Ruiz, F. (2001). Gnathostomiasis en vertebrados silvestres de la Laguna de Tres Palos, Acapulco, Guerrero. Tesis de Licenciatura.Facultad de Ciencias, UNAM, México, DF.

Bertoni-Ruiz, F., García-Prieto, L., Osorio-Sarabia, D., and Leon-Regagnon V. (2005). A new species of *Gnathostoma* (Nematoda: Gnathostomatidae) in *Procyon lotor hernandezii* from Mexico. J Parasitol 91(5), 1143–1149.

Chai, J.Y., Han, E.T., Shin, E.H., Park, J.H., Chu, J.P., Hirota, M., Nakamura-Uchiyama, F., and Nawa, Y. (2003). An outbreak of gnathostomiasis among Korean emigrants in Myanmar. Am J Trop Med Hyg 69(1), 67–73.

Chandler, A.C., and Reed, C.P. (1961). Introduction to Parasitology with Special Reference to the Parasites of Man, 10th ed. New York, London: John Wiley and Sons, 502.

Chandler, A.C. (1925). The heminthic parasites of cats in Calcutta and the relation of cats to human helminthic infections. Indian J Med Res 13, 213.

Chandler, A.C. (1932). Notes on the helminth parasites of the opossum (*Didelphis virginiana*) in Southeast,Texas with descriptions of four new species. Proc. U.S. Natl Museum Art 16, 1–15.

Chandler, A.C. (1942). The helminthes of raccoons in East Texas. J Parasitol 28, 255–268.
Chappuis, F., Farinelli T., and Loutan L. (2001). Ivermectin treatment of a traveler who returned from Peru with cutaneous gnathostomiasis. Clin Infect Dis 33, 17–19.
Chitchang, S. (1987). The efficacy of albendazole in the treatment of human gnathostomiasis. J Postgrad Med Thai 5(1), 39–43 (in Thai).
Daengsvang, S. (1980). A Monograph on the Genus *Gnathostoma* and Gnathostomiasis in Thailand. Tokyo: SEAMIC/IMFJ, 1–87.
Daengsvang, S. (1981). Gnathostomiasis in Southeast Asia. Southeast Asia J Trop Med Public Health 12(3), 319–332.
Daengsvang, S. (1983). The world's first-story of *Gnathostoma spinigerum* infection in the young Thai woman. J Parasitol Trop Med Assoc Thailand 6(1), 1–10.
Diaz Camacho, S.P., Willms, K., de la Cruz Otero, M. del C., Zazueta, M.L., Bayliss, S., Castro Velazquez, R., Osuna, I., Bojorquez, A., Torres, E.H., and Sanchez Gonzales, S. (2003). Acute outbreak of gnathostomiasis in a fishing community in Sinaloa, Mexico. Parasitol Int 52(2), 133–140.
Diaz Camacho, S.P., Willms, K., Zazueta, M.L., de la Cruz Otero, M., del C., Nawa, Y., and Akahane, H. (2002). Morphology of *Gnathostoma* spp isolated from natural hosts in Sinaloa, Mexico. Parasitol Res 88, 639–645.
Díaz-Camacho, S.P., Zazueta, M., Ponce, E., Osuna, I., Castro, R., Baquera, J., Flores, A., Willms, K., Nawa, Y., and Akahane, H. (1998). Clinical manifestations and immunodiagnosis of Gnathostomosis en Sinaloa, México. Am J Trop Med Hyg 59(6), 908–915.
Feinstein, R.J., and Rodriguez-Valdez, J. (1984). Gnathostomiasis or *larva migrans profundus*. J Am Acad Dermatol 1, 738–740.
Grobusch, M.P., Bergmann, F., Teichmann, D., and Klein E. (2000). Cutaneous gnathostomiasis in a woman from Bangladesh. Int J Infect Dis 4(1), 51–54.
Gyi, K. (1960). Intra-ocular gnathostomiasis. Br J Ophthalmol 44(1), 42–45.
Hadidjaja, P., Margono, S.S., and Moeloek, F.A. (1979). *Gnathostoma spinigerum* from the cervix of a woman in Jakarta. Am J Trop Med Hyg 28(1), 161–162.
Han, E.T., Park, J.H., Shin, E.R., and Chai, J.Y. (2002). Surface ultrastructure of the advanced third stage larvae of *Gnathostoma nipponicum* found in the republic of Korea. Proceedings of First International Meeting of Gnathosomosis, Mexico, 78.
Hennies, F., Jappe, U., Kapaun, A., and Enk, A. (2006). Gnathostomiasis: import from Laos. J Dtsch Dermatol Ges 4(5), 414–416.
Heydon, G.M. (1929). Creeping eruption or larva migrans in North Queensland and a note on the worm *Gnathostoma spinigerum* (Owen). Med J Aust 1, 583.
Hoan, N.H., Pham, Y., Luyen, T.V., Minh, L., Mai, T.X., and Xuan, L.T. (2001). Cerebromyelitis caused by *Gnathostoma* sp., a review of 4 cases detected at Choray Hospital. H.C.M.C. Med Magaz 1(5), 100–102.
Kamala, D., Rohela, M., and Anuar, A.K. (1997). Blurring of vision due to VLM. An abstract of Joint International Tropical Medicine Meeting, 25–27 August 1997, Siam City Hotel Bangkok, 121.
Kaminsky, C.A., De Kaminsky, A.R., Constantini, S.E., and Abulafia, J. (1989). Eosinophilic migratory nodular panniculitis (human gnathostomiasis). Med Cutan Ibero Lat Am 17(3), 158–162.
Khin, T. (1968). Intra-ocular gnathostomiasis. Br J Ophthalmol 52(1), 57–60.
Kim, Y.K. (1973). A study on *Gnathostoma*. 1. An investigation into the geographical distribution of larvae on the second-third stage in Gyengsang Nam do. Bull Fusan Natl Univ 15, 111–116.

Kiriratana, C. (1989). Acute appendicitis and intraabdominal abscess caused by *Gnathostoma* infestation: report of 2 cases. Bull Dept Med Serv 14, 163–167 (in Thai with English abstract).
Koga, M., and Ishii, Y. (1987). Surface morphology of the advanced third-stage larvae of *Gnathostoma doloresi*—an electron microscopic study Jpn J Parasitol 36, 231–235.
Koga, M., Akahane, H., and Ishii, Y. (1991). Surface ultraestructure of adults and eggs of *Gnathostoma spinigerum* (Nematoda,Gnathostomatidae). Trans Am Microscopical Soc 110, 315–320.
Koga, M., Ishibashi, J., Ishi, Y., and Nishimura, T. (1988). Scanning electron microscopic comparisons among the early and advanced third-stage larvae of *Gnathostoma hispidum* and the *Gnathostoma* larvae obtained from loaches. Jpn J Parasitol 37, 220–226.
Komiya, Y. (1965). "Changchiang edema": the review of research. Igaku no Ayumi 52, 78–79 (in Japanese).
Kraivichian, P., Kulkumthorn, M., Yingyourd, P., Akarabovorn, P., and Paireepai, C. (1992). Albendazole for the treatment of human gnathostomiasis. Trans R Soc Trop Med Hyg 86(4), 418–421.
Kraivichian, P., Nuchprayoon, S., Sitichalernchai, P., Chaicumpa, W., and Yentakam, S. (2004). Treatment of cutaneous gnathostomiasis with ivermectin. Am J Trop Med Hyg 71(5), 623–628.
Le Rouk, P.L. (1958). *Gnathostoma spinigerum* in lion in Northern Rhodesia. Trans R Soc Trop Med Hyg 52, 14.
Leon-Regagnon, V., Osorio-Sarabia, D., Garcia-Prieto, L., Lamothe-Argumedo, R., Bertoni-Ruiz, F., and Oceguera-Figueroa, A. (2005). New host records of the nematode *Gnathostoma* sp. in Mexico. Parasitol Int 54(1), 51–53.
Lertanekwattana, S., Waikagul, J., Pilomvarakorn, S., and Saepeung, K. (2004). Gnathostomiasis: two cases admitted to Nong Khai Hospital Thailand. Southeast Asia J Trop Med Public Health 35(suppl 1), 100–103.
Levinsen, G.M.R., (1889). Om en ny Rundorm hos Mennesket, *Cheiracanthus siamensis* n. sp. Videnskabel. Meddel.fra den naturhist. Foren. I Kjöbenhaven 323–326. (Quoted from Centr Bakteriol Parasitenk 8, 182, 1890.)
Lie, K.J. (1949). A case of gnathostomiasis in Indonesia. Doc Ned Indones Morb Trop 1, 368–370.
Magaña, M., Messina, M., Bustamante, F., and Cazarin J. (2004). Gnathostomiasis: clinicopathologic study. Am J Dermatopathol 26(2), 91–95.
Maleewong, W., Loahabhan, P., Wongkham, C., Intapan, P., Morakote, N., and Khamboonroung, C. (1992). Effects of albendazole on *Gnathostoma spinigerum* in mice. J Parasitol 78(1), 125–126.
Margono, S.S., Indris, K.N., and Brodjonegoro, N.S. (1978). Another case of human gnathostomiasis in Indonesia. Southeast Asia J Trop Med Public Health 9(3), 406–408.
Martínez-Cruz, J.M., Bravo-Zamudio, R., Aranda-Patraca, A., and Martínez-Marañón, R. (1989). La gnathostomiasis en México. Salud Púb Mex 31, 541–549.
Mitter, S.N. (1910). *Gnathostoma spinigerum* in a domestic cat. J Trop Vet Sc 5, 284–285.
Miyazaki, I., and Dunn, F.L. (1965). *Gnathostoma malaysiae* sp. n. from rats on Tioman Island, Malaysia (Nematoda: Gnathostomidae). J Parasitol 51 (3), 382–384.
Miyazaki, I. (1960). On the genus *Gnathostoma* and human gnathostomiasis, with special reference to Japan. Exp Parasitol 9, 338–370.
Miyazaki, I. (1991). An Illustrated Book of Helminthic Zoonoses. Tokyo: International Medical Foundation of Japan.

Montero, E., Montero, J., Jose Rosales, M., and Mascaro, C. (2001). Human gnathostomosis in Spain: first report in humans. Acta Trop 78(1), 59–62.

Moore, D.A.J., McCroddan, J., Dekumyoy, P., and Chiodini, P. (2003). Gnathostomiasis: an emerging disease. Emerg Infect Dis 9 (6), 647–650.

Morishita, K., and Faust, E.C. (1925). Two new cases of human creeping disease (gnathostomiasis) in China with a note on the infection in reservoir hosts in the China area. J Parasitol 11, 158.

Nawa, Y. (1991). Historical review and current status of gnathostomiasis in Asia. Southeast Asian J Trop Med Public Health 22(suppl), 217–219.

Ngarmamonpirat, C., Waikagul, J., Petmitr, S., Dekumyoy, P., Rojekittikhun, W., and Anantapruti, M.T. (2005). Analysis of sequence variation in *Gnathostoma spinigerum* mitochondrial DNA by single-strand conformation polymorphism analysis and DNA sequence. Parasitol Int 54 (1), 65–68.

Nomura, Y., Nagakura, K., Kagei, N., Tsutsumi, Y., Araki, K., and Sugawara, M. (2000). Gnathostomiasis possibly caused by *Gnathostoma malaysiae*. Tokai J Exp Clin Med 25 (1), 1–6.

Nontasut, P., Bussratid, V., Chullawichit, S., Charoensuk, N., and Visetsuk, K. (2000). Comparison of ivermectin and albendazole treatment for gnathostomiasis. Southeast Asian J Trop Med Public Health 31(2), 374–377.

Ogata, K., Nawa, Y., Akahane, H., Díaz-Camacho, S.P., Lamothe-Argumedo, R., and Cruz-Reyes, A. (1998). Short report: Gnathostomiasis in Mexico. Am J Trop Med Hyg 58(3), 316–318.

Olds, R.J. (1952). *Gnathostoma spinigrum* (Owen, 1836) in cats in North Queensland. Australia Vet J 29, 124.

Ollage, W. (1985). Gnathostomiasis. VII. Guayaquil, Ecuador: Monografía del Colegio Iberoamericano de Dermatología.

Ollage, W., Ollage, J., Guevara de Veliz, A., and Peñaherreras, S. (1984). Human gnathostomiasis in Ecuador (nodular migratory eosinophilic panniculitis: first finding of the parasite in South America. Int J Dermatol 23, 647–651.

Ollage, W., Ollage, J., Guevara de Veliz, A., Peñaherreras, S., Von Buchwald, C., Mancheno, J., and Tanca, J. (1981). Paniculitis nodular migratoria eosinofílica (Gnathostomiasis profunda y superficial) primer hallazgo del parásito en América del Sur. Nuestra Med 7, 9–23.

Owen, R. (1836). Anatomical descriptions of two species of *Entozoa* from the stomach of a tiger (*Felis tigris* Linn.), one of which forms a new genus of Nematoidea, Gnathostoma. Proc Zool Soc Lond 22, 123–126

Pelaez, D., and Pérez-Reyes, R. (1970). Gnathostomiasis humana en América. Rev Latinoamer Microbiol 12, 83–91.

Punyagupta, S. (1969). Two newly recognized clinical manifestations due to *Gnathostoma* infection in man. In: Seminar on Some Development in Parasitology in Thailand (in Thai), organized by the Parasitology and Tropical Medicine Association of Thailand, April 1969, Bangkok.

Rahman, M.M., and Moula, M.R. (2006). Gnathostomiasis: a rare nematode infection. Mymensingh Med J 15(1), 105–107.

Rojas-Molina, N., Pedraza-Sanchez, S., Torres-Bibiano, B., Meza-Martinez, H., and Escobar-Gutierrez A. (1999). Gnatostomosis, an emerging foodbome zoonotic disease in Acapulco, México. Emerg Infect Dis 5, 264–266.

Rojekittikhun, W., and Buchachart, K. (2002). The infectivity of frozen *Gnathostoma spinigerum* encysted larvae in mice. J Trop Med Parasitol 25(2), 79–82.

Rojekittikhun, W. (2005). Gnathostomiasis in Thailand. In: Asian Parasitology. Vol 1: Food-Borne Helminthiasis in Asia. Arizono, N., Chai, J.Y., Nawa, Y., and Takahashi, Y., eds. Japan: FAP Journal, 241–259.

Ruiz, J.M. (1952). Sobre un novo *Gnathostoma* assinalado no Brasil (Nematoda: Gnathostomatidae). Men Inst Butantan 24:37–44.

Samarasinghe, S., Perera, B.J., and Ratnasena, B.G. (2002). First two cases of gnathostomiasis in Sri Lanka. Ceylon Med J 47(3), 96–97.

Scholz, T., Uhlirova, M., and Ditrich, O. (2003). Helminth parasites of cats from the Vientiane province, Laos, as indicators of the occurrence of causative agents of human parasitoses. Parasite 10(4), 343–350.

Setasuban, P., Punsri, W., and Muennoo, C. (1981). Effects of temperature, chemicals and some native Thai food upon the viability of the infective stage of *Gnathostoma spinigerum*. J Parasit Trop Med Assoc Thailand 4(2), 77–85 (in Thai with English abstract).

Setasuban, W., Hiranyacharttada, P., Pubampen, S., Prachasitthisak, Y., Rojekittikhun, W., and Dekumyoy, P. (1992). Effects of Cobalt-60 gamma radiation on *Gnathostoma spinigerum* larvae. In: The 4th Nuclear Science and Technology Conference; 1992 Oct. 20–22; Bangkok, Thailand. Bangkok: Office of Atomic Energy for Peace, 66–79.

Shaikh, H., Hug, M.M., Rahman, M.S., Hag, S., and Mollah, M.A.U. (1968). Trial with Promintic in the treatment of gnathostomiasis in cat. Pakistan J Vet Sc 2, 200.

Sohn, W.M., and Lee, S.H. (1998). The first discovery of larval *Gnathostoma hispidum* (Nematoda: Gnathostomatidae) from a snake host, *Agkistrodon brevicaudus*. Korean J Parasitol 36 (2), 81–89.

Sohn, W.M., Kho, W.G., and Lee, S.H. (1993). Larval *Gnathostoma nipponicum* found in the imported Chinese loaches. Korean J Parasitol 31 (4), 347–352.

Suntharasamai, P., Riganti, M., Chittamas, S., and Desakorn, V. (1992). Albendazole stimulates outward migration of *Gnathostoma spinigerum* to the demis in man. Southeast Asian J Trop Med Public Health 23(4), 716–722.

Tapchaisri, P., Nopparatana, C., Chaicumpa, W., and Setasuban P. (1991). Specific antigen of *Gnathostoma spinigerum* for immunodiagnosis of human gnathostomiasis. Int J Parasitol 21, 315–319.

Travassos, L. (1925). Contibuicoes para o conhecimento da fauna helmintologica Brasilerira. Sobre as especies brasileiras do generoGnathostoma Owen, 1936. Sci Med 3, 508–517.

Tubangui, M.A. (1947). A summary of the parasitic worms reported from the Philippines. Philip J Sci 76, 225.

Villar de Cipriani, E. (2003). Paniculitis migratoria eosinofilica en el Peru. *Gnathostoma* como agente causal. Rev Peru Med Exp Salud Pública 20(4), 220–223.

Vinh, N.Q. (2001). Clinical features of gnathostomiasis in adult. H.C.M.C. Med Magaz 1(5), 100–102.

Waikagul, J., Anantapruti, M.T., and Visiassuk, K. (1994). A preliminary study on the efficacy of mebendazole, thiabendazole, praziquantel and ivermectin against gnathostomiasis in rats. J Trop Med Parasitol 17(2), 76–81 (in Thai with English abstract).

Waikagul, J., Nuamtanong, S., Anantaphruti, M.T., Pubampen, S., Visiadsuk, K., and Tawatsin, A. (1997). Sequential mebendazole treatment against *Gnathostoma spinigerum* in rat. J Trop Med Parasitol 20(1), 40–46.

Wharton, L.D. (1917). The intestinal worms of dogs in the Philippines Islands. J Parasitol 4, 80–82.

Witenberg, G., Jacoby, J., and Steckelmacher, S. (1960). A case of ocular gnathostomiasis. Ophtalmologia 119, 114.

Xuan, L.T., Hoa, P.T.L., Dekumyoy, P., Hoan, N.H., Khuong, L.H., Van, T.T.H., Mai, V.T.C., Tu, L.X., and Hien, T.V. (2004). *Gnathostoma* infection in South Vietnam. Southeast Asian J Trop Med Public Health 35(suppl 1), 97–99.

Xuan, L.T., Rojekittikhun, W., Punpoowong, B., Trang, L.N., and Hien, T.V. (2002). Case report: intraocular gnathostomiasis in Vietnam. Southeast Asian J Trop Med Public Health 33(3), 485–489.

8 Angiostrongyliasis

J.H. Cross and E.R. Chen

Animal life is a major source of food for many of the world's populations, and the preparation and the eating of the animals is highly variable. In some societies, animal meat may be cooked, smoked, pickled, fermented, or eaten raw. Unfortunately, there are often parasitic infections in the animals that are transmitted to humans, especially when animal products are eaten raw or poorly cooked. There are many of these zoonotic parasites, and one that is emerging and spreading throughout the world is the rat-lung worm, *Angiostrongylus cantonensis.* The helminth in the natural rodent hosts is not pathogenic, when only a few worms are involved; however, when humans acquire the parasite after eating the infected molluscan intermediate or paratenic host, angiostrongyliasis leading to eosinophilic meningitis may develop and cause severe illness.

History

Angiostrongylus cantonensis is one of 20 metastrongyle nematodes that reside in the vascular system of animals. Species of *Angiostrongylus* have been reported from animals since 1886 when *A. vasorum* was found in the pulmonary vessels of dogs in Europe. Other species of the parasite have been reported over the years from Europe, Africa, North and South America, Australia, and the Asian-Pacific basin and is found in rodents, canines, felines, and insectivores (Table 8.1). At the present time only two species, *A. cantonensis* and *A. costarecensis,* have been found to cause disease in humans; however, two other species. *A. mackerrasae* and *A. malaysiensis* may have the potential of being pathogenic in humans (Prociv et al., 2000).

Angiostrongylus cantonensis was first reported as *Pulmonema cantonensis* found in the lung of rats in South China in 1933 (Chen, 1935). Matsumoto (1937) found it in rats on Taiwan, and Yokogawa (1937) described it as a new species, *Haemostrongylus ratti. Pulmonema* and *Haemostrongylus* were subsequently synonymized to *Angiostrongylus* by Dougherty (1946). Mackerras and Sanders (1955) described the life cycle of *A. cantonensis* in rats, but it was later determined that these authors actually worked with *A. mackerrasae* (Bhaibulaya, 1968, 1975).

TABLE 8.1. *Angiostrongylus* species.

Parasite	Host	Geographic Areas
A. vasorum	Dogs	Europe, South America, Australia
A. raillieti	Crab eating dog	Brazil
A. tateronae	Jerboa	West Africa
A. ondatrae	Muskrat	Russia
A. cantonensis	Rats	Asia, Pacific
A. ten	Marten	Japan
A. gubernaculatus	Badger, skunk	United States
A. blarini	Shrew	United States
A. soricis	Shrew	Poland
A. chabaudi	Wildcat	Italy
A. sciuri	Squirrel	Turkey
A. michiganensis	Shrew	United States
A. sandarasae	Rodent	East Africa
A. mackerrasae	Rats	Australia
A. dujardini	Rodent	France
A. schmidti	Rice rat	United States
A. malaysiensis	Rats	Southeast Asia
A. costaricensis	Rats	Central and South America
A. minutus	Mole	Japan
A. siamensis	Rats	Thailand

A high prevalence of eosinophilic meningitis in humans was reported from the Caroline Islands in the Pacific in 1948 (Bailey, 1948) and from Tahiti in 1960 (Franco et al., 1960). The first report of human infection with *A. cantonensis*, however, was reported by Nomura and Lin in 1945 (Beaver and Rosen, 1964) when 10 actively moving immature worms were recovered from the cerebrospinal fluid (CSF) of a 15-year-old boy from Taiwan. The importance of the parasitosis was not recognized internationally, however, until Rosen and coworkers (1961, 1962) reported the findings of the nematode in the brain of a man who died in Hawaii. At about the same time, Prommindaroj et al. (1962) reported finding the parasite in the eye of a Thai male, and Alicata (1962) reported *A. cantonensis* as a pathogen in humans. In subsequent years, the parasite has been reported in rodents, definitive hosts, molluscan intermediate hosts, paratenic hosts (frogs, crabs, prawn, planaria), and humans worldwide. In 1986, Ubelaker changed the name of the helminth to *Parastrongylus cantonensis*. This change, however, has yet to be generally accepted since the disease is recognized internationally as angiostrongyliasis.

Parasite/Biology

The morphological features of *A. cantonensis* have been described by Alicata and Jindrak (1970) and Bhaibulaya (1979). The nematode has been classified as a member of the superfamily Metastrongyloidea, family Angiostrongylidae. The family members have filariform bodies that taper slightly at both ends. In *A. cantonensis*, there are three lips around the mouth, one dorsal with two

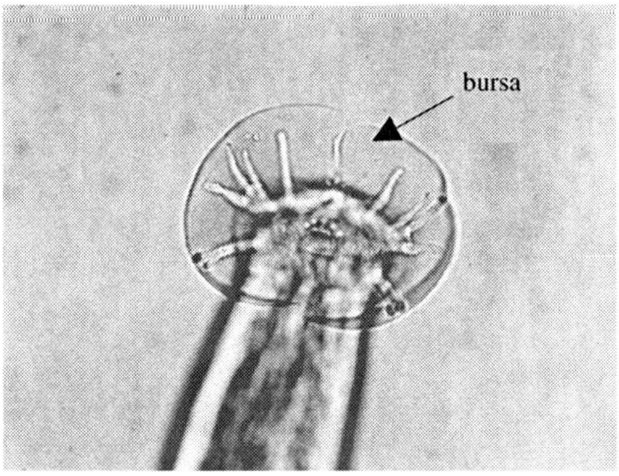

FIGURE 8.1. Posterior end of male *Angiostrongylus cantonensis* showing the bursa.

submedium papillae and two subventral lips each with two submedium papillae. The mouth opens directly into the esophagus followed by the intestine, which extends to the end of the body. The body is long and thin with a smooth transparent cuticle and transverse striae.

The males measure 20 to 25 mm in length and 0.32 to 0.42 in width. The caudal bursa of the male is small, well develop, kidney shaped, and single lobed (Fig. 8.1). The bursal rays have a ventral ray branched at a point two thirds of the length into a small ventroventral and a large lateroventral ray. Lateral rays arise from a common trunk; the anterolateral ray is thickened more than the others and projected like a thumb. The mediolateral ray and posterolateral ray usually originate from a common trunk. The posterolateral ray is normally shorter than the mediolateral ray and sometimes reduced to a stump. The external ray is simple and arises from between the lateral and dorsal rays. The dorsal is viable, emerging as a short trunk, terminating in several small digitations. Spicules are equal, slender, and with conspicuous striations (Fig. 8.2). The spicule measures 1.00 to 1.46 μm. A gubernaculum is present.

A female *A. cantonensis* has uterine tubules that wind spirally around a blood-filled intestine, which can be seen through the transparent cuticle as a barber's pole pattern. This is more apparent in the living parasite. A single thin-walled

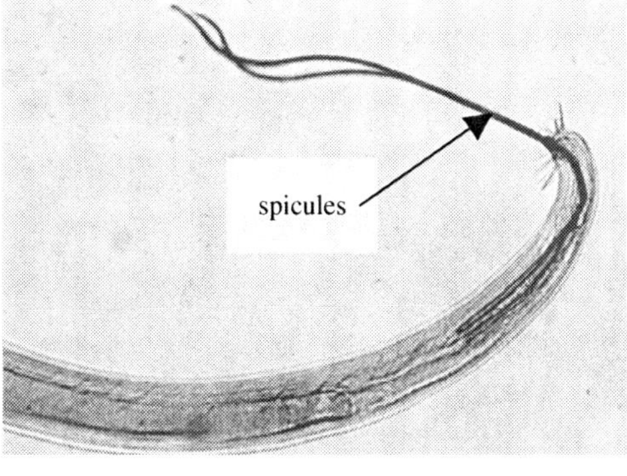

FIGURE 8.2. Posterior end of male *Angiostrongylus cantonensis* showing the bursa and extended copulatory spicules.

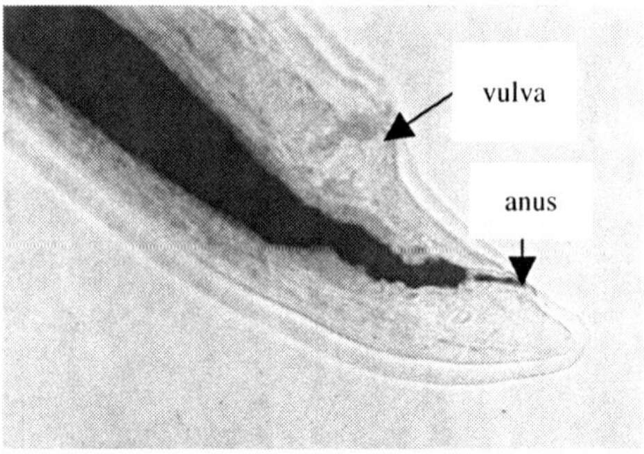

FIGURE 8.3. Posterior end of female *Angiostrongylus cantonensis* showing the vulva and anus.

vagina commences at the junction of the uterine tubules, extending posteriorly and opening at the vulva. The vulva and anus are close together near the posterior end of the worm. The anus is located about 0.05 mm and the vulva 0.2 mm from the tail end (Fig. 8.3). The females measure 22 to 34 mm in length and a maximum width of 0.34 to 0.56 mm (Fig. 8.4).

Adult male and female *A. cantonensis* live in the branches of the pulmonary artery of rats (Fig. 8.5). Following copulation, female worms deposite eggs that are carried by the blood to capillaries of the lung where they embryonate in 5 to 6 days (Fig. 8.6). The first stage larva hatches from the egg and breaks through the capillaries into the lung alveolus. The larvae subsequently migrate into the bronchioles up to the trachea, are swallowed, pass through the digestive tract, and out in the feces (Fig. 8.7). The first-stage larvae are eaten or penetrate the body of a molluscan intermediate hosts and migrate to the muscle tissue (Fig. 8.8). The larvae molt into second-stage larvae in 7 to 9 days and into the infective third-stage larvae (Fig. 8.9) in 12 to 16 days. Many species of the mollusks may serve as intermediate hosts, and when eaten by a rodent, the larvae are released from the molluscan tissue by digestion in the gastrointestinal tract. The released larvae penetrate the intestinal wall and are picked up in the blood and carried to the liver, heart, and lungs, and may reach the central nervous system in 1 to 2 days. The worms enter the neural parenchyma and molt into the fourth stage in 4 to 6 days.

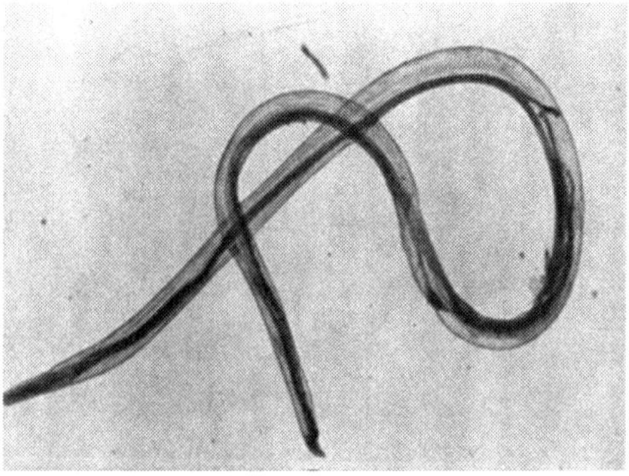

FIGURE 8.4. Adult female *Angiostrongylus cantonensis* recovered from the pulmonary artery of a rat.

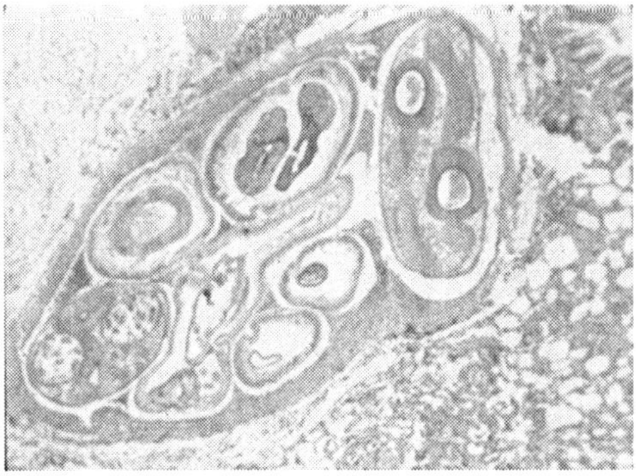

FIGURE 8.5. Section of adult *Angio-strongylus cantonensis* in the pulmonary artery of a rat.

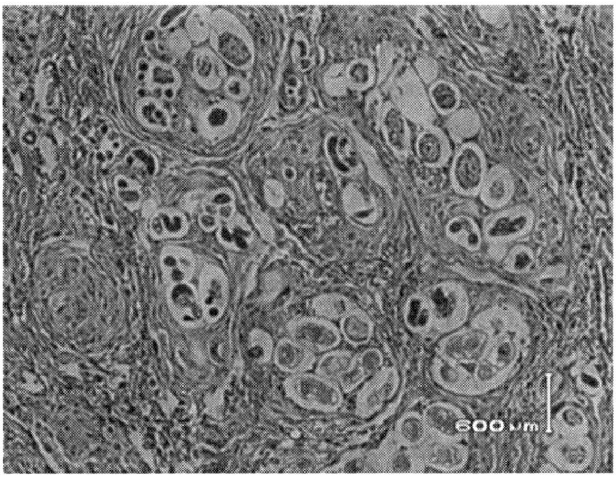

FIGURE 8.6. Section of a rat lung showing *Angiostrongylus cantonensis* eggs and larvae in the lung parenchyma.

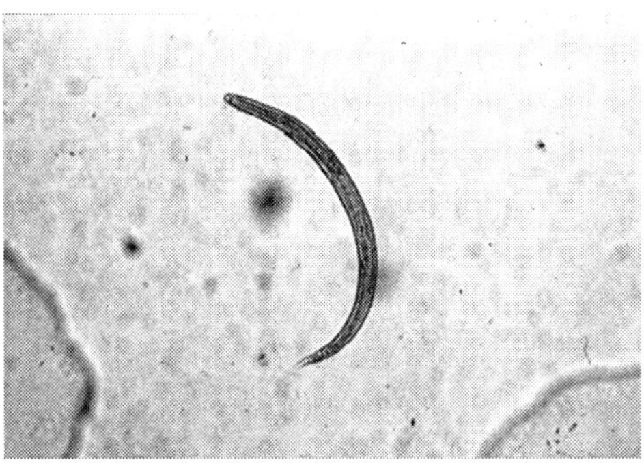

FIGURE 8.7. First stage larva of *Angiostrongylus cantonensis* recovered from an infected rat's feces.

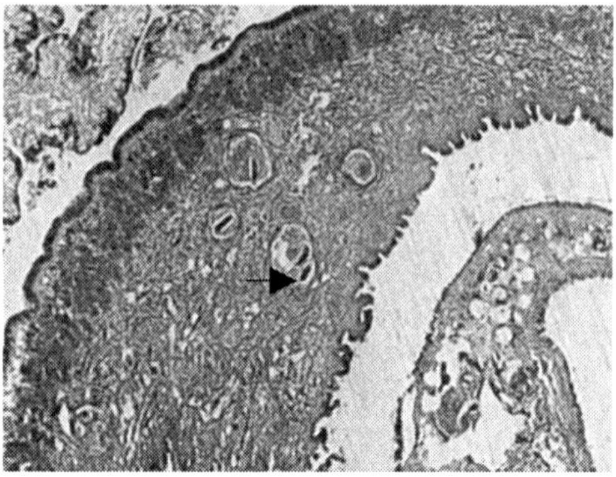

FIGURE 8.8. Developing larvae of *Angiostrongylus cantonensis* in snail muscle.

The final molt occurs in 7 to 9 days, and the young worms move into the subarachnoid space. They remain there for 10 to 14 days before invading the cerebral vein and migrate to the heart and lungs and eventually reach the pulmonary artery. Sexual maturity occurs in the artery and females deposit eggs. The eggs mature quickly and first-stage larvae (Fig. 8.7) may be found in the rat feces 6 to 7 weeks after the rat eats the molluscan intermediate host. The prepatent period is between 42 to 45 days in rats. The life cycle of the parasite is shown in Fig. 8.10.

Geographic Distribution

Angiostrongylus cantonensis has been reported in rats and humans from most parts of the world (Alicata and Jindrak, 1970; Kliks and Palumbo, 1992; Prociv et al., 2000). It has been suggested that the parasite originated in Madagascar and was carried eastward to Asia by the giant African snail, *Achatina fulica*, a major intermediate host. It has also been suggested that the snail was introduced to some areas of Asia and the Pacific Basin by the Japanese military as a food source, and consequently it has been called the Japanese snail by some indigenous populations. Many Taiwanese in southern Taiwan eat the snail cooked and uncooked, and there are indications that the parasite can be picked up by simply handling the snail (Wan and Weng, 2004).

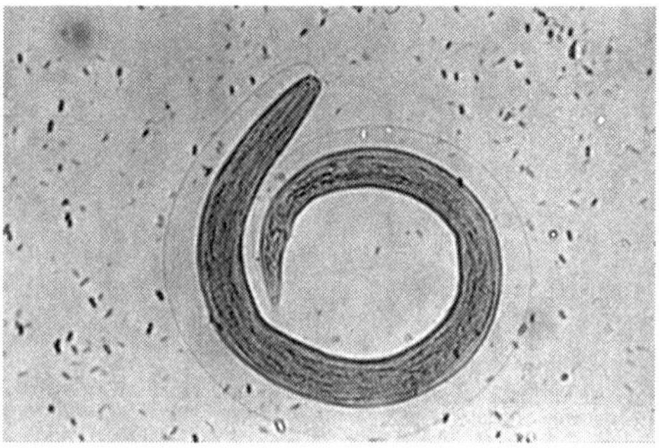

FIGURE 8.9. Third-stage larva of *Angiostrongylus cantonensis* digested out of snail muscle.

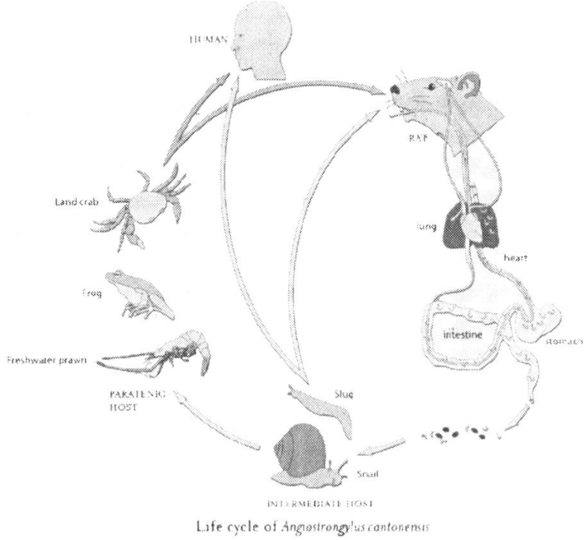

FIGURE 8.10. Life cycle of *Angiostrongylus cantonensis*. (From the Centers for Disease Control and Prevention.)

As indicated in Figure 8.11, the parasite today seems to be spreading from Asia and the Pacific Basin to the Western Hemisphere. Usually, once the parasite is found in rats and snails in an area, infections and eosinophilic meningitis eventually occur in humans.

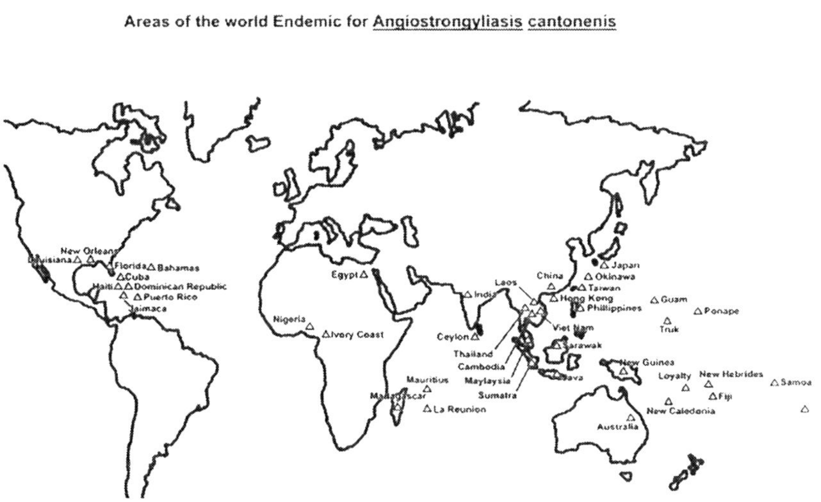

FIGURE 8.11. Map of the world indicating areas reporting *Angiostrongylus cantonensis*.

Disease

The natural rat host will tolerate infection with *A. cantonensis* when less than 100 worms are involved; however, infection of more than 100 worms may lead to death of the animals. Rats and monkeys, on the other hand, may develop immunity to infection experimentally when given nonlethal immunizing doses of third-stage larvae (Heyneman and Lim, 1967; Cross, 1979). Monkeys were able to tolerate infections of hundreds or thousands of third-stage larvae after being given the immunizing infections (Cross, 1979).

When infective-stage larvae are ingested by humans after eating infected mollusks or a paratenic host, the parasites are digested from the vector tissue and enter the intestinal tissue causing enteritis. Passage of the worm through the liver may also cause hepatomegaly. Cases may be benign and self-limiting, but when many worms are involved, severe central nervous system (CNS) symptoms can develop. The symptoms may be abrupt and persistent (Yii, 1976) and may be attributed partly to the destruction and inflammation of nerve fibers. Severe headache, stiff neck, nausea, vomiting, and fever as well as myalgia, pain, and paresthesia in the skin of the trunk and the limbs may develop (Punyagupta et al., 1975; Cross, 1978). There are numerous reports of ocular symptoms especially from Indonesia (Widagdo et al., 1997), Sri Lanka (Dissanaike et al., 2001), Vietnam (Thu et al., 2002), Okinawa (Toma et al., 2002), Thailand (Eamsobhana, 2005), India (Malhotra et al., 2006) and Taiwan (Liu et al., 2006; Wang et al., 2006). It is possible that the first report of human angiostrongyliasis was from Sri Lanka in 1925, when a larval worm was recovered from the eye of a patient with iritis (Dissanaike and Cross, 2004). There also was a report of sensorinneural hearing loss in a Thai patient with eosinophilic meningitis (Chotmongkol et al., 2004a).

Generalized weakness and flaccid paralysis of the extremities have been reported and, at times, coma. When the developing worms migrate from the neural tissue to the surface of the brain and cord and enter the subarachnoid space, they may cause inflammation of the meninges and the production of eosinophils. Many worms may die in the CNS and provoke more reactions and disease. The incubation of the disease is highly variable, 1 day to several weeks, depending on the number of parasites involved. Table 8.2 lists the major symptoms and signs in 114 cases of eosinophilic meningitis in Taiwan (Yii, 1976). The severity of the disease is related to the number of parasites involved and at an autopsy carried out on Taiwan, a large numbers of worms were recovered (Yii et al., 1968).

Pathogenesis

Rats usually suffer little from infection with *A. cantonensis* except when large numbers of worms are involved. There may be cellular infiltration and edema when the parasite enters the gastric mucosa and small necrotic foci in the liver and lungs with eosinophilic and seropurulent pleurisy. Small foci of edema and hemorrhage may develop in the CNS along with granulomatous reactions.

TABLE 8.2. Major symptoms and clinical signs in cases of eosinophilic meningitis on Taiwan.

Symptoms	Signs
Headache	Abnormal patellar reflex
Nausea, vomiting	Neck stiffness
Somnolence, lethargy	Abnormal Achilles reflex
Fever	Absence abdominal wall reflex
Constipation	Hepatomegaly
Malaise	Kernig's sign
Anorexia	Abnormal biceps reflex
Abdominal pain	Eye muscle paralysis

The molted sheaths of the worm can cause granulomas with giant cells and monocytes with abscesses developing especially in heavy infections. As the infection progresses, the subarachnoid spaces become dilated with hemorrhage resulting from dilated cerebral veins. Hemorrhage also occurs in the nerve roots of the cranial and spinal nerves. Parasites in the lung arteries may cause embolism and hypertrophy with cellular infiltration in the arteries and bronchi with thrombosis in the arteries. Thrombi may encase parasite eggs and larvae along with nodular formation in the lungs parenchyma (Alicata and Jindrak, 1970).

In humans, a gastroenteritis and hepatomegaly may be experienced after eating infected intermediate hosts. Cough, rhinorrhea, sore throat, malaise, and fever may develop when the worms pass through the lungs, and when they reach the CNS, symptoms of eosinophilic meningitis and eosinophilic pleocytosis develop. There are a few reports, however, where the patients exhibited peripheral eosinophilia yet did not develop eosinophilic pleocytosis (Tsai et al., 2001; Lindo *et al.*, 2004).

It appears that dead parasites cause more pathology than living worms, and in the few autopsies carried out, both living and dead worms have been found. Living fourth- and fifth-stage worms have been recovered from subarachnoid spaces and on the surface of the brain and the cord. Upon gross examination, few lesions can be seen; however, there may be congestion and hemorrhage with thickening of the basal portion of the leptomeninges (Fig. 8.12). Histologically,

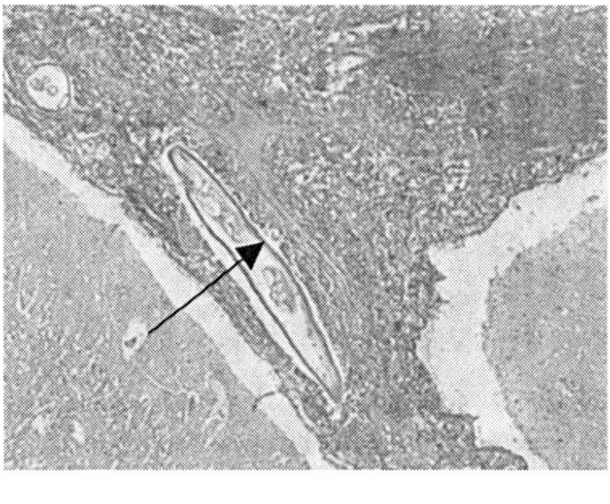

FIGURE 8.12. Section of brain showing *Angiostrongylus cantonensis* (arrow) in the leptomeninges.

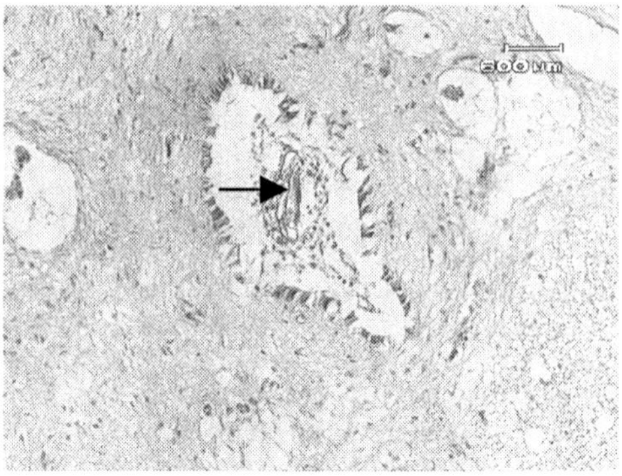

FIGURE 8.13. Section of spinal cord showing *Angiostrongylus cantonensis* (arrow) in the central canal.

the worms are not concentrated and tissue sections from different parts of the brain and the cord have to be examined in order to find them. Dead worms, the sheath, and worm fragments provoke the inflammatory reaction, eosinophils, and granulomatous reaction, including giant cells and focal necrosis (Fig. 8.13). Tracks made by the migrating worms may be seen in the tissue with the presence of glial scars containing hemosiderin, hemorrhage, eosinophils, and Charcot-Leyden crystals. There may be arterial and venous dilation in the subarachnoid spaces. Nerve cells adjacent to worms and tracks may show critical chromatolysis and axonal swelling. Similar changes occur in the spiral cord (Punyagupta, 1979).

An autopsy was carried out on a 5-year-old child in Taiwan who died with eosinophilic meningitis. Tissues were examined and sections of the parasite were detected in the brain (Fig. 8.14), spinal cord (Fig. 8.15) and lungs (Fig. 8.16).

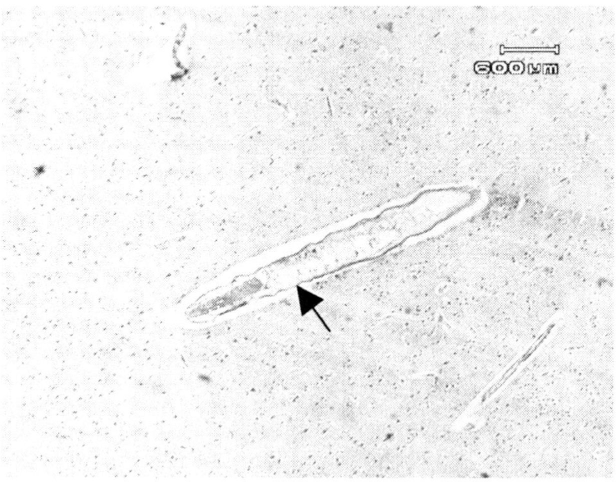

FIGURE 8.14. Section (arrow) of *Angiostrongylus cantonensis* in the parenchyma of the brain showing little inflammatory reaction.

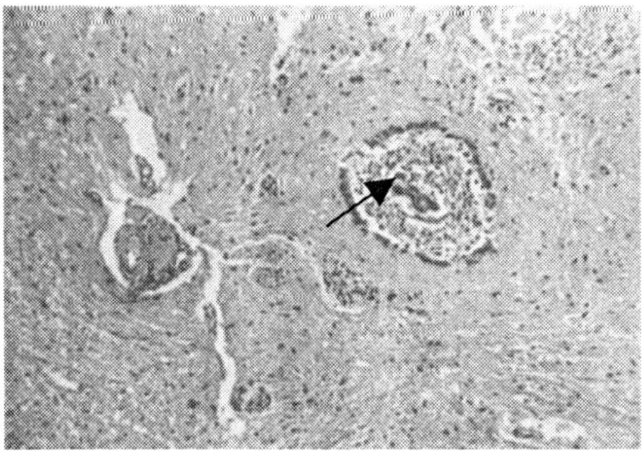

FIGURE 8.15. Section of *Angiostrongylus cantonensis* in the central canal of the spinal cord (arrow) surrounded by an intense inflammatory reaction.

The leptomeninges were infiltrated with lymphocytes, pigment-laden macrophages, and mononuclear cells (Fig. 8.17). Foreign-body–type giant cells were found close to degenerative worms (Fig. 8.18). Cerebral and cerebellar sulci sections revealed sections of viable worms (Fig. 8.19). Lymphocytes surrounded cerebral and meningeal vessels. Numerous glial elements and glitter cells were also in these regions. There was also perivascular cuffing by lymphocytes. Foreign-body reaction surrounded dead worms and recently dead parasites provoked a mixed reaction of mononuclear cells, lymphocytes, and eosinophils. Over 500 immature and young adult worms were recovered from the CNS, and developing eggs were found in the uterus of some female worms. The lungs were congested, and macrophages with pigments were found in the alveolus. Proteinaceous material and neutrophils infiltrated around the bronchi and bronchioles. Focal hemorrhages were seen and some small blood vessels contained thrombi. Living worms were found in pulmonary vessels (Figs. 8.20 and 8.21) (Yii et al., 1968). Worms in the lungs have also been reported from another Taiwan patient (Hung and Chen, 1988) as well as patients in Australia (Cooke-Yarborough et al., 1999), Jamaica (Lindo et al., 2004), and in Thailand (Eamsobhana, 2005).

The parasite has also been found to infect primates with the pathological findings similar to those reported from humans. Monkeys given as many as 10,000 larvae

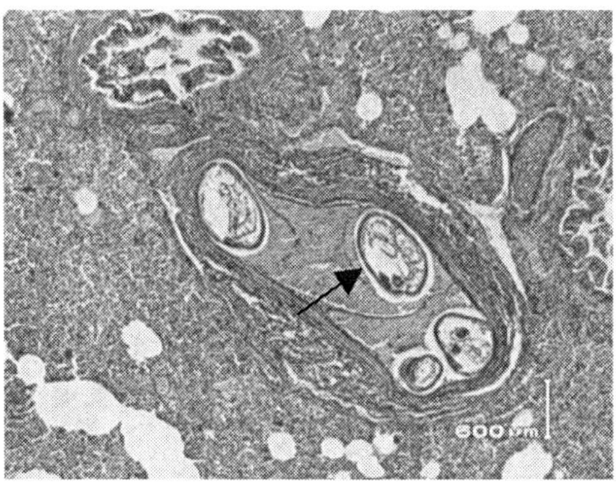

FIGURE 8.16. Section of *Angiostrongylus cantonensis* (arrow) in a pulmonary artery of a human.

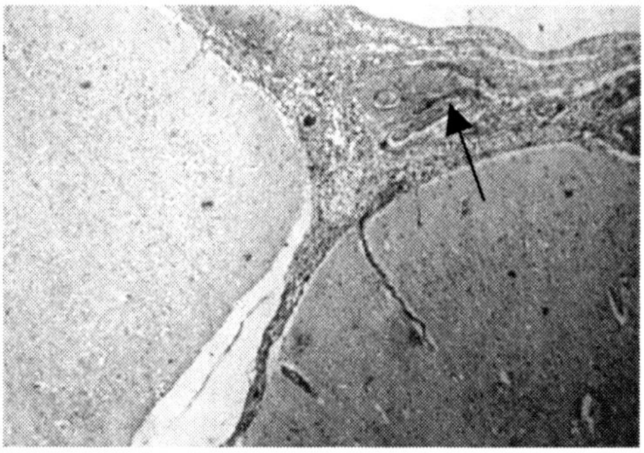

FIGURE 8.17. Section of *Angiostrongylus cantonensis* (arrow) in the leptomeninges surrounded by inflammatory cells.

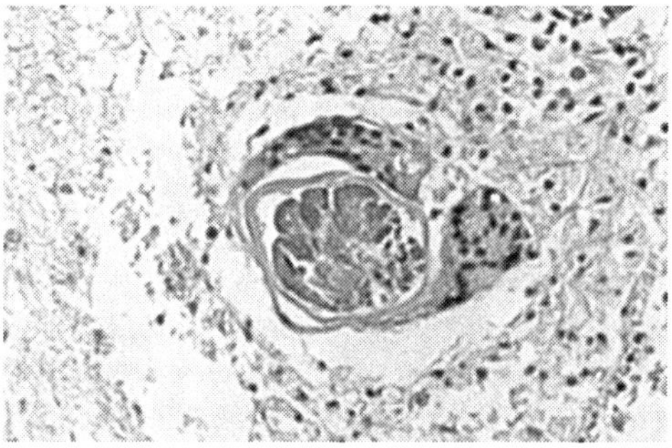

FIGURE 8.18. Section of the spinal cord showing a degenerated *Angiostrongylus cantonensis* surrounded by a foreign-body giant cell.

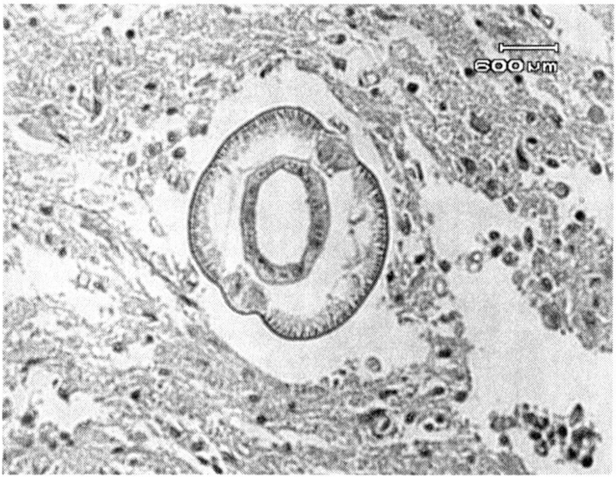

FIGURE 8.19. Section of human brain showing *Angiostrongylus cantonensis* without inflammatory reaction.

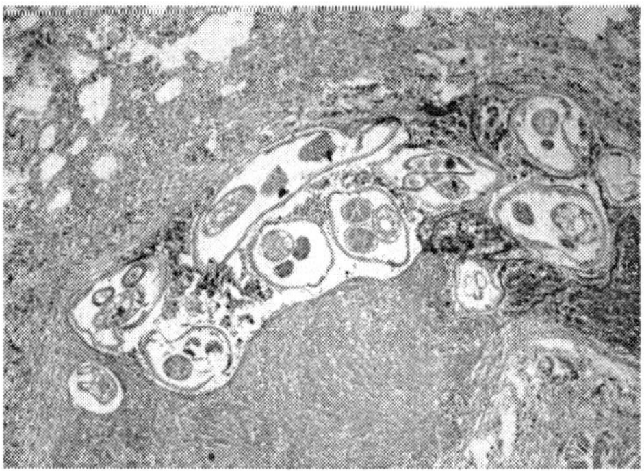

FIGURE 8.20. Human lung showing multiple section of *Angiostrongylus cantonensis* in a pulmonary artery.

usually died, and at necropsy little tissue reaction was seen around the worms in the CNS of animals examined early (10 days) in the infection (Fig. 8.22). However, cellular infiltration around worms consisting of eosinophils, plasma cells, lymphocytes, giant cells, and perivascular hemorrhage were found at 29 days postinfection (Fig. 8.23). The reactions were associated with dead worms rather than living worms. In early infections, the worms were still in the third stage, while those found later were in the fourth larval stage or young adults (Cross, 1979).

Diagnosis

A confirmed diagnosis for *A. cantonensis* eosinophilic meningitis is rare. In areas endemic for the parasitosis, diagnosis is usually presumptive based on the symptoms of headache, nausea, vomiting, fever, neck stiffness, paresthesia, diplopia, and strabismus, and a history of contact or ingestion of an intermediate or paratenic host. Cerebro spinal fluid (CSF) may demonstrate eosinophilic pleocytosis, hemorrhage, and occasional xanthochromia. The confirmed diagnosis is based on the recovery of larval stages of the parasite in CSF (Fig. 8.24) (Kuberski *et al.*, 1979) or from ocular chambers (Fig. 8.25). Recovery of worms from the

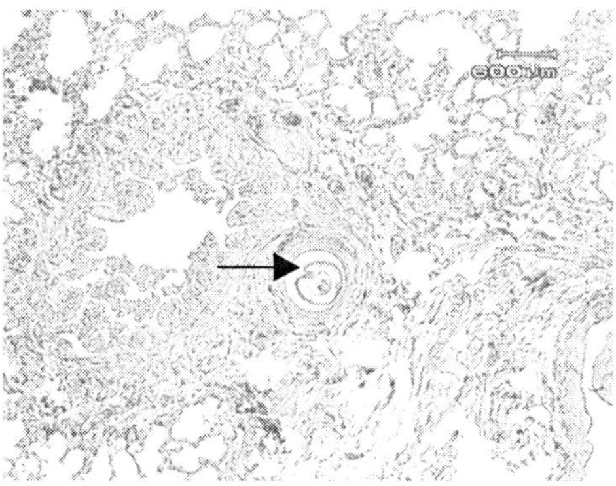

FIGURE 8.21. Human lung with section of *Angiostrongylus cantonensis* (arrow) in a pulmonary vessel.

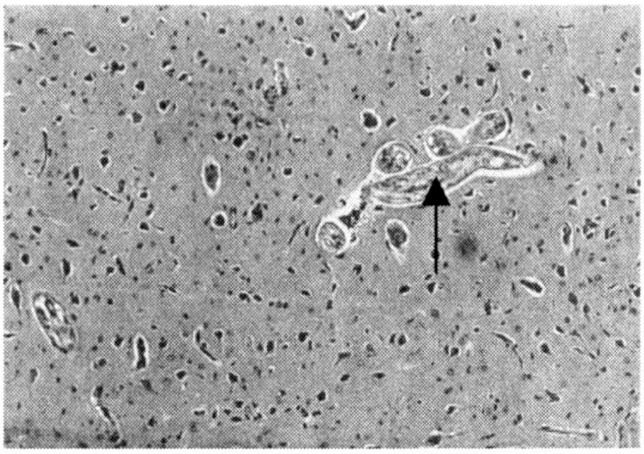

FIGURE 8.22. Section of *Angiostrongylus cantonensis* in the brain parenchyma of a monkey 10 days post-infection showing little reactions.

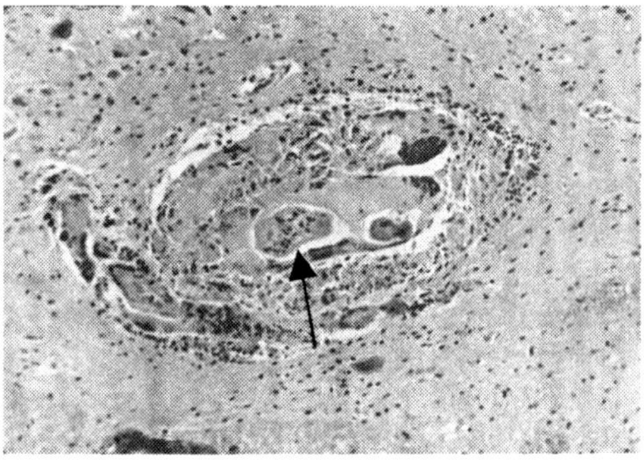

FIGURE 8.23. Section of a monkey brain 29 days after infection with *Angiostrongylus cantonensis* showing an intense foreign-body giant cell reaction around degenerating worms.

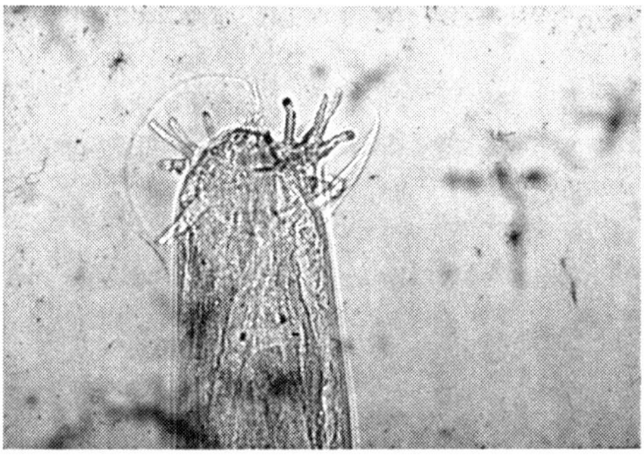

FIGURE 8.24. Young adult male *Angiostrongylus cantonensis* recovered from the spinal fluid of a Taiwanese child. (Courtesy of Dr. K.P. Hwang.)

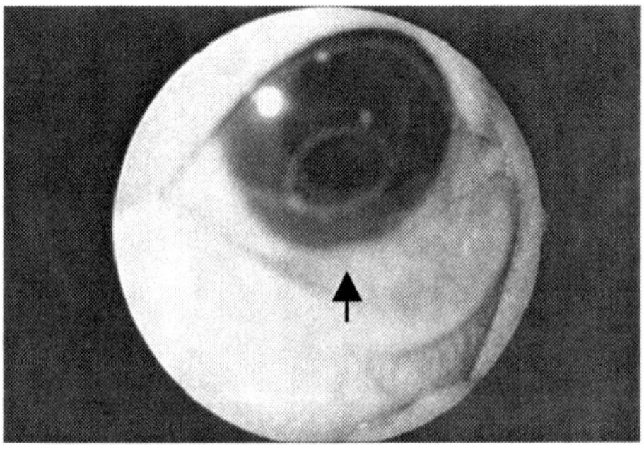

FIGURE 8.25. Larva of *Angiostrongylus cantonensis* (arrow) in the eye of a Taiwanese child. (Courtesy of Dr. K.P. Hwang.)

CSF has improved (Hwang and Chen, 1991). Serological tests such as the enzyme-linked immunosorbent assay (ELISA) has been found to be satisfactory (Cross, 1978; Cross and Chi, 1982; Jaroonesama et al., 1985), and a number of improved immunological tests have been developed for both antibody and antigen detection (Chye et al., 2004; Maleewong et al., 2001; Eamsobhana and Tungtrongchitr, 2005). A dot-blot ELISA using blood dried on filter paper has proven to be convenient for handling field samples for epidemiological surveys (Eamsobhana and Tungtrongchitr, 2005; Eamsobhawa et al., 2006). Antigens from *A. cantonensis* can also be detected in serum by immuno–polymerase chain reaction (PCR) (Chye et al., 2004). Computed tomography (CT) and magnetic resonance imaging (MRI) may reveal the presence of lesion in the meninges (Jin et al., 2005).

Eosinophilic meningitis may be caused by other infectious agents as well as malignancies. Paragonimiasis, schistosomiasis, neurocysticercosis, and gnasthostomiasis should be considered in the differential diagnosis of eosinophilic meningitis.

Treatment

Most cases of angiostrongyliasis are mild and self-limiting, with symptoms abating in 4 to 6 weeks. Treatment is supportive or symptomatic using analgesics with corticosteroids (Pien and Pien, 1999) and frequent removal of CSF to relieve increasing intracranial pressure and to relieve headaches. Specific anthelminthic treatment is controversial since dead worms are considered by some to cause more pathological changes. However, mebendazole and albendazole have been reported effective in treating children (Hwang and Chen, 1991), and a combination of albendazole and corticosteroids has also been reported as effective treatment (Chotmongkol et al., 2006). A recent study has suggested that a Chinese herbal medicine, yin-chen extract, in combination with albendazole may be effective in managing eosinophilic meningitis or eosinophilic meningoencephalitis (Lai, 2006). Worms in the eye usually require surgical removal (Kumar et al., 2005); however, paralysis of the parasite with intracameral preservative-free lidocaine provides easy removal of the worm (Mehta et al., 2006).

Epidemiology

Angiostrongylus cantonensis is found in most tropical areas with warm moist environments where rodent definitive hosts and molluscan intermediate hosts abound. *Rattus rattus* and *R. norvegicus* are the most common definitive host, but other species of rats found in rural and forested areas such as *R. exulans*, *R. diardii*, *R. coxinga*, *R. argentiventer*, *R. losea*, *R. jalorensis*, *R. tiomanicus*, *R. mindanensis*, and *Bandicota indica*, *B. savilei*, and *B. malobarica* are also reported to be natural hosts. It is quite possible that all species of *Rattus* or *Bandicota* are susceptible to infection. Although other species of mammals can be infected, the parasite is usually unable to complete its development and the "abnormal" host commonly dies when a large number of worms are involved. The worm has been reported to cause death to primates in zoos in the United States (Gardiner et al., 1990; Aguilar et al., 1999) and Australia (Carlisle et al., 1998).

Rats, humans, and other accidental animal hosts acquire *A. cantonensis* infection by eating molluscan intermediate hosts harboring third-stage larvae of the parasite. Terrestial and some aquatic snails and slugs are the primary sources of infection. The infective larvae are encysted in the tissues of the hosts (Fig. 8.8) and the larvae released upon ingestion and digestion of the tissue. Most species of mollusks are susceptible and are capable of transmitting the worm; however, the intensity of the infection is variable. The giant African snail, *Achatina fulica,* is considered a major source of infection, and, according to Alicata and Jindrak (1970), the dispersion of the parasite is associated with the spread of the snail throughout Asia and the Pacific Basin. Once established in a country, local mollusks acquire the parasite and rats become infected by eating the infected mollusks. *Angiostrongylus cantonensis*–infected mollusks were probably introduced worldwide, initially by ship traffic or by migrating human populations that include the snail as part of their diet (Kliks and Palumbo, 1992; Cross, 2004). It is also believed that heavy ship traffic during World War II may have been a means of dispersal of snails throughout Asia and the Pacific Islands. The hitch-hiking snails could have been hiding in heavy equipment moving around the war zones.

Achatina fulica is a large terrestrial snail (Fig. 8.26) with an enormous reproduction potential, and when introduced to an area, becomes a serious problem to agriculture. The snail may be eaten cooked, and, when mixed with various seasoning, eaten uncooked. It is also possible for larvae to be released during preparation of the snail when released larvae contaminate the knives or chopping blocks that are subsequently used for cutting other foods. Some populations also use snails and juice from snails for medicinal purposes.

Other species of snails throughout the world that are known vectors of *A. cantonensis* include *Cryptozona bristalis, Bradybaenae similaris, Macrochlamys resplendens, Subulina octona, Pila ampullacea, P. polita, P. scutata* (Fig. 8.27), *Ampullarium caniculatus, Ampularia gigas* and the rice-paddy snail, *Cipangopaludina chinensis*

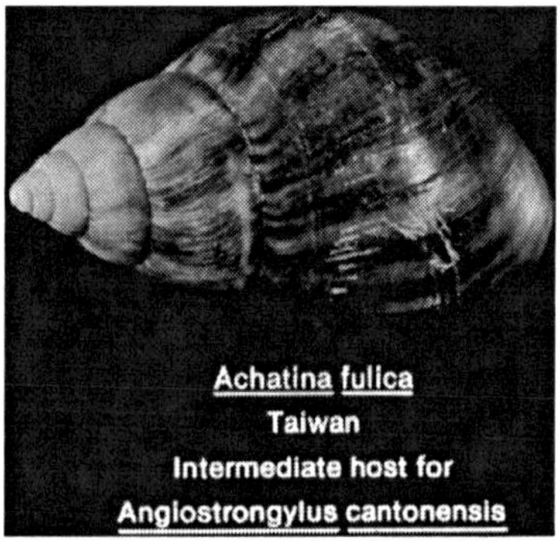

FIGURE 8.26. *Achatina fulica* an important intermediate host and vector for *Angiostrongylus cantonensis* in the Asia Pacific area.

FIGURE 8.27. Baskets of *Pila* sp. snails. An important intermediate host of *Angiostrongylus cantonensis* being sold in a market in Thailand.

(Fig. 8.28). Many species of snails have been experimentally infected reflecting the widespread susceptibility of snails to infection (Richards and Merritt, 1967; Wallace and Rosen, 1969). Slugs in various endemic areas are also known vectors. *Vaginalus plebeius* (Fig. 8.29), *Veronicella alte, Decerocerus laeve, Parmarion martens, and Microparmarion malayanus* have been found to be infected or susceptible to infection and able to serve as intermediate hosts.

A large number of animals are known paratenic hosts. Land crabs, coconut crabs, and freshwater prawns are known paratenic hosts and are eaten raw (Alicata, 1962). Frogs have been considered a paratenic host (Ash, 1968), and one human case of eosinophilic meningitis has recently been reported in a man who ate, on a dare, two uncooked green tree frog legs in Louisiana in the United States

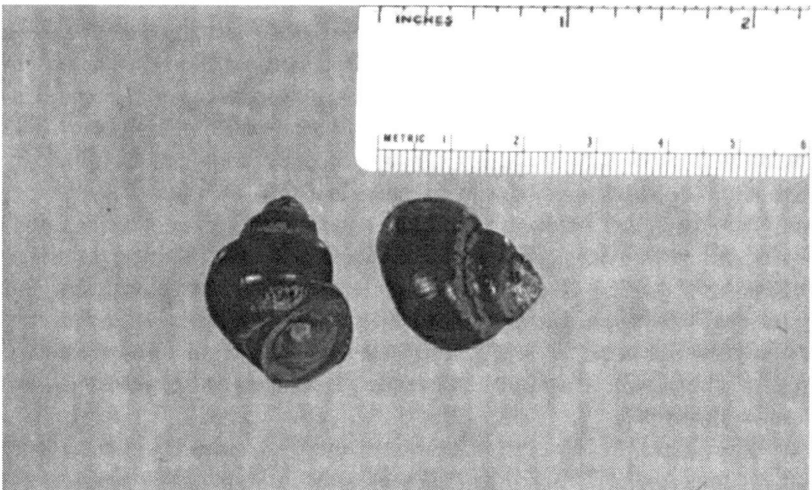

FIGURE 8.28. Rice paddy snail (*Cipangopaludina chinensis*) responsible for an infection of *Angiostrongylus cantonensis* in a Taiwan child. The child ate the snails directly from a rice paddy.

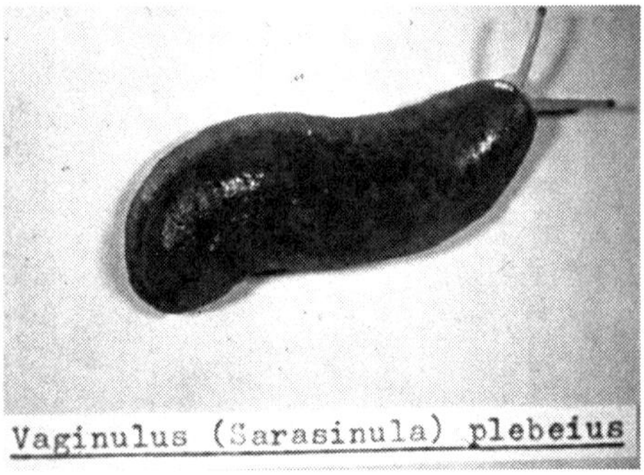

FIGURE 8.29. The slug *Vaginalus plebeius*, an intermediate host and vector for *Angiostrongylus cantonensis*.

(Cuneo et al., 2006). Eating raw liver of a toad has also been implicated in infections in Japan (Kinjo et al., 1975). Juice from crushed crustaceans is also a source of infection. Planarians are also considered a source of infection in some areas. These flatworms are predaceous on dead or dying snails and slugs and subsequently acquire third-stage larvae. Humans become infected when eating the planaria on uncooked vegetables. In New Caledonia, there is an increase of human infections in the cool season, a time when there is an influx in planaria in vegetable gardens (Ash, 1976). The tiny organisms are unseen and accidentally ingested when eating uncooked vegetables. Human infections have also been reported in people that ate raw or partially cooked monitor lizards in Thailand (Eamsobhana and Tungtrongchit, 2005), Sri Lanka (Hidelaratchi et al., 2005), and India (Panackel et al., 2006).

In Taiwan, most eosinophilic meningitis is seen in children and most have been associated with eating *A. fulica* meat. The snail is usually heavily infected and consequently causes more disease. In Thailand, on the other hand, most infections occurred in adult males who ate Pila spp snails (Fig. 8.27), a poor host that contain few infective stage larvae (Eamsobhana and Tungtrongchitr, 2005). The symptoms therefore are milder. Thai men acquire the infections primarily at the time of drinking alcohol with friends. Thai men also feel that the snails have some medicinal value. *Achatina fulica* has become a commercial value in Taiwan and other places and are raised commercially for the meat used in the preparation of escargot (Fig. 8.30). In a preliminary study on Taiwan, snail meat in a market in southern Taiwan was fed to laboratory rats. Some of the meat had been cooked and some uncooked. Every rat fed with the uncooked meat developed infections with *A. cantonensis,* while the animals fed cooked meat were negative at necropsy.

In 1976, three U.S. Marines on survival training in Okinawa, Japan, ate five to 10 *A. fulica* uncooked. Several other Marines tasted the snails while others ate only cooked snails. Three weeks after eating the raw snails, the three Marines became ill and were hospitalized. They had symptoms of eosinophilic meningitis and were serologically positive. The hospital course was uneventful, although one required ventricular drainage and repeated lumbar punctures. The Marines who ate the

FIGURE 8.30. Empty shells of *Achatina fulica* that have been raised and provided the meat for export for the preparation of escargot.

cooked snails and those who only tasted the snails had no ill effects (Cross, 1978). Group infections have been reported such as the event in Samoa when six Koreans were sold, as a prank, living *A. fulica* by the native Samoans. The Koreans ate the snails raw and became severely ill and one died (Kliks et al., 1982). The first case of eosinophilic meningitis in the United States occurred when a child in New Orleans, Louisiana, was dared by his sister to eat a live snail (New et al., 1995).

Water may also be a source for infection. Terrestrial mollusks falling into water may drown and release larvae into the water (Alicata and Jindrak, 1970; Crook et al., 1971). There is also a report from Taiwan reporting an outbreak of eosinophilic meningitis by drinking vegetable juice probably contaminated with a small intermediate or paratenic host or infective larvae in molluscan mucus deposited on the vegetables (Tsai et al., 2004). Another interesting report from Taiwan was a boy who acquired the disease while raising *A. canaliculatus* as pets (Wan and Weng, 2004), and in mainland China 18 people who ate these snails raw also developed the disease (Wang et al., 1999). There are an increasing number of cases of angiostrongyliasis reported from China (Chen et al., 2005), and many of the patients had eaten snails from one particular restaurant in Beijing specializing in the dish. The snails involved were called Amazon snails (*Ampularia gigas*) that originated from South America and introduced into China in 1980. A total of 137 cases were reported by the Xinhua News Agency in September 2006.

Impact and Issues

There are relatively few problems associated with angiostrongyliasis. Human infections are sporadic and are rarely seen even in endemic areas.

In April 2003, 12 of 23 tourists who visited Jamaica met the case definition of eosinophilic meningitis. No parasites were found in the CSF, but 11 of the 12 had positive antibody titers. The source of infection was not determined, but the tourists all had eaten a salad in a restaurant (Slom et al., 2002). Events such as this can make an impact on the tourist industry. Cases of eosinophilic meningitis have been reported in the Jamaican population, with at least one death and the parasite has been isolated in rats and snails (Lindo et al., 2002).

Angiostrongylus cantonensis has also had an impact on zoos. The death of primates in the Audubon Park and Zoological Garden in New Orleans, Louisiana, was of great concern to the veterinarians (Gardiner et al., 1990; Aguilar et al., 1999). Death was also reported in a gibbon from the Miami, Florida, Metrozoo. The monkey had been in the zoo since 1963, and the parasite was recovered from the CNS (Duffy et al., 2004). A lemur in a zoo in New Iberia, Lousiana, was also found to be infected (Kim et al., 2002). If there is an increase of the parasite in the rat population, other wildlife species such as the wood rat and opossums may acquire fatal infections (Kim et al, 2002). The parasite has also been found in a miniature horse in Louisiana (Costa et al., 2000). It is anticipated that *A. cantonensis* will continue to remain endemic in the southern United States, and the spread of the parasitosis in nature will undoubtedly lead to human infection and disease.

Only one other species of *Angiostrongylus* is known to cause disease in humans. *A. costaricensis* is in Latin America and is known to cause eosinophilic granuloma in the human intestines (Morera and Cespedes, 1971; Cross, 1998). The definitive hosts of the parasite are cotton rats (*Sigmodon hispidus*) and black rats (*Rattus* sp.). Infections have also been reported in animals in the Miami, Florida, Metrozoo (Miller et al., 2006). Slugs (*Vaginulus plebeius*) serve as the intermediate host. Humans acquire the infection accidentally by ingesting infected slugs or vegetations that have slug mucus trails containing infectious-stage larvae. The larvae from the slugs penetrate the intestinal wall of the rat and mature in lymph nodes and lymphatic vessels. The young worms migrate to the mesenteric arteries and lay eggs. The female worms lay eggs in the intestinal wall. They hatch and the larvae migrate into the intestinal lumen. In humans, eggs and any hatched larvae are usually destroyed by the cellular reaction and the parasite products are not found in the feces. The inflammatory reaction results into the formation of a mass and in some cases may partially or completely obstruct the intestines. Most lesions are found in the appendix. Surgical removal is the treatment, and anthelminthics are not recommended. Most cases involved children, and while the parasitosis is reported throughout Latin America, Costa Rica reports a significant number of cases each year (Sun, 1998). A parasitologic diagnosis is not available and the serological diagnosis presents several drawbacks. However, PCR technologically is reported to be an alternative (da Silva et al., 2003; Caldeira et al., 2003).

Unsolved Problems

There is a need to determine the geographic distribution of *A. cantonensis* in wildlife populations of rats and mollusks. This requires surveys (especially for rat infections), support funds, and man hours. Rats would have to be trapped and killed and the heart and lungs examined for worms. Once found in rats, surveys should be carried out to determine infection in snails and paratenic hosts.

Public health workers and physician in endemic areas must be made aware of the parasite, its hosts, and the dangers to the health of populations. Programs should be developed to make the indigenous populations aware of the problems and to warn them about eating mollusks and paratenic hosts uncooked. More specific and sensitive diagnostic techniques must be developed. It is difficult to make a parasitological diagnosis, as worms are not easily detected antemortem. Serological and molecular tools are available, but are only used at a few universities and research centers in Asia, and the tests are not always conclusive.

It would be interesting to determine if strains of the parasite exist. In a study of strains of *A. cantonensis* from Asia and Hawaii, a new species, *A. malaysiensis*, was discovered (Cross, 1979). There are now four different species of *Angiostrongylus* found in Southeast Asia and Australia: *A. cantonensis* (Chen, 1935), *A. mackerrasae* (Bhaibulaya, 1968), *A. malaysiensis* (Bhabulaya and Cross, 1972; Cross and Bhaibulaya, 1974), and *A. siamensis* (Ohbayashi et al., 1979). The four species are found essentially in the same intermediate and definitive host; however, only *A. cantonensis* has been confirmed to be the cause of human eosinophilic meningitis. It is possible, however, that *A. mackerrasae* in Australia (Prociv and Carlisle, 2001) and *A. malaysiensis* in Malaysia (Ambu et al., 1997) may infect humans.

Control

The control of *Angiostrongylus* species in nature is very difficult. However, rodent control in areas close to human communities is recommended. The elimination of molluscan hosts of the parasite near housing and vegetable gardens should be routine. Populations in endemic areas should be made aware of the parasite, the host the means of transmission, and the possibility of illness. The habit of eating uncooked snails and paratenic hosts should be discouraged. Vegetation should be examined for snails, slugs, or planaria and washed thoroughly if eaten raw; however, cooking vegetables is recommended. Natural water should not be drunk unless boiled. It is difficult to change eating habits of the population that have existed for generations, but it would be necessary to prevent infection with *A. cantonensis*.

The spread of *A. cantonensis* has been attributed to the introduction of *A. fulica* accidentally to Asia and the Pacific Basin. This may have contributed to the dispersal of the snail initially, but in most recent times it is believed the spread has been considered to be via ocean shipping. Snails are known to be in ship cargo and can be carried ashore. However, the transporting of rats via ship is probably the most accepted means of dispersal of the parasite. Therefore, snail and rodent control measures should be employed on ships.

Summary

Angiostrongylus cantonensis is a metastrongyle nematode considered an important cause of eosinophilic meningitis in areas endemic for the parasite. Adult worms are found in the pulmonary vessels of *Rattus* and *Bandicotta* species, and larvae from eggs deposited in the lung migrate to the intestine and pass in the feces. The larvae are ingested by snail and slug intermediate hosts and develop into the infective stage of the worm. When intermediate hosts are eaten by a definitive host, the larvae migrate to the brain. Young adult worms that develop after a few weeks migrate from the brain to the pulmonary vessels, where they complete development and reproduce.

Humans become infected by eating an intermediate or paratenic host (planaria, amphibians, crustaceans, lizards) raw or poorly cooked. The larvae are digested from the animal tissue, and like the rodents, the larvae migrate to the brain and cause disease. The important signs and symptoms are headache, stiff neck, nausea, vomiting, myalgia, paresthesia, and eosinophilia pleocytosis, and on rare occasions the parasite may complete development in humans with worms found in the lungs at autopsy. There are reports of *A. cantonensis* invading the eye. The disease may persist for several weeks and is usually limiting.

The parasite was first reported in rats in China but was thought to have originated in Africa and spread eastward, possibly carried by the giant African snail, *Achatina fulica*. Human infection was first reported from Taiwan in 1945. It was not until 1961, however, when the parasite was found in the brain of a man at autopsy in Hawaii. It was not too long before the parasite and disease was reported throughout the Far East and the Pacific Basin. The disease is now recognized in many parts of the world and has reached the Western Hemisphere. Most reports of the parasite and disease are from Taiwan and Thailand with increasing reports from the China mainland. The parasite has reached the United States, with reports of infection in rats, molluscan intermediate hosts, some abnormal host animals, and two human cases of eosinophilic meningitis.

The cause of clinical disease is questionable. Some physicians believe that reaction to dead worms are responsible for the pathology and do not recommend anthelminthic treatment. There are a few physicians, however, who report successful treatment with mebendazole, or albendazole along with a steroid. Worms found in the eye are removed surgically.

The parasite is on the move. The dispersal is possibly due to rats and snails that are stowaways in cargo aboard sea-going ships. When the ships reach port, the rats jump ship and the snails in cargo are moved ashore. Control of the parasite in nature is nearly impossible, but community control measures could be implemented. Changes in eating habits can be difficult, but populations in endemic area should be made aware of the dangers of eating snails and paratenic hosts raw. The populations should also be informed of hazards of accidentally eating slugs and planaria that are often found in gardens and on vegetables.

References

Aguilar, R.F., Topham, K., Heatley, J.J. et al. (1999). Neural angiostrongyliasis in non-human primates: diagnosis, treatment and control of an outbreak in Southern Louisiana. Proc Am Assoc Zoo Vet 272–276.

Alicata, J.E. (1962). *Angiostrongylus cantonensis* (Nematoda: Metastrongylidae) as a causative agent of eosinophilic meningoencephalitis of man in Hawaii and Tahiti. Can J Zool 40, 5–8.

Alicata, J.E., and Jindrak, K. (1970). Angiostrongylosis in the Pacific and Southeast Asia. Springfield, IL: Charles C. Thomas.

Ambu, S., Rain, A.N., Mak, J.W., Maslah, D., and Maidah, S. (1997). Detection of *Angiostrongylus malaysiensis* circulating antigen using monoclonal antibody-based enzyme-linked immunosorbent assay (MAB-ELISA). Southeast Asian J Trop Med Public Health 28(suppl), 143–147.

Ash, L.R. (1968). The occurrence of *Angiostrongylus cantonensis* in frogs of New Caledonia with observations on paratenic hosts of metastrongyles. J Parasitol 54, 432–436.

Ash, L.R. (1976). Observation on the role of mollusks and planarians in the transmission of *Angiostrongylus cantonensis* infection to man in New Caledonia. Rev Biol Trop 24, 163–174.

Bailey, C.A. (1948). An epidemic of eosinophilic meningitis, a previously undescribed disease occurring on Ponape, Eastern Carolines. Naval Med Res Enst Rpt No 7, 1–32.

Beaver, P.C., and Rosen, L. (1964). Memorandum on the first report of Angiostrongyliasis in man by Nomura and Lin, 1945. Am J Trop Med Hyg 13, 589–690.

Bhaibulaya, M. (1968). A new species of *Angiostrongylus* in an Australian, rat *Rattus fuscipes*. Parasitology 58, 789–799.

Bhaibulaya, M. (1975). Comparative studies of the life history of *Angiostrongylus macherrasae* (Bhaibulaya, 1968) and *Angiostrongylus cantonensis* (Chen, 1935). Int J Parasitol 5, 7–20.

Bhaibulaya, M. (1979). Morphology and taxonomy of major *Angiostrongylus* species of Eastern Asia and Australia. In: Cross, J.H., ed. Studies on Angiostrongyliasis in Eastern Asia and Australia. NAMRU-2-SP-44. MS Naval Medical Research unit No2 Taipei, Taiwan: 4–13.

Bhaibulaya, M., and Cross, J.H. (1972). *Angiostrongylus malaysiensis* (Nematode: Metastrongylidae), a new species of rat-lungworm from Malaysia. Southeast Asian J Trop Med Public Health 2, 527–533.

Caldeira, R.L., Carvalho, G.S., Mendonca, C., Groeff-Teixeira, C., Silva, M., Ben, R., Maurer, R., Lima, W.S., and Lenzi, H.L. (2003). Molecular differentiation of *Angiostrongylus costariciensis, A. cantonensis* and *A. vasorum* by Polymerase chain reaction–restriction fragment length polymorphism. Mem Iros Oswaldo Cruz, Rio de Janeiro 98, 1039–1043.

Campbell, B.C., and Little, M.D. (1988). The findings of *Angiostrongylus cantonensis* in rats in New Orleans. Am J Trop Med Hyg 38, 568–573.

Carlisle, M.S., Prociv, P., Grennan, J., et al. (1998). Cerebrospinal angiostrongyliasis in five captive tamrins (*Sanguinus spp*). Aust Vet J 76, 167–170.

Chen, H.T. (1935). Un nouveau nematode pulmonaire, *Pulmonema cantonensis* n.g., n. sp. des rats de Canton. Ann Parasitol 13, 312–317.

Chen, X.G., Li, H., and Lun, Z.R. (2005). Angiostrongyliasis, Mainland China. Emerg Infect Dis 11, 1645–1647.

Chotmongkol, V., Sawadpanitch, K., Sawanyawesuth, K., Louhawilal, S., and Limpawattana, P. (2006). Treatment of eosinophilic meningitis with combination of prednisolone and albendazole. 74, 1122–1124.

Chotmongkol, V., Wongjitrat,C.L., Sawadpanit, K., and Sawanyawisuth, K. (2004a). The treatment of eosinophilic meningitis with a combination of albendazole and corticosteroid. Southeast Asian J Trop Med Public Health 35, 172–174.

Chotmongkol, V., Yimtae, K., and Itapan, P.M. (2004b). Meningitis associated with sensorineural hearing loss. J Laryngol Otol 118, 57–58.

Chye, S.M., Lim, S.R., Chin, Y.L., Chung, L.Y., and Yin, C.M. (2004). Immuno-PCR for detection of antigen to *Angiostrongylus cantonensis* circulating fifth-staged worms. Clin Chem 50, 51–57.

Cooke-Yarborough, C.M., Kornberg, A.J., Hogg, G.G., Spratt, D.M., and Forsyth, J.R. (1999). A fatal case of Angiostrongyliasis in an 11 month old infant. Med J Aust 170, 541–543.

Costa, L.R.R., McClure, J.J., Cnider, T.G., and Stewart, T.B. (2000). Verminous meningoencephalomyelitis by *Angiostrongylus (Parastrongylus) cantonensis* in an American miniature horse. Equine Vet Educ 12, 2–6.

Crook, J.R., Fulton, S.E., and Supanwong, K. (1971). The infectivity of third-stage *Angiostrongylus cantonensis* shed from drowned *Achatina fulica* and the effect of chemical agents on infectivity. Trans R Soc Trop Med Hyg 65, 602–605.

Cross, J.H. (1977). Angiostrongyliasis. In: Stanford, CT: eds. Appleton & Lange, Pathology of Infectious Disease. Connor, D., Chandler, F., Schwatz, D., Manz, H., and Lack, E., 1307–1314.

Cross, J.H. (1978). Clinical manifestations and laboratory diagnosis of eosinophilic meningitis syndrome associated with angiostrongyliasis. Southeast Asian J Trop Med Public Health 9, 161–170.

Cross, J.H., ed. (1979). Experimental studies on *Angiostrongylus* species and strains in monkeys and laboratory animals. In: Cross, J.H., ed. Studies on Angiostrongyliasis in Eastern Asia and Australia. NAMRU-2-SP-44, Taipei, Taiwan, 118–127.

Cross, J.H. (1998). Angiostrongylosis. In: Palmer, S.R., Saulsby, L., and Simpson, D.H., eds. Zoonoses. Oxford University Press, 773–781.

Cross, J.H. (2004). Iterant *Angiostrongylus (Parastrongyliasis) cantonensis*. Proceedings of the European Multicolloqum on Parasitology, Valencia, Spain, 165–174.

Cross, J.H., and Bhaibulaya, M. (1974). Validity of *Angiostrongylus malaysiensis*. Bhaibulaya and Cross. Southeast Asian J Trop Med Public Health 5, 374–378.

Cross, J.H., and Chi, J.H.C. (1982). ELISA for the detection of *Angiostrongylus cantonensis* antibodies in patients with eosinophilic meningitis. Southeast Asian J Trop Med Public Health 13, 73–77.

Cuneo, P., Clement, S., and Sokol, T. (2006). Eosinophilic meningitis and angiostrongyliasis. Louisiana Morbid Rep 17:1.

de Silva, A.C., Graeff-Teixeira, C., and Zoho, A. (2003). Diagnosis of angiostrongyliasis by PCR from sera of patients. Rev Inst Med Trop San Palo 45, 295–297.

Dissanaike, A.S., Ihalamulla, R.L., Naotunne, T.S., Sexarathna, T., and Withana, D.S. (2001). Third report of ocular Parastrongyliasis (Angiostrongyliasis) from Sri Lanka. Parasitologia 43, 95–97.

Dissanaike, A.S., and Cross, J.H. (2004). Ocular parastrongyliasis (angiostrongyliasis): probable first report of human infection in a patient in Ceylon (Sri Lanka). Parasitologia 46, 315–316.

Dougherty, E.C. (1946). The genus *Aelurostrongylus* Cameron, 1927 (Nematoda: Metastrongylidae) and its relatives with description of Parafilaroides gen. Nov., and *Angiostrongylus gubernaculatus* sp. nov. Proc Helm Soc Wash 13, 16–26.

Duffy, M.S., Miller, C.L., Kinsella, J.M., and deLahunta, A. (2004). *Parastrongylus cantonensis* in a non-human primate, Florida. Emerg Infect Dis 10, 2207–2210.

Eamsobhana, P., Yoolex, A., Punthuprapasa, P., and Suvouttho, S. (2004). ELISA mimicking immunoblot analysis for the specific diagnosis of human parastrongyliasis. Southeast Asian J Trop Med Public Health 35:

Eamsobhana, P., and Tungtrongchitr, A. (2005). Angiostrongyliasis in Thailand. In: Arizowo, R., Chai, J.Y., Nawa, Y., Takahashi, Y., and Yano, A., eds. Asian Parasitology. Japan: FAP Journal, 183–198.

Eamsobhana, P., Ongrotchanakun, J., Yoolek, A., Punthuprapasa, P., Monkong, N., and Dekumyoy, P. (2006). Multi-immunodot for rapid differential diagnosis of eosinophilic meningitis due to parasitic infections. J Helminth 88, 249–254.

Franco, R., Bories, S., and and Couzin, B. (1960). A propos de 142 cas de meningite a eosinophiles observes a Tahiti et en Nouvell-Caledonie. Med Trop 20, 41–55.

Gardiner, C.H., Wells, S., Gutter, A.E., Fitzgerald, L., Anderson, D., Harris, R., and Nichols, D. (1990). Eosinophilic meningoencephalitis due to *Angiostrongylus cantonensis* as the cause of death in captive non-human primates. Am J Trop Med Hyg 42, 70–74.

Heyneman, D., and Lim, B.L. (1967). *Angiostrongylus cantonensis*: proof of direct transmission with its epidemiological implication. Science 158, 1057–1058.

Hidelaratchi, M.D., Riffsy, M.T., and Wijesekera, J.C. (2005). A case of eosinophilic meningitis following monitor lizard consumption, exacerbated by anthelminthics. Ceylon Med J 50, 84–86.

Hung, T.P., and Chen, E.R. (1988). Angiostrongyliasis (*Angiostrongylus cantonensis*). In: Vinken, P., Bruyn, G., and Klawans, H., eds. Handbook of Clinical Neurology. New York: Elsevier, 8, 545–562.

Hwang, K.P., and Chen, E.R. (1991). Clinical studies on angiostrongyliasis cantonensis among children in Taiwan. In Proceedings of the 33rd SEAMEO-TROPMED. Seminar on Emerging Problems in Food-borne Parasitic Zoonoses, November 1990, Chiang Mai, Thailand, 194–199.

Jaroonvesama, N., Charoenlarp, K., Buranasin, P., Zaraspe, G.G., and Cross, J.H. (1985). ELISA testing in cases of clinical angiostrongyliasis in Thailand. Southeast Asian J Trop Med Public Health 116, 110–112.

Jin, E., Ma, D., Liang, Y., Ji, A., and Gan, S. (2005). MRI findings on eosinohpilic myelomeningoencephalitis due to *Angiostrongylus cantonensis*. Clin Radiobiol 60, 242–250.

Kim, D.Y., Stewart, T.B., Bauer, B.W., and Mitchell, M. (2002). *Parastrongylus (Angiostrongylus) cantonensis* now endemic in Louisiana wildlife. 88, 1024–1026.

Kinjo, K., Sueyoshi, S., Ota, M., Ohama, M., and Nakamoto, M. (1975). Two cases of eosinophilic meningoencephalitis, Angiostrongyliasis caused by swallowing of liver of a toad, *Bufo asiaticus*. Abstracts of the 46th annual meeting, Okinawa-ken Ishikai-igakkai, (In Alicanta and Jindrak 1970) 7–8.

Kliks, M.M., Kroenke, K., and Hardman, J.M. (1982). Eosinophilic radiculomyeloencephalitis: an angiostrongyliasis outbreak in American Samoa related to the ingestion of *Achatina fulica* snails. Am J Trop Med Hyg 31, 1114–1122.

Kliks, M.M., and Palumbo, N.F. (1992). Eosinophilic meningitis beyond the Pacific basin: the global dispersion of a peridomestic zoonosis caused by *Angiostrongylus cantonensis*, the nematode lungworm of rats. Soc Sci Med 34, 199–212.

Kuberski, T., Bart, R.D., Briley, J.M., and Rosen, L. (1979). Recovery of *Angiostrongylus cantonensis* from cerebrospinal fluid of a child with eosinophilic meningitis. J Clin Microbiol 9, 629–631.

Kumar, V., Kyprianou, I., and Keenan, J.M. (2005). Ocular Angiostrongyliasis: removal of a live nematode from the anterior chamber. Eye 19, 229–230.

Lai, S.C. (2006). Chinese herbal medicine Yin-Chin-extract as an adjunct to anthelminthic albendazole used against *Angiostrongylus cantonensis*–induced eosinophilic-meningitis meningoencephalitis. Am J Trop Med Hyg 75, 556–562.

Lindo, J.F., Waugh, C., Hall, J., Cunningham-Myrie, C., Ashley, D., Robinson, D.G., Holtz, T., and Robinson, R.D. (2002). Enzootic *Angiostrongylus cantonensis* in rats and snails after an outbreak of human eosinophilic meningitis in Jamaica. Emerg Infect Dis 8, 324–326.

Lindo, J.F., Escoffery, C.T., Reid, B., Codrington, C., Cunningham-Myrie, C., and Eberhard, M. (2004). Fatal autochthonous eosinophilic meningitis in a Jamaican child caused by *Angiostrongylus cantonensis*. Am J Trop Med Hyg 70, 425–428.

Liu, I.H., Chung, Y.M., Chen, S.J., and Cho, W.I. (2006). Necrotizing retinitis induced by *Angiostrongylus cantonensis*. Am J Ophthalmol 141, 577–579.

Mackerras, M.J., and Sandars, D.F. (1955). The life history of the rat lungworm. *Angiostrongylus cantonensis* (Chen, Nematoda: Metastrongyldiae). Aust J Zool 3, 1–25.

Maleewong, W., Sombatsawat, P., Intapan, P.M., Wongkham, C., and Chotmongkol, C. (2001). Asia Pac J Allergy Immunol 19, 267–273.

Malhotra, S., Mehta, D., Arora, R., Chauhan, D., Ray, S., and Jain, M. (2006). Ocular Angiostrongyliasis in a child. First case report from India. J Trop Pediatr 52, 223–225.

Matsumoto, T. (1937). On nematode found in the lung, especially in the pulmonary artery of the wild rat (in Japanese) J Formosan Med Assoc 36, 2620–2623.

Mehta, D.K., Arora, A., Chouhan, D., Shroff, D., and Narula, R. (2006). Chemoparalysis for the removal of a line intraocular worm in ocular angiostrongyliasis. Clin Exp Ophthalmol 34, 493–495.

Miller, C.L., Kinsella, J.M., Garner, M.M., Evans, S., Gulleti, P.A., and Schmidt, R.E. (2006). Endemic infections of *Parastrongylus* (= *Angiostrongylus*) *costaricensis* in two species of non-human primates, raccoons and opposums from Miami, Florida. J Parasitol 9, 406–408.

Morera, P., and Cespedes, R. (1971). *Angiostrongylus costaricensis* n. sp. (Nematoda: Metastrongylidae) A new lungworm occurring in man in Costa Rica. Rev Biol Trop (Costa Rica) 18, 173–185.

New, D., Little, M.D., and Cross, J.H. (1995). *Angiostrongylus cantonensis* infection from eating raw snails. N Engl J Med 332, 1105–1106.

Nomura, S., and Lin, P.H. (1945). First case report of human infection with *Haemostrongylus ratti*. Yokogawa Taiwan no Ikai 3, 589–592 (in Japanese).

Ohbayashi, M., Kamiya, M., and Bhaibulaya, M. (1979). Studies on parasitfauna of Thailand. I. Two new metastrongylid nematodes. *Angiostrongylus siamensis* sp.n and *Thaistrongylus harinsuti* gen. et sp.n (Metastrongylidae: Angiostrongylidae) from wild rats. Jpn J Vet Res 27, 5–10.

Panackel, C., Vishad, Cherian, G., Vijayakumar, K., and Sharma, R.N. (2006). Eosinophilic meningitis due to *Angiostrongylus cantonensis*. Indian J Med Microbiol 24, 220–221.

Pien, F.D., and Pien, B.C. (1999). *Angiostrongylus cantonensis* eosinophilic meningitis. Int J Infect Dis 3, 161–163.

Prociv, P., Spratt, D.M., and Carlisle, M.S. (2000). Nuero-angiostrongyliasis: unsolved issues. Int J Parasitol 30, 1295–1303.

Prociv, P., and Carlisle, M.S. (2001) The spread of *Angiostrongylus cantonensis* in Australia. Southeast Asian J Trop Med Public Health 32(suppl 2), 126–128.

Prommindaroj, K., Leelawong, S.N., and Pradatsundarasar, A. (1962). Human angiostrongyliasis of the eye in Bangkok. Am J Trop Med Hyg 11, 759–761.

Punyagupta, S., Tuttijudat, P., and Bunnag, T. (1975). Eosinophilic meningitis in Thailand. Clinical studies of 484 typical cases probably caused by *Angiostrongylus cantonensis*. Am J Trop Med Hyg 24, 921–931.

Punyagupta, S. (1979). Angiostrongyliasis: clinical features and human pathology. In: Cross, J.H., ed. Studies of Angiostrongyliasis in Eastern Asia and Australia. U.S. Naval Medical Research Unit No. 2, special publication 44. Taipei, Taiwan. R.O.C., 138–150.

Richards, C.S., and Merritt, J.W. (1967). Studies on *Angiostrongylus cantonensis* in molluscan intermediate hosts. J Parasitol 53, U.S. Naval Medical Research Unit No 2 382–388.

Rosen, L., Chapell, R., Lacquer, G.L., Wallace, G.D., and Weinstein, P.P. (1962). Eosinophilic meningoencephalitis caused by metastrongylid lungworm of rats. JAMA 179, 620–624.

Rosen, L., Laigret, J., and Bories, S. (1961). Observations on an outbreak of eosinophilic meningitis on Tahiti, French Polynesia. Am J Hyg 74, 26–42.

Slom, T.J., Cortese, M.M., Gerber, S.I., Jones, R.C., Holtz, T.H., Lopez, A.S., Zambrano, C.H., Sakolvaree, Y., Chaicumpa, W., Herwalt, B.L., and Johnson, S. (2002). An outbreak of eosinophilic meningitis caused by *Angiostrongylus cantonensis* in travelers returning from the Caribbean. N Engl J Med 346, 668–675.

Sun, T. (1998). Parasitic Disorders: Pathology Diagnosis and Management, 2nd ed. Baltimore: Williams & Wilkins.

Thu, T.P., Nguyen, N.X., Lan, L.T., and Kuchie, M. (2002). Ocular Angiostrongyliasis cantonensis in a female Vietnamese patient: case report. Klin Monatshl Augenheilkd 219, 892–895.

Toma, H., Matsumura, Oshiro, C., Hidaka, T., and Sato, Y. (2002). Ocular angiostrongyliasis without meningitis symptoms in Okinawa, Japan. J Parasitol 88, 211–213.

Tsai, H.C., Liu, Y.C., Kunin, C.M., Lee, S.J.L., Chen, Y.S., Lin, H.H., Tsai, T.H., Lin, W.R., Huang, C.K., Yen, H.H., Tsai, T.H., Lin, W.R., Huang, C.K., Yen, M.H., and Yen, C.M. (2001a). Eosinophilic meningitis caused by *Angiostrongylus cantonensis*: report of 17 cases. Am J Med 111, 109–114.

Tsai, H.C., Liu, Y.C., and Kunin, C.M. (2001b). Eosinophilic meningitis caused by *Angiostrongylus cantonensis* with eating raw snails: correlation of brain magnetic resonance imaging scans with clinical findings. Am J Trop Med Hyg 68, 281–285.

Tsai, H.C., Lee, S.J., Huang, C.K., Yen, C.M., Chen, E.R., and Liu, Y.C. (2004). Outbreak of eosinophilic meningitis associated with drinking raw vegetables juice in Southern Taiwan. Am J Trop Med Hyg 71, 222–226.

Ubelaker, J.F. (1986). Systematics of species referred to the genus Angiostrongylus. J Parasitol 72, 237–244.

Wallace, G.D., and Rosen, L. (1969). Studies on eosinophilic meningitis. V. Molluscan Hosts of *Angiostrongylus cantonensis* on Pacific Islands. Am J Trop Med Hyg 18, 206–216.

Wan, K.S., and Weng, W.C. (2004). Eosinophilic meningitis in a child raising snails as pets. Acta Trop 90, 51–53.

Wang, X., Huang, H., Dong, Q., Lin, Y., Wang, Z., and Li, F. (1999). A clinical research for eosinophilic meningoencephalitis caused by Angiostrongyliasis. Zhonghua Nei Ke Za Zhi 38, 326–328.

Widagdo, S., Lokollo, D.M., and Margono, S.S. (1997). Ocular Angiostrongyliasis in Semorong, Central Java. Am J Trop Med Hyg 26, 72–74.

Wang, L.C., Wang, I.H., and Jou, J.R. (2006). Optic neuritic secondary to *Angiostrongylus cantonensis* infection. Ocul Immunol Inflamm 14, 189–191.

Yii, C.Y. (1976). Clinical observations on eosinophilic meningitis and meningoencephalitis caused by *Angiostrongylus cantonensis* in Taiwan. Am J Trop Med Hyg 25, 233–249.

Yii, C.Y., Chen, C.Y., Fresh, J.W., Chen, T., and Cross, J.H. (1968). Human angiostrongylosis involving the lungs. Chin J Microbiol 1, 148–150.

Yokogawa, S. (1937). A new species of nematode found in the lungs of rats. *Haemostrongylus ratti* n.sp. now. Trans Nat Hist Soc Formosa 27, 247–250.

II
Plant-Borne Parasites

9
Plant-Borne Trematode Zoonoses: Fascioliasis and Fasciolopsiasis

Santiago Mas-Coma, María Dolores Bargues, and María Adela Valero

There are six plant-borne trematode species known affecting humans: *Fasciola hepatica, F. gigantica*, and *Fasciolopsis buski* (Fasciolidae), *Gastrodiscoides hominis* (Gastrodiscidae), *Watsonius watsoni*, and *Fischoederius elongatus* (Paramphistomidae). Whereas *F. hepatica* and *F. gigantica* are hepatic, the other four species are intestinal parasites. The gastrodiscid causes a secondary disease, while *W. watsoni* and *F. elongatus* have been only accidentally detected in humans (Mas-Coma et al., 2005). The fasciolids cause important zoonoses that appear to be emerging or reemerging in many countries as a consequence of many phenomena related to both environmental changes and man-made modifications, similarly as in other snail-transmitted helminthiases. Well known since long ago and with considerable information about the disease mainly in animals (Dalton, 1999), there is a new emphasis on the disease in humans after multidisciplinary research in recent years.

Fascioliasis

Etiology

F. hepatica is distributed in all continents, whereas *F. gigantica* appears restricted to Africa and Asia. Both fasciolids follow a similar two-host life cycle, in which freshwater snails of the family Lymnaeidae act as intermediate or vector hosts and a broad spectrum of mammals, mainly herbivorous large-size species, act as definitive hosts, including humans.

The Adult Stage

Intraspecific Characterization

The morphology of flukes found in biliary canals and gall bladder has been the subject of numerous studies and is well known (see review in Mas-Coma and Bargues, 1997). However, recent studies on intraspecific variation of adults and eggs have provided new light, thanks to modern phenotyping techniques, alone or combined, as mathematical modelling for ontogeny analysis and the computer image analysis system (CIAS) for morphometric studies, together with measurement standardization (Valero et al., 1996, 2005).

The CIAS facilitates morphometrically characterizing fluke features that were not measurable by the standard microscopic techniques. This becomes very useful in the case of fasciolids, in which many structures present complicated forms, as the ceca, testes, or ovary. An example has been the study of uterine size (Valero et al., 2001a). Bolivian sheep and cattle liver fluke populations proved to have a uterine size smaller than that of European populations. Although this may be attributable to intraspecific variation, these uterus differences between populations of highland and lowland flukes were tentatively related to high-altitude influences. High-altitude hypoxia conditions could be the origin of a reduced egg production by the flukes. Moreover, although the uterus in digeneans has traditionally not been considered as a storage organ but mainly an organ adapted to the developmental time of the eggs (in fasciolids, eggs are laid unembryonated, the miracidium beginning its development in eggs once in freshwater), in the Northern Bolivian Altiplano, climatic conditions, freshwater body characteristics, and lymnaeid ecology enable fascioliasis transmission to take place throughout the year (Fuentes et al., 1999; Mas-Coma et al., 1999c), so that egg storage is *a priori* not needed as in the Northern Hemisphere latitudes where fascioliasis transmission is typically seasonal (Valero et al., 2001a).

These new methods have enabled more accurate analyses of fluke development, both in experimentally and naturally infected hosts. A good example has been the characterization of the crowding effect in the rat model (Valero et al., 2006a). Results showed that when the burden increases, the maximum values of fluke measurements decrease. The crowding effect is manifest when fluke measurements approximate their maximums in the advanced chronic stage. The prepatent period and egg production decrease when the burden increases. This means that measurements of eggs per gram of feces tend to underestimate the fluke burden. This study demonstrated how to quantify the fascioliasis experimental rat model crowding effect on adult growth, prepatent period, and egg production. This quantification may be of great interest in epidemiological studies and in experimental research on the in vivo actions of different anthelmintic drugs and vaccines, pathology, immunology, and resistance studies.

Similar studies also enable accurate comparisons of flukes and eggs from different host species from the same endemic area. Exhaustive morphometric comparisons of *F. hepatica* adults and eggs from sheep, cattle, pigs, and donkeys of the Bolivian Altiplano, as well as of *F. hepatica* adults and eggs experimentally obtained in Wistar rats infected with Altiplanic sheep, cattle, and pig isolates, were made. Comparative statistical analysis of the allometries showed that fluke adult populations from sheep, cattle, and pigs significantly differ in the functions of (1) body length versus body width, and (2) body length versus distance between the posterior end of the body and the ventral sucker. Statistical analysis of *F. hepatica* egg size showed characteristic morphometric traits in each definitive host species. In experimentally infected rats, fluke adult allometry and egg morphometry did not vary depending on the Altiplanic definitive host species isolate. This study revealed that the definitive host species decisively influences the size of *F. hepatica* adults and eggs, and that this influence does not persist in a heterologous host (Valero et al., 2001b). Eggs shed by both naturally and experimentally infected murid rodents (wild *Mus musculus* and *Rattus rattus* from Corsica island, and *R. norvegicus* Wistar laboratory strain) were smaller in size than those shed by naturally infected cattle from the same region (Valero et al., 2002).

The modern methods also allow for significant comparisons between different fluke populations. Slight but significant differences between sheep liver flukes from the Bolivian Altiplano and Spain were found (Valero et al., 1999). Results did not show differences in egg size, and in adults only a smaller size in the majority of the parameters in the Bolivian material was found. The minor differences can be accounted for by geographical variation related to altitude effects or genetic isolation and suggest that the Bolivian population is a recent isolate from Iberian populations.

Interspecific Characterization

The above-mentioned phenotyping methods have been recently used to get a significant differentiation of both fasciolids. Using flukes from the same host to avoid host influences, the quantification of the different size and shape of *F. hepatica* and *F. gigantica* has been achieved for the first time in natural populations (Periago et al., 2006). Linear measurements, areas and ratios of gravid adults, and eggs of *F. hepatica* (from France and Spain) and *F. gigantica* (from Burkina Faso) were analyzed by CIAS and an allometric model. All the measurements proved to overlap in the two fasciolids, apart from the distance between the ventral sucker and the posterior end of the body, body roundness, and body length/body width ratio.

These results may be useful in *Fasciola* species identification in countries where both species coexist. The overlapping distribution of both fasciolids has even become the basis of an already long controversy on the taxonomic identity of the *Fasciola* species occurring in Far East countries, especially Japan, Taiwan, the Philippines, and Korea, in which a wide range of morphological types is detected. Some resemble *F. hepatica*, whereas others resemble *F. gigantica*, with intermediate forms also occurring and involving phenomena such as abnormal gametogenesis, diploidy, triploidy, and mixoploidy, parthenogenesis, and hybridization events between different genotypes (see review in Mas-Coma and Bargues, 1997). The recent finding of an aspermic triploid, necessarily asexually reproducing, liver fluke isolate in the United Kingdom suggests that facultative gynogenesis may be widespread in this parasite (Fletcher et al., 2004).

The confirmation of the existence of intermediate forms in Asia has also recently been done by applying the new phenotyping techniques (Ashrafi et al., 2006a). Fasciolids from naturally infected bovines from the human endemic province of Gilan, Iran, were studied and compared with *F. hepatica* and *F. gigantica* standard populations from areas where both species do not coexist. Although morphometric values somewhat overlapped, there were clear differences in allometric growth.

Results revealed that Iranian *F. hepatica*–like specimens are larger than the *F. hepatica* standard and Iranian *F. gigantica*–like specimens are longer and narrower than the *F. gigantica* standard, but with a smaller body area. Measurements that permit a specific differentiation in allopatric populations overlap in the specimens from Gilan, thus proving the presence of intermediate forms. Moreover, this study showed that simple, traditional microscopic measurements may be sufficient for the morphometric characterization of fasciolids, even in areas where intermediate forms are present.

Genetic Markers

Different molecular techniques and DNA markers are useful for epidemiological and diagnostic studies, as well as for intraspecific variation analyses of fasciolids.

Most studies on fasciolid proteins have concentrated on isoenzymes. Only a very few studies considered individual or population-level variation. The same isoenzymes of *F. hepatica* were detected regardless of the host species (cattle, sheep, goats), but densities of some isoenzyme bands differed according to the host (Blair, 1993). Profiles of whole-body proteins and excretory/secretory products obtained with isoelectric focusing differed among worms from different hosts (Lee et al., 1992). Random amplified polymorphic DNA (RAPD) markers applied to *F. hepatica* showed that the majority of genetic diversity occurred within, rather than between, hosts and was also greater within than between populations. Individual cows were infected by numerous genetically different liver flukes, suggesting the influence of mainly migrations and transportation of definitive hosts (Semyenova et al., 2003).

The whole mitochondrial genome of *F. hepatica* has been recently sequenced, and will be suitable for studies of variation (Le et al., 2001). Part of the mitochondrial DNA of *F. hepatica* showed length heterogeneity, suggesting differences among individual mitochondrial genomes (Zurita et al., 1988). Sequences of nicotinamide adenine dinucleotide dehydrogenase (NADH) subunit I (NDI) and cytochrome c oxidase subunit I (COI) showed that the Japanese *Fasciola* forms were more closely related to *F. gigantica* than to *F. hepatica* (Itagaki et al., 1998), contrary to previous information (Itagaki et al., 1995). Moreover, a high intraspecific variation (8.3%) in an NDI fragment of *F. gigantica* and a low one (0.2%) in a COI fragment of *F. hepatica* and the Japanese form of *Fasciola* were found (Itagaki et al., 1998). Restriction fragment length polymorphism (RFLP) patterns were analyzed for the whole mitochondrial DNA of *F. hepatica* from Australia, *F. gigantica* from Malaysia, and *Fasciola* sp. from Japan after digestion with three different four-base-cutting endonucleases (Hashimoto et al., 1997). The mtDNA digestion patterns differed markedly between the three fasciolids. For each enzyme there were some bands specific for each geographical isolate with the Japanese *Fasciola* sp. sharing more bands with *F. gigantica* than with *F. hepatica*. However, given the variation observed within the short COI region, there are likely to be considerable differences in RFLP patterns even between quite closely related forms. A total of 25 to 28 nucleotide differences were detected in a COI fragment between *F. hepatica* and *Fasciola* sp./*F. gigantica*, and only four to five between *Fasciola* sp. and *F. gigantica*. Moreover, intraspecific variation was found at one nucleotide site between different specimens from the same *F. gigantica* population from Malaysia (Hashimoto et al., 1997).

In regard to nuclear ribosomal DNA, restriction endonuclease maps of ribosomal RNA (rRNA) genes were distinct for *F. hepatica* and *F. gigantica*, Japanese *Fasciola* sp. being identical to *F. gigantica*. No intraspecific variations in the maps of *F. hepatica* or of *F. gigantica* were detected, but length heterogeneity was noted in the intergenic spacer, even within individual worms (Blair and McManus, 1989). A total of six differences was detected between *F. hepatica* from sheep of Ipswich and *F. gigantica* from cattle of Malaysia in a 28S rRNA gene D1 domain fragment (Barker et al., 1993).

In an rDNA ITS-2 fragment, a single nucleotide difference between *F. hepatica* from Mexico and the same fluke species from Australia, Hungary, and New Zealand was found, but no differences between *F. gigantica* from Indonesia and Malaysia were found. *Fasciola hepatica* and *F. gigantica* differed at six nucleotide sites among 213 nucleotides compared, and *Fasciola* sp. from Japan differed at seven nucleotide positions from *F. hepatica* and in one from *F. gigantica* (Adlard et al., 1993). Similar results were found in the almost complete, 362–base pair (bp)-long ITS-2 sequence, although no differences were found between Japanese *Fasciola* sp. and *F. gigantica* (Hashimoto et al., 1997).

The ITS-2 sequences of the seven Japanese triploid fasciolids were divided into two distinct types: *F.*sp.I, almost identical to that of *F. hepatica* from Uruguay, and *F.*sp.II, similar to *F. gigantica* from Indonesia and Japan. No intraspecific variation was detected between two *F. hepatica* specimens from Uruguay, but two different sequences were obtained from individuals of *F. gigantica* from Zambia and a third sequence from *F. gigantica* from Indonesia, which was identical to that from Malaysia. It was concluded that the Japanese triploid form of *Fasciola* may be a hybrid between *F. hepatica* and *F. gigantica* because the NDI and COI sequences of *F.*sp.I were almost identical to those of *F. gigantica* from Zambia but not to *F. hepatica* from Uruguay (Itagaki and Tsutsumi, 1998). The ITS-2 sequences have also been used to characterize the liver flukes from mainland China. *Fasciola hepatica, F. gigantica*, and an intermediate genotype, including polymorphism among ITS-2 copies within the same fluke individual, were found (Huang et al., 2004). The complete ITS-2 sequence of *F. hepatica* from Spain and Bolivia proved to be identical but differed from other geographic origins at least in one position. Similarly, the complete ITS-1 and ITS-2 showed no nucleotide differences between flukes from Spain and Bolivia, suggesting that the Bolivian flukes were derived from Iberian ones, probably because of a recent introduction having taken place at the time of the Spanish colonization or somewhat later (Mas-Coma et al., 2001).

According to molecular clock estimations based on ITS sequences, the relatively few ITS nucleotide differences between different regions of the world suggest a recent geographical diffusion of *F. hepatica* from Europe into the other continents (Mas-Coma et al., 2003). It needs to be taken into account that

F. hepatica and *F. gigantica* are evolutionarily relatively recent species having diverged about 19 million years ago, according to sequence analysis of cathepsin-L–like cysteine proteases (Irving et al., 2003).

Five among six microsatellite markers appear polymorphic in *F. hepatica* from Bolivia. No genetic differentiation between sampling sites or between definitive host species (sheep, cattle, pig) was found (Hurtrez-Bousses et al., 2004). The current large livestock import-export trade is undoubtedly in the background of this large genetic variability of a hermaphroditic fluke that preferentially out-crosses. This reproduction mode also explains the existence of hybrids, as shown in given Japanese triploid forms that presented ITS-2 sequences almost identical to those of *F. hepatica* and NDI and COI sequences almost identical to those of *F. gigantica* (Itagaki and Tsutsumi, 1998).

Life Cycle and Transmission

The Definitive Mammal Hosts

Fasciolid flukes show a reduced specificity at the adult-stage level, presenting a broad spectrum of definitive mammal host species (Boray, 1982; Mas-Coma and Bargues, 1997). Moreover, *F. hepatica* has a large capacity to adapt to new definitive hosts, which gives it a high geographic spreading potential. This species has succeeded in expanding from the European original area thanks to the exportation of European livestock to actually colonize the five continents where it has adapted to other autochthonous mammal species such as camelids in Africa, aukenids in South America, and marsupials in Australia (Mas-Coma et al., 2003).

Studies on isoenzymes, protein sequences, and mitochordrial DNA (mtDNA) sequences suggest that fasciolids are able to develop a capacity for definitive host species selection (Miller et al., 1993; Panaccio and Trudgett, 1999; Spithill et al., 1999). This finding raises the question of liver fluke circulation within an endemic area (Valero et al., 2001b). In Bielorrusia, for example, it has been found in seven different wild animal species (Shimalov and Shimalov, 2000). The quick capacity of *F. hepatica* to adapt to new definitive host species is illustrated by examples of the black rat in Corsica island (Valero et al., 1998, 2002; Mas-Coma et al., 2003), the nutria in France (Menard et al., 2001) and the pig in Andean countries (Mas-Coma et al., 1997; Valero and Mas-Coma, 2000; Valero et al., 2001a,b). In all these cases, the newly acquired hosts play an important role as reservoir hosts in fascioliasis transmission, contributing to the spread of the disease.

Lymnaeid Intermediate Hosts

Fasciolid specificity at the snail host level is a crucial factor for the transmission and spread of the disease. The fluke larval stages not only develop in freshwater snails of the family Lymnaeidae, they require given lymnaeid species. *Fasciola hepatica* uses mainly small-sized species of the broad *Galba/Fossaria* group, with species as stagnicolines being able to transmit it in given circumstances. *Fasciola gigantica* uses species of the genus *Radix*. A few species have been noted as being able to transmit both fasciolids, such as the American species *Pseudosuccinea columella*. A list of vector species is given in Bargues et al. (2001).

The expansion of *F. hepatica* from its European origin to other continents appears related to the spreading of its original European vector *Galba truncatula* and secondarily to *Pseudosuccinea columella*, as well as to its adaptation to other authochthonous lymnaeids in the colonized areas. The smaller geographical distribution of *F. gigantica* seems to be related to the lesser diffusion capacity of their intermediate snail hosts, the African *Radix natalensis* and the Eurasian *R. auricularia*.

Galba truncatula has spread into other continents most probably together with livestock exportation (i.e., in mud attached to the feet of sheep and cattle). The expanding potential of *G. truncatula* is also related to its capacity for ecological niche widening, as observed on Corsica island (Gil-Benito et al., 1991). Studies showed that *G. truncatula* is distributed throughout the insular periphery (coastal zones) as well as in the inland, up to 1500 m altitude, in both reservoir habitats (permanent presence and renewal of water) and invasion habitats (only seasonal presence of water), including numerous different types of biotopes (Oviedo et al., 1992). Several atypical habitats may be understood as a consequence of the influences of the insularity phenomenon. This fact is in turn related to the extraordinary distribution of the disease on the island.

Pseudosuccinea columella is also related to the spread of fascioliasis. Originally from Central America, the Caribbean, and the southern part of North America, this rapidly colonizing, more aquatic, more heat-tolerant species is today present in South America, Europe, Africa, Australia, New Zealand, and even Tahiti. In Brazil, *P. columella* appears to be the only lymnaeid present in many fascioliasis areas. Interestingly, a strain of *P. columella* resistant to fasciolid infection has been recently found in Cuba (Gutierrez et al., 2003a,b; Fernandez Calienes et al., 2004). This find opens research possibilities to look for the genes responsible for resistance and for future applications in control strategies.

The adaptation to many different local lymnaeid species has allowed *F. hepatica* to spread in the Americas, Asia, Hawaii, Papua New Guinea, Philippines, Japan, Australia, and New Zealand (Mas-Coma and Bargues, 1997; Bargues et al., 2001). The reports of other lymnaeids (*L. palustris, L. turricula, Omphiscola glabra, Catascopia occulta, Radix ovata*) and planorbid species *Planorbis leucostoma* as alternate or facultative natural host species transmitting *F. hepatica* in Europe are extremely rare, because such a possibility only exists when lymnaeid infection occurs during the first days of the snail's life (Dreyfuss et al., 1994, 2002; Bargues et al., 2001). In the laboratory, *O. glabra, L. palustris*, and *L. fuscus*, and even *L. stagnalis, Radix peregra*, and *Myxas glutinosa*, can be infected only if the miracidium penetrates very young snails, and a high lymnaeid mortality is obtained (Oviedo et al., 1996; Dreyfuss et al., 2000, 2002). In Egypt, a finding of the planorbid *Biomphalaria alexandrina* naturally infected with *F. gigantica* (Farag and El Sayad, 1995) still needs confirmation.

Classification and Genetic Characterization of Lymnaeids

The Lymnaeidae are involved in systematic-taxonomic confusion (Bargues et al., 2001). Numerous lymnaeids show an interspecific morphologic and anatomic uniformity that usually causes considerable difficulties in specimen classification at the species level, sometimes even impeding it, as in the most important *Galba/Fossaria* group (e.g., Oviedo et al., 1995). Intraspecific shell variation is considerable within lymnaeids, although a genetic component in shell shape has been shown in some lymnaeid populations (Samadi et al., 2000). A large range of situations can be found in lymnaeids, from heterogeneous, polymorphic populations (Rudolph and Burch, 1989; Jarne and Delay, 1990a; Coutellec-Vreto et al., 1994) to completely homogeneous, monomorphic populations (Jabbour-Zahab et al., 1997; Meunier et al., 2001; Trouve et al., 2001), a phenomenon related to both selfing and crossing capacities of these freshwater snails (Jarne and Delay, 1990b; Jarne et al., 1993). As a consequence, lymnaeid-fasciolid interrelationships are far from being sufficient. The necessity for tools enabling species distinction and population characterization within lymnaeids is evident.

The DNA markers prove to be the best tools (Bargues and Mas-Coma, 2005). Concerning mtDNA, sequences of the large subunit (16S) enabled distinguishing between several lymnaeid species and analyzing their phylogenetic relationships (Remigio and Blair, 1997a; Remigio, 2002). However, mtDNA sequences have not been used to advance lymnaeid systematics and taxonomy.

The 18S rDNA sequence differentiates between species belonging to different genera and subgenera, and sometimes even between species of the same genus. It is, however, not useful when comparing populations (Bargues and Mas-Coma, 1997; Bargues et al., 1997). When locating nucleotide differences in the 18S rRNA secondary structure, modified positions mainly appear in helix E10–1 of the variable region V2. The small region of this helix and the obtained phylogenetic cladograms allow species groupings, thus distinguishing supraspecific entities and suggesting 18S usefulness for the definitive, supraspecific taxonomic reorganization of Lymnaeidae. Moreover, 18S rRNA gene results have also showed an applied interest in the distinction between fasciolid transmitter and nontransmitter lymnaeid species groups, as well as between lymnaeids transmitting *F. hepatica* from those transmitting *F. gigantica* (Bargues and Mas-Coma, 1997). However, studies showed that levels of intraspecific divergence of the V1 and V2 were not appreciably different (1%) from the interspecific when comparing different lymnaeid species and would therefore question the validity of the 18S rDNA marker for lymnaeid taxonomy and phylogeny (Stothard et al., 2000).

In recent years, rDNA ITS sequences have furnished most of the valuable information for lymnaeid systematics, taxonomy, and population characterization. The ITSs have enabled distinction between four systematically problematic, very closely related North American stagnicoline species (Remigio and Blair, 1997b).

The usefulness of both ITSs was appreciated in Bolivia, where ITS sequencing of two inhabiting American species *L. viatrix* and *L. cubensis* (Ueno et al., 1975) proved that those were in fact the extreme morphs of a large intraspecific shell variability within only one species and that this species was neither of the above mentioned species but rather the European *G. truncatula* (Mas-Coma et al., 2001). As with the liver fluke, ITS results suggest that fascioliasis transmitting snails of the Altiplano were introduced from Europe, most probably imported by Spanish colonizers (Mas-Coma et al., 2001). Additionally, isoenzyme and microsatellite analyses proved that all lymnaeid populations inhabiting the Bolivian endemic area are monomorphic, a clonicity related to selfing reproduction processes deriving from a foundational original population (Jabbour-Zahab et al., 1997; Meunier et al., 2001). Moreover, differences detected in comparative studies between Bolivian lymnaeids and *G. truncatula* from Europe were always nonsignificant or nonexistent. Thus, *G. truncatula*, as *F. hepatica*, has followed a process of adaptation from the European lowlands to the Bolivian highland. The initial founding snail individual or a few individuals imported from Europe that have given rise by selfing to the numerous monomorphic populations today inhabiting the Bolivian Altiplano endemic area were most probably very susceptible snails. These snails would have genetically transmitted their high susceptibility to their descendants by almost absolute predomination of autofecundation, suggesting a large and homogeneous susceptibility of all Altiplanic *G. truncatula* populations (Mas-Coma et al., 2001).

The ITS-2 is a useful tool for resolving supraspecific, specific, and population relationships in Lymnaeidae and an excellent marker for systematic and taxonomic purposes (Bargues and Mas-Coma, 2005). Numerous populations of different lymnaeid species and subspecies from Europe, Morocco, Bolivia, and the United States and belonging to different genera/subgenera taxa were sequenced by Bargues et al. (2001, 2003) and Mas-Coma et al. (2001). When comparing the sequences, several populations originally classified as belonging to different species showed identical ITS-2 sequences, and other populations originally classified as pertaining to the same species presented different ITS-2 sequences. Sometimes the sequence differences were very few, suggesting intraspecific variability. However, sometimes differences detected among populations classified as pertaining to the same species were numerous, sufficient to consider different species involved. Moreover, the number of sequence differences between species sometimes appeared lower than that between populations of the same species (Bargues et al., 2001, 2003). After sequence comparisons and phylogenetic studies, snail specimens analyzed were reexamined to ensure correct classification by specialists; lymnaeid species studied were systematically revisited, several synonymies were proposed, and the taxonomic validity and relationships of genera and subgenera involved were established. Subsequently, the analysis of genetic distances and sequence differences found between the distinct populations and taxa studied facilitated distinguishing the upper limit to be expected within a single species and determining how different sister species can be expected to be at the ITS-2 sequence level. The information furnished by ITS-2 is of applied interest concerning molluscan host specificity. The ITS-2 phylogenies distinguish between lymnaeids transmitting fasciolids and those not transmitting these flukes, as well as between those transmitting *F. hepatica* and those transmitting *F. gigantica* (Bargues et al., 2001).

The ITS-1 has been used for only relatively few numbers of species (Remigio and Blair, 1997b; Mas-Coma et al., 2001; Bargues et al., 2006). The ITS-1 studies confirmed the results and conclusions previously reached with ITS-2. The slightly higher percentage of nucleotide differences suggests that ITS-1 may evolve somewhat faster than ITS-2 in the Lymnaeidae. Consequently, the ITS-1 may offer a valuable marker for taxon differentiation and relationships in the Lymnaeidae, not only at genera and species levels, but also for subspecies and other information for populations.

Single nucleotide mutations in both ITSs have been shown to be useful to distinguish between *P. columella* populations susceptible and resistant to *F. hepatica* (Gutierrez et al., 2003b).

Intramolluscan Larval Development

The literature concerning the larval development of *F. hepatica* and *F. gigantica* is numerous (see review by Mas-Coma and Bargues, 1997). In recent years, studies have shown that larval development is related to the spread of fascioliasis, thanks to the capability of fasciolids to colonize and adapt to new environments, even the extreme characteristics of inhospitality of a very high altitude. The development of *F. hepatica/G. truncatula* from the Northern Bolivian Altiplano (3800–4100 m) was compared to that of the European homologue. Several aspects of fluke development were similar to those in Europe, such as embryonation time, infection percentages, prepatent period, and infectivity of metacercariae. However, certain aspects appeared to favor transmission, such as the longer cercarial shedding period and the greater cercarial production, both aspects related to the high survival capacity of infected lymnaeid snails (Mas-Coma et al., 2001).

Lymnaeid survival in experiments with *F. hepatica* and *G. truncatula* from the northern Bolivian Altiplano is worth mentioning. The longevity of the experimentally parasitized Bolivian mollusks was considerably longer than in the European *G. truncatula*, and even longer than that of other American lymnaeids such as *Lymnaea viatrix* and *L. bulimoides*. The capacity of Altiplanic lymnaeids to survive for more than 4 months after the end of the shedding period is surprising, as in Europe snails die during the shedding period, or immediately after the end of shedding. Moreover, the absence of survival differences between parasitized and nonparasitized mollusks suggests a better parasite–host adaptation in Bolivia (Mas-Coma et al., 2001).

In experimentally infected *G. truncatula* from different geographic origins by the same fluke isolate, *F. hepatica* produced more larval stages when infecting lymnaeid snails from another place (Gasnier et al., 2000). This interesting capacity may undoubtedly be one reason for its fast geographical spreading power.

For the facilitation of field transmission studies, attempts have been made to develop a molecular probe for the sensitive and specific detection of *F. hepatica* in lymnaeid snails. The value of several *F. hepatica* 18S rRNA sequences have been tested by Shubkin et al. (1992) and Rognlie et al. (1994), with different results. An assay using the reverse-transcriptase polymerase chain reaction (RT-PCR) to amplify a 18S rRNA region, followed by hybridization to a *F. hepatica*–specific probe, was able to detect infected snails immediately after miracidial exposure and throughout the parasite's development period, although the assay was not species-specific. Another DNA probe was the squash-blot method (Heussler et al., 1993, 1998). The high sensitivity of a DNA probe that does not cross-hybridize with other trematodes (Kaplan et al., 1995) was improved by using chemiluminescent detection (Kaplan et al., 1997). Similar results were obtained by applying a specific repeated, noncoding, and very abundant short DNA sequence (Kozak-Cieszczyk et al., 2002).

Epidemiology

Classification of Human Fascioliasis

We now know that human fascioliasis is an important public health problem in many countries on different continents (Chen and Mott, 1990; WHO, 1995b; Mas-Coma et al., 1999a,b). Human fascioliasis reports have increased in 51 countries on five continents (Esteban et al., 1998; Mas-Coma et al., 1999a,b). Estimations are that 2.4 million people (Rim et al., 1994) up to 17 million people (Hopkins, 1992) or even higher are affected, depending on hitherto unknown situations in many countries, mainly in Asia and Africa (Mas-Coma, 2004a).

An epidemiological classification of human infection situations has been proposed (Mas-Coma et al., 1999a), as follows:

1. Authochthonous, isolated, nonconstant cases: Humans acquire the infection in an area where they live and where animal fascioliasis is also present; these human cases appear sporadically, without any constancy.
2. Imported cases: Human cases are diagnosed in a zone lacking the parasite, even in animals, who were infected in an area where transmission occurs.
3. Hypoendemic: Prevalence is less than 1%, and the arithmetic mean intensity is less than 50 eggs per gram of feces (epg). High epg numbers occur only in sporadic cases. Human participation in transmission through egg shedding may be neglected. Hygiene-sanitation characteristics usually including latrines and waste or sewage disposal facilities, and outdoor defecation is not commonly practiced.
4. Mesoendemic: prevalence is between 1% and 10%, and 5- to 15-year-old children may present higher prevalences (holoendemic). The arithmetic mean intensity in human communities is usually between 50 and 300 epg. Individual high epg numbers can be found, although intensities over 1000 epg are rare. Human subjects may participate in transmission through egg shedding. Hygiene-sanitation characteristics may or may not include latrines and waste or sewage disposal facilities, and outdoor defecation may be practiced.
5. Hyperendemic: Prevalence is more than 10%, and 5- to 15-year-old children usually present higher prevalences (holoendemic). The arithmetic mean intensity

in human communities is usually more than 300 epg. Very high individual epg numbers are encountered, with intensities over 1000 epg being relatively frequent. Human subjects significantly participate in transmission through egg shedding. Hygiene-sanitation characteristics do not include the use of latrines. There are no proper waste or sewage disposal facilities. Indiscriminate defecation is commonly practiced.

6. Epidemics in nonhuman endemic areas but in animal endemic areas: Outbreaks occur in zones where previous human reports have been isolated and are sporadic. Such outbreaks usually involve very few subjects infected from the same contamination source (family or small group reports; contaminated wild, home-grown, or commercially grown watercress or other metacercariae-carrying vegetables).
7. Epidemics in human endemic areas: Outbreaks appear in zones presenting human endemics. A larger number of subjects may be concerned, usually related to previous climatic conditions having favored both the parasite and the snail life cycles. Epidemics can take place in hypoendemic, mesoendemic, and hyperendemic areas.

Human Endemic Areas

Andean countries (Bolivia, Peru, Chile, Ecuador), the Caribbean area (Cuba), northern Africa (Egypt), western Europe (Portugal, France, and Spain), and the Caspian area (Iran and neighboring countries) present the major health problems, including human endemic areas (Mas-Coma, 2004a). It is worth mentioning that high prevalences in humans do not seem to be necessarily related to high prevalences in livestock. Prevalences detected in some communities of the northern Bolivian Altiplano were of up to 72% and 100% in coprological and serological surveys, respectively (Hillyer et al., 1992; Bjorland et al., 1995; Mas-Coma et al., 1995, 1999c; Esteban et al., 1997a,b, 1999; O'Neill et al., 1998), and intensities reached up to more than 5000 epg in children (Esteban et al., 1997a,b, 1999). Although more prevalent and intense in children (with a peak in the 9- to 11-year age group), adult subjects were also infected, with prevalences reaching 40% and arithmetic mean intensities up to 752 epg (Esteban et al., 1997a,b, 1999). Prevalence and intensity situations found in Peru and Egypt were only somewhat lower (Esteban et al., 2002, 2003). Concerning gender, in human hyperendemic areas, prevalences and intensities appeared to be higher in females (Esteban et al., 1999, 2002, 2003).

In South America, well-known human hyperendemic areas are the Northern Altiplano in Bolivia (Mas-Coma et al., 1995, 1999c), the Puno Altiplano, the Cajamarca valley, and the Mantaro valley in Peru (Knobloch et al., 1985; Esteban et al., 2002; Marcos Raymundo et al., 2004). In Cuba, more than 10,000 people were infected in 1947–1948 (Mitterpak, 1968), an outbreak involved more than 1000 subjects in 1983 (Gonzalez et al., 1985, 1987; Diaz et al., 1990), a new outbreak involving 81 subjects occurred in 1995 (Perez et al., 1997), and patients were continuously diagnosed (Millan et al., 2000).

France is considered an important human endemic area (Anonymous, 1988). The first large modern epidemic of human fascioliasis occurred in 1956 (Coudert and Triozon, 1958). Between 1950 and 1983, 3297 cases from published reports have been cataloged (Gaillet et al., 1983). Most cases were reported from the areas of Lyon, Bretagne Nord–Pas de Calais, and Sud-Ouest. Recent reports on Sud-Ouest France refer to more than 300 cases (Giap, 1987; Ripert et al., 1987). The paper by Danis et al. (1985), which reported on 5863 human cases recorded from nine hospitals between 1970 and 1982, demonstrated that published data largely underestimate the true situation. The French Mediterranean island of Corsica maintains a low hypoendemia (Gitard et al., 1965; Gil-Benito et al., 1991). The disease is also important in Portugal, with the northern part of the country as a marked endemic area. Sampaio Silva et al. (1996) referred to 1011 cases diagnosed in Porto between 1970 and 1992. In Spain, human fascioliasis appears to be underestimated and mainly distributed in the northern part (Sorribes et al., 1990). Moreover, imported cases were recently added to authochthonous ones (Turrientes et al., 2004).

Concerning Africa, numerous human cases have been detected in many governorates of Egypt (Curtale et al., 2000, 2003a,b; Haseeb et al., 2002; Esteban et al., 2003). Initial estimates of 830,000 subjects affected in the Nile Delta region (WHO, 1995b) probably underestimate the true situation if the high prevalences reaching 18% to 19% of the total population in villages (Esteban et al., 2003) are taken into account.

In the Middle East, Iran is worth mentioning. Human cases are mainly concentrated in Gilan, at the Caspian Sea, where several large epidemics, including thousands of human cases, were reported from the end of the 1980s and during the 1990s (Massoud, 1990, 1993; Ashrafi et al., 2004). In Mazandaran, fascioliasis has recently been shown to be a large human health problem (Moghaddam et al., 2004). The recent detection of a 1.8% human prevalence in a village in Eastern Turkey (Yilmaz and Gödekmerdan, 2004) suggests that the endemic area around the Caspian Sea may be widespread.

In eastern Asia, cases in Japan and Korea are sporadic, but recent information on Vietnam is troublesome (Mas-Coma, 2004b). Only occasional cases of human fascioliasis were reported in Vietnam until the 1990s, but over 500 human cases have been diagnosed between 1997 and 2000 (De et al., 2003).

Epidemiological Patterns

The life cycle of fasciolids is dependent on environmental characteristcs. Despite these restrictions, fascioliasis has become the vector-borne disease presenting the widest latitudinal, longitudinal, and altitudinal distribution known (Mas-Coma, 2004a). Fascioliasis is unique in being capable of giving rise to human endemic areas from below sea level (as besides the Caspian Sea) up to the very high altitude (as in Bolivia, Peru, Ecuador, and Venezuela) (Mas-Coma et al., 2003).

Fascioliasis in human hypo- to hyperendemic areas presents a very wide spectrum of transmission and epidemiological patterns related to the large diversity of environments, including different human endemic/epidemic situations; different human demographics, races, diets, habits, traditions, and religions; different domestic and wild mammal reservoir species; different lymnaeid transmitting species, zones in both the Northern and Southern hemispheres, altitudes from −27 m up to 4200 m, hot and cold weather, seasonal and yearly constant temperatures, scarce to pronounced annual rainfall, low and high mean annual potential evapotranspiration; and from the lack of a dry period to the lack of a wet period through different dryness/humidity rates. From the landscape point of view, these areas range from altiplanos to valleys, from islands to mainlands, from natural to artificial irrigations, from lakes to lagoons, from large rivers to small streams, and from permanent to temporal water bodies (Mas-Coma et al., 2003).

From results obtained in the human endemic areas of Europe, South America, Africa, and Asia, it may be concluded that known epidemiological patterns of fascioliasis may not always explain the transmission characteristics in a given area. This means that when dealing with an endemic zone not previously studied, known patterns must always be taken into account merely as the starting base, before control measures can be considered. A classification of epidemiological patterns has recently been proposed to offer a baseline for future research (Mas-Coma, 2005):

Class A: A very high altitude pattern related to only *F. hepatica* transmitted by imported *G. truncatula* in Andean countries following transmission throughout the year; within this category, two subpatterns may be distinguished according to physiographic and seasonal characteristics:
  a: The altiplanic pattern, with transmission throughout the whole year, e.g., in the Northern Bolivian Altiplano and the Puno Altiplano
  b: The valley pattern, with seasonality and prevalences and intensities related to altitude, e.g., in the valleys of Cajamarca and Mantaro

Class B: A Caribbean insular pattern, with reduced but repeated outbreaks in human hypoendemic areas and lymnaeid species other than the main vector species being involved in the transmission, e.g., the Pinar del Rio Province in Cuba

Class C: A pattern related to Afro-Mediterranean lowlands, including overlapping *F. hepatica* and *F. gigantica* and several *Galba-Fossaria* and *Radix* lymnaeids together with secondary transmitting *Pseudosuccinea*, and where seasonality is typical, e.g., the Behera Governorate in the Nile Delta region in Egypt

Class D: A pattern related to Caspian surrounding areas, including human hypoendemic areas in which large epidemics occur, occasionally involving up to 10,000 people and with overlapping of *F. hepatica* and *F. gigantica* and several *Galba-Fossaria, Radix* and stagnicoline lymnaeids, e.g., the area of Rasht and Bandar-e Anzali in the Gilan province in Iran

The parasite distribution appears irregular within a human endemic area. The transmission foci have a patchy distribution and are linked to the presence of appropriate water collections, and human prevalences in schoolchildren appear to be related to the distance to water bodies containing lymnaeids (Mas-Coma et al., 1999c).

Human Infection Sources

Human contamination takes place by ingestion of infective metacercariae. Metacercarial infectivity is dependent on storage time, being lower when metacercariae are older: the maximum longevity was 31 and 48 weeks using doses of 20 and 150 metacercariae per rat, respectively, although in the latter case only a very low percentage was viable. Moreover, metacercarial viability and infectivity did not show differences between isolates from different reservoir species, demonstrating that flukes from secondary reservoirs such as pigs and donkeys involve the same potential risk as those from the main ones, that is, sheep and cattle (Valero and Mas-Coma, 2000).

Although it has been assumed that human contamination takes place through ingestion of metacercariae attached to freshwater vegetables (mainly watercress), up to nine human contamination sources may be identified (Mas-Coma, 2004a): ingestion of (1) freshwater wild plants, (2) freshwater cultivated plants, (3) terrestrial wild plants, (4) terrestrial cultivated plants, (5) food dishes and soups made with contaminated water, and (6) infected raw liver; and drinking of (7) beverages made from local plants and (8) contaminated water, and (9) washing of kitchen utensils or other objects with contaminated water. All of them may be related to human endemic areas. Traditional local beverages and foods appear to be quite involved, such as alfalfa juice in Peru (Marcos et al., 2006) or zeitoon-parvardeh appetizer and delar paste in Iran (Ashrafi et al., 2006b).

Climate and Environment

In fascioliasis transmission, climatic factors are crucial. The yearly definitive host infection incidence of fascioliasis has been related to air temperature, rainfall, and potential evapo-transpiration. Several climatic forecast indices have been successfully applied to animal fascioliasis but only recently to human infection (Fuentes et al., 1999). When applied to the northern Bolivian Altiplano, analyses showed that the very high altitude climatic characteristics of this region markedly differ from those of European fascioliasis endemic lowland areas. In the Altiplano, the temperature has no marked seasonal character; there are large variations in temperature within a daily 24-hour period; the rainfall distribution is seasonal, with a long dry season coinciding with the lowest minimum temperatures and a long wet season in which rainfall is concentrated; the evapo-transpiration is very high, temporary water bodies are of short duration, mainly in the arid period, and the solar radiation is high not only because of altitude but also because of the lack of trees and shrubs. Following needed modifications introduced in climate diagrams to fit the conditions of this endemic area, the Mt index and the water budget–based system index (Wb-bs) were also modified for high altitude and low latitude (Fuentes et al., 1999). Values of both modified indices showed optimum transmission during the December to March period. Moreover, the modified Wb-bs index allowed for the classification of transmission into low-, moderate-, and high-risk areas.

Detectable environmental factors useful for fascioliasis evaluation can be observed from space-borne platforms. Remote Sensing (RS) and Geographic Information Systems (GIS) have already been used for animal fascioliasis (Malone and Yilma, 1999), but never for human endemic areas. In GIS for animal fascioliasis, surface hydrology, vegetation indices, and temperature have proved to be very useful. The first attempt to apply these technologies to a human fascioliasis endemic area was that of Fuentes and Malone (1999) in Chile. Annual normalized difference vegetation index (NDVI) values were calculated for each region, using specialized computer software to extract values from satellite images. Based on different NDVI classes, a map of risk of fascioliasis transmission was made for each of the administrative zones, with the differentiation of four risk levels (zero, low, moderate, and high). Studies were undertaken to analyze whether a GIS predicting model would be viable and useful in the northern Bolivian Altiplano human endemic zone (Mas-Coma et al., 1999c). The prediction capacity of the remote sensing map based on NDVI data (Fuentes et al., 2001) appeared to be higher than that from forecast indices based only on climatic data (Fuentes et al., 1999). An overlap between real ranges of human fascioliasis prevalence and predicted ones (transmission risk through NDVI) is worth mentioning. The NDVI data maps represent a further step on the way to a GIS based on various parameters that could accurately fit real epidemiological and transmission situations of human fascioliasis in high-altitude endemic areas in Andean countries (Fuentes, 2004). A GIS forecast model to conduct an epidemiological analysis of human and animal fasciolosis in the central part of the Andes has recently been proposed (Fuentes et al., 2005). This model, through the classification of transmission into low-, moderate-, and high-risk areas, facilitates identifying those areas requiring the implementation of control activities.

Climate change and man-made modifications of the landscape may also influence the spread of fascioliasis. Outbreaks of human fascioliasis in the western coast countries of South America may be expected in relation to climatic anomalies associated with the El Niño–southern oscillation phenomenon (Githeko et al., 2001). The colonization by *F. hepatica* of man-made irrigation systems giving rise to human disease has been recently described. In the large Nile Delta irrigation area, human fascioliasis is unexpectedly emerging after many years of successful control measures against schistosomiasis (Curtale et al., 2000). In Peru, the Asillo zone is a man-made irrigation area built in the 1956–1974 period, to which both liver fluke and lymnaeid snails have quickly adapted. This refers to the high health risk of water resource constructions and puts a question mark on several irrigation projects proposed for the Peruvian and Bolivian Altiplanos (Esteban et al., 2002).

Clinics and Pathology

The following clinical periods can be distinguished: incubation phase (from the ingestion of metacercariae to the appearance of the first symptoms), invasive or acute phase (fluke migration to the bile ducts), latent phase (maturation of the parasites and the start of oviposition), and obstructive or chronic phase (Mas-Coma and Bargues, 1997; Mas-Coma et al., 1999b, 2000). In humans, the incubation period has not been accurately determined (only "a few" days, 6 weeks, 2 to 3 months, or even more—Chen and Mott, 1990; Bouree and Thiebault, 1993) and may vary considerably depending on the number of metacercariae ingested and the individual host response. In the acute phase, the symptomatology is due mainly to mechanical destruction of liver tissue and the abdominal peritoneum by the migrating larvae causing localized or generalized toxic and allergic reactions lasting 2 to 4 months (Mas-Coma and Bargues, 1997). The latent phase can last for months or years, and the proportion of asymptomatic individuals in this phase is unknown, being often discovered during family screening after a patient is diagnosed (Arjona et al., 1995); these persons may have prominent eosinophilia suggestive of infection (Gil-Benito et al., 1991), gastrointestinal complaints, or one or more relapses of the acute symptoms (Chen and Mott, 1990). A chronic or obstructive phase may develop after months or years of infection. Adult flukes in the bile ducts cause inflammation, hyperplasia of the epithelium, and thickening and dilatation of the duct and gall bladder walls. The resulting cholangitis and cholecystitis, combined with the large body of the flukes, are sufficient to cause obstruction. This phase includes biliary colic, epigastric pain, fatty food intolerance, nausea, jaundice, pruritus, and right upper-quadrant abdominal tenderness, among others. Lithiasis of the bile duct or the gall bladder is frequent. The bile duct and the gall bladder may contain blood mixed with bile (hemobilia), blood clots, and fibrinous plugs (Chen and Mott, 1990; Mas-Coma and Bargues, 1997; Mas-Coma et al., 1999b, 2000).

Despite the decrease of the prevalence from children to adult subjects, the results demonstrate that in high endemic zones adult subjects either maintain the parasites acquired when young or can be newly infected as the consequence of inhabiting a zone of high infection risk (Esteban et al., 1999). It must be taken into account that the life span of the adult fluke in humans is between 9 and 13.5 years (Mas-Coma and Bargues, 1997). Such a picture suggests that, in those areas, the majority of adult subjects should be in the chronic phase, acute lesions by repetitive infections being superimposed on chronic disease with relative frequency. Thus, the acute phase may be prolonged and overlap with both the latent and the obstructive phase. Pathological studies on those subjects in the advanced chronicity stage present in human endemic areas are needed.

The laboratory rat model offers a useful approach for pathological studies in the advanced chronic period (Valero et al., 2000, 2003). In experimentally infected Wistar rats, the presence of gallstones increased with infection time, the lithogenic induction by infection becoming manifest in situations of the advanced chronicity stage more than 100 days postinfection. The relative risk of gallstone disease increased when the number of flukes per rat and the rat weight also increased. The presence of gallstones was strongly associated with the number of flukes located in the bile duct. The risk of pigment stones appears to depend mainly on factors that favor bile duct obstruction (cholangitis, fluke body development versus time, intensity of infection). Situations of undiagnosed cases, as in subjects presenting undistinguishable symptoms or in those retaining their infection for a long time because of nontreatment or of repetitive reinfections, usually in human endemic areas of developing countries, imply a higher lithiasis risk. A high gallstone risk may be expected in subjects inhabiting human hyperendemic areas where very high egg outputs detected in humans suggest that liver fluke burdens may also be very high (Valero et al., 2003).

The synergistic capacity of fasciolids in co-infection with other pathogenic agents is well known, immunological responses to pathogenic antigens being markedly suppressed and concomitant infection being exacerbated following fascioliasis infection (Brady et al., 1999). Interestingly, the parasitological spectrum of protozoan and helminthic species found in the inhabitants of the human fascioliasis endemic areas, the multiparasitisms, and the associations between liver fluke infection and infection by other pathogenic parasites, all appear to be similar in the different human endemic zones (Esteban et al., 1997a,b, 1999, 2002, 2003). Additionally, a recent experimental study with rats demonstrated an association between bacterobilia by *Escherichia coli* (45% of cases), *Enterococcus faecalis* (45%), and *Klebsiella pneumoniae* (10%), and the duration of fasciolid infection, intensity of fasciolid infection, and liver damage, and supported the fact that the obstruction caused by advanced chronic fasciolosis may be related to biliary sepsis. These results lead to a reconsideration of treatment features in human disease; therapeutic strategies should not only include a parasitic treatment but also consider the possibility of bacterial co-infection (Valero et al., 2006b). These synergistic associations of fascioliasis with other pathogens are believed to be responsible for the high morbidity and mortality rates of Aymara children inhabiting the northern Bolivian Altiplano (Mas-Coma et al., 1995).

Diagnosis

There are several types of techniques for human fascioliasis diagnosis, although some suggestive clinical presentation aspects may be useful. Direct parasitological techniques, indirect immunological tests, and other noninvasive diagnostic techniques are presently used for human fascioliasis diagnosis. Quantitative coprological analyses are important in epidemiological surveys, as well as in posttreatment (and future postvaccination) monitoring. Besides eggs in coprological analyses, adults and eggs also may be found elsewhere by means of other invasive techniques: obtaining duodenal fluid, duodenal and biliary aspirates; surgery (laparotomy, cholecystectomy, sphincterotomy); and histological examination of liver or other organ biopsy materials. Serological, intradermal, and stool antigen detection tests have been developed. Immunological techniques have the advantages of being applicable during all periods of the disease, but fundamentally during the invasive or acute phase, as well as to the other situations in which coprological techniques may present problems. At any rate, immunological techniques offer other types of problems related mainly to sensibility and specificity. Different serological tests have been used for human diagnosis. Almost all these techniques concern the detection of circulating antibodies and only a few are designed to detect circulating antigens and immune complexes. Several serological techniques have recently proved to be useful for monitoring posttreatment evolution. Noninvasive diagnostic techniques that can be used for human diagnosis are radiology, radioisotope scanning, ultrasound, computed tomography, and magnetic resonance (see reviews in Esteban et al., 1998, and Hillyer, 1999).

Human infection by *F. hepatica* and *F. gigantica* cannot be differentiated by clinical, pathological, coprological, or immunological methods. This is a problem in overlapping areas because this differential diagnosis is very important owing to the different pathological, transmission, and epidemiological characteristics of the two fasciolids. To distinguish between *F. hepatica* and *F. gigantica*, a simple and rapid PCR-RFLP assay, using the common restriction enzymes *Ava*II and *Dra*II, has recently been described. It is based on a 618-bp-long sequence of the 28S rRNA gene recently obtained from populations in South America, Europe, and Africa. This sequence showed no intraspecific variations within each species and a few nucleotide differences between both fasciolids. This assay provides unambigous results and may be useful for both individual subject diagnosis and epidemiological surveys of humans and animals in endemic regions of sympatry in Africa and Asia (Marcilla et al., 2002). A similar PCR-RFLP assay using restriction endonucleases *Hsp*92II and *Rca*I has been recently applied to differentiate between Chinese liver flukes (Huang et al., 2004).

Present efforts are concentrated in obtaining purified excretory/secretory antigens or recombinant molecules to improve serological tests, owing to the problems of the parasitological diagnosis because of the delay in its usefulness in the acute phase (coprological examination positive only after 3 to 4 months postinfection), intermittent egg output dynamics, very low or even absence of egg shedding in cases of only one or a few adult flukes and old chronic or ectopic infections, "false" fascioliasis related to eggs in transit after ingestion of infected liver from domestic animals, or flukes unable to attain maturity in human subjects in nonhuman endemic areas (Esteban et al., 1998; Mas-Coma et al., 1999b).

Cysteine proteinases are secreted by the adult and juvenile forms (Dalton et al., 2003; Law et al., 2003) and are highly antigenic in both animals (Cornelissen et al., 2001; Neyra et al., 2002) and humans (Cordova et al., 1997). Several cysteine proteinases offer highly sensitive and specific markers for human fascioliasis serodiagnosis for *F. hepatica* (Sampaio Silva et al., 1996; Cordova et al., 1997, 1999; O'Neill et al., 1998, 1999; Strauss et al., 1999; Rokni et al., 2002; Espinoza et al., 2007) as well as for *F. gigantica* infection (Ikeda, 1998; Intapan et al., 1998, 2004; Maleewong et al., 1999; Tantrawatpan et al., 2005). *Fasciola hepatica* recombinant cysteine proteinases produced in yeast (O'Neill et al., 1999) or in *E. coli* (Carnevale et al., 2001) have been used in enzyme-linked immunosorbent assay (ELISA) methods for human infection diagnosis, offering results similar to native antigens. Although not yet evaluated for diagnosis in humans, a recent ultrasensitive capture ELISA test useful for both serodiagnosis and coproantigen detection appears promising (Mezo et al., 2003, 2004).

Treatment

Many drugs have been used to treat fascioliasis in human patients. Dehydroemetine was considered the therapy of choice a few decades ago, but toxic manifestations let bithionol become the choice for years, despite its long treatment course. The lack of consensus ended when it was proved that appropriate doses of triclabendazole were highly efficient (Esteban et al., 1998; Mas-Coma et al., 1999b). Triclabendazole for human use (Egaten®) is at present the drug of choice for human fascioliasis caused by both *F. hepatica* and *F. gigantica* (Savioli et al., 1999).

This drug is better adsorbed if administered after meals (Lecaillon et al., 1998). The recommended dosage is two separate regimens of 10 mg/kg. A cure rate of 79.2% when first used and 100% after a second round of therapy was found by Apt et al. (1995) in Chile, and 79.4% and 93.9%, respectively, by El-Morshedy et al. (1999) in Egypt. Triclabendazole appears to keep its efficiency in human endemic areas after years (Talaie et al., 2004). However, Millan et al. (2000) found that 8% of Cuban patients, with *F. hepatica* infection refractory to previous chemotherapy with other antihelmintics, were still shedding eggs in feces on day 60 posttherapy with two doses of 10 mg/kg administered after food 12 hours apart, although all these failure cases responded to a third triclabendazole dose.

Although the possibility of reinfection after treatment because of living in an endemic area cannot be ignored, the risk of the appearance of resistance to triclabendazole cannot be forgotten, taking into account the veterinary use of triclabendazole (Fasinex®) for livestock treatment in endemic areas since long ago, the tradition of human self-treatments with Fasinex® owing to the general availability of this drug, and the appearance of triclabendazole resistance in animals in different countries (Overend and Bowen, 1995; Lane, 1998; O'Brien, 1998; Mitchell et al., 1998; Moll et al., 2000; Gaasenbeek et al., 2001; Vara-del-Rio et al., 2005). The colonization abilities related to domestic animal management and export/import seem to be at the base of the present expansion of triclabendazole resistance, a serious problem as this is the only drug currently available for human use.

Strategies to minimize resistance development in human endemic areas include selective treatment of only infected subjects detected in previous surveys (Curtale et al., 2005). The use of synergistic drug combinations has been proposed at the veterinary level (Fairweather and Boray, 1999), although this approach has the risk of building up multiple drug resistance (Gaasenbeek et al., 2001). It must be considered that there are no drug alternatives for human treatment at present; drugs such as bithionol and others are no longer commercially available (Millan et al., 2000). Nitazoxanide, recently marketed in Mexico and having been reported to be effective against fascioliasis in human patients in treatments of

7-day duration (Rossignol et al., 1998; Favennec et al., 2003), and myrrh (Mirazid®), registered in Egypt (Massoud et al., 2001), still require more studies on efficacy and tolerability. The high fasciolicidal activity recently detected in artemisinin derivatives is encouraging (Keiser, 2006).

Prevention and Control

Human infection was always related to animal endemics, so that prevention and control measures recommended were always the same to be applied for veterinary fascioliasis, at the levels of domestic animals and snails, and in the field (see reviews by Roberts and Suhardono, 1996; Spithill et al., 1999; Torgerson and Claxton, 1999). However, recent studies on human endemic areas have shown that traditional epidemiological patterns of fascioliasis may not always explain the transmission characteristics in a given area, and that control measures must consider the results of the ecoepidemiological studies undertaken in the zone concerned (Mas-Coma et al., 1999b).

The prevention of human fascioliasis may be achieved by strict control of the contamination sources in each place. Unfortunately, potassium permanganate, which had been suggested to be the most effective preventive tool for killing metacercariae attached to leaves and vegetables used in salads, has been shown to have no effectivity on metacercarial viability, even at the very high doses of 300, 600, and 1200 mg/L (Ashrafi et al., 2006).

Contamination risks should not be restricted only to ingestion of freshwater vegetables, as already mentioned. In many human hyperendemic areas of the Americas, people do not have a history of eating watercress or other freshwater plants (Hillyer and Apt, 1997; Esteban et al., 2002). In the Nile Delta region, people living in houses where piped water is present showed a higher infection risk (Curtale et al., 2003a). In the Egyptian locality of Tiba, where a 18.0% prevalence was initially found, human infection has markedly decreased after the construction and utilization of the so-called washing units, in which the water is appropriately filtered (Mas-Coma, 2004a).

Moreover, one should not think that the possibility of human contamination is restricted to rural areas. Thanks to the transport of vegetables (both aquatic and terrestrial) from rural endemic zones to cities, plants carrying metacercariae can be sold in noncontrolled city markets, giving rise to urban infection, as already detected in many countries (Mas-Coma, 2004a).

Fasciolopsiasis

Etiology and Life Cycle

The giant Asian intestinal fluke *Fasciolopsis buski* is one of the largest digeneans (body: 2–10/0.8–3 cm; eggs: 130–140/80–85 μm) infecting humans (Kumar, 1980). It inhabits the duodenum and jejunum and can also be found in much of the intestinal tract, including the stomach, in moderate and heavy infections (Graczyk et al., 2001).

The pig is the only important reservoir host, although it harbors few flukes (usually only 3 to 12) (Rim, 1982), and produces less eggs per adult than in humans. There is still disagreement as to whether there is only one or several strains of *F. buski*, more or less adapted to either humans or animals. Infections vary, with places in which the parasite is common in humans but absent in pigs; places with relatively important prevalences in pigs but very rare infection in humans; and other areas where infection rates among humans and pigs are almost equal. Although traditions, diet, and customs have been mentioned to explain these differences (Rim, 1982), studies refer to morphological and microtopographical strain variations in different geographical areas (Roy and Tandon, 1993). Mice, rats, monkeys, and dogs were experimentally refractory, and guinea-pigs only partially susceptible, but rabbits were susceptible (Malviya, 1985), as were squirrel monkeys (Kuntz and Lo, 1967). This fluke has been rarely reported in rhesus monkeys (Hartman, 1961), and there are no reports incriminating monkeys in endemic areas.

Contamination of the definitive host is by ingestion of plants or water carrying metacercariae. Metacercariae excyst in the duodenum and attach to the intestinal wall to grow to mature flukes in 3 months. The adult produces a large number of eggs in humans (13,000–26,000, mean 16,000 worm/day) (Hsieh, 1960). Eggs are unembryonated when shed in the feces and must reach freshwater to continue the cycle. The miracidium development period is 16 to 77 days, with a mean of 22 days, the optimum being 27° to 30°C water temperature and a 6.5 to 7.2 pH range (Barlow, 1925). Sporocysts, mother and daughter rediae and cercariae develop in the snail. Cercariae emerge from the rediae after 25 to 30 days but are released from the snail after an average of 49 days; cercariae do require a maturation period in snail tissues (Barlow, 1925). Other incubation periods recorded are 46 to 59 days at 22° to 24°C, 85 to 100 days at 18° to 22°C (Lo, 1967), and 31 days after exposure to the miracidia (Buckley, 1939). A short prepatent period of only 21 days has been reported in *S. trochoideus* and *H. umbilicalis*, in which *F. buski* causes 100% snail mortality because of mechanical damage to the ovotestis (Graczyk et al., 2000).

Cercarial emergence is dependent on light, with great variation in daily emergence patterns (Preet and Prakash, 2001). Cercariae swim in water until encystment occurs on a substratum, mainly aquatic plants and debris. Cercarial survival in water vary from 64 to 72 days (Gilman et al., 1982; Nguyen Van Tho, 2002b). Important to transmission is the consumption of various water plants, among which are water caltrop (*Trapa natans* in China, *T. bispinosa* in Taiwan, and *T. bicornis* in Bangladesh and Thailand), water chestnut (*Eliocharis tuberosa*), water hyacinth (*Eichhornia* sp.), water bamboo (*Zizania* sp.), water lotus (*Nymphaea lotus*), water lily (*Nymphaea* sp.), watercress, gankola (*Otelia* sp.), and water morning glory (*Ipomoea aquatica*). Up to 200 metacercarial cysts may be found on the surface of one water caltrop, but the usual number is about 15 to 20. Cercariae may also encyst on the water surface. The number of floating metacercariae is 3.6% of the total.

It was found that 10.3% to 12.8% of the patients and 35.1% to 40% of the infested pigs were possibly infected by drinking water (Weng et al., 1989).

Small planorbids of the genera *Segmentina, Hippeutis*, and *Gyraulus* act as snail hosts, that is, *S. hemisphaerula* (syn. *S. coenosus, S. nitidella, S. calathus, S. largeillierti*), *H. cantori* (syn. *H. smackeri*), *H. umbilicalis*, and *G. convexiusculus* in China, Vietnam, and Taiwan (Hsu, 1964; Wang et al., 1977; Nguen Tkhi Le, 1978). In India (Assam), the snail hosts are *S. trochoideus, S. hemisphaerula*, and probably also *H. umbilicalis* (Tripathi et al., 1973). *S. hemisphaerula, S. trochoideus*, and *Gyraulus chinensis* are snail hosts in Thailand (Manning and Ratanarat, 1970; Nguen Tkhi Le, 1978), *S. hemisphaerula* in Taiwan (Hsieh, 1960), *H. umbilicalis* in Laos (Ditrich et al., 1992), *S. trochoideus* and *H. umbilicalis* in Bangladesh (Gilman et al., 1982; Graczyk et al., 2000), and *Indoplanorbis exustus* in India (Preet and Prakash, 2001).

Geographical Distribution

Fasciolopsiasis is known in many Asian countries (WHO, 1995b). In China, infections have been reported from 10 provinces in both humans and pigs, with prevalences up to 85% in Chekiang and Kiangsi provinces. In other areas of China prevalence varies from less than 1% to 5% (Weng et al., 1989; Gai et al., 1995; Lo and Lee, 1996). Recent nationwide surveys suggest a decrease in fasciolopsiasis (Yu et al., 1994; Xu et al., 1995). In Taiwan, the most higly endemic area is in southern Tainan Hsier (Lee, 1986; Lee et al., 1989; Lo and Lee, 1996), where infection rates reached 48% (Hsieh, 1960) and a high of 61% in Pa-Weng village (Lee, 1972). A prevalence of 25% was found in schoolchildren (Shyu et al., 1984). In Liuying area, the intensity per person was estimated to be about 10 worms (Hsieh, 1960).

In India, 60% of the people were found infected with 1 to 57 adult worms per person in Assam (Buckley, 1939). Infections in pigs and humans have been reported in Calcuta, and Bombay, where Shah et al. (1966) found a prevalence of 29% in the city with *F. buski* absent in pigs, as in Assam. New human foci were detected in Maharastra (Manjarumkar and Shah, 1972). In Bangladesh, surveys showed 39.2% and 8.6% infection in two villages (Muttalib and Islam, 1975). Prevalences in schoolchildren reached 50% in an endemic focus (Gilman et al., 1982).

In Vietnam, early studies reported infection in both Asians and Europeans (Fournier, 1954; Harinasuta et al., 1984), and both humans and pigs have been found positive (Yoshihara et al., 1999). In Thailand, the main endemic area is the central part of the country, with an estimate of 100,000 persons infected among 500,000 (Manning and Ratanarat, 1970; Manning et al., 1971). A total of 13% of 1,500 people from three provinces were infected (Sadun and Maiphoom, 1953). In areas such as Pak Hai, 100% of the indigenous population is likely to be infected. A lower prevalence of 10% was found by Bunnag et al. (1983). A prevalence of 7.1% has been reported in northeastern Thailand (Wiwanitkit et al., 2002).

Human reports are also known from Laos (Waikagul, 1991; Giboda et al., 1991). In Cambodia, prevalences of 0.04% in humans and 5% in pigs have been reported in Phnom-Penh (Waikagul, 1991). It has also been reported from Borneo and Sumatra, Indonesia (Rim, 1982; Hadidjaja et al., 1982; Margono, 2003), where a 27.0% prevalence was reported in residents of Kalimantan, and 20.3% outside that area, including the highest prevalence in the 5- to 14-year-old age group (56.8%) and decreasing with age, and a male/female ratio of 1.4:1 (Handoyo et al., 1986). The parasite has also been detected in Kampuchea, the Philippines, Singapore (Waikagul, 1991), Burma (Rim, 1982), and Malaysia (Shekhar, 1991).

Reports of infections in Korean and Japanese subjects indicate that the parasite is contracted outside of Korea and Japan (Rim, 1982). Reports in the U.S., Venezuela, Australia, Guatemala, Israel, and Cuba may be associated with immigrants from the Far East (Basnuevo and Seuc-Chiu, 1950; Greenberg et al., 1994) or to misidentification of fecal eggs (Schubert and Granz, 1981).

Epidemiology

The disease is underreported and most prevalent in remote rural places and semi-urban areas (Lee, 1972; Shah et al., 1973; Idris et al., 1980; Rahman et al., 1981; Gilman et al., 1982; Bunnag et al., 1983; Harinasuta et al., 1984; Gai et al., 1995), mainly in schoolchildren in which the number of worms per child can exceed 800 (Gilman et al., 1982; Weng et al., 1989). A higher prevalence among females (16%) than males (11%) was reported in coprological surveys in Thailand (Sadun and Maiphoom, 1953). The prevalence varied from 8% among adults to 15% in 5- to 14-year-old children. The severity of infection increased in the 10- to 14-year age group and decreased in older groups. Similar gender and age relationships were found in Thailand (Manning and Ratanarat, 1970), Bangladesh (Rahman et al., 1981), and Taiwan (Hsieh, 1960). Prevalences detected in children were 57% in mainland China (Lee, 1972; Weng et al., 1989), 25% in Taiwan (Shyu et al., 1984), 60% in India (Muttalib and Islam, 1975), 50% in Bangladesh (Rahman et al., 1981), and 10% in Thailand (Bunnag et al., 1983).

This disease occurs focally, is widespread, and is linked to freshwater habitats with stagnant or slow moving waters; it is associated with common social and agricultural practices and promiscuous defecation (Kuntz and Lo, 1967; Gilman et al., 1982; Cross, 1984; Weng et al., 1989). In pigs, it is seasonal, with a peak in June–September declining thereafter to a low level during winter and early spring (November–March), and it is absent during January and February, at least in northeastern India (Roy and Tandon, 1992).

Contamination is by ingestion of metacercarial cysts through consumption of raw or undercooked aquatic plants, drinking or using raw water, and handling or processing water-derived plants (Kuntz and Lo, 1967; Gilman et al., 1982; Weng et al., 1989). Human infection occurs when the hull or skin of infected plants is peeled off between the teeth. Cultivation of aquatic, metacercariae-carrying plants for consumption on a large scale and the pollution of the areas in which they are grown with human and animal (mainly pig) feces are important spreading factors (Kuntz and Lo, 1967; Manning and Ratanarat, 1970; Cross, 1984). Prevalences among people living near water caltrop plantations was much higher than in villages distant from plantations (Sadun and Maiphoom, 1953; Hsieh, 1960).

Fasciolopsiasis is aggravated by social and economic factors, such as poverty, malnutrition, and a free-food market associated with lack of food inspection, poor sanitation, other helminthiasis, and declining economic conditions (Muttalib and Islam, 1975; Gilman et al., 1982; Cross, 1984; Weng et al., 1989; Yu et al., 1994; WHO, 1995a). Differences of incidence in the same area are due to factors such as economic status, educational background, standard of health, and mode of living (Jaroonvesama et al., 1980).

The main reservoir is the pig (Kuntz and Lo, 1967; Weng et al., 1989; Roy and Tandon, 1992). Fresh aquatic green fodder and raw water used to raise pigs are the main infection sources for farm animals (Weng et al., 1989; D'Souza et al., 2001). Different pig infection rates have been reported: 30% in India (Tripathi et al., 1973); 10% in China (Hsu, 1964); 52% in Taiwan (Hsieh, 1960).

Clinics and Pathology

Morbidity in endemic areas is high, and the disease can be fatal, depending on the worm burden (Lee, 1972; Idris et al., 1980; Gilman et al., 1982; Bunnag et al., 1983; Lee et al., 1986; Weng et al., 1989). Human morbidity and mortality are difficult to assess, because quantitative data exclusive for *F. buski* infections are not available (Graczyk et al., 2001). Fasciolopsiasis is considered a main factor for a persistent poor nutritional status in children (Lee, 1972; Shah et al., 1973; Gilman et al., 1982).

Flukes cause extensive intestinal and duodenal erosions, ulceration, hemorrahage, abscesses, and catarrhal inflammation. Absorption of toxic and allergic worm metabolites cause ascites and both general and facial edema, for example, cheek and orbital edema (Jaroonvesama et al., 1986). Pathological changes may be traumatic, obstructive, and toxic, especially in heavy infections, in which worms disturb the secretion of intestinal juices, cause excess mucus secretion, and obstruct the passage of food. Feces are profuse, yellow-brown, and contain pieces of undigested food due to malabsorption (Jaroonvesama et al., 1986).

Light infections are usually asymptomatic except for diarrhea, alternating with periods of constipation and abdominal pain. They may include anemia with eosinophilia, headache, dizziness, stomach ache, gastric pain, loose stools (Gilman et al., 1982), asthenia, pallor, malnutrition, protuberant abdomen, and abdominal distention (Chandra, 1976). Moderate and heavy infections are associated with malnutrition, severe epigastric and abdominal pain, diarrhea or bowel obstruction, poor appetite, mild abdominal colic, vomiting, fever, nausea (occurring especially in the morning and resolving after the first meal), acute ileus, anasarca, marked eosinophilia and leukocytosis, and a significant lowering in serum vitamin

B_{12} content (Rahman et al., 1981; Gilman et al., 1982; Handoyo et al., 1986; Jaroonvesama et al., 1986). Generalized toxic and allergic symptoms, usually in the form of edema, particularly of the face, abdominal wall, and lower extremities, may appear in heavy infections. Generalized abdominal pain and ascites are common, in addition to poor appetite, bitemporal headache, giddiness, a low-grade fever, nonpalpable liver and spleen, nausea, and vomiting (Viranuvatti et al., 1953).

The disease becomes evident only in massive infections. Patients purged of the worms usually recover completely, although advanced, heavy infections can be fatal. Mortality has been reported in heavily infected children owing to profound intoxication (Sadun and Maiphoom, 1953; Viranuvatti et al., 1953; Yu and Mott, 1994). However, little, if any, evidence has been found to suggest that this parasite is harmful to humans, if present in less than massive numbers (Plaut et al., 1969; Jaroonvesama et al., 1986).

Diagnosis and Treatment

Diagnosis is by coprology by examining the eggs, or occasionally by examination of expelled adult worms vomited or passed in feces (Gilman et al., 1982; Rim, 1982; Weng et al., 1989; Le et al., 2004). The clinical picture is usually not distinctive, although highly suggestive in endemic areas. Obtaining the 18S rDNA sequence from a fluke vomited by a Vietnamese child has recently been used to verify the diagnosis, and only two nucleotide substitutions were found in comparison with the sequences available in the GenBank (Le et al., 2004).

Tetrachloroethylene proved to be effective (Shah et al., 1966), and was suggested as the drug of choice for mass treatment some time ago (Plaut et al., 1969; Chandra, 1976). Niclosamide was less effective (Suntharasamai et al., 1974; Idris et al., 1980). Thiabendazole, mebendazole, levamisole, and pyrantel pamoate were ineffective (Rabbani et al., 1985). Praziquantel proved its efficacy even in severe fasciolopsiasis, so that a single dose of 15 mg/kg was proposed as the treatment of choice (Bunnag et al., 1983; Harinasuta et al., 1984; Handoyo et al., 1986; Lee, 1986; Taraschewski et al., 1986), although this drug could not save the life of a heavily infected 20-year-old girl (Gupta et al., 1999). Triclabendazole, oxyclozanide, and rafoxanide were recently evaluated in pigs. Respective efficacies were 97.1%, 93.3%, and 83.2%, with no side effects (Datta et al., 2004).

Prevention and Control

Control can be achieved by pharmacological treatment of people and pigs, preventing reinfection, and modern pig farming (Kuntz and Lo, 1967; Muttalib and Islam, 1975; Cross, 1984; Gai et al., 1995). Individual prevention is simply by avoiding the consumption of raw, water-derived food. However it is extremely difficult to enforce this in view of century-old traditions. Such measures demand fundamental changes in the eating habits, customs, and economic conditions of the people (Rim, 1982; Graczyk et al., 2001). Infections follow a familial trend, as food preparation and eating habits are passed from one generation to the next (Gai et al., 1995). In addition, water plants are a common food source because they are cheap and readily available (Gilman et al., 1982; Weng et al., 1989; Gai et al., 1995).

Prevention of pollution of the ponds where aquatic plants are cultivated becomes crucial. The use of human feces as fertilizers, promiscuous defecation, and washing of pig feces into neighboring water bodies should be avoided. Plants used to feed pigs must always be checked for metacercariae (Nguyen Van Tho, 2002a). Dried aquatic plants are not dangerous because desiccation and direct solar radiation kills metacercariae (Lo, 1967; Rim, 1982; Weng et al., 1989). Prevention might also be accomplished by immersing plants and fruits in boiling water for a few minutes (Weng et al., 1989). Moreover, metacercariae are killed in 1.0% HCl in 18 days, 2% acetic acid in 9 days, 3.0% acetic acid in 6 days, 5% salt solution in 3 hours, by soybean sauce in 30 minutes, and 10% cane sugar in 3 days (Komiya, 1964).

Educational programs should stress the importance of thoroughly cooking aquatic plants and boiling water where treated water is not available (Cross, 1984; Weng et al., 1989; Gai et al., 1995). In the past, fasciolopsiasis was highly endemic in Taiwan, but the number of human cases decreased in the 1980s, thanks to aggressive educational programs (Lee, 1986; Lee et al., 1989). Unfortunately, this tendency was not maintained, because many cultures still enjoy eating raw food (Lee et al., 1989; Weng et al., 1989).

Fasciolopsiasis can be controlled along with other food-borne parasitoses (Mott et al., 1995; WHO, 1995a,b). However, despite control programs, fasciolopsiasis still remains a public health problem in endemic areas (WHO, 1995a,b). Where it was thought to be controlled, as in Uttar Pradesh, and where no cases were detected since the 1990s, there are now reports of recent reemerging infections (Bhatti et al., 2000; Muralidhar et al., 2000).

Acknowledgements. Fascioliasis research studies whose results are summarized here were supported by funding from: STD Programme of the Commission of the European Community (DG XII) (Contract No. TS3-CT94–0294), Brussels, EU; Programme of Scientific Cooperation with Latin America, Instituto de Cooperación Iberoamericana, Agencia Española de Cooperación Internacional, Madrid, Spain; Project PDP B2/181/125, WHO, Geneva, Switzerland; project Nos. PB87–0623, UE96–0001, PB96–0401-C02–02, PM97–0099, BOS2002–01978 and SAF2006–09278, Spanish Ministry of Education and Science, Madrid; the Red de Investigación de Centros de Enfermedades Tropicales—RICET (project Nos. C03/04 and PI050574) and project No. PI030545, FIS, Spanish Ministry of Health, Madrid; French-Spanish Acciones Integradas 68/240 (Area 4), 91/89 and HF-121/90; financial aid to the Valencia-Paris VI Interuniversity Agreement; project No. 3006/1999 of the Dirección General de Cooperación para el Desarrollo, Generalitat Valenciana, Valencia, Spain; and by Patronat Sud-Nord of the Universidad de Valencia.

The numerous scientists from Valencia (Spain), Banyuls-sur-Mer, Perpignan, Montpellier, Marseille, and Corsica (France), La Paz (Bolivia), Lima, Puno, and Cajamarca (Peru), Mérida (Venezuela), Baton Rouge (Louisiana, U.S.), Roma (Italy), Cairo and Damanhour (Egypt), Tehran, Rasht, and Bandar-e Anzali (Iran), and Quito (Ecuador) who participated in field or laboratory studies are gratefully acknowledged. Special thanks are given to Dr. L. Savioli, Dr. D. Engels, and Dr. A. Montresor of WHO (Geneva, Switzerland) for the logistic facilities provided.

References

Adlard, R.D., Barker, S.C., Blair, D., and Cribb, T.H. (1993). Comparison of the second internal transcribed spacer (ribosomal DNA) from populations and species of Fasciolidae (Digenea). Int J Parasitol 23, 422–425.

Anonymous. (1988). Parasitic diseases: hepatic distomiasis caused by *Fasciola hepatica.* Wkly Epidemiol Rec (WHO) 63, 109–111.

Apt, W., Aguilera, X., Vega, F., Miranda, C., Zulantay, I., Perez, C., Gabor, M., and Apt, P. (1995). Treatment of human chronic fascioliasis with triclabendazol: drug efficacy and serologic response. Am J Trop Med Hyg 52, 532–535.

Arjona, R., Riancho, J.A., Aguado, J.M., Salesa, R., and Gonzalez-Macias, J. (1995). Fascioliasis in developed countries: a review of classic and aberrant forms of the disease. Medicine (Baltimore) 74, 13–23.

Ashrafi, K., Massoud, J., Houlakuei, K., Mahmoodi, M., Jo-Afshani, M.A., Valero, M.A., Fuentes, M.V. Khoubbane, M., Bargues, M.D., and Mas-Coma, S. (2004). Evidence suggesting that *Fasciola gigantica* may be the most prevalent causal agent of fascioliasis in the endemic province of Gilan, northern Iran. Iranian J Public Health 33, 31–37.

Ashrafi, K., Valero, M.A., Panova, M., Periago, M.V., Massoud, J., and Mas-Coma, S. (2006a). Phenotypic analysis of adults of *Fasciola hepatica, Fasciola gigantica* and intermediate forms from the endemic region of Gilan, Iran. Parasitol Int. 55,249–260

Ashrafi, K., Valero, M.A., Massoud, J., Sobhani, A.R., Solaymani-mohammadi, S., Conde, P., Khoubbane, M., Bargues, M.D., and Mas-Coma, S. (2006b). Plant-borne human contamination by fascioliasis Am J Trop Med Hyg 75, 295–302.

Bargues, M.D., and Mas-Coma, S. (1997). Phylogenetic analysis of lymnaeid snails based on 18S rDNA sequences. Mol Biol Evol 14, 569–577.

Bargues, M.D., and Mas-Coma, S. (2005). Reviewing lymnaeid vectors of fascioliasis by ribosomal DNA sequence analyses. J Helminthol 79, 257–267.

Bargues, M.D., Mangold, A.J., Muñoz-Antoli, C., Pointier, J.P., and Mas-Coma, S. (1997). SSU rDNA characterization of lymnaeid snails transmitting human fascioliasis in South and Central America. J Parasitol 83, 1086–1092.

Bargues, M.D., Vigo, M., Horak, P., Dvorak, J., Patzner, R.A., Pointier, J.P., Jackiewicz, M., Meier-Brook, C., and Mas-Coma, S. (2001). European Lymnaeidae (Mollusca: Gastropoda), intermediate hosts of trematodiases, based on nuclear ribosomal DNA ITS-2 sequences. Inf Gen Evol 1, 85–107.

Bargues, M.D., Horak, P., Patzner, R.A., Pointier, J.P., Jackiewicz, M., Meier-Brook, C., and Mas-Coma, S. (2003). Insights into the relationships of Palaearctic and Nearctic lymnaeids (Mollusca: Gastropoda) by rDNA ITS-2 sequencing and phylogeny of stagnicoline intermediate host species of *Fasciola hepatica.* Parasite 10, 243–255.

Bargues, M.D., Artigas, P., Jackiewicz, M., Pointier, J.P., and Mas-Coma, S. (2006). Ribosomal DNA ITS-1 sequence analysis of European stagnicoline Lymnaeidae (Gastropoda). Heldia 6, 57–68.

Barker, S.C., Blair, D., Garret, A.R., and Cribb, T.H. (1993). Utility of the D1 domain of nuclear 28S rRNA for phylogenetic inference in the Digenea. Syst Parasitol 26, 181–188.

Barlow, C.H. (1925). The life cycle of the human intestinal fluke, *Fasciolopsis buski* (Lankester). Am J Hyg Monogr Ser 4, 1–98.

Basnuevo, J.G., and Seuc-Chiu, A. (1950). [*Fasciolopsis buski* in Cuba; two cases treated with chloroquine, gentian violet, hexylresorcinol and tetrachloroethylene]. Rev Cubana Med Trop Parasitol 6, 91–96.

Bhatti, H.S., Malla, N., Mahajan, R.C., and Sehgal, R. (2000). Fasciolopsiasis, a re-emerging infection in Azamgarh (Uttar Pradesh). Indian J Pathol Microbiol 43, 73–76.

Bjorland, J., Bryan, R.T., Strauss, W., Hillyer, G.V., and McAuley, J.B. (1995). An outbreak of acute fascioliasis among Aymara Indians in the Bolivian Altiplano. Clin Infect Dis 21, 1228–1233.

Blair, D. (1993). Molecular variation in fasciolids and *Paragonimus*. Acta Trop 53, 277–289.

Blair, D., and McManus, D.P. (1989). Restriction enzyme mapping of ribosomal DNA can distinguish between fasciolid (liver fluke) species. Mol Biochem Parasitol 36, 201–208.

Boray, J.C. (1982). Fascioliasis. In: Hillyer, G.V., Hopla, C.E., eds. Handbook Series in Zoonoses. Section C. Parasitic Zoonoses. Volume 3. Boca Raton, FL: CRC Press, 71–88.

Bouree, P., and Thiebult, M. (1993). Fasciolose à *Fasciola hepatica* en Basse Normandie de 1980 à 1990. Bull Soc Fr Parasitol 11, 79–84.

Brady, M.T., O'Nneill, S.M., Dalton, J.P., and Mills, K.H. (1999). *Fasciola hepatica* supresses a protective Th1 response against *Bordetella pertussis*. Infect Immun 67, 5372–5378.

Buckley, J.J.C. (1939). Observations on *Gastrodiscoides hominis* and *Fasciolopsis buski* in Assam. J Helminthol 17, 1–12.

Bunnag, D., Radomyos, P., and Harinasuta, T. (1983). Field trial on the treatment of fasciolopsiasis with praziquantel. Southeast Asian J Trop Med Public Health 14, 216–219.

Carnevale, S., Rodriguez, M.I., Guarnera, E.A., Carmona, C., Tanos, T., and Angel, S.O. (2001). Immunodiagnosis of fasciolosis using recombinant procathepsin L cysteine proteinase. Diagn Microbiol Infect Dis 41, 43–49.

Chandra, S.S. (1976). A field study on the clinical aspects of *Fasciolopsis buski* infections in Uttar Pradesh. Med J Armed Forces India 32, 181–189.

Chen, M.G., and Mott, K.E. (1990). Progress in assessment of morbidity due to Fasciola hepatica infection: a review of recent literature. Trop Dis Bull 87 (4), R1-R38.

Cordova, M., Herrera, P., Nopo, L., Bellatin, J., Naquira, C., Guerra, H., and Espinoza, J.R. (1997). *Fasciola hepatica* cysteine proteinases: immunodominant antigens in human fascioliasis. Am J Trop Med Hyg 57, 660–666.

Cordova, M., Reategui, L., and Espinoza, J.R. (1999). Immunodiagnosis of human fascioliasis with *Fasciola hepatica* cysteine proteinases. Trans R Soc Trop Med Hyg 93, 54–57.

Cornelissen, J.B., Gaasenbeek, C.P., Borgsteede, F.H., Holland, W.G., Harmsen, M.M., and Boersma, W. (2001). Early immunodiagnosis of fasciolosis in ruminants using recombinant *Fasciola hepatica* cathepsin L-like protease. Int J Parasitol 31, 728–737.

Coudert, J., and Triozon, F. (1958). Recherche sur l'épidémiologie de la distomatose humaine à *Fasciola hepatica*. A propos d'une épidémie récente de 500 cas. Rev Hyg Méd Soc 6, 840–864.
Coutellec-Vreto, M.A., Guiller, A., and Daguzan, J. (1994). Allozyme variation in some populations of the freshwater snails *Lymnaea peregra, L. auricularia* and *L. stagnalis* (Gastropoda: Pulmonata). J Moll Stud 60, 393–403.
Cross, J.H. (1984). Changing patterns of some trematode infections in Asia. Arzneimittelforschung 34, 1224–1226.
Curtale, F., Hammoud, E.S., El Wakeel, A., Mas-Coma, S., and Savioli, L. (2000). Human fascioliasis, an emerging public health problem in the Nile Delta, Egypt. Res Rev Parasitol 60, 129–134.
Curtale, F., Hassanein, Y.A.E., El Wakeel, A., Mas-Coma, S., and Montresor, A. (2003a). Distribution of human fascioliasis by age and gender among rural population in the Nile Delta, Egypt. J Trop Pediatr 49, 264–268.
Curtale, F., Mas-Coma, S., Hassanein, Y.A.E., Barduagni, P., Pezzotti, P., and Savioli, L. (2003b). Clinical signs and household characteristics associated with human fascioliasis among rural population in Egypt: a case-control study. Parassitologia 45, 5–11.
Curtale, F., Hassanein, Y.A.E., and Savioli, L. (2005). Control of human fascioliasis by selective chemotherapy: Design, cost and effect of the first public health, school-based intervention implemented in endemic areas of the Nile Delta, Egypt. Trans R Soc Trop Med Hyg 99, 599–609.
Dalton, J.P., ed. (1999). Fasciolosis. Wallingford, Oxon, UK: CAB International.
Dalton, J.P., Neill, S.O., Stack, C., Collins, P., Walshe, A., Sekiya, M., Doyle, S., Mulcahy, G., Hoyle, D., Khaznadji, E., Moire, N., Brennan, G., Mousley, A., Kreshcenko, N., Maule, A.G., and Donnelly, S.M. (2003). *Fasciola hepatica* cathepsin L-like proteases: biology, function, and potential in the development of first generation liver fluke vaccines. Int J Parasitol 33, 1173–1181.
Danis, M, Nozais, J.P., and Chandenier, J. (1985). La distomatose à *Fasciola hepatica*. II: La fasciolose humaine en France. Action Vet 907, VII-VIII.
Datta, S., Mukerjee, G.S., and Ghosh, J.D. (2004). Comparative efficacy of triclabendazole, oxyclozanide and rafoxanide against *Fasciolopsis buski* in naturally infected pigs. Indian J Anim Health 43, 53–56.
De, N.V., Murrell, K.D., Cong, L.D., Cam, P.D., Chau, L.V., Toan, N.D., and Dalsgaard, A. (2003). The food-borne trematode zoonoses of Vietnam. Southeast Asian J Trop Med Public Health 34(suppl 1), 12–34.
Diaz, J., Pina, B., Lastre, M., Rivera, L., and Perez, O. (1990). Fascioliasis humana epidémica. Cuba 1983. VI. Estudio clínico de 40 niños del Hospital Provincial de Sagua La Grande. GEN 44, 385–388.
Ditrich, O., Nasincova, V., Scholz, T., and Giboda, M. (1992). Larval stages of medically important flukes (Trematoda) from Vientiane province, Laos. Part II. Cercariae. Ann Parasitol Hum Comp 67, 75–81.
Dreyfuss, G., Moukrim, A., Rondelaud, D., and Vareille-Morel, C. (1994). Field observations concerning infection of *Lymnaea palustris* by *Fasciola hepatica*. J Helminthol 68, 115–118.
Dreyfuss, G., Abrous, M., and Rondelaud, D. (2000). The susceptibility of *Lymnaea fuscus* to experimental infection with *Fasciola hepatica*. J Parasitol 86, 158–160.
Dreyfuss, G., Vignoles, P., Arous, M., and Rondelaud, D. (2002). Unusual snail species involved in the transmission of *Fasciola hepatica* in watercress beds in central France. Parasite 9, 113–120.

D'Souza, P.E., Jagannath, M.S., and Prasanna, K.M. (2001). A note on the occurrence of *Fasciolopsis buski* in pigs. Cheiron 30, 178–179.

El-Morshedy, H., Farghaly, A., Sharaf, S., Abou-Basha, L., and Barakat, R. (1999). Triclabendazole in the treatment of human fascioliasis: a community-based study. Eastern Mediterranean Health J 5, 888–894.

Espinoza, J.R., Maco, V., Marcos, L., Saez, S., Neyra, V., Terashima, A., Samalvides, F., Gotuzzo, E., Chavarry, E., Huaman, C., Bargues, M.D., Valero, M.A., and Mas-Coma, S. (2007). Evaluation of Fas2-ELISA for the serological detection of *Fasciola hepatica* infection in humans. Am J Trop Med Hyg 76, 977–982.

Esteban, J.G., Flores, A., Aguirre, C., Strauss, W., Angles, R., and Mas-Coma, S. (1997a). Presence of very high prevalence and intensity of infection with *Fasciola hepatica* among Aymara children from the Northern Bolivian Altiplano. Acta Trop 66, 1–14.

Esteban, J.G., Flores, A., Angles, R., Strauss, W., Aguirre, C., and Mas-Coma, S. (1997b). A population-based coprological study of human fascioliasis in a hyperendemic area of the Bolivian Altiplano. Trop Med Int Health 2, 695–699.

Esteban, J.G., Bargues, M.D., and Mas-Coma, S. (1998). Geographical distribution, diagnosis and treatment of human fascioliasis: a review. Res Rev Parasitol 58, 13–42.

Esteban, J.G., Flores, A., Angles, R., and Mas-Coma, S. (1999). High endemicity of human fascioliasis between Lake Titicaca and La Paz valley, Bolivia. Trans R Soc Trop Med Hyg 93, 151–156.

Esteban, J.G., Gonzalez, C., Bargues, M.D., Angles, R., Sanchez, C., Naquira, C., and Mas-Coma, S. (2002). High fascioliasis infection in children linked to a man-made irrigation zone in Peru. Trop Med Int Health 7, 339–348.

Esteban, J.G., Gonzalez, C., Curtale, F., Muñoz-Antoli, C., Valero, M.A., Bargues, M.D., El Sayed, M., El Wakeel, A., Abdel-Wahab, Y., Montresor, A., Engels, D., Savioli, L., and Mas-Coma, S. (2003). Hyperendemic fascioliasis associated with schistosomiasis in villages of the Nile Delta, Egypt. Am J Trop Med Hyg 69, 429–437.

Fairweather, I., and Boray, J.C. (1999). Fasciolicides: efficacy, action, resistance and its management. Vet J 158, 81–112.

Farag, H.F., and El Sayad, M.H. (1995). *Biomphalaria alexandrina* naturally infected with *Fasciola gigantica* in Egypt. Trans R Soc Trop Med Hyg 89, 36.

Favennec, L., Jave Ortiz, J., Gargala, G., Lopez Chegne, N., and Ayoub, A. (2003). Double blind, randomized, placebo-controlled study of nitazoxanide in the treatment of fascioliasis in adults and children from northern Peru. Aliment Pharmacol Ther 17, 265–270.

Fernandez Calienes, A., Fraga, J., Pointier, J.P., Yong, M., Sanchez, J., Coustau, Ch., Gutierrez, A., and Theron, A. (2004). Detection and genetic distance of resistant populations of *Pseudosuccinea columella* (Mollusca: Lymnaeidae) to *Fasciola hepatica* (Trematoda: Digenea) using RAPD markers. Acta Trop 92, 83–87.

Fletcher, H.L., Hoey, E.M., Orr, N., Trudgett, A., Fairweather, I., and Robinson, M.W. (2004). The occurrence and significance of triploidy in the liver fluke, *Fasciola hepatica*. Parasitology 128, 69–72.

Fournier, J. (1954). Services pratiques. Saigon: Rapp Inst Pasteur.

Fuentes, M.V. (2004). Proposal of a Geographical Information System for modelling zoonotic fascioliasis transmission in the Andes. Parasitol Latinoamer 59, 51–55.

Fuentes, M.V., and Malone, J.B. (1999). Development of a forecast system for fascioliasis in central Chile using remote sensing and climatic data in a Geographic Information System. Res Rev Parasitol 59, 129–134.

Fuentes, M.V., Malone, J.B., and Mas-Coma, S. (2001). Validation of a mapping and predicting model for human fasciolosis transmission in Andean very high altitude endemic areas using remote sensing data. Acta Trop 79, 87–95.

Fuentes, M.V., Sainz-Elipe, S., Nieto, P., Malone, J.P., and Mas-Coma, S. (2005). Geographical Information Systems risk assessment models for zoonotic fasciolosis in the South American Andes region. Parassitologia 47, 151–156.
Fuentes, M.V., Valero, M.A., Bargues, M.D., Esteban, J.G., Angles, R., and Mas-Coma, S. (1999). Analysis of climatic data and forecast indices for human fascioliasis at very high altitude. Ann Trop Med Parasitol 93, 835–850.
Gaasenbeek, C.P.H., Moll, L., Cornelissen, J.B.W.J., Vellema, P., and Borgsteede, F.H.M. (2001). An experimental study on triclabendazole resistance of *Fasciola hepatica* in sheep. Vet Parasitol 95, 37–43.
Gai, L., Ma, X., Fu, Y., and Huang, D. (1995). Relationships between the rate of parasitic infection and the knowledge of prevention. Chin J Parasitol Parasit Dis 13, 269–272.
Gaillet, P., Liance, M., Rivollet, D., and Houin, R. (1983). Situation de la fasciolose humaine en France, enquête retrospective portant sur les 30 derniéres années. Bull Soc Fr Parasitol 1, 79–82.
Gasnier, N., Rondelaud, D., Abrous, M., Carreras, F., Boulard, C., Diez-Baños, P., and Cabaret, J. (2000). Allopatric combination of *Fasciola hepatica* and *Lymnaea truncatula* is more efficient than sympatric ones. Int J Parasitol 30, 573–578.
Giap, L.H. (1987). Distomatose hépatique à *Fasciola hepatica*. Bourdeaux, France: Laboratoire de Parasitologie, Université de Bordeaux.
Giboda, M., Ditrich, O., Scholz, T., Viengsay, T., and Bouaphanh, S. (1991). Current status of food-borne parasitic zoonoses in Laos. Southeast Asian J Trop Med Public Health 22 (suppl), 56–61.
Gil-Benito, A., Ciolkovitch, A., Mas-Coma, S., and Quilici, M. (1991). Enquête sur la Distomatose à *Fasciola hepatica* en Corse. Mediterr Med Marseille 403, 21–25.
Gilman, R.H., Mondal, G., Maskud, M., Alam, K., Rutherford, E., Gilman, J.B., and Khan, M.U. (1982). Endemic focus of *Fasciolopsis buski* in Bangladesh. Am J Trop Med Hyg 31, 796–802.
Gitard, R., Coquilhat, P., Silicany, V., Blanc, B., and Nicoli, R.M. (1965). La Distomatose humaine à *Fasciola hepatica* Linnaeus, 1758 en Corse. Bull Soc Pathol Exot 3, 471–474.
Githeko, A.K., Lindsay, S.W., Confalonieri, U.E., and Patz, J.A. (2001). Climate change and vector-borne diseases: a regional analysis. Bull WHO 78, 1136–1147.
Gonzalez, J.F., Perez, O., Rodriguez, G., Arus, E., and Lastre, M. (1985). Fasciolasis humana epidémica, Cuba 1983. VI Estudio Clínico de 44 adultos de Hospital General de Fomento. GEN 39, 276–281.
Gonzalez, J.F., Perez, R., Perez, O., Gonzalez de la Torre, R., Lastre, M., Brito, E., and Diaz, J. (1987). Fasciolasis humana epidémica, Cuba 1983. II. Estudio epidemiológico. GEN 41, 53–57.
Graczyk, T.K., Alam, K., Gilman, R.H., Mondal, G., and Ali, S. (2000). Development of *Fasciolopsis buski* (Trematoda: Fasciolidae) in *Hippeutis umbilicalis* and *Segmentina trochoideus* (Gastropoda: Pulmonata). Parasitol Res 86, 324–326.
Graczyk, T.K., Gilman, R.H., and Fried, B. (2001). Fasciolopsiasis: is it a controllable food-borne disease? Parasitol Res 87, 80–83.
Greenberg, Z., Giladi, L., Bashary, A., and Zahavi, H. (1994). Prevalence of intestinal parasites among Thais in Israel. Harefuah 126, 507–509.
Gupta, A., Xess, A., Sharma, H.P., Dayal, V.M., Prasad, K.M., and Shahi, S.K. (1999). *Fasciolopsis buski* (giant intestinal fluke): a case report. Indian J Pathol Microbiol 42, 359–360.

Gutierrez, A., Pointier, J.P., Yong, M., Sanchez, J., and Theron, A. (2003a). Evidence of phenotypic differences between resistant and susceptible isolates of *Pseudosuccinea columella* (Gastropoda: Lymnaeidae) to *Fasciola hepatica* (Trematoda: Digenea) in Cuba. Parasitol Res 90, 129–134.

Gutierrez, A., Pointier, J.P., Fraga, J., Jobet, E., Modat, S., Perez, R.T., Yong, M., Sanchez, J., Loker, E.S., and Theron, A. (2003b). *Fasciola hepatica:* identification of molecular markers for resistant and susceptible *Pseudosuccinea columella* snail hosts. Exp Parasitol 105, 211–218.

Hadidjaja, P., Dahri, H.M., Roesin, R., Margono, S.S., Djalins, J., and Hanefiah, M. (1982). First autochthonous case of *Fasciolopsis buski* infection in Indonesia. Am J Trop Med Hyg 31, 1065.

Handoyo, I., Ismuljowono, B., Darwis, F., and Rudiansyah, (1986). A survey of fasciolopsiasis in Sei Papuyu village of Babirik subdistrict, Hulu Sungei Utara Regency, South Kalimantan Province. Trop Biomed 3, 113–118.

Harinasuta, T., Bunnag, D., and Radomyos, P. (1984). Efficacy of praziquantel on fasciolopsiasis. Arzneimittelforschung 34, 1214–1215.

Hartman, H.A. (1961). The intestinal fluke (*Fasciolopsis buski*) in a monkey. Am J Vet Res 22, 1123–1126.

Haseeb, A.N., El Shazly, A.M., Arafa, M.A.S., and Morsy, A.T.A. (2002). A review on fascioliasis in Egypt. J Egypt Soc Parasitol 32, 317–354.

Hashimoto, K., Watanobe, T., Liu, C.X., Init, I., Blair, D., Ohnishi, S., and Agatsuma, T. (1997). Mitochondrial DNA and nuclear DNA indicate that the Japanese *Fasciola* species is *F. gigantica.* Parasitol Res 83, 220–225.

Heussler, V., Kaufman, D., Strahm, J., Liz, J., and Dobbelaere, D. (1993). DNA probes for the detection of *Fasciola hepatica* in snails. Mol Cell Probes 7, 261–267.

Heussler, V., Kaufman, D., Glaser, I., Ducommun, D., and Dobbelaere, D. (1998). A DNA probe for the detection of *Dicrocoelium dendriticum* in ants of *Formica* spp. and *Lasius* spp. Parasitol Res 84, 505–508.

Hillyer, G.V. (1999). Immunodiagnosis of human and animal fasciolosis. In: Dalton, J.P., ed. Fasciolosis. Wallingford, Oxon, UK: CAB International Publishing, 435–447.

Hillyer, G.V., and Apt, W. (1997). Food-borne trematode infections in the Americas. Parasitol Today 13, 87–88.

Hillyer, G.V., Soler de Galanes, M., Rodriguez-Perz, J., Bjorland, J., Silva de Lagrava, M., Ramirez Guzman, S., and Bryan, R.T. (1992). Use of the Falcon Assay Screening Test – Enzyme-Linked Immunosorbent Assay (FAST-ELISA) and the Enzyme-Linked Immunoelectrotransfer Blot (EITB) to determine the prevalence of human Fascioliasis in the Bolivian Altiplano. Am J Trop Med Hyg 46, 603–609.

Hopkins, D.R. (1992). Homing in on helminths. Am J Trop Med Hyg 46, 626–634.

Hsieh, H.C. (1960). Studies on the epidemiology of *Fasciolopsis buski* in South Taiwan. Formosan Sci 14, 95–120.

Hsu, P.J. (1964). [A survey of fasciolopsiasis of pigs in Kwantung province.] Acta Vet Zootech Sinica 7, 143–150.

Huang, W.Y., He, B., Wang, C.R., and Zhu, X.Q. (2004). Characterisation of *Fasciola* species from Mainland China by ITS-2 ribosomal DNA sequence. Vet Parasitol 120, 75–83.

Hurtrez-Bousses, S., Durand, P., Jabbour-Zahab, R., Guegan J.F., Meunier, C., Bargues, M.D., Mas-Coma, S., and Renaud, F. (2004). Isolation and characterization of microsatellite markers in the liver fluke (*Fasciola hepatica*). Mol Ecol Notes 4, 689–690.

Idris, M., Rahman, K.M., Muttalib, M.A., and Azad Khan, A.K. (1980). The treatment of fasciolopsiasis with niclosamide and dichlorophen. J Trop Med Hyg 83, 71–74.
Ikeda, T. (1998). Cystatin capture enzyme-linked immunosorbent assay for immunodiagnosis of human paragonimiasis and fascioliasis. Am J Trop Med Hyg 59, 286–290.
Intapan, P.M., Mallewong, W., Wongkham, C., Tomanakarn, K., Ieamviteevanich, K., Pipitgool, V., and Sukolapong, V. (1998). Excretory-secretory antigen components of adult *Fasciola gigantica* recognized by infected human sera. Southeast Asian J Trop Med Public Health 29, 579–583.
Intapan, P.M., Sadee, P., Wongkham, C., and Maleewong, W. (2004). Development of rapid agglutination test using *Fasciola gigantica* specific antigen for serodiagnosis of human fascioliasis. Southeast Asian J Trop Med Public Health 35(suppl 1), 313–317.
Irving, J.A., Spithill, T.W., Pike, R.N., Whisstock, J.C., and Smooker, P.M. (2003). The evolution of enzyme specificity in *Fasciola* spp. J Mol Evol 57, 1–15.
Itagaki, T., and Tsutsumi, K. (1998). Triploid form of *Fasciola* in Japan: genetic relationships between *Fasciola hepatica* and *Fasciola gigantica* determined by ITS-2 sequence of the nuclear rDNA. Int J Parasitol 28, 777–781.
Itagaki, T., Tsutsumi, K.I., Sakamoto, T., Tsutsumi, Y., and Itagaki, H. (1995). Characterization of genetic divergence among species within the genus *Fasciola* by PCR-SSCP. Jpn J Parasitol 44, 244–247.
Itagaki, T., Tsutsumi, K.I., Ito, K., and Tsutsumi, Y. (1998). Taxonomic status of the Japanese triploid forms of *Fasciola*: comparison of mitochondrial ND1 and COI sequences with *F. hepatica* and *F. gigantica*. J Parasitol 84, 445–448.
Jabbour-Zahab, R., Pointier, J.P., Jourdane, J., Jarne, P., Oviedo, J.A., Bargues, M.D., Mas-Coma, S., Angles, R., Perera, G., Balzan, C., Khallaayoune, K., and Renaud, F. (1997). Phylogeography and genetic divergence of some lymnaeid snails, intermediate hosts of human and animal fascioliasis, with special reference to lymnaeids from the Bolivian Altiplano. Acta Trop 64, 191–203.
Jarne, P., and Delay, B. (1990a). Population genetics of *Lymnaea peregra* (Müller) (Gastropoda: Pulmonata) in Lake Geneva. J Moll Stud 56, 317–322.
Jarne, P., and Delay, B. (1990b). Inbreeding depression and self-fertilization in *Lymnaea peregra* (Gastropoda: Pulmonata). Heredity 64, 169–175.
Jarne, P., Vianey-Liaud, M., and Delay, B. (1993). Selfing and crossing in hermaphrodite freshwater gastropods (Basommatophora): where, when and why. Biol J Linn Soc 49, 99–125.
Jaroonvesama, N., Charoenlarp, K., Areekul, S., Aswapokee, N., and Leelarasmee, A. (1980). Prevalence of *Fasciolopsis buski* and other parasitic infections in residents of three villages in Sena district, Ayudhaya Province, Thailand. J Med Assoc Thai 63, 493–499.
Jaroonvesama, N., Charoenlarp, K., and Areekul, S. (1986). Intestinal absorption studies in *Fasciolopsis buski* infection. Southeast Asian J Trop Med Public Health 17, 582–586.
Kaplan, R.M., Dame, J.B., Reddy, G.R., and Courtney, C.H. (1995). A repetitive DNA probe for the sensitive detection of *Fasciola hepatica* infected snails. Int J Parasitol 25, 601–610.
Kaplan, R.M., Dame, J.B., Reddy, G.R., and Courtney, C.H. (1997). The prevalence of *Fasciola hepatica* in its snail intermediate host determined by DNA probe assay. Int J Parasitol 27, 1585–1593.
Keiser, J. (2006). Fasciolicidal activity of artemisinin derivatives. In: XI International Congress of Parasitology (ICOPA XI), Glasgow 6–11 August 2006, UK: CD-ROM.

Knobloch, J., Delgado, A.E., Alvarez, G.A., Reymann, U., and Bialek, R. (1985). Human fascioliasis in Cajamarca/Perú. I. Diagnostic methods and treatement with praziquantel. Trop Med Parasitol 36, 88–90.

Komiya, Y. (1964). *Fasciolopsis buski*. In: K. Morishita, K., Komiya, Y., and Matsubayashi, H., ed. Progress of Medical Parasitology in Japan. Tokyo: Meguro Parasitological Museum, 1, 277–285.

Kozak-Cieszczyk, M., Wedrychowicz, H., and Wielguszewska, I. (2002). A PCR-based test for the detection of *Fasciola hepatica* in the snail intermediate host—*Galba truncatula*. In: XXXII Annual Meeting of the European Society for New Methods in Agricultural Research (ESNA), Conference Materials, Warsaw.

Kumar, V. (1980). The digenetic trematodes, *Fasciolopsis buski, Gastrodiscoides hominis* and *Artyfechinostomum malayanum*, as zoonotic infections in South Asian countries. Ann Soc Belge Med Trop 60, 331–339.

Kuntz, R.E., and Lo, C.T. (1967). Preliminary studies on *Fasciolopsis buski* (Lankester, 1857) (giant Asian intestinal fluke) in the United States. Trans Am Microsc Soc 86, 163–166.

Lane, G. (1998). Anthelminthic resistance. Vet Rec 143, 332.

Law, R.H., Smooker, P.M., Irving, J.A., Piedrafita, D., Pointing, R., Kennedy, N.J., Whisstock, J.C., Pike, R.N., and Spithill, T.W. (2003). Cloning and expression of the major secreted cathepsin B-like protein from juvenile *Fasciola hepatica* and analysis of immunogenicity following liver fluke infection. Infect Immun 71, 6921–6932.

Le, T.H., Blair, D., and McManus, D.P. (2001). Complete DNA sequence and gene organization of the mitochondrial genome of the liver fluke, *Fasciola hepatica* L. (Plathelminthes; Trematoda). Parasitology 123, 609–621.

Le, T.H., Nguyen, V.D., Phan, B.U., Blair, D., and McManus, D.P. (2004). Case report: unusual presentation of *Fasciolopsis buski* in a Vietnamese child. Trans R Soc Trop Med Hyg 98, 193–194.

Lecaillon, J.B., Gobdillon, J., and Campestrini, J. (1998). Effect of food on bioavailability of triclabendazole in patients with fascioliasis. Br J Clin Pharmacol 45, 601–604.

Lee, C.G., Zimmerman, G.L., and Bishop, J.K. (1992). Host influence on the banding profiles of whole-body protein and excretory-secretory product of *Fasciola hepatica* (Trematoda) by isoelectric focusing. Vet Parasitol 41, 57–68.

Lee, H.H. (1972). *Fasciolopsis buski* infection among children of Liu-ying Primary School in Tainan Hsien south Taiwan. Chin J Microbiol 5, 110–114.

Lee, H.H. (1986). Studies on epidemiology and treatment of fasciolopsiasis in southern Taiwan. Kaohsiung J Med Sci 2, 21–27.

Lee, H.H., Shyu, L.Y., and Chen, E.R. (1989). Experimental control study of fasciolopsiasis in Taiwan. Kaohsiung J Med Sci 5, 335–344.

Lo, C.T. (1967). Life history of the snail, *Segmentina hemisphaerula* (Benson), and its experimental infection with *Fasciolopsis buski* (Lankester). J Parasitol 53, 735–738.

Lo, C.T., and Lee, K.M. (1996). Intestinal parasites among Southeast Asian laborers in Taiwan during 1993–1994. Zhonghua Yi Xue Za Zhi, Taipei 57, 401–404.

Maleewong, W., Wongkham, C., Intapan, P.M., and Pipitgol, V. (1999). *Fasciola gigantica*-specific antigens: purification by a continuous-elution method and its evaluation for the diagnosis of human fascioliasis. Am J Trop Med Hyg 61, 648–651.

Malone, J.B., and Yilma, J.M. (1999). Predicting outbreaks of fasciolosis: from Ollerenshaw to satellites. In: Dalton, J.P., ed. Fasciolosis. Wallingford, Oxon, UK: CAB International, 151–183.

Malviya, H.C. (1985). The susceptibility of mammals to *Fasciolopsis buski*. J Helminthol 59, 19–22.
Manjarumkar, P.V., and Shah, P.M. (1972). Epidemiological study of *Fasciolopsis buski* in Palghar Taluk. Indian J Public Health 16, 3–6.
Manning, G.S., and Ratanarat, C. (1970). *Fasciolopsis buski* (Lankester, 1857) in Thailand. Am J Trop Med Hyg 19, 613–619.
Manning, G.S., Brockelman, W.Y., and Yiyant, V. (1971). An analysis of the prevalence of *Fasciolopsis buski* in central Thailand using catalyitic models. Am J Epidemiol 93, 354–360.
Marcilla, A., Bargues, M.D., and Mas-Coma, S. (2002). A PCR-RFLP assay for the distinction between *Fasciola hepatica* and *F. gigantica*. Mol Cell Prob 16, 327–333.
Marcos Raymundo, L.A., Maco Flores, V., Terashima, A., Samalvides, F., Miranda, E., Tantalean, M., Espinoza, J.R., and Gotuzzo, E. (2004). Hiperendemicidad de Fasciolosis humana en el Valle del Mantaro, Perú: factores de riesgo de la infeccion por *Fasciola hepatica*. Rev Gastroenterol Peru 24, 158–164.
Marcos, L., Maco, V., Samalvides, F., Terashima, A., Espinoza, J.R., and Gotuzzo, E. (2006). Risk factors for *Fasciola hepatica* infection in children: a case-control study. Trans R Soc Trop Med Hyg 100, 158–166.
Margono, S.S. (2003). Important human helminthiasis in Indonesia. In: Crompton, D.W.T., Montresor, A., Nesheim, M.C., and Savioli, L., eds. Controlling Disease due to Helminth Infections. Geneva: World Health Organization, 3–14.
Mas-Coma, S. (2004a). Human fascioliasis. In: Cotruvo, J.A., Dufour, A., Rees, G., Bartram, J., Carr, R., Cliver, D.O., Craun, G.F., Fayer, R., and Gannon, V.P.J., eds. World Health Organization (WHO), Waterborne Zoonoses: Identification, Causes and Control. London: IWA Publishing, 305–322.
Mas-Coma, S. (2004b). Human fascioliasis: epidemiological patterns in human endemic areas of South America, Africa and Asia. Southeast Asian J Trop Med Public Health 35 (suppl 1), 1–11.
Mas-Coma, S. (2005). Epidemiology of fascioliasis in human endemic areas. J Helminthol 79, 207–216.
Mas-Coma, S., and Bargues, M.D. (1997). Human liver flukes: a review. Res Rev Parasitol 57, 145–218.
Mas-Coma, S., Angles, R., Strauss, W., Esteban, J.G., Oviedo, J.A., and Buchon, P. (1995). Human fasciolasis in Bolivia: a general analysis and a critical review of existing data. Res Rev Parasitol 55, 73–93.
Mas-Coma, S., Rodriguez, A., Bargues, M.D., Valero, M.A., Coello, J.R., and Angles, R. (1997). Secondary reservoir role of domestic animals other than sheep and cattle in fascioliasis transmission in the Northern Bolivian Altiplano. Res Rev Parasitol 57, 39–46.
Mas-Coma, S., Esteban, J.G., and Bargues, M.D. (1999a). Epidemiology of human fascioliasis: a review and proposed new classification. Bull WHO 77, 340–346.
Mas-Coma, S., Bargues, M.D., and Esteban, J.G. (1999b). Human Fasciolosis. In: Dalton, J.P., ed. Fasciolosis. Wallingford, Oxon, UK: CAB International, 411–434.
Mas-Coma, S., Angles, R., Esteban, J.G., Bargues, M.D., Buchon, P., Franken, M., and Strauss, W. (1999c). The Northern Bolivian Altiplano: a region highly endemic for human fascioliasis. Trop Med Int Health 4, 454–467.
Mas-Coma, S., Bargues, M.D., Marty, A.M., and Neafie, R.C. (2000). Hepatic Trematodiases. In: Meyers, W.M., Neafie, R.C., Marty, A.M., and Wear, D.J., eds. Pathology of Infectious

Diseases, Vol. 1 Helminthiases. Washington, DC: Armed Forces Institute of Pathology and American Registry of Pathology, 69–92.

Mas-Coma, S., Funatsu, I.R., and Bargues, M.D. (2001). *Fasciola hepatica* and lymnaeid snails occurring at very high altitude in South America. Parasitology 123, S115-S127.

Mas-Coma, S., Bargues, M.D., Valero, M.A., and Fuentes, M.V. (2003). Adaptation capacities of *Fasciola hepatica* and their relationships with human fascioliasis: from below sea level up to the very high altitude. In: Combes, C., and Jourdane, J., eds. Taxonomy, Ecology and Evolution of Metazoan Parasites, vol 2. Perpignan: Perpignan University Press, 81–123.

Mas-Coma, S., Bargues, M.D., and Valero, M.A. (2005). Fascioliasis and other plant-borne trematode zoonoses. Int J Parasitol 35, 1255–1278.

Massoud, A., El Sisi, S., Salama, O., and Massoud, A. (2001). Preliminary study of therapeutic efficacy of a new fasciolicidal drug derived from *Commiphora molmol* (myrrh). Am J Trop Med Hyg 65, 96–99.

Massoud, J. (1990). Fascioliasis outbreak of man and drug test (Triclabendazol) in Caspian littoral, northern part of Iran, 1989. Bull Soc Fr Parasitol 8 (suppl 1), 438.

Massoud, J. (1993). Present status of human fascioliasis in Iran. In: Food-borne trematodes, World Health Organization, WHO Manual, Manila, Mimeogr Rep Sch/SG/93 w, p. 19.

Menard, A., Agoulon, A., L'Hostis, M., Rondelaud, D., Collard, S., and Chauvin, A. (2001). *Myocastor coypus* as a reservoir host of *Fasciola hepatica* in France. Vet Res 32, 499–508.

Meunier, C., Tirard, C., Hurtrez-Bousses, S., Durand, P., Bargues, M.D., Mas-Coma, S., Pointier, J.P., Jourdane, J., and Renaud, F. (2001). Lack of molluscan host diversity and the transmission of an emerging parasitic disease in Bolivia. Mol Ecol 10, 1333–1340.

Mezo, M., Gonzalez-Warleta, M., and Ubeira, F.M. (2003). Optimized serodiagnosis of sheep fascioliasis by fast-D protein liquid chromatography fractionation of *F. hepatica* excretory-secretory antigens. J Parasitol 89, 843–849.

Mezo, M., Gonzalez-Warleta, M., Carro, C., and Ubeira, F.M. (2004). An ultrasensitive capture ELISA for detection of *Fasciola hepatica* coproantigens in sheep and cattle using a new monoclonal antibody (MM3). J Parasitol 90, 845–852.

Millan, J.C., Mull, R., Freise, S., and Richter, J. (2000). The efficacy and tolerability of triclabendazole in Cuban patients with latent and chronic *Fasciola hepatica* infection. Am J Trop Med Hyg 63, 264–269.

Miller, C.M.D., Howell, M.J., and Boray, J.C. (1993). Host effects of glutathione s-tranferase activity in *Fasciola hepatica*. Int J Parasitol 23, 1073–1076.

Mitchell, G.B., Maris, L., and Bonniwell, M.A. (1998). Triclabendazole-resistant liver fluke in Scottish sheep. Vet Rec 143, 399.

Mitterpak, J. (1968). Zu den Besonderheiten der Epizootologie von Fasziolose in den tropischen Bedingungen von Kuba. Wiad Parazytol 14, 503–507.

Moghaddam, A.S., Massoud, J., Mahmoodi, M., Mahvi, A.H., Periago, M.V., Artigas, P., Fuentes, M.V., Bargues, M.D., and Mas-Coma, S. (2004). Human and animal fascioliasis in Mazandaran province, northern Iran. Parasitol Res 94, 61–69.

Moll, L., Gaasenbeek, C.P.H., Vellema, P., and Borgsteede, F.H.M. (2000). Resistance of *Fasciola hepatica* against triclabendazole in cattle and sheep in the Netherlands. Vet Rec 91, 153–158.

Mott, K.E., Nuttall, I., Desjeux, P., and Cattand, P. (1995). New geographical approaches to control of some parasitic zoonoses. Bull WHO 73, 247–257.

Muralidhar, S., Srivastava, I., Aggarwal, P., Jain, N., and Sharma, D.K. (2000). Fasciolopsiasis, a persisting problem in eastern U.P.: a case report. Indian J Pathol Microbiol 43, 69–71.

Muttalib, M.A., and Islam, N. (1975). *Fasciolopsis buski* in Bangladesh—a pilot study. J Trop Med Hyg 78, 135–137.

Neyra, V., Chavarry, E., and Espinoza, J.R. (2002). Cysteine proteinases Fas1 and Fas2 are diagnostic markers for *Fasciola hepatica* infection in alpacas (*Lama pacos*). Vet Parasitol 105, 21–32.

Nguen Tkhi Le (1978). [A study of the biology of the trematode *Fasciolopsis buski* (Lankester, 1857) in Vietnam]. Med Parazitol Moskva 47, 81–84.

Nguyen Van Tho (2002a). Vectors of the pig intestinal fluke *Fasciolopsis buski*. Khoa Hoc Ky Thuat Thu Y (Vet Sci Tech) 9, 38–42.

Nguyen Van Tho (2002b). Evolution cycle of *Fasciolopsis buski* in the mollusc-intermediate host and in nature. Khoa Hoc Ky Thuat Thu Y (Vet Sci Tech) 9, 43–48.

O'Brien, D.J. (1998). Fasciolosis: a threat to livestock. Irish Vet J 51, 539–541.

Ollerenshaw, C.B. (1971). The influence of climate on the life cycle of *Fasciola hepatica* in Britain with some observations on the relationship between climate and the incidence of fasciolasis in the Netherlands. In: Facts and Reflections, Parasitology Department, Centraal Diergeneeskundig Institut, Lelystad, 41–63.

O'Neill, S.M., Parkinson, S.M., Strauss, W., Angles, R., and Dalton, J.P. (1998). Immunodiagnosis of *Fasciola hepatica* (Fascioliasis) in a human population in the Bolivian Altiplano using purified cathepsin L cysteine proteinase. Am J Trop Med Hyg 58, 417–423.

O'Neill, S.M., Parkinson, S.M., Dowd, A.J., Strauss, W., Angles, R., and Dalton, J.P. (1999). Immunodiagnosis of human fascioliasis using recombinant *Fasciola hepatica* cathepsin L1 cysteine proteinase. Am J Trop Med Hyg 60, 749–751.

Overend, D.J., and Bowen, F.L. (1995). Resistance of *Fasciola hepatica* to triclabendazole. Australian Vet J 72, 275–276.

Oviedo, J.A., Mas-Coma, S., Dominici, J.L., and Roche, B. (1992). Distribution and ecology of *Lymnaea truncatula* (Müller, 1774) (Gastropoda: Basommatophora: Lymnaeidae), intermediate host of fasciolosis, in the island of Corsica (France). In: EMOP VI, The Hague, 7–11 September 1992, abstracts, 218.

Oviedo, J.A., Bargues, M.D., and Mas-Coma, S. (1995). Lymnaeid snails in the human fascioliasis high endemic zone of the Northern Bolivian Altiplano. Res Rev Parasitol 55, 35–43.

Oviedo, J.A., Bargues, M.D., and Mas-Coma, S. (1996). The intermediate snail host *of Fasciola hepatica* on the Mediterranean island of Corsica. Res Rev Parasitol 56, 217–220.

Panaccio, M., and Trudgett, A. (1999). Molecular Biology. In: Dalton, J.P., ed. Fasciolosis. Wallingford, Oxon, UK: CAB International, 449–464.

Perez, A., Celestino, M., Millan, J.C., and Finlay, C.M. (1997). Brote de *Fasciola hepatica* en humanos: estudio epidemiológico. In: XIII Congreso Latino Americano de Parasitología (La Habana, Cuba), Resúmenes, 122–123.

Periago, M.V., Valero, M.A., Panova, M., and Mas-Coma, S. (2006). Phenotypic comparison of allopatric populations of *Fasciola hepatica* and *Fasciola gigantica* from European and African bovines using a computer image analysis system (CIAS). Parasitol Res, 99, 368–378.

Plaut, A.G., Kampanart-Sanyakorn, C., and Mannig, G.S. (1969). A clinical study of *Fasciolopsis buski* infection in Thailand. Trans R Soc Trop Med Hyg 63, 470–478.

Preet, S., and Prakash, S. (2001). Cercarial emergence in *Fasciolopsis buski* (Lankerster). J Parasit Dis 25, 108–110.

Rabbani, G.H., Gilman, R.H., Kabir, I., and Mondel, G. (1985). The treatment of *Fasciolopsis buski* infection in children: a comparison of thiabendazole, mebendazole, levamisole, pyrantel pamoate, hexylresorcinol and tetrachloroethylene. Trans R Soc Trop Med Hyg 79, 513–515.

Rahman, K.M., Idris, M., and Khan, A.K.A. (1981). A study of fasciolopsiasis in Bangladesh. J Trop Med Hyg 84, 81–86.

Remigio, E.A. (2002). Molecular phylogenetic relationships in the aquatic snail genus *Lymnaea*, the intermediate host of the causative agent of fascioliasis: insights from broader taxon sampling. Parasitol Res 88, 687–696.

Remigio, E.A., and Blair, D. (1997a). Molecular systematics of the freshwater snail family Lymnaeidae (Pulmonata: Basommatophora) utilising mitochondrial ribosomal DNA sequences. J Moll Stud 63, 173–185.

Remigio, E.A., and Blair, D. (1997b). Relationships among problematic North American stagnicoline snails (Pulmonata: Lymnaeidae) reinvestigated using nuclear ribosomal DNA internal transcribed spacer sequences. Can J Zool 75, 1540–1545.

Rim, H.J. (1982). Fasciolopsiasis. In: G.V. Hillyer, G.V., Hopla, C.E. (Ed.), Handbook Series in Zoonoses. Section C. Parasitic Zoonoses, vol 3. Boca Raton, FL: CRC Press, 89–97.

Rim, H.J., Farag, H.F., Sornmani, S., and Cross, J.H. (1994). Food-borne trematodes: ignored or emerging? Parasitol Today 10, 207–209.

Ripert, C., Tribouley, J., Luang Dinh Giap, G., Combe, A., and Laborde, M. (1987). Epidémiologie de la fasciolose humaine dans le sud ouest de la France. Bull Soc Fr Parasitol 5, 227–230.

Roberts, J.A., and Suhardono, (1996). Approaches to the control of fasciolosis in ruminants. Int J Parasitol 26, 971–981.

Rognlie, M.C., Dimke, K.L., and Knapp, S.E. (1994). Detection of *Fasciola hepatica* in infected intermediate hosts using RT-PCR. J Parasitol 80, 748–755.

Rokni, M.B., Massoud, J., O'Neill, S.M., Parkinson, M., and Dalton, J.P. (2002). Diagnosis of human fasciolosis in the Gilan province of Northern Iran: application of cathepsin L-ELISA. Diag Micr Inf Dis 44, 175–179.

Rossignol, J.F., Abaza, H., and Friedman, H. (1998). Successful treatment of human fascioliasis with nitazoxanide. Trans R Soc Trop Med Hyg 92, 103–104.

Roy, B., and Tandon, V. (1992). Seasonal prevalence of some zoonotic trematode infections in cattle and pigs in the north-east montane zone in India. Vet Parasitol 41, 69–76.

Roy, B., and Tandon, V. (1993). Morphological and microtopographical strain variations among *Fasciolopsis buski* originating from different geographical areas. Acta Parasitol 38, 72–77,

Rudolph, P.H., and Burch, J.B. (1989). Electrophorectic analysis of enzymes in three species of *Stagnicola* (Pulmonata: Lymnaeidae). J Med Appl Malacol 1, 57–64.

Sadun, E.H., and Maiphoom, C. (1953). Studies on the epidemiology of the human intestinal fluke, *Fasciolopsis buski* (Lankester) in Central Thailand. Am J Trop Med Hyg 2, 1070–1084.

Samadi, S., Roumegoux, A., Bargues, M.D., Mas-Coma, S., Yong, M., and Pointier, J.P. (2000). Morphological studies of lymnaeid snails from the human fascioliasis endemic zone of Bolivia. J Moll Stud 66, 31–44.

Sampaio-Silva,M.L., Correia, da Costa, J.M., Viana da Costa, A.M., Pires, M.A., Lopes, S.A., Castro, A.M., and Monjour, L. (1996). Antigenic components of excretory-secretory products of adult *Fasciola hepatica* recognized in human infections. Am J Trop Med Hyg 54, 146–148.

Savioli, L., Chistulo, L., and Montresor, A. (1999). New opportunities for the control of fascioliasis. Bull WHO 77, 300.

Schubert, S., and Granz, W. (1981). [Problems in the differential diagnosis of intestinal diseases in patients who have traveled in the tropics]. Z Gesamte Med 36, 230–233.

Semyenova, S.K., Morozova, E.V., Chrisanfova, G.G., Asatrian, A.A., Movsessian, S.O., and Ryskov, A.P. (2003). RAPD variability and genetic diversity in two populations of liver fluke, *Fasciola hepatica*. Acta Parasitol 48, 125–130.

Shah, A., Gadgil, R.K., and Manohar, K.D. (1966). Fasciolopsiasis in Bombay. A preliminary communication. Indian J Med Sci 20, 805–811.

Shah, P.M., Udani, P.M., Manjarumkar, P.V., and Naik, P.A. (1973). *Fasciolopsis buski* infestation in children. Indian Pediatr 10, 721–724.

Shekhar, K.C. (1991). Epidemiological assessment of parasitic zooonoses in Malaysia. Southeast Asian J Trop Med Publ Health 22, 337–339.

Shimalov, V.V., and Shimalov, V.T. (2000). Findings of *Fasciola hepatica* Linnaeus, 1758, in wild animals in Bielorussian Polesye. Parasitol Res 86, 342.

Shubkin, C.D., White, M.W., Abrahamsen, M.S., Ronglie, M.C., and Knapp, S.E. (1992). A nucleid acid-based test for detection of *Fasciola hepatica*. J Parasitol 78, 817–821.

Shyu, L.Y., Lee, H.H., and Chen, E.R. (1984). A preliminary study on epidemiology of fasciolopsiasis in Tainan Hsien, south Taiwan. Chin J Microbiol Immunol 17, 118–120.

Sorribes, J., Aubary, M.J., Bargues, M.D., and Mas-Coma, S. (1990). Revisión de los casos humanos de Fascioliasis en España entre los años 1970 y 1989. In: Congreso de Zoonosis. 1° Interautonómico. Reunión de la International Zoonoses Association (IZA) (Valencia, Spain), Resúmenes, 186.

Spithill, T.W., Smooker, P.M., and Copeman, D.B. (1999). *Fasciola gigantica*: epidemiology, control, immunology and molecular biology. In: Dalton, J.P., ed. Fasciolosis. Wallingford, Oxon, UK: CAB International, 465–525.

Stothard, J.R., Bremond, P., Andriamaro, L., Loxton, N.J., Sellin, B., Sellin, E., and Rollinson, D. (2000). Molecular characterization of the freshwater snail *Lymnaea natalensis* (Gastropoda: Lymnaeidae) on Madagascar with an observation of an unusual polymorphism in ribosomal small subunit genes. J Zool 252, 303–315.

Strauss, W., O'Neill, S.M., Parkinson, M., Angles, R., and Dalton, J.P. (1999). Diagnosis of human fascioliasis: detection of anti-cathepsin L antibodies in blood samples collected on filter paper. Am J Trop Med Hyg 60, 746–748.

Suntharasamai, P., Bunnag, D., Tejavanij, S., Harinasuta, T., Migasena, S., Vutikes, S., and Chindanond, D. (1974). Comparative clinical trials of niclosamide and tetrachloroethylene in the treatment of *Fasciolopsis buski* infection. Southeast Asian J Trop Med Public Health 5, 556–559.

Talaie, H., Emami, H., Yadegarinia, D., Nava-Ocampo, A.A., Massoud, J., Azmoudeh, M., and Mas-Coma, S. (2004). Randomized trial of a single, double and triple dose of 10 mg/kg of a human formulation of triclabendazole in patients with fascioliasis. Clin Exp Pharmacol Physiol 31, 777–782.

Tantrawatpan, C., Maleewong, W., Wongkham, C., Wongkham, S., Intapan, P.M., and Nakashima, K. (2005). Serodiagnosis of human fascioliasis by cystatin capture enzyme-linked immunosorbent assay with recombinant *Fasciola gigantica* capthepsin L antigen. Am J Trop Med Hyg 72, 82–86.

Taraschewski, H., Mehlhorn, H., Bunnag, D., Andrews, P., and Thomas, H. (1986). Effects of praziquantel on human intestinal flukes (*Fasciolopsis buski* and *Heterophyes heterophyes*). Zentralbl Bakteriol Mikrobiol Hyg 262, 542–550.

Torgerson, P., and Claxton, J. (1999). Epidemiology and control. In: Dalton, J.P., ed. Fasciolosis. Wallingford, Oxon, UK: CAB International, 113–149.

Tripathi, J.C., Srivastava, H.D., and Dutt, S.C. (1973). A note on experimental infection of *Helicorbis coenosus* and pig with *Fasciolopsis buski*. Indian J Anim Sci 43, 647–649.

Trouve, S., Degen, L., Meunier, C., Tirard, C., Hurtrez-Boussees, S., Durand, P., Guegan, J.F., Goudet, J., and Renaud, F. (2001). Microsatellites in the hermaphroditic snail, *Lymnaea truncatula*, intermediate host of the liver fluke, *Fasciola hepatica*. Mol Ecol 9, 1662–1663.

Turrientes, M.C., A. Saenz de Sta. Maria, A., Ceballos, E., Diaz, M., Bareno, M., Muro, A., Pardo, J., and Lopez-Velez, R. (2004). Fasciolosis importada y autóctona. Enf Emerg 6, (119), 188.

Ueno, H., Arandia, R., Morales, G., and Medina, G. (1975). Fascioliasis of livestock and snail host for *Fasciola* in the Altiplano region of Bolivia. Nat Inst An Health Q 15, 61–67.

Valero, M.A., and Mas-Coma, S. (2000). Comparative infectivity of *Fasciola hepatica* metacercariae from isolates of the main and secondary reservoir animal host species in the Bolivian Altiplano high human endemic region. Folia Parasitol 47, 17–22.

Valero, M.A., Marcos, M.D., and Mas-Coma, S. (1996). A mathematical model for the ontogeny of *Fasciola hepatica* in the definitive host. Res Rev Parasitol 56, 13–20.

Valero, M.A., Marcos, M.D., Fons, R., and Mas-Coma, S. (1998). *Fasciola hepatica* development in experimentally infected black rat, *Rattus rattus*. Parasitol Res 84, 188–194.

Valero, M.A., Marcos, M.D., Comes, A.M., Sendra, M., and Mas-Coma, S. (1999). Comparison of adult liver flukes from highland and lowland populations of Bolivian and Spanish sheep. J Helminthol 73, 341–345.

Valero, M.A., Varea, M.T., and Marin, R. (2000). *Fasciola hepatica*: lithogenic capacity in experimentally infested rats and chemical determination of the main stone components. Parasitol Res 86, 558–562.

Valero, M.A., Panova, M., and Mas-Coma, S. (2001a). Developmental differences in the uterus of *Fasciola hepatica* between livestock liver fluke populations from Bolivian highland and European lowlands. Parasitol Res 87, 337–342.

Valero, M.A., Darce, N.A., Panova, M., and Mas-Coma, S. (2001b). Relationships between host species and morphometric patterns in *Fasciola hepatica* adults and eggs from the Northern Bolivian Altiplano hyperendemic region. Vet Parasitol 102, 85–100.

Valero, M.A., Panova, M., Comes, A.M., Fons, R., and Mas-Coma, S. (2002). Patterns in size and shedding of *Fasciola hepatica* eggs by naturally and experimentally infected murid rodents. J Parasitol 88, 308–313.

Valero, M.A., Santana, M., Morales, M., Hernandez, J.L., and Mas-Coma, S. (2003). Risk of gallstone disease in advanced chronic phase of fascioliasis: an experimental study in a rat model. J Infect Dis 188, 787–793.

Valero, M.A., Panova, M., and Mas-Coma, S. (2005). Phenotypic analysis of adults and eggs of *Fasciola hepatica* by computer image analysis system. J Helminthol 79, 217–225.

Valero, M.A., De Renzi, M., Panova, M., Garcia-Bodelon, M.A., Periago, M.V., Ordoñez, D., and Mas-Coma, S. (2006a). Crowding effect on adult growth, pre-patent period and egg shedding of *Fasciola hepatica*. Parasitology 133, 453–463.

Valero, M.A., Navarro, M., Garcia-Bodelon, M.A., Marcilla, A., Morales, M., Garcia, J.E., Hernandez, J.L., and Mas-Coma, S. (2006b). High risk of bacterobilia in advanced experimental chronic fasciolosis. Acta Trop 100,17–23.

Vara-del-Rio, M.P., Villa, H., Martinez-Valladares, M., Cruz-Rojo, M.A., and Rojo Vazquez, F.A. (2005). Genetic heterogeneity of *F. hepatica* isolates from sheep in northwest of Spain. Acta Parasitol Port 12, 399–400.

Viranuvatti, V., Stitinimankarn, T., and Tansurat, P. (1953). A fatal case of infection with *Fasciolopsis buski* in Thailand. Ann Trop Med Parasitol 47, 132–133.

Waikagul, J. (1991). Intestinal fluke infections in Southeast Asia. Southeast Asian J Trop Med Public Health 22(suppl), 158–162.

Wang, P., Zhang, K., Wu, F., and Yao, T. (1977). [Studies on the life-history of *Fasciolopsis buski* (Lankester, 1857) with consideration of its seasonal infection in pigs]. Acta Zool Sinica 23, 88–96.

Weng, Y.L., Zhuang, Z.L., Jiang, H.P., Lin, G.R., and Lin, J.J. (1989). Studies on ecology of *Fasciolopsis buski* and control strategy of fasciolopsiasis. Zhongguo Ji Sheng Chong Xue Yu Ji Sheng Chong Bing Za Zhi 7, 108–111.

Wiwanitkit, V., Suwansaksri, J., and Chaiyakhun, Y. (2002). High prevalence of *Fasciolopsis buski* in an endemic area of liver fluke infection in Thailand. Med Gen Med 4(3), 6.

World Health Organization. (1995a). Foodorne trematode infections. Bull WHO 73, 397–399.

World Health Organization. (1995b). Control of foodborne trematode infections. WHO Tech Rep Ser 849, 1–157.

Xu, L., Jiang, Z., Yu, S., Xu, S., Chang, J., Wu, Z., Xu, J., Zhang, X., Chen, Z., Zhang, B., *et al.* (1995). [Characteristics and recent trends in endemicity of human parasitic diseases in China]. Zhongguo Ji Sheng Chong Xue Yu Ji Sheng Chong Bing Za Zhi 13, 214–217.

Yilmaz, H., and Gödekmerdan, A. (2004). Human fasciolosis in Van province, Turkey. Acta Trop 92, 161–162.

Yoshihara, S., Hung, N.P., Hung, N.H., and Loc, C.B. (1999). Helminths and helminthiosis of pigs in the Mekong Delta, Vietnam with special reference to ascariosis and *Fasciolopsis buski* infection. Jpn Agric Res Q 33, 193–199.

Yu, S., Xu, L., Jiang, Z., Xu, S., Han, J., Zhu, Y., Chang, J., Lin, J., and Xu, F. (1994). [Report of the first nationwide survey of the distribution of human parasites in China. 1. Regional distribution of parasitic species]. Zhongguo Ji Sheng Chong Xue Yu Ji Sheng Chong Bing Za Zhi 12, 241–247.

Yu, S.H., and Mott, K.E. (1994). Epidemiology and morbidity of food-borne intestinal trematode infections. Trop Dis Bull 91 (7), R125-R152.

Zurita, M., Bieber, D., Ringold, G., and Manssur, T.E. (1988). cDNA cloning and gene characterization of the mitochondrial large subunit (LSU) rRNA from the liver fluke *Fasciola hepatica*. Evidence of heterogeneity in the fluke mitochondrial genome. Nucleic Acids Res 16, 7001–7012.

III
General Aspects of Infection

10
Immunology of the Infection

Haruhiko Maruyama and Yukifumi Nawa

What motivates people to study the immunology of parasitic infections, especially those described in the preceding chapters of this book? One of the most practical purposes would be the development of a diagnostic kit to detect parasite antigens or specific antibodies in clinical samples. For decades, fish-borne and plant-borne zoonoses have been limited for the most part to populations living in low- and middle-income countries who have specific food consumption customs. However, along with the growing international market and improved transportation system, populations at risk, including those in developed countries, are expanding enormously, and people as well as physicians are not familiar with food-born parasitic zoonoses (Chai et al., 2005). In such situations, convenient clinical kits would greatly facilitate rapid diagnosis of exotic diseases.

For the development of diagnostic tools, parasite excretory/secretory (ES) products seem to provide promising target molecules. These products are highly antigenic (Harrison and Parkhouse, 2001), and the performance of immunodiagnostic measures is greatly improved when ES products are employed. Because ES products usually contain a fairly large amount of proteases (Tort et al., 1999; Sajid and McKerrow, 2002), much work has been done on parasite proteases not only for practical purposes, but also for the elucidation of the biology of parasites.

Development of an effective vaccine is also a driving force behind a number of immunological studies. In the case of *Fasciola* infections, strong demand from global livestock industries has so encouraged vaccine development that promising vaccine candidates have now been emerging (Spithill and Dalton, 1998; Dalton et al., 2003a). In experimental animals, anti-*Clonorchis* vaccines have been studied (E.G. Lee et al., 2006; J.S. Lee et al., 2006).

Yet man shall not live by bread alone. To think again about food-borne parasitic zoonoses, one might notice that worms that cause fish-borne and plant-borne zoonoses are extraordinarily versatile, and that immunology, in spite of a vast body of research, cannot give clear answers to questions concerning how these parasites get along with host immune responses. They have complex life cycles that involve invertebrates, vertebrates, and even plants, and their specificity for the final host is usually broad. Moreover, food-borne parasitic infections are often chronic with repeated or recurrent reinfections (Chai et al., 2005). Induction of protective immunity appears difficult against many of food-borne worms. All of these features indicate that food-transmitted parasites are well adapted to make use of the wildlife food web to maintain generations, and, consequently, they have evolved highly sophisticated strategies for dealing with the defense mechanisms of the host. From the point of immunology, these zoonotic parasites are difficult but valuable study subjects.

This chapter provides an overview of immune responses in parasitic infections in rodent model systems, and describes immunological phenomena that occur in fish-borne and plant-borne helminth infections. Then, we focus on the following key issues in the field: (1) biological significance of parasite-derived proteases, (2) significance of immune responses in host defense, and (3) encystment as an immune evasion mechanism.

Overview of Immune Responses Against Parasitic Worms

Parasitic worms are never destroyed directly by T cells or cytokines. Upon infection, innate cells at the front line, such as epithelial cells, macrophages, and granulocytes, recognize the entry, and initiate T-cell differentiation by stimulating them with cytokines or antigen presentation. Antigen-specific T cells then start to secrete an array of cytokines, which in turn stimulate innate cells to eliminate parasites (Fig. 10.1). Known effector cells that directly kill or expel parasitic

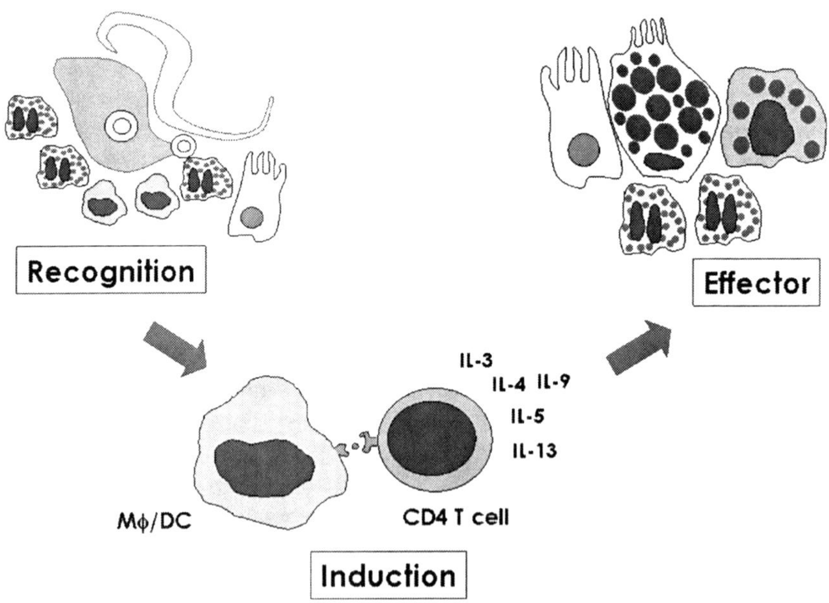

FIGURE 10.1. Immune responses to parasitic worms. Immune responses can be divided into three phases. Innate cells recognize the entry of parasite (recognition phase), and they induce antigen-specific T cells, which secrete various cytokines (induction phase). Cytokines activate innate cells, which eliminate parasitic worms (effector phase).

worms are macrophages, eosinophils, mucosal mast cells, and intestinal epithelial cells. This section discusses immune effector mechanisms and their regulation, which is knowledge derived mostly from experimental infections of mice and rats with intestinal nematodes.

An immune response is initially stimulated on entry of a parasite into the host body. However, we start the description of the immune response with the stage of parasite elimination—the effector stage. Because worm expulsion is often employed as an indicator for an immune response, it is important to know what cells are effective against these worms. It has been long recognized that infections with parasitic worms elicit almost stereotyped responses regardless of the nature of parasites (Finkelman et al., 1997, 2004). Although flukes, tapeworms, and roundworms are only distantly related phylogenetically, they induce similar ("hallmark") responses including eosinophilia, elevated immunoglobulin E (IgE) production, mucosal mastocytosis, and goblet cell hyperplasia.

Immune Effector Mechanisms

Eosinophils

Peripheral blood eosinophilia is one of the clinical signs of parasitic infections. Because eosinophils migrate to tissue-invading parasites and surround them, eosinophils have been proposed to be important in the destruction of parasites. However, their contribution to the control of parasitic infections has long been controversial. The advent of genetic engineering has made it possible to produce mice with forced expression of a certain protein (transgenic mice) or mice with a disrupted gene (knockout mice). For example, interleukin-5 (IL-5) is a key cytokine for the growth and activation of eosinophils, and IL-5 transgenic and knockout mice have been used to evaluate the role of eosinophils in parasite infections (Ovington and Behm, 1997). *Nippostrongylus brasiliensis* and *Strongyloides venezuelensis* are widely used intestinal nematodes in these model systems. In mice and rats, their larvae start migrating through the connective tissues right after percutaneous or subcutaneous infection, and then they migrate to the lungs and subsequently the intestinal mucosa. During tissue migration, eosinophils infiltrate around migrating larvae. However, a substantial portion of larvae are permitted to reach the lungs without being killed, indicating that eosinophils are not able to destroy nematode larvae effectively in a primary infection. Further, the number of tissue-migrating larvae recovered from the lungs in a primary infection is not affected by anti–IL-5 monoclonal antibody treatment (Korenaga et al., 1991). On the other hand, in IL-5 transgenic mice, in which eosinophils are highly activated, the number of larvae that reach the lungs are much smaller than in normal wild type mice (Shin et al., 1997; Daly et al., 1999; El-Malky et al., 2003). Thus eosinophils appear to become able to kill nematodes when activated by IL-5.

In contrast to the results of intestinal nematode infections, in a primary infection with *Angiostrongylus cantonensis* in mice, larvae are completely eliminated even in nontransgenic mice. Because the number of larvae is significantly increased in IL-5 receptor knockout mice, eosinophils in *A. cantonensis* infected mice are fully capable of killing nematode larvae in a primary infection (Sugaya et al., 1997b). The difference between *N. brasiliensis*/*S. venezuelensis* and *A. cantonensis* is that the mouse is an adequate host for the former, but an unsuitable host for the latter. It appears that the state of eosinophil activation determines the host specificity early in the infection. Related to this, an interesting finding was obtained when IL-5 transgenic mice were infected with *Toxocara canis*. In contrast to *N. brasiliensis*/*S. venezuelensis* infections, *T. canis* larvae were recovered in similar numbers from tissues of IL-5–transgenic mice having activated eosinophils and from nontransgenic mice (Sugane et al., 1996). For *T. canis* larvae, ingestion by an unsuitable host must be accommodated for their survival strategy (Fig. 10.2). Among food-borne parasitic zoonoses, *Anisakis*, *Gnathostoma,* and *A. cantonensis* infection in humans could be equated to *A. cantonensis* type (protective eosinophil response), whereas sparganosis with *S. erinaceieuropaei* could be of the *T. canis* type (eosinophil response ineffective).

Mucosal Responses in Nonmammalian Animals

Host specificity depends on granular cells in nonmammalian animals, at least in some cases. In teleostean fish, mast cells and eosinophilic granule cells, together with rodlet cells, massively accumulate in the parasitized gills or intestinal tract (Reite, 2005; Reite and Evensen, 2006). As early as 1967, Kearn (1967) demonstrated that host-finding and host-specificity of monogenean skin parasites of the fish is determined by the nature of the mucous substance of the skin of host fish. In the chicken, implanted adult worms of *S. venezuelensis* are eliminated by heterophils infiltrated into the mucosal tissues (Maruyama et al., 2003). Possibly, the physicochemical nature of glycoconjugates present on intestinal mucosal surface and the responses of bone marrow–derived granular cells are determining factors for this host-parasite relationship.

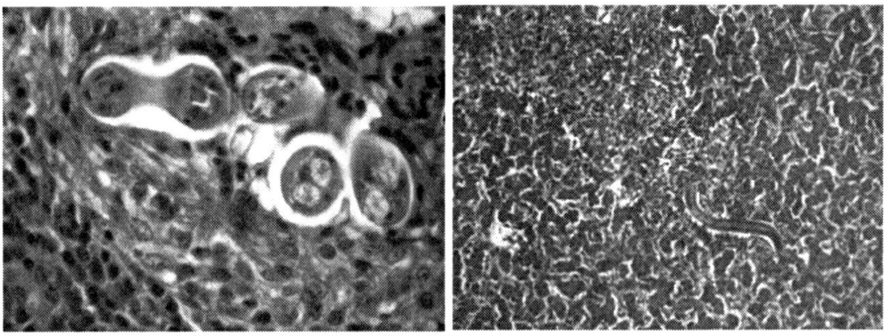

FIGURE 10.2. *T. canis* larvae in backyard chicken liver. Tissue section of backyard chicken liver infected with *T. canis*. A Japanese family that ate slices of the liver get infected with *T. canis* and presented with visceral larva migrans syndrome.

Mucosal Mast Cells

In mucosal tissues in mice and rats, a unique granular cell population, the mucosal mast cell (Fig. 10.3) appears in parasitic infections. As in eosinophils, the mucosal mast cell response has been recognized as a hallmark of worm infection; however, its role in protective immunity was for a long time obscure. Nawa and Korenaga (1983) first recognized that after concurrent infection with *N. brasiliensis* and *Strongyloides ratti* in rats, these parasites were separately expelled with different kinetics from the host. Interestingly, the expulsion of *N. brasiliensis* and that of *S. ratti* was closely associated with intestinal goblet cell hyperplasia in the former and mucosal mastocytosis in the latter infection (Nawa and Korenaga, 1983). Subsequently a crucial role for mucosal mast cells in the expulsion of *S. ratti* was clearly demonstrated in experiments with mast cell–deficient *W/Wv* mice, which have mutation in stem cell factor (SCF) receptor (Nawa et al., 1985; Abe and Nawa, 1987). Since then, an absolute requirement of mucosal mast cells for the expulsion of *Strongyloides* nematodes has been demonstrated repeatedly. Mice lacking in critical genes for mucosal mast cell responses also fail to expel *Strongloides* adult worms (Lantz et al., 1998; Fukao et al., 2002).

As for the mechanism of the expulsion by mucosal mast cells, it has been shown that glycosaminoglycans in mast cell granules, namely chondroitin sulfate

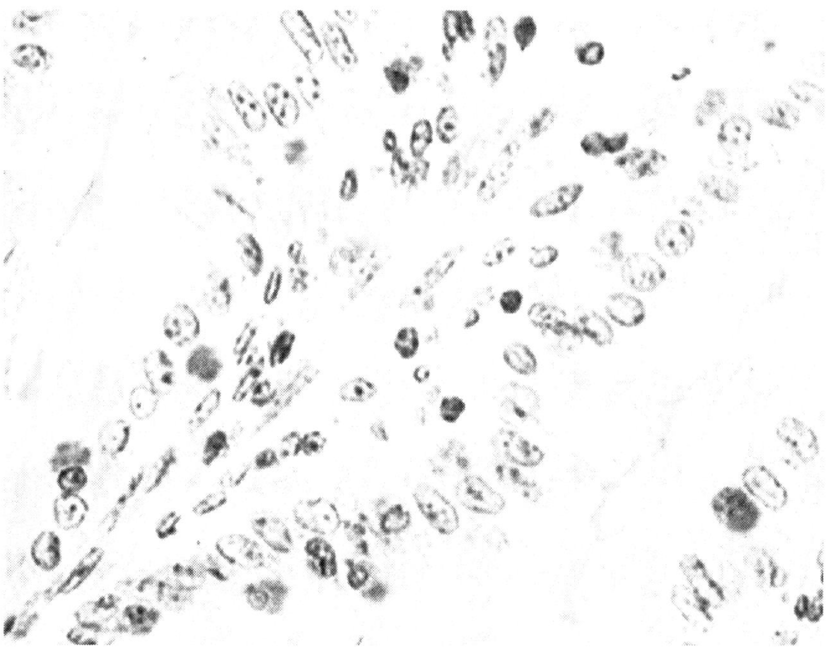

FIGURE 10.3. Mucosal mast cell. Tissue section of mouse intestine infected with Strongyloides venezuelensis. Mucosal mast cells contain granules with glycosaminoglycans, which are stained blue. (Alcian blue/Safranin O staining)

A and E, inhibit reentry of *S. venezuelensis* into the mucosal epithelium by blocking the attachment of adult worms to epithelial cell surface (Maruyama et al., 2000). Because *S. venezuelensis* adult worms continue exiting and reentering the epithelial layer, inhibition of reentry results in worm expulsion (Fig. 10.4). Glycosaminoglycans are highly sulfated polysaccharides and it has been demonstrated that any sulfated sugars, whatever the cell source, inhibit the mucosal invasion of *S. venezuelensis*. In hamsters, *S. venezuelensis* is expelled by sulfated mucin of goblet cells (Shi et al., 1994, 1995), and sulfated mucins of reserpine-treated rats (Ishikawa et al., 1995) and normal mouse large intestine (Maruyama et al., 2002) inhibit mucosal invasion of *S. venezuelensis* worms. Because sulfated sugars expel intestinal worms by making the mucosal surface slippery, worms that require tight adhesion to the mucosal surface would be expelled by mucosal mast cells and goblet cells with sulfated mucins. As Nawa et al. (1994) predicted, the physicochemical nature at the intestinal mucosa can be a determining factor for the establishment of an intestinal parasite.

Epithelial Cells

Following the discovery that rats expel both *N. brasiliensis* and *S. ratti* in concurrent infection but with different kinetics (Nawa and Korenaga, 1983; see above), a series of experiments were carried out to determine the critical factors in goblet cell mucins in the expulsion of *N. brasiliensis*. Surprisingly, it turned out that in rats goblet cell hyperplasia, which is a sign of parasitic infections, was not essential, but the biochemical structure of the mucin sugars was a critical factor for expulsion (Ishikawa et al., 1993, 1994). Ishikawa et al. suggested that the importance of sialated terminal sugars in inhibiting the establishment of *N. brasiliensis* is undisputed, although the precise mechanisms remain unclear. In *N. brasiliensis* infection in mice, it has been shown that IL-13 and the downstream signal transducer and activator of transcription-6 (STAT6) are critical for goblet cell proliferation and expulsion of *N. brasiliensis* (Urban et al., 1997).

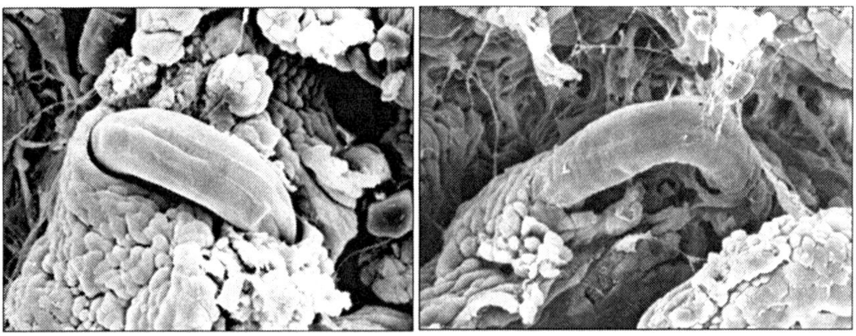

FIGURE 10.4. *S. venezuelensis* in intestinal mucosa. Adult worms exit and reenter the intestinal epithelial layer. Inhibition of reentry results in the worm expulsion

Mouse intestinal whipworm, *Trichuris muris*, is another often employed mouse-parasite model for investigations on mucosal immunology. This worm has been used for more than 30 years in experimental research, especially on T-helper-2 (Th2) responses (see below). However, no clear explanation has been given as to how this nematode is expelled. As the expulsion of *T. muris* is tightly associated with goblet cell hyperplasia, most study has been directed at finding goblet cell–derived inhibitory factors (Artis et al., 2004). Recently, Cliffe et al. (2005) demonstrated that an increase in the rate of epithelial cell turnover in the large intestine acts like an "epithelial escalator" to expel *Trichuris*. Also, the rate of epithelial cell movement is under immune control, in which IL-13 elevates the rate of epithelial cell turnover to displace the parasites, whereas chemokine CXCL10, also known as interferon-γ (IFN-γ)–induced protein 10, was inhibitory to cell turnover (Cliffe et al., 2005).

Adaptive T-Cell Responses

In contrast to helminth infections, in viral and bacterial infections, eosinophils, mucosal mast cells, and goblet cells do not play a major role in immunity. Because all these cells are induced and activated by T-cell cytokines, it is essential to know what kind of T cells secrete what kind of cytokines in infection. For nearly two decades, two lineages of functional CD4 T cells have been recognized, namely Th1 and Th2, providing a framework for understanding T-cell biology (Mosmann and Coffman, 1989; Infante-Duarte and Kamradt, 1999). Th1 cells secrete IFN-γ, whereas Th2 cells secrete IL-4, IL-5, IL-9, and IL-13. Because Th2 cytokines activate eosinophils, mucosal mast cells, and goblet cells, pathology in parasitic worm infections can be explained by the induction of Th2 cells. Although Th17 and regulatory T cell populations have been assigned to the adaptive T-cell lineages (Weaver et al., 2006), the roles of these cell populations have not been clarified yet in Th2 cell differentiation (Fig. 10.5).

Cytokine Milieu

A major question in parasite immunology is how and why Th2 cells are induced in worm infections. It is widely accepted that the differentiation and expansion of naive CD4 T cells into Th1 or Th2 are influenced by the cytokine environment in which T-cell priming occurs (Murphy and Reiner, 2002). Because IL-4 promotes Th2 differentiation in vitro, a responsible cellular source of IL-4 has been sought in parasitic infections. Among various cell types that express IL-4, eosinophils are demonstrated to be the most prevalent IL-4–producing cells that appear in the lungs of mice infected with *N. brasiliensis* (Shinkai et al., 2002; Voehringer et al., 2004). However, recently, it has become apparent that this is not so simple. It is becoming clear that Th2 responses can be induced by multiple pathways, and partitioned into several subsystems. For example, although IL-4/IL-13 was earlier thought essential for Th2 differentiation (Takeda et al., 1996), another line of evidence now indicates that IL-4/IL-13 is not essential for Th2 cell differentiation in vivo, because both IL-4R– and STAT6-deficient mice generate substantial

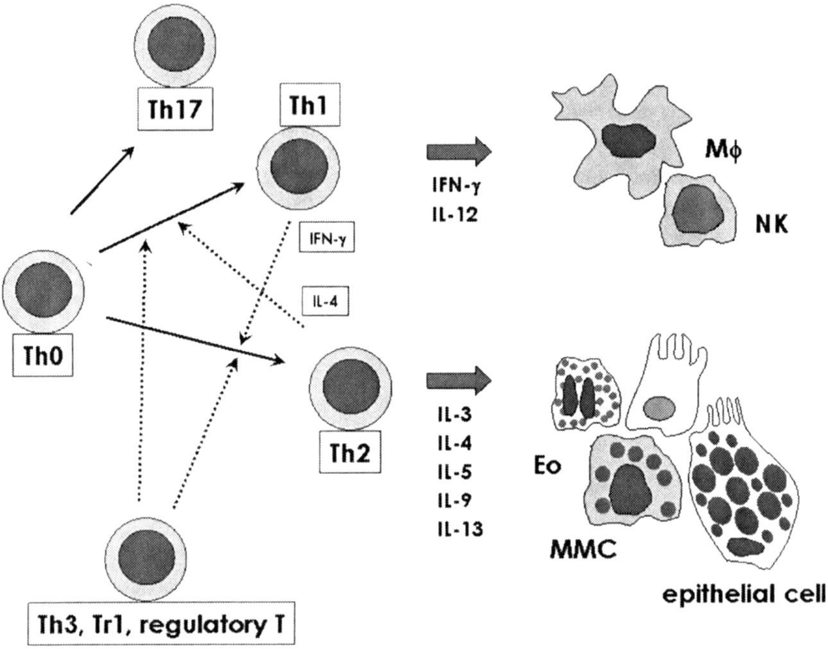

FIGURE 10.5. T-cell lineages in parasitic infections. The native CD4 T cell is a multipotent precursor that differentiates into Th1, Th2, Th17, and regulatory T cells (Th3 and Tr1) depending on signals from the innate immune system. Th1 and Th2 are cross-inhibitory, in which IL-4 from Th2 cells inhibit differentiation of Th1, and IL-12 from Th1 cells blocks Th2 differentiation. Th3, Tr1, and naturally occurring regulatory T cells are inhibitory for Th1 and Th2 responses. Dotted arrow: inhibitory effects.

numbers of Th2 cells (Noben-Trauth et al., 1997; Jankovic et al., 2000; Mohrs et al., 2001). Innate hematopoietic cell-derived IL-4/IL-13 is unnecessary for Th2 cell differentiation in lymph nodes but is required for effector cell recruitment and tissue responses (Voehringer et al., 2006). A major reason for this revision is that the indicator employed for Th2 responses was not appropriate. Furthermore, a recent series of investigations show that IL-18, which is known as a proinflammatory cytokine, plays a key role for the switching of Th1 and Th2 responses and is dependent on its cytokine milieu (Yoshimoto and Nakanishi, 2006). Interleukin-18 and IL-12 synergistically induce T cells, B cells, natural killer (NK) cells, macrophages, and dendritic cells (DCs) to produce IFN-γ (Nakanishi et al., 2001). In contrast, IL-18 with IL-2 induces such Th2 cytokines as IL-3, IL-4, IL-9, and IL-13 in T cells, basophils, and mast cells (Sasaki et al., 2005).

Antigen-Presenting Cell

Dendritic cells (DCs) are professional antigen-presenting cells that have the ability to stimulate naive T cells to differentiate into Th1 and Th2 subsets (Banchereau and Steinman, 1998; Shortman and Liu, 2002). In vivo depletion of DC abolishes eosinophilic inflammation and goblet cell hyperplasia in an asthma model, and endogenous or adoptively transferred CD4 Th2 cells into a DC-depleted animal do not produce IL-4, IL-5, or IL-13 (van Rijt et al., 2005). Regulatory pathways and signaling events by which the DC instructs Th2 cell differentiation remain poorly defined; however, recent studies demonstrated the importance of NF-κB1 and Ym1 expression within DC (Artis et al., 2005; Arora et al., 2006). Ym1 is a lectin that binds heparin/heparan sulfate with a shared homology to chitinases of lower organisms, though it does not have enzymatic activity (Chang et al., 2001). Macrophages activated by Th2 cytokines in vivo produce high levels of Ym1 (Chang et al., 2001; Welch et al., 2002) and Fizz1 (Raes et al., 2002; Nair et al., 2003). In antigen-presenting cells in nematode infected mice, acidic mammalian chitinase and Fizz2, another Fizz family member, are also upregulated (Nair et al., 2005). Fizz2 is a goblet-cell–specific immune-effector molecule in the intestinal nematode infection (Artis et al., 2004), and is another example of the interchangeability between inflammatory cell and epithelial cell, as was discussed above on sulfated glycans in mast cells and goblet cells.

Immunological Aspects of Diseases

Parasites with Minimal Tissue Invasion

Intestinal Flukes

About 70 species of intestinal flukes (trematodes) have been known to infect humans (Yu and Mott, 1994), mainly in Asian countries (Liu and Cheng, 1996; Chai, 2005; Waikagul and Radomyos, 2005). Parasite occurrence depends on the presence of intermediate hosts, reservoir hosts, and risky culinary habits of local people. A majority of intestinal flukes are fish-borne zoonoses, but aquatic plants, snails, frogs, insects, and snakes can also be a source of infection with some species of intestinal flukes

Because the morbidity of intestinal fluke infections is relatively low, and because patients are relatively easily diagnosed by stool examination and easily treated by trematodicidal drugs, systematic research work on pathology and immunology on intestinal fluke infections in humans is largely lacking. Recently, Toledo et al. (2006) made an extensive collective review on immunology and pathology of six selective families of relatively common intestinal trematodes—Brachylaimidae, Diplostomidae, Echinostomatidae, Gymnophallidae, Heterophydae, and Paramphistomiade—in their definitive hosts. As was proposed by Toledo et al. (2006), the knowledge derived by the nematode/rodent model is insufficient for the understanding of immunological as well as pathological events in chronic parasite infections. In this context, the use of intestinal flukes as experimental models may provide new insights into host responses that facilitate intestinal helminth resistance (Toledo and Fried, 2005).

An interesting point for future investigation in conjunction with immunology is the narrow host specificity of some intestinal flukes. For example, *F. buski* can develop into mature adult worms in pigs and rabbits (Wu, 1947), but is unable to develop in goats, sheep, cows, donkeys, dogs, cats, guinea pigs, rats, and mice (Wu, 1947; Li, 1980). It is not known how *F. buski* is able to discriminate between these animals to show such a remarkable host specificity.

Like intestinal nematode infections, intestinal flukes also induce villous atrophy, crypt hyperplasia, intestinal mastocytosis, and goblet cell hyperplasia, as has been observed in *M. yokogawai*, *Neodiplostomum seoulense*, and *Echinostoma hortense* infection in laboratory rodents (Chai, 2005), though whether those cells play a role in protection or expulsion remains unclear. While no significant increases in the number of Paneth cells, goblet cells, and mast cells were observed in *E. caproni* infections, there was a marked increase in those cells in *E. trivolvis* infection (Fujino and Fried, 1993). *E. caproni* may cause some level of immunosuppression and thereby parasitize the intestine for a longer period. Although goblet cell mucin seems important for expulsion of *E. trivolvis*, Frazer et al. (1999) noted marked goblet cell response in Recombination-activating gene (RAG)-2-deficient mice that could not expel *E. trivolvis*.

Clonorchiasis and Opisthorchiasis (Liver Flukes)

Clonorchis sinensis, distributed in Far East Asian countries, and *Opisthorchis viverrini*, distributed in the Indochina Peninsula, are two important liver flukes, of which infections occur by ingesting freshwater fishes; 30 million people are supposed to be affected with these parasites (World Health Organization, 1995; Crompton, 1999). Adult worms dwell in the bile duct, causing chronic hepatobiliary disease. Because of their relatively small size and their long life span, morbidity primarily depends on the number of worms and the duration of infection. With ongoing exposure, reinfection readily occurs following curative treatment, and cumulative infections result in significant morbidity and a predisposition to cholangiocarcinoma (for clonorchiasis, Chen et al., 1994; for opisthorchiasis, Sithithaworn and Haswell-Elkins, 2003). Epidemiological studies suggest that humans do not develop resistance to reinfection or superinfection with *C. sinensis*, and that reinfection occurs whenever reexposure happens throughout life in those accustomed to consuming undercooked fish in endemic areas (Seo et al., 1981; Hong et al., 1994). A comparable situation is seen in *Opisthorchis viverrini* infection in that the prevalence of liver cirrhosis and cholangiocarcinoma is extremely high in endemic areas (Sithithaworn et al., 1994; Sithithaworn and Haswell-Elkins, 2003), although a possible protective role for IgA and IgM antibody is suggested from a seroepidemiological study on egg-positive and -negative opisthorchiasis cases (Akai et al., 1994).

In spite of a failure to develop an effective protective immunity against reinfection/superinfection, potent antibody production is induced by liver fluke infections. As usual for trematodes (Sajid and McKerrow, 2002), excretory/secretory products are the major source of antigenic components, and cysteine proteinases are a major and reliable antigen for immunodiagnosis for clonorchiasis (Na et al., 2002; Nagano et al., 2004). Specific antibody against a 17-kd antigen, cysteine proteinase, of *C. sinensis* cross-reacts with *O. viverrini* (Chung et al., 2000) and these two liver flukes share several antigenic components (Choi et al., 2003). In contrast to the apparent situation with humans, potent protective immunity can be induced in experimental infection in rats (Chung et al., 2004) and mice (Kwon, 1987). Protective immunity against *C. sinensis* was successfully induced in rats by a DNA vaccine encoding cysteine protease (J.S. Lee et al., 2006). At the moment, the precise mechanisms responsible for protective immunity in rats remain unclear. Since *C. sinensis* dwells in the bile duct lumen, it might be expelled by a mechanism similar to the one that operates against intestinal nematodes in the gut, in which mast cells and goblet cells play key roles as selective effectors against intestinal nematodes (Nawa et al., 1994). Related to this, histochemical changes in the nature of biliary mucins have been reported in *C. sinensis* infestation (Sheung-To and Gibson, 1970).

In *O. viverrini* infections in laboratory animals, Silian golden hamsters have been commonly used as a model animal because of the similarities of pathological finding in chronic infections to those of humans, along with the development of partial protective immunity (Flavell, 1982). Using the *O. viverrini*–hamster model, the effect of immune status on the severity of pathological changes have been studied in detail (Flavell and Flavell, 1986; Wongratanacheewin et al., 1987). Recently, *O. viverrini* antigen was found to induce Toll-like receptor 2 in a macrophage cell line (Pinlaor et al., 2005). Because cytokine expression profile analyses in *O. viverrini*–infected hamsters have just recently begun (Jittimanee et al., 2007), it is hoped that identification of each cytokine involved in the immune response and pathology of opisthorchiasis will occur in the near future.

Intestinal Capillariasis

Intestinal capillariasis is a life-threatening disease because once humans are infected with this nematode, it may multiply in the gut (autoinfection), resulting in severe mucosal damage, which is observed clinically as watery diarrhea, malabsorption syndrome, and protein-losing enteropathy. Since the first outbreak recorded in northern Luzon, the Philippines, the disease has been gradually spreading around the world (Hong and Cross, 2005).

Due to the rapid progress of clinical symptoms by autoinfection in humans, protective immunity in the gut appears not to operate against this nematode. Although the definitive host for *C. philippinensis* in natural life is thought to be fish-eating birds, the parasite does not appear to be fatal to birds (Cross and Basaca-Sevilla, 1983). On the other hand, Mongolian gerbil was found experimentally to be a definitive host for this parasite, allowing autoinfection (Cross et al., 1978). The Mongolian gerbil is a suitable experimental model not only for rodent strongyloidiasis (Horii et al., 1994) but also for human strongyloidiasis in causing autoinfection (Nolan et al., 1993). Recently, a Mongolian gerbil was found to have deficiency to T-independent antigens (Mohanty and Ravindran, 2002). Whether such immunodeficiency has some relation to high susceptibility of Mongolian gerbils to various parasites (Nawa, 2002) should also be examined.

Because symptomatic capillariasis cases can be easily diagnosed by stool examination, and, once diagnosis is made, the patient can be easily treated by anthelminthic benzimidazole derivatives, as a consequence, little attention has been paid to the immunology of intestinal capillariasis. Banzon et al. (1975) reported that patients' sera show remarkable cross-reactivity to *Trichinella spiralis* and *Trichuris vulpis* antigens but not with *Schistosoma* antigens, illustrating close phylogenetic relationship between *Capillaria* and whipworms. Later on, using this cross-reactivity, Intapan et al. (2006) developed a serodiagnosis for intestinal capillariasis by immunoblot analysis using *T. spiralis* antigen as a probe. For the diagnosis for asymptomatic stage of capillariasis, El Dib et al. (2004) reported the usefulness of sandwich enzyme-linked immunosorbent assay (ELISA) to detect coproantigen.

Parasites with Extensive Tissue Migration

Anisakidosis

Anisakidosis is primarily an acute gastrointestinal disease caused by infection with larvae of subfamily Anisakinae, a nematode parasites of marine mammals. Although worms belonging to four genera, *Anisakis*, *Pseudoterranova*, *Contracaecum*, and *Phoconema*, are known as the causative agents for human diseases (Garcia, 2001), the vast majority of human cases are infection with *Anisakis simplex*. The Anisakinae larvae reside in the muscles and the visceral organs of marine fish, and human infection occurs by ingesting raw or undercooked fish. Thus, the majority of anisakidosis patients have been found in Japan, where people frequently consume sushi and sashimi dishes, although an apparent increase in the number of patients during the 1970s and 1980s is mainly due to popularization of gastrointestinal (GI) endoscopy (Oshima, 1987; Nawa et al., 2005). After being ingested, the larvae preferentially penetrate into the stomach wall, causing acute abdominal pain, nausea, and vomiting within a few minutes to several hours (gastric anisakidosis). Due to acute onset and severity of the disease, most of gastric cases are diagnosed and the worms are extirpated by GI endoscopy at the emergency wards. More than 90% of anisakidosis cases are of this type. Based on clinical observations, gastric anisakidosis can be divided into two types: the fulminant form and the mild form. In the fulminant form, mucosa is edematous with spotted bleeding or small erosions. Those changes are thought to be the results of a local anaphylactic response (Ohtaki and Ohtaki, 1989). In Japan, Nishino et al. (1990) reported that about 60% of apparently healthy adults were skin test positive and about 40% were indirect hemagglutination antibody (IHA) test positive against *Anisakis* antigen. Using more sensitive and specific antigen-capture ELISA, about 10% of healthy adults in Japan were found to be seropositive against *Anisakis* antigen (Takahashi et al., 1993). The significance of presensitization on clinical manifestation of GI anisakidosis has been stressed with reference to immediate hypersensitivity (Alonso et al., 1997).

Although the frequency is far less than that of gastric anisakidosis, occasionally the larvae pass through stomach and invade the intestinal mucosa (intestinal anisakidosis). Diagnosis for intestinal anisakidosis is often problematic because of the difficulties of detection of worms by GI endoscopy. In addition in Japan, a problem exists with regard to differential diagnosis of intestinal anisakidosis from another emerging larva migrans due to infection with *Spirurina* type X, which causes abdominal symptoms, including intestinal obstruction, similar to intestinal aniskidosis (Yoshikawa and Ishizaka, 2005). Only one laboratory in Tokyo can perform specific immunodiagnosis for *Spirurina* type X (Goto et al., 1998). In extraordinary cases, larvae accidentally migrate into the peritoneal cavity or other visceral organs (extra-GI anisakidosis) to form chronic eosinophilic granulomas. The presence of larvae is disclosed by histopathological examinations for biopsy specimens removed during surgery for neoplasma (Yoshimura, 1990). Pleurisy due to migration of *Anisakis* larva has been reported rarely in Japan (Matsuoka et al., 1994).

Apart from actual *Anisakis* infection in the GI tract, which is almost exclusively seen in Japan, the relationship between *Anisakis* and allergy has become a hot topic in Europe, especially in Spain, where anisakidosis is still endemic (Fernandez de Corres et al., 1996; Valinas et al., 2001; Daschner and Pascual, 2005). Because of its food-borne nature, *Anisakis* is assumed to be a potent allergen, causing gastroallergic diseases (Lopez-Serrano et al., 2000; Audicana et al., 2002). Interestingly, some researchers have emphasized the importance of the presence of live worms for sustaining sensitivity (Sastre et al., 2000; Garcia et al., 2001), while others regard even a dead worm or frozen ES product as sufficient to cause allergic reaction (Audicana et al., 1997; Audicana et al., 2002; Falcao et al., 2002). Since transient passage of worms, either dead or alive, would cause GI symptoms such as abdominal pain, nausea, and vomiting, it is difficult to discriminate Anisakis-related food allergy from actual infection with larvae (Daschner et al., 2000).

In wildlife, although both porpoises and whales are known as the definitive host for *Anisaxis simplex* (Kuramochi et al., 1996), mucosal damage of minke whales was very mild in spite of the presence of large numbers of fully mature adult worms, whereas severe erosion or ulceration is seen with mostly immature worms in the gastric mucosa of porpoises (Uchida et al., 1998). Immunological reaction in porpoises might cause gastric lesions and, at the same time, would cause growth retardation of the parasite.

A major problem in exploring the relationship between *Anisakis* and allergy resides in the high degree of cross-reactivity of *Anisakis* allergen with that of other nematodes (Iglesias et al., 1996) or even that of house dust mites (Johansson et al., 2001; Bernardini et al., 2005) or of certain insects (Pascual et al., 1997). As mentioned as above, many Japanese are sensitized with *Anisakis* allergen and also with house dust mite allergen (Japanese Ministry of Health, Labor, and Welfare, 2006). Thus, analysis of results of epidemiological surveys on the relationship between *Anisakis* and allergy is an extremely complicated task. A recombinant protein of an IgE-reactive antigen (Arrieta et al., 2000) is claimed to be useful for the diagnosis of *Anisakis* allergy (Lorenzo et al., 2000).

As is usual in helminth parasite infections, Th2 type responses such as peripheral blood eosinophilia and parasite-specific IgE response occur in anisakidosis

(Buendia, 1997; Pichler, 1999). Recently, however, using a murine model for investigations on *Anisakis* allergy, anaphylaxis induced by parasitic proteins displayed a mixed Th1/Th2 pattern (Baeza et al., 2005). Since *Anisakis* is a multicellular organism composed of complicated components, it is not surprising that helminth parasites stimulate a mixed Th1/Th2 responses, though it makes it difficult to study *Anisakis* allergy from the immunological point of view.

Gnathostomiasis

Among over 10 species of the genus *Gnathostoma* (Daengsvang, 1980), five species, *G. spinigerum*, *G. hispidum*, *G. nipponicum*, *G. doloresi*, and *G. binucleatum*, were found responsible for causing cutaneous larva migrans (Nawa, 1991). While the first four species are endemic in Asia, *G. binucleatum* is responsible for the outbreak of gnathostomiasis in Latin America (Koga et al., 1999; Diaz-Camacho et al., 2002). These five species can be divided into two groups on the basis of the clinical features induced in humans (Nawa and Nakamura-Uchiyama, 2004). *G. spinigerum* and *G. binucleatum* cause long-lasting recurrent infection in a relatively deeper part of the skin of the peripheral site of the body, whereas *G. hispidum*, *G. nipponicum*, and *G. doloresi* larvae preferentially migrate into the surface skin of the trunks with a short duration (<3 months) (Fig. 10.6). Such

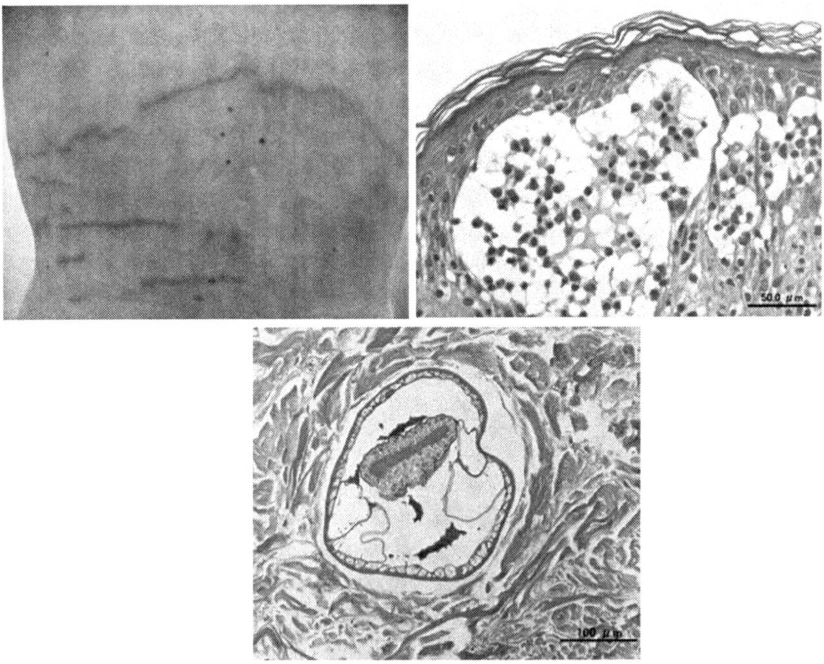

FIGURE 10.6. Creeping eruption in gnathostomiasis. Skin eruption made by a *G. doloresi* in a middle-aged man. Tissue section of a biopsy specimen shows eosinophilic infiltration in epidermis.

differences in clinical manifestation are probably in some way related to the antigenicity/immunogenicity of the two groups; *G. spinigerum* and *G. binucleatum* are supposed to be less immunogenic and adapt well to the human host, while short duration of the disease caused by the latter three species are presumably related to immune-mediated eradication of the larvae (Nawa and Nakamura-Uchiyama, 2004). Conversely, this could be a smart strategy of *Gnathostoma* larvae, because Koga and Ishii (1990) found that *G. hispidum* larvae could survive for years encysted in the muscles of experimentally infected rats. In nonmammalian hosts like fish, amphibians, reptiles, and even birds, *Gnathostoma* larvae remain encysted for long period without signs of an inflammatory response (Miyazaki, 1991).

Although definitive diagnosis of gnathostomiasis depends on detection of larvae removed from skin lesions, the detection rate is very low because of the rapid migrations of *Gnathostoma* larvae in the host tissue. Furthermore, the clinical manifestation of cutaneous gnathostomiasis is somewhat similar to cutaneous larva migrans caused by animal hookworm larvae (Nakamura-Uchiyama et al., 2002b) or by type X larvae of *Spirurina* (Taniguchi et al., 1994; Yoshikawa and Ishizaka, 2005). Thus, specific antibody detection by ELISA has been developed for diagnosis for gnathostomiasis (Suntharasamai et al., 1985) and has been in use for a number of years. For practical purposes, crude somatic extract of adult worms has been frequently used because of its easy availability, although efforts to detect species-specific antigens for *G. spinigerum* (Anantaphruti, 1989a,b), the species complex of *Gnathostoma,* appear to share potent common or cross-reactive antigens. *G. doloresi* antigen prepared in Japan has proved useful in ELISA tests used to detect *G. binucleatm* in Ecuador (Mimori et al., 1987) and in Mexico (Ogata et al., 1998), where the latter is endemic. In fact, adult worm extract of three *Gnathostoma* species showed extremely high antigenic similarities in ELISA reactivity against Mexican and Japanese patients (Ishiwata et al., 2003). Advantages of using purified adult (Nopparatana et al., 1991) or larval (Rojekittikhun et al., 1993; Uparanuknaw et al., 1999) somatic antigens, larval ES antigen (Saksirisampant et al., 2001), or cuticle antigen (Dantrakool et al., 2001) has been reported, but these have limited availability. Recently, ES antigen of *G. binucleatum* larvae was produced, and the major antigenic components were found to be metalloproteinases (Cabballero-Garcia et al., 2005). Molecular cloning of a gene encoding matrix metalloproteinase–like protein from *G. spinigerum* has been accomplished (Uparanukraw et al., 2001). Since matrix metalloproteinases of nematodes are generally important for tissue invasion (Tort et al., 1999), the biological role of metalloproteinases should be investigated.

Paragonimiasis

Paragonimiasis is a subacute to chronic inflammatory lung disease caused by infection with lung flukes, the *Paragonimus* species. Human infection occurs mainly by ingesting freshwater crustaceans (crabs/crayfishes) contaminated with metacercariae (infective stage) of the parasites. Wild boar was proven as the paratenic (transport or alternate intermediate) host for *P. westermani* (Miyazaki,

1991), and infection by ingesting raw meat of wild boars is still common in Japan (Nakamura-Uchiyama et al., 2002a).

Paragonimiasis patients almost exclusively exhibit peripheral blood eosinophilia and elevated total IgE in the serum (Nakamura-Uchiyama et al., 2002a), showing that *Paragonimus* worms stimulate a Th2 response. The highest eosinophilia observed so far was in a 5-year-old boy infected with *P. westermani* and whose WBC was about 100,000/μL, more than 90% of which were eosinophils (Kan et al., 1995). A high degree of eosinophilia in association with a high level of cytokines is often observed in pleural effusions of paragonimiasis patients (Taniguchi et al., 2001; Matsumoto et al., 2002). In spite of such potent induction of a Th2 type response, there is no clear-cut evidence that any protective immunity is induced by *Paragonimus* infection. In fact, once *Paragonimus* worms reach in the lungs of definitive mammalian hosts (humans, cats, dogs), they can survive for many years. Although infection frequency is extremely low, among the over 300 cases of paragonimiasis referred to the authors' laboratory in Miyazaki, Japan, two cases of repeat paragonimiasis was detected: one a 65-year-old man who had a previous history of paragonimiasis 30 years previously, and the other a 35-year-old man who had a twice previous history of paragonimiasis at the age of 16 and 27. In both cases, the patients do not have immunodeficiency and were successfully treated with praziquantel at the dose of 75 mg/kg/day for 3 days.

Evasion by *Paragonimus* from the host's defense can be explained, at least in part, by the immunosuppressive effects of its excretory-secretory proteases (see below). Since the earlier work by Yamakami (1986), an array of cysteine proteinases and thiol proteinases have been identified at various maturation stages of *P. wesetermani* (Chung et al., 1997a; Choi et al., 2006) as a part of the attempts to understand the host–parasite interplay in paragonimiasis. The biological significance of *Paragonimus*- and other parasite-derived proteinases is discussed later in the chapter. Apart from biological functions, cysteine proteases of *Paragonimus* were proven as the major diagnostic antigens (Ikeda et al., 1996; E.G. Lee et al., 2006).

While *P. westermani* can evade from immune defense in humans, cats, and dogs, to become fully mature adult worms, they remain juvenile stage in wild boars, a paratenic host in human epidemiology in Japan (Miyazaki, 1991). Whether any immunological processes are involved in the developmental arrest of *P. westermani* worms in wild boars is unknown, but is of interest.

Fascioliasis

Fascioliasis is caused by infection with the liver fluke, *Fasciola* sp. Infection in humans occurs by ingesting aquatic plants contaminated with the metacercariae, and as an unique route, by ingestion of bovine liver as sashimi (Taira et al., 1997). Clinically, fever and abdominal pain are associated with marked eosinophilia and elevation of total IgE level in serum, showing that *Fasciola* induces a typical Th2-type response in host animals. The egg detection rate is extremely low in human fascioliasis, so that immunodiagnostic measures have been developed for diagnosis for this disease. Similar to those of *Paragonimus*, the major diagnostic antigens of *Fasciola* are cysteine proteases in ES (see below).

Fascioliasis is one of a very few food-borne zoonotic helminths that induces a potent protective immunity in the natural definitive hosts, cattle and sheep (Ross, 1967), and also in experimental hosts such as rats (Rajasekariah et al., 1979). Destruction of newly excysted juveniles occurred within 1 to 5 hours after challenge infection in immune rats with a high correlation to specific IgG1 antibody and the local accumulation of eosinophils (van Milligen et al., 1999). Along with the development of immunodiagnostic measures and the identification and purification of antigens, extensive work has been done on the development of vaccine against *Fasciola* sp. (Emery, 1996; Spithill et al., 1997; Spithill and Dalton, 1998; Smith, 1999; Knox et al., 2001; Dalton and Mulcahy, 2001; Dalton et al., 2003a,b).

Several molecules have been found to be effective experimentally as candidates for a vaccine. While diagnostic antigens require high specificity and sensitivity, vaccine candidate antigens are not necessarily so specific, because they should elicit a broad protection spectrum, even interspecies or -genera. In this regard, Hillyer and his group (Hillyer and Serrano, 1982; Hillyer et al., 1988) demonstrated cross-protection against *Schistosoma mansoni* using *F. hepatica* tegument antigen. Subsequently, *S. mansoni* fatty acid binding protein, Sm14, was found to be the potential vaccine candidate for not only schistosomiasis but also fascioliasis due to its cross-reactivity (Tendler et al., 1995, 1996; Muro et al., 1997). A possible counterpart molecule of Sm14 was isolated from *F. hepatica* (Hillyer, 1995). A similar molecule has been found in *Clonorchis sinensis* (Lee and Yong, 2004). Paramyosin has been proposed as a vaccine candidate for *Fasciola* and for *S. japonicum* (Chen et al., 2000; Cancela, 2004), though their intergeneric cross-reactivity has not been explored.

Apart from structural proteins, several enzymes are also identified as the vaccine candidates. Among them, glutathione S-transferase was proposed as the potent vaccine candidate for *Fasciola* and its antigenic epitope was identified by a three-dimensional model (Sexton et al., 1994; Rossjohn et al., 1997). Field trials have been carried out on cattle (Morrison, 1996) and sheep (Paykari et al., 2002) with promising results. Catepsin L–like proteinase could also induce protective immunity (Dalton et al., 1996, 2003b; Mulcahy and Dalton, 2001), and its antigenic epitope was identified within the propeptide sequence (Harmsen et al., 2004). Interestingly, *Fasciola* catepsin L itself induces a Th1 response (Bentancor et al., 2002), while it can also suppress Th1 responses against bacteria (Brady, 1999).

While most of these vaccine candidate molecules were identified from somatic or ES products of adult worms, van Milligen et al. (2000) reported that newly excysted juvenile worms contain potent antigen that can successfully induce protective immunity. Recently, Espino and Hillyer (2004) demonstrated the immunoprophylactic potential of a recombinant saposin-like proteins (SAPLIPs) of *F. hepatica*, and considered this molecule as a vaccine candidate (Espino et al., 2005; Torres and Espino, 2006). The SAPLIPs are a diverse family of lipid-interacting proteins widely distributed among protozoa to mammals, and are involved in membrane-permeability (Bruhn, 2005), being identified also in various other parasites such as *Clonorchis sinensis* (Lee et al., 2002) and *Ancylostoma caninum* (Don et al., 2007), and are considered to play an important role for blood-feeding by the parasite. Antibodies against such molecules conceptually could prevent nutrient uptake, thereby resulting in host protection.

While various vaccine candidate molecules have been identified and successful results were obtained at the experimental level, a major bottleneck for practical application is the requirement for the adjuvant to obtain long-lasting effective protective immunity. To overcome adjuvant problems, the DNA vaccine approach was developed as an effective way of antigen delivery to host animals. DNA vaccines of glutathione S-transferase (Smooker et al., 1999), fatty acid binding protein and cathepsin L (Smooker et al., 2001), and saposin (Espino et al., 2005) of *Fasciola* have been tested in laboratory animals with promising results.

Sparganosis (Spirometrosis)

Spirometra erinaceieuropaei (also called *Sparganum mansoni*) is a tapeworm parasitizing the intestine of dogs and cats. Its larva (plerocercoid) looks like a white tape of about 10 to 20 cm (up to 70 cm) in length, and resides in the connective tissues, muscles, or viscera of various amphibians, reptiles, birds, and mammals. Infection in humans occurs in two ways: by ingesting cyclops contaminated with procercoids, or by ingesting raw or undercooked meat of those animals contaminated with plerocercoids. The disease is usually called sparganosis. Patients can be seen worldwide, especially in East and Southeast Asian countries. In Japan, sashimi of frogs, snakes, and backyard chicken are the major sources of human infection. Some 500 cases have been reported in Japan (Yoshida, 2006), over 100 cases in Korea (Min, 1990), and 34 cases (predominantly ocular cases) in Thailand (Wiwanitkit, 2005), but this is probably an underestimate.

In the human body, larvae preferentially appear in the subcutaneous tissues of the anterior chest, the abdominal wall, or the inguinal region, and form slow-growing migratory nodular lesions without causing pain or redness. Larva may occasionally migrate into an unexpected site of the body such as the pleural cavity (Ishii et al., 2001) or the central nervous system (CNS) (Nobayashi et al., 2006), causing unusual or even fatal manifestations. A total of 11 cases of cerebral sparganosis have been reported in Japan (Nobayashi et al., 2006) and 16 cases in Korea (Min, 1990). Diagnosis is based on histological identification of the worm or serologic tests. Surgical removal of the worm is the only practical treatment. If head and neck portions of the worms have been left behind, the worms can regenerate from this residual portion.

For diagnostic purposes, crude somatic extracts of sparganum have been used as the antigen for ELISA, although more purified antigens are being sought. Cho et al. (1992) analyzed antigenic components in ES preparations and found protease activities. Among those, 31- to 36-kd proteins have been shown to be highly specific and to be useful for the diagnosis of human sparganosis (Morakote and Kong, 1993). Interestingly, the carbohydrate moieties of those antigens vary depending on the intermediate/paratenic hosts utilized (Yang, 2004).

Related to pathological changes or host–parasite interactions, proteolytic enzymes secreted by parasites are thought to play important roles in invasion or migration through host tissue. Three species of neutral serine proteases (Kong et al., 1994a) and three species of cysteine proteases have been identified in *Sparaganum* (Kong et al., 1997). Among those, the 27-kd cathepsin L–like enzyme has IgG cleaving activity (Kong et al., 1994b) and IgE provoking activity (Kong et al., 1997), and is expressed only in the coracidium and plerocercoid stages, but not detected in eggs and adult worms (Kong et al., 2000). Plerocercoids of spirometric tapeworms produce a growth factor, which interacts with growth hormone (GH) receptors and mimics many of the biological actions of GH (Phares, 1987). Surprisingly this GH activity is coexpressed with neutral cysteine proteinase activity of the 27.5-kd cathepsin L–like enzyme (Phares and Kubik, 1996). In terms of biological functions, ES products from plerocercoids of *S. erinaceieuropaei* also affect phagocytic cells in mice. Tumor necrosis factor-α (TNF-α), IL-1β, inducible nitric oxide synthase (iNOS), and chemokine production by lipopolysaccharide (LPS)-stimulated macrophages is suppressed by ES products of *S. erinaceieuropaei* (Fukumoto et al., 1997; Miura et al., 2001; Dirgahayu et al., 2004). Molecular cloning of parasite-derived immunosuppressant will facilitate our understanding of immune system of host animals and their responses to parasite infection.

Angiostrongyliasis

Angiostrongylus cantonensis, also known as a rat lung worm, is a blood-dwelling nematode, of which adult worms live in the pulmonary arteries of wild rats. Infective stage (L3) of this parasite resides in snails and slugs, and after being ingested, they migrate into CNS where they become young adults. In rats, a permissive (suitable) host, they can gain entry into the cerebral veins and reach pulmonary arteries, where they become fully mature adults. In contrast, in nonpermissive hosts, such as mice, guinea pigs, rabbits, or humans, the parasites in the CNS are unable to migrate further, and are eventually killed by the host defense, primarily by eosinophils in the CNS (Yoshimura et al., 1994). Humans are, together with mice, typical nonpermissive hosts for *A. cantonensis*, making this parasite the most famous pathogen for eosinophilic meningoencephalitis (Lo Re and Gluckman, 2003), and *A. cantonensis* infection in mice is an excellent model for understanding pathology and immunology of human angiostrongyliasis.

A protective role of eosinophils against intracranial worms in nonpermissive hosts has been clearly demonstrated by a series of experiments using mainly an *A. cantonensis*–infected mouse model by Yoshimura et al. (1994). Intracranial worms stimulate $CD4^{+}$(Th2) cells to produce IL-5 (Sugaya et al., 1997a), which leads to bone marrow eosinopoiesis, followed by peripheral eosinophilia (Sugaya and Yoshimura, 1988). Interleukin-5 transgenic mice, which have constitutive eosinophilia with activated eosinophils, showed significant resistance against this parasite (Sugaya et al., 2002). Eosinophils accumulate in cerebrospinal fluid (CSF) and adhere to, and damage, intracranial worms by releasing various toxic substances contained in their granules (Yoshimura et al., 1994). Unfortunately for nonpermissive host animals, such deleterious effects of eosinophils also cause damage of the central nervous tissues of the host (Yoshimura et al., 1994), resulting in eosinophilic meningitis.

In terms of diagnostic purposes, it is practically difficult to detect intracranial *A. cantonensis* worms in humans, so that immunodiagnostic methods for sera or CSF of patients have been developed (Cross and Chi, 1982; Chen, 1986). Cross-reactivity is a large problem with these tests, and to improve specificity and sensitivity, various antigenic components have been purified and their clinical applicability has been evaluated (Shih and Chen, 1991; Nuamtanong, 1996; Chye et al., 2000; Maleewong et al., 2001). In addition to these diagnostic antigens, Maki and Yanagisawa (1986) identified carboxyl and thiol proteases in *A. cantonensis* adult worms. Recently, Lee and Yen (2005) found matrix metalloproteinase secreted by the larvae, suggesting its role in invasion through the intestine of host animals. The diagnostic potential of these proteinases is yet to be determined.

After a primary infection in the permissive host like rats, *A. cantonensis* adult worms developed in the pulmonary artery can survive for long periods. It is not known how adult worms evade immunological attack. At least their evasion mechanism is not dependent on host antigenic mimicry or disguise with host antigen, because preimmunization with host antigen of nonpermissive host animals did not affect the growth/maturation of implanted immature worms from the same host (Yoshimura et al., 1980). In spite of persistent infection with adult worms in the pulmonary artery after a primary infection, those worm-harboring rats show acquired resistance to young adult worms of challenge infection with marked strain difference (Yoshimura et al., 1979). A parasite-specific IgE response (Yoshimura and Yamagishi, 1976) and mast cell hyperplasia in lungs (Tiengkamol and Brockelman, 1982) were induced by *A. cantonensis* infection in rats. Interestingly, strain differences in acquired resistance in rats was correlated well with the antigen-specific IgE responsiveness, but not with IHA antibody responses (Yoshimura et al., 1979), suggesting the possible importance of the Th2 response in the acquired resistance in the lungs of rats.

Key Issues in the Field

Parasite Proteases

Excretory/Secretory Products and Proteases

Helminth parasites release a wide variety of molecules into their surrounding environment, host tissues, and organs. These molecules are collectively designated as excretory/secretory (ES) products or antigens (Hillyer, 1986; Harrison and Parkhouse, 2001), which consist of cuticular and tegumental surface components, intestinal epithelial cells, and secretions released by intestinal mucosa and specialized excretory/secretory organs.

Among various components in ES products, proteases are a major molecular species, both quantitatively and qualitatively (Tort et al., 1999). Proteases catalyze the cleavage of amide linkages in proteins and oligomeric peptides, and are classified as serine, threonine, aspartate, and metalol- and cysteine proteases, depending on the structure of catalytic active center. As in free-living organisms, parasitic worms deploy proteases to accomplish housekeeping functions such as digestion, hatching from eggs, and molting. In addition, parasite proteases are required for excystation, tissue penetration, and invasion, and possibly immune evasion. It is now established that many parasite proteases are promising chemotherapeutic or vaccine targets (Selzer et al., 1999). For example, synthetic cysteine protease inhibitor successfully prevents infection with *Schistosoma mansoni* (Abdulla et al., 2007).

Proteases in Parasite Life

By far the majority of reported parasite proteases belong to the papain superfamily of cysteine proteinases, though this number might be biased by polymerase chain reaction (PCR) amplification (Sajid and McKerrow, 2002). However, large amounts of serine protease inhibitors contained in the body fluid of the mammalian host might be one of the major reasons why internal parasites are equipped with an array of cysteine proteases.

Among them, the most studied are the *Fasciola* cysteine proteases. To date 18 cathepsin family enzymes were identified in *F. hepatica* (Irving et al., 2003), and their biological function has been analyzed (Dalton et al., 2003b). Newly hatched metacercaria of *F. hepatica* utilize cysteine proteases for tissue invasion (Berasain et al., 1997), and some *Fasciola* proteases have been suggested to have immunosuppressive effects by cleaving host immunoglobulin (Smith et al., 1993a) or by preventing eosinophil adhesion to juvenile worms (Carmona et al., 1992). They are also involved in the virulence of the fluke (Collins et al., 2004). O'Neill et al. (2001) reported that *F. hepatica* cathepsin L suppresses *Bordetella pertussis*–specific IFN-γ production in vivo. Cysteine proteases of minute intestinal flukes also play a role in nutrition uptake (Choi et al., 1998a,b, 1999).

Paragonimus westermani produces and releases various proteinases depending on their developmental stage (Song and Dresden, 1990; Chung et al., 1997a; Park et al., 2002). Cysteine proteases of *P. westermani* are considered to play important roles in nutrient uptake (Choi et al. 2006), metacercarial excystment (Chung et al., 1995; Intapan and Maleewong, 2001; Ikeda, 2003), and tissue invasion (Na et al., 2006). Cysteine proteases from *P. westermani* metacercariae also have immunosuppressive effects in mice (Hamajima et al., 1994), and varying degrees of capacity to cleave human IgGs (Chung et al., 1997b). However, this cannot be interpreted as evidence for an immunosuppressive role in vivo (Sajid and McKerrow, 2002), although human eosinophil function is modified by *Paragonimus* proteinases (Min et al., 2004; Shin et al., 2005).

Apart from cysteine proteases, matrix metalloproteinases (MMPs) play a key role in invasion of host tissues, as seen in the infective stage larvae of various nematodes such as *Ostertagia ostertagii* (De Maere et al., 2005), *A. cantonensis* (Lai et al., 2005), *Strongyloides stercoralis* (Gomez Gallego et al., 2005), *Gnathostoma binucleatum* (Caballero-Garcia et al., 2005), and *Ancylostoma caninum* (Williamson et al., 2006). In *S. venezuelensis*–infective larvae, an MMP activity at 40 kd is tightly associated with the ability of percutaneous infection (Maruyama et al., 2006) (Fig. 10.7).

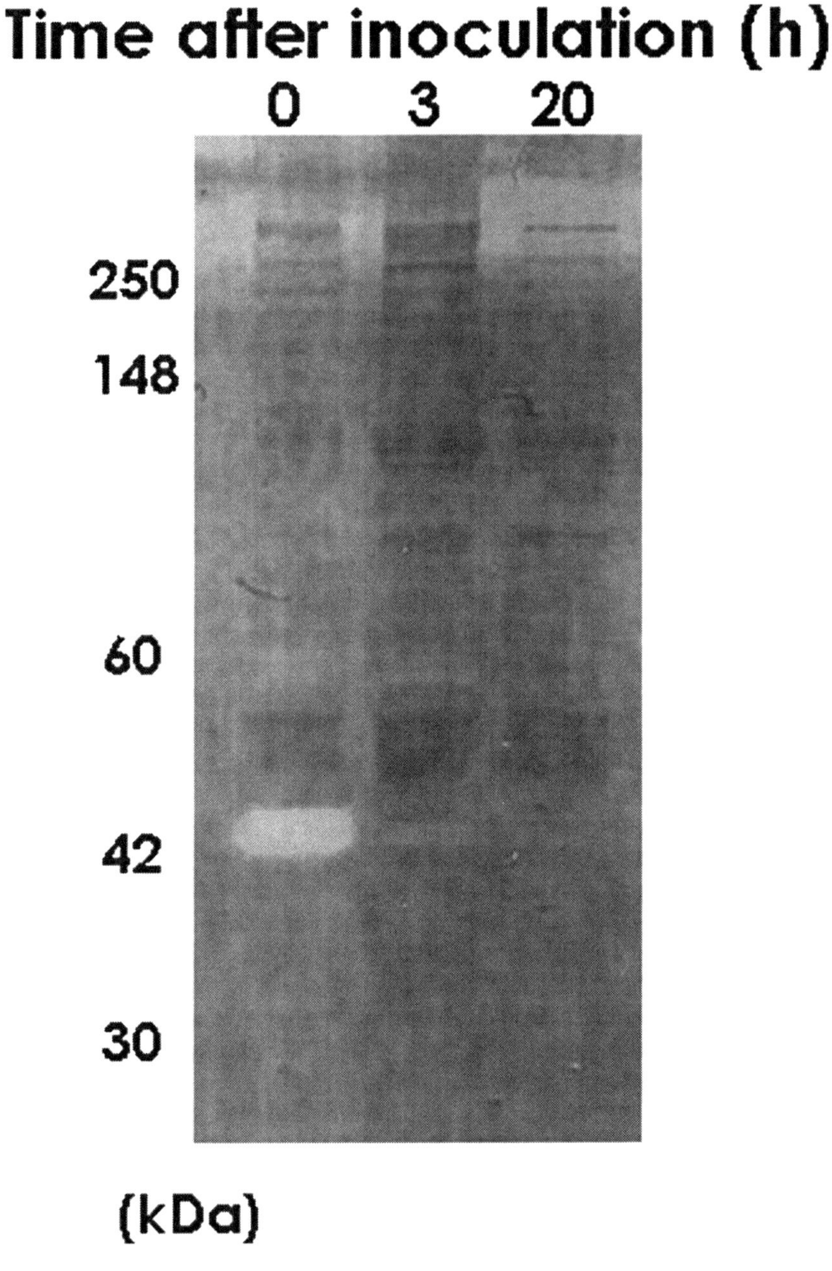

FIGURE 10.7. Matrix metalloproteinase in *S. venezuelensis* larvae. Infective larvae of *S. venezuelensis* have a matrix metalloproteinase activity at 40 kd, which disappears rapidly after the entry into the host. The ability of skin penetration also disappears within 20 hours of infection (Maruyama et al., 2006).

Proteases as Diagnostic Antigens

As pointed out above, proteases released in ES products are often useful as antigens for immunodiagnosis (Tort et al., 1999). The ES products of *Fasciola* sp. contain various antigenic components with variable sensitivity and specificity (Sampaio Silva et al., 1996). Among the various antigens, cysteine proteinases that were purified from ES products are proven to be highly specific diagnostic antigens in ELISA (Yamasaki et al., 1989; Cordova et al., 1997). Cathepsin L of *F. hepatica*, which is a major ES antigen (Smith et al., 1993b; Cordova et al., 1997) has been cloned, and recombinant cathepsin L has been applied successfully to immunodiagnosis in animal infections (Yadav et al., 2005; Sriveny et al., 2006) as well as for humans (Wongkham et al., 2005). Besides *Fasciola,* cysteine proteases of lung flukes (Yang et al., 2004; E.G. Lee et al., 2006) and liver flukes (Nagano et al., 2004) have been used in diagnosis.

To circumvent laborious purification steps, Ikeda (1998) developed the cystatin-capture ELISA technique, in which cysteine proteinases in crude ES are captured by a specific cysteine proteinase inhibitor, cystatin, coated on the wells of ELISA. Sensitivity and specificity in the diagnosis for human fascioliasis and paragonimiasis were greatly improved (Ikeda 1998; Tantrawatpan et al., 2005).

Because it is becoming increasingly difficult, especially in developed countries, to obtain enough quantities of fresh worm antigens for immunodiagnosis, a panel of recombinant ES protease antigens may in the future replace crude antigen preparations for clinical diagnosis and screening.

Immune Evasion in Mammalian Hosts

The gut lumen must be a relatively peaceful place for many food-borne parasites. After being swallowed by a definitive host, infective forms of flukes (metacercaria), tapeworms (plerocercoid, cysticercus, cysticercoid, etc.), and nematodes remain in the gut or its derivatives without invading mucosal tissues and invoking little or only moderate host reactions. In humans, intestinal flukes and *Diphyllobothrium* are included in this group, eliciting little immune response.

The gastrointestinal tract, however, is not a perfect sanctuary because some of the antigens in the gastrointestinal tract are absorbed as proteins or protein fragments and reach lymph nodes. In addition, dendritic cells present in the lamina propria extend processes across the epithelium and take up parasite antigens to induce immune responses (Rescigno et al., 2001; Chieppa et al., 2006). In most of these cases, parasitic worms evoke intense immune responses. However, in spite of an immune response, they survive and reproduce successfully (Ovington and Behm, 1997). What then, are their strategies to avoid the consequences of hostile immune reactions?

It is true that parasites, especially those that penetrate the gut wall to travel around the host's tissues, evoke strong immune responses and intense pathological changes, characterized by elevated production of eosinophils, IgE, mucosal mastocytosis, and goblet cell hyperplasia. However, some parasites do counteract or mitigate immune effector functions. An array of substances in ES products contribute to immune evasion strategies of the parasites through mechanisms such as shedding of surface-bound ligands and cells, and modulation of innate and acquired immune functions (Lightowlers and Rickard, 1988). Recent studies clearly show that helminth infections suppress immune responses by inducing regulatory T cells (Wilson et al., 2005; Kitagaki et al., 2006; Metwali et al., 2006), because removal of regulatory T cells results in the clearance of parasites (Taylor et al., 2005); such immunomodulation by parasites can contribute to their persistence in vivo (Babu et al. 2006). Macrophage functions are also suppressed in parasitic infections. For example, TNF-α, IL-1β, iNOS, and chemokine production by LPS-stimulated macrophages are suppressed by ES products of *S. erinaceieuropaei* (Fukumoto et al., 1997; Miura et al., 2001; Dirgahayu et al., 2004).

While a few parasites are eliminated by Th2-induced cellular responses, the vast majority of parasites remain unaffected. Migrating larvae continue their way in the face of eosinophilia, and massively accumulated mucosal mast cells are ineffective (e.g., against *N. brasiliensis*). Conversely *S. venezuelensis* is not affected by goblet cells induced by *N. brasiliensis* infection (Nawa et al., 1994). Apart from immune-mediated worm expulsion seen in intestinal nematode infections such as *N. brasiliensis*, *S. ratti*, *S. venezuelensis,* and *T. spiralis* infection in rodent models (Finkelman et al., 1997, 2004), it is rather difficult to find a clear example of immune-mediated worm expulsion, especially in food-borne parasitic zoonoses. Thus, it is tempting to speculate that parasitic worms might induce Th2 responses in order to suppress potentially dangerous Th1-type inflammation (see Fig. 10.5).

In schistosome infections, vaccination of mice with irradiated *Schistosoma mansoni* larvae confers high levels of immunity that is mediated by Th1 cells (Smythies et al., 1992; Wynn et al., 1995, 1996). In *Fasciola* infection in cattle, Th1-type responses characterized by IgG2 and IFN-γ production confer high resistance (Mulcahy et al., 1998). The murine cysticercosis model further illustrates the critical role of the Th1-type response. In susceptible BALB/c mice, *Taenia crassiceps* induces strong Th2 responses and the infection persists; however, STAT6-KO BALB/c mice that produce high levels of IgG2a, IFN-γ, and nitric oxide upon infection control the *T. crassiceps* infection very effectively (Rodriguez-Sosa et al., 2002). In contrast, genetically resistant C57BL/6 $\times$ 129Sv/Ev mice become susceptible to *T. crassiceps* infection when Th1 development is impaired by *STAT4* gene disruption (Rodriguez-Sosa et al., 2004). Parasites may be mimicking immunologically intestinal epithelial cells. Intestinal epithelial cells regulate mucosal homeostasis by guiding mucosal dendritic cells to promote the polarization of T cells toward a classical noninflammatory Th2 response, even after exposure to a Th1-inducing pathogen (Rimoldi et al., 2005).

Immune Evasion in Intermediate/Paratenic Host—Encystment

When researchers scrutinize fish, crab, or meat of wild animals, they sometimes find encysted parasites. Most are metacercariae of trematodes, bladder worms of cestodes, or nematode larvae. Inside the cyst, parasites appear to survive for long periods under stable conditions; they become immunologically invisible, waiting quietly until their host is eaten by another (usually definitive) host. This is an extremely smart strategy for a parasite to maintain its life cycle. Intense immune reactions against parasites can result in severe damage not only to worms but to host tissues as well. When the intermediate host harboring encysted larval parasites is ingested by the next host, a large number of larval worms can excyst simultaneously and infect the final host, increasing their chances of successful infection and their probability of finding a mate and reproducing. To understand the mechanisms employed for the formation of the cyst is to know the secret of the evolutionary success of these clever and highly successful worms.

Metacercaria of Flukes

Except for schistosomes, almost all digenean trematode parasites enter the final host orally as the metacercaria stage. During metamorphosis from a cercaria to metacercaria, typically a cyst wall is formed from the outer layer of the cercaria. By morphological and histochemical study, the cyst wall of *F. hepatica* metacercariae is formed from the cystogenic cells of cercariae (Dixon, 1966; Dixon and Mercer, 1967). The formation of the cyst wall has also been studied in *Notocotylus attenuatus* (Southgate, 1971). Harada and Suguri (2001) reported that, by direct observation and histochemical study using *Cercaria shikokuensis*, four layers of cyst wall of metacercariae are derived from distinct cell groups of the secretory gland cells of cercariae. Recently, transformation from cercaria to metacercaria has been achieved in vitro in *Echinostoma cinetorchis* (Park et al., 2006). In that study, 0.5 × RPMI 1640 medium with 10% fetal bovine serum gave the highest encystment rate and the development ratio of normal metacercariae, suggesting the importance of the exposure of cercariae to low osmotic conditions.

In the crustacean intermediate hosts, *Paragonimus* metacercariae seem to encyst without an inflammatory reaction. However, it is reported that miracidia of *P. westermani* can survive and develop in the snail host, *Semisulcospira libertina*, only when the defense system of the snail is suppressed by a preceding infection with other trematode parasites (Hamajima et al., 1989). It is well known that snails show innate resistance against invading pathogens via their hemocytes and humoral factors in their hemolymph (Loker and Bayne, 1986). Lie et al. (1981) reported that, after *Echinostoma paraensei* infection, *Biomphalaria glabrata* lose their ability to encapsulate homologous and heterologous trematode larvae, though the snails retain their ability to encapsulate larval nematode, *Angiostrongylus malaysiensis*. Such a phenomenon is termed *interference* (Lie, 1982). Noda (1990) reported that *E. paraensei*–infected *B. glabrata* lose their encapsulation capacity to *A. cantonensis* at 10 days postexposure, indicating the requirement of a longer incubation period to manifest interference to nematodes. Since metacercariae produce/release several proteinases, those enzymes may play a role in the prevention of the host's reaction. Recently, Doanh and Le (2005) found *Paragonimus* metacercariae of two different species concurrently in the same mountainous crab. Whether any interference could occur in crustacean hosts needs further study.

Cyst Around the Nematode Larvae

In the case of nematodes, the most famous cyst formation is seen in muscle larvae of *Trichinella spiralis* (Ritterson, 1966; Kwiatkiewicz-Pauw et al., 1994; Wranicz et al., 1998). When *T. spiralis* larvae invade muscle cells, they modify myocytes in such a way that myocytes transform into nurse cells that serve parasites food and accommodation (Montgomery et al., 1995). Nurse cells then start to producing type IV collagen to form a cystic capsule around the larva (Haehling et al., 1995). Discovery of non–cyst-forming *Trichinella* sp. (Garkavi, 1972), which was termed later as *T. pseudospiralis* (Pozio et al., 1992), brought about intensive molecular analyses of the mechanisms of cystogenesis by comparing capsulating and non-encapsulating species (Nagano and Takahashi, 2005).

We can easily see another example of the encystment of a nematode in the case of *Anisakis simplex* third-stage larvae in the visceral organs of fish. As was reviewed earlier in this section (see above), *Anisakis* larvae in vitro produce and release immunologically or biochemically active materials in ES, including proteases (Raybourne et al., 1986; Kennedy et al., 1988). While infected humans suffer from vigorous granulomatous reaction around larvae (Yoshimura, 1990), and even anaphylactic reactions (Alonso et al., 1997), fish seem to tolerate large numbers of larvae, indicating that encapsulation certainly prevents vigorous immune responses. Although precise mechanisms for the encystment of *Anisakis* larvae remain unsolved, it has been reported that infected fish, *Chub mackerel*, produces a protein factor with apoptosis-inducing activity through its H_2O_2-producing function (Jung et al., 2000). This protein, designated as apoptosis-inducing protein (AIP), is secreted by fish cells and concentrated in the capsule cavity around the larvae, and possibly suppresses larvae's activities (Jung et al., 2000). In the case of *Anisakis*, the host forms a capsule around the larvae to prevent their migration from the abdominal region into various viscera. What component of the nematode elicits AIP production remains unknown.

As in *Anisakis*, *Gnathostoma* larvae are usually found encysted in the muscles or viscera of intermediate/paratenic host fish (Daengsvang, 1980). In addition, *G. doloresi* (Mako and Akahane, 1985; Ishiwata et al., 1997) and *G. hispidum* (Sohn and Lee, 1998) larvae can be found encysted in snakes. In Mexico, *G. binucleatum* larvae are also found encysted in the muscles of fish-eating birds (Diaz-Camacho et al., 2002). Thus, in spite of their invasive or migratory nature in humans, *Gnathostoma* larvae generally encyst without visible harm to their intermediate/ paratenic hosts. *G. spinigerum* and *G. binucleatum* cause long-lasting migratory skin lesions over several years in humans, probably by using their immune evasion mechanisms. On the other hand, human disease caused by other three species, *G. doloresi, G. hispidum,* and *G. nipponicum,* persists for only 2 to 3 months. Although the larvae of those short-lasting species are presumed to be killed by immune reactions, they may survive in humans because of factors related to encystment. In fact, Koga and Ishii (1990) found that *G. hispidum* larvae could survive many years encysted in the muscles of experimentally infected rats.

Conclusions: Animals Within Animals

Along with a series of beautiful experiments that denied spontaneous generation of flies, Francesco Redi in Tuscany dissected various kinds of domestic and wild animals one after another to describe more than 100 parasites in fishes, reptiles, birds, and mammals (Fig. 10.8), probably wondering what these creatures were. Since that time our knowledge about parasites has grown enormously. Today we know that parasites are animals that have evolved from free-living species, and that they enter our body either orally or percutaneously, with no need of spontaneous generation. We know how to prevent parasitic infections, and are equipped with effective antiparasite drugs, which people in Redi's days never knew.

However, if Redi could ask us questions such as how these animals get along with the strong immune responses of their hosts, or why our immune system responds similarly against different animal parasites, we would not be able to satisfy him. Our present knowledge of parasite immunology largely depends on laboratory data mostly obtained from clean experiments with inbred mice and established parasite strains. It is well known that different strains of mice respond to the same parasite differently, and that the same strain of mice responds differently to different strains of parasites from the same species (Bellaby et al., 1996; Johnston et al., 2005). It is thus possible that our understanding of the immune system and immune responses against parasites has been biased and oversimplified by our choice of laboratory hosts and parasite strains (laboratory adapted in most cases). To understand deeply the immunology of parasitic infections, especially that of food-borne parasites, we must be brave enough to get into the biological diversity of host and parasite, as the Italian physician practiced 300 years ago.

FIGURE 10.8. Pieces of Francesco Redi's work. Francesco Redi (1626–1697) described a large number of animal parasites while he was working as a court doctor for the Great Duke of Tuscany. He published "Esperienze Intorno alla Generazione degli Insetti" in 1668, in which the oldest known figure of Fasciola adult worm appears. In "Osservazioni Intorno agli Animali Viventi che si Trovano Negli Animali Viventi" in 1684, he dissected *A. lumbricoides*.

References

Abdulla, M. H., Lim, K. C., Sajid, M., McKerrow, J. H., and Caffrey, C. R. (2007). *Schistosomiasis mansoni*: novel chemotherapy using a cysteine protease inhibitor. PLoS Med 4, 14.

Abe, T., and Nawa Y. (1987). Reconstitution of mucosal mast cells in W/W^v mice by adoptive transfer of bone marrow-derived cultured mast cells and its effects on the protective capacity to *Strongyloides ratti*-infection. Parasite Immunol 9, 31–38.

Akai, P. S., Pungpak, S., Kitikoon, V., Bunnang, D., Befus, A. D. (1994). Possible protective immunity in human opisthorchiasis. Parasite Immunol 16, 279–288.

Alonso, A., Daschner, A., and Moreno-Ancillo, A. (1997). Anaphylaxis with *Anisakis simplex* in the gastric mucosa. N Engl J Med 337, 350–351.

Anantaphruti, M. T. (1989a). ELISA for diagnosis of gnathostomiasis using antigens from *Gnathostoma doloresi* and *G. spinigerum*. Southeast Asian J Trop Med Public Health 20, 297–304.

Anantaphruti, M. T. (1989b). Demonstration of species-specific antigens of *Gnathostoma spinigerum*, a preliminary attempt. Southeast Asian J Trop Med Public Health 20, 305–312.

Arora, M., Chen, L., Paglia, M., et al. (2006). Simvastatin promotes Th2–type responses through the induction of the chitinase family member Ym1 in dendritic cells. Proc Natl Acad Sci USA 103, 7777–7782.

Arrieta, I., del Barrio, M., Vidarte, L., et al. (2000). Molecular cloning and characterization of an IgE-reactive protein from *Anisakis simplex*: Anis 1. Mol Biochem Parasitol 107, 263–268.

Artis, D., Kane, C. M., Fiore, J., et al. (2005). Dendritic cell-intrinsic expression of NF-κB1 is required to promote optimal Th2 cell differentiation. J Immunol 174, 7154–7159.

Artis, D., Wang, M. L., Keilbaugh, S. A., et al. (2004). RELMβ/FIZZ2 is a goblet cell-specific immune-effector molecule in the gastrointestinal tract. Proc Natl Acad Sci USA 101, 13596–13600.

Audicana, L., Audicana, M. T., Fernandez de Corres, L., Kennedy, M. W. (1997). Cooking and freezing may not protect against allergenic reactions to ingested *Anisakis simplex* antigens in humans. Vet Rec 140, 235.

Audicana, M.T., Ansotegui, I.J., de Corres, L.F., and Kennedy, M.W. (2002). *Anisakis simplex*: dangerous-dead and alive? Trends Parasitol 18, 20–25.

Babu, S., Blauvelt, C. P., Kumaraswami, V., and Nutman, T. B. (2006). Regulatory networks induced by live parasites impair both Th1 and Th2 pathways in patent lymphatic filariasis: implications for parasite persistence. J Immunol 176, 3248–3256.

Baeza, M. L., Conejero, L., Higaki, Y., et al. (2005). Anisakis simplex allergy: a murine model of anaphylaxis induced by parasitic proteins displays a mixed Th1/Th2 pattern. Clin Exp Immunol 142, 433–440.

Banchereau, J., and Steinman, R.M. (1998). Dendritic cells and the control of immunity. Nature 392, 245–252.

Banzon, T. C., Lewert, R. M., and Yogore, M. G. (1975). Serology of *Capillaria philippinensis* infection: reactivity of human sera to antigens prepared from *Capillaria obsignata* and other helminthes. Am J Trop Med Hyg 24, 256–263.

Bellaby, T., Robinson, K., and Wakelin, D. (1996). Induction of differential T-helper-cell responses in mice infected with variants of the parasitic nematode Trichuris muris. Infect Immun 64, 791–795.

Bentancor, A., Piacenza, L., and Carmona, C. (2002). Immunization with cathepsin L proteinases CL1 and CL2 secreted by *Fasciola hepatica* elicit a preferential type 1 response based on IgG2a antibodies in rats. J Helminthol 76, 199–205.

Berasain, P., Goni, F., McGonigle, S., et al. (1997). Proteinases secreted by *Fasciola hepatica* degrade extracellular matrix and basement membrane components. J Parasitol 83, 1–5.

Bernardini, R., Mistrello, G., Novembre, E., et al. (2005). Cross-reactivity between IgE-binding proteins from *Anisakis simplex* and *Dermatophagoides pteronyssinus*. Int J Immunopathol Pharmacol 18, 671–675.

Brady, M. T., O'Neill, S. M., Dalton, J. P., and Mills, K. H. (1999). *Fasciola hepatica* suppresses a protective Th1 response against Bordetella pertussis. Infect Immun 67, 5372–5378.

Bruhn, H. (2005). A short guided tour through functional and structural features of saposin-like proteins. Biochem J 389, 249–257.

Buendia, E. (1997). *Anisakis*, anisakidosis, and allergy to *Anisakis*. Allergy 52, 481–482.

Caballero-Garcia, M. de L., Almeyda-Artigas, R. J., Mosqueda-Cabrera, M. A., and Jimenez-Cardoso, E. (2005). *Gnathostoma binucleatum*: excretion-secretion antigen analysis obtained from advanced third-stage larvae in in vitro culture. Exp Parasitol 110, 140–145.

Cancela, M., Carmona, C., Rossi, S., Frangione, B., Goni, F., and Berasain, P. (2004). Purification, characterization, and immunolocalization of paramyosin from the adult stage of *Fasciola hepatica*. Parasitol Res 92, 441–448.

Carmona, C., Smith, A., Dowd, A., and Dalton, J. P. (1992). A *Fasciola hepatica* cathepsin L proteinase prevents the adherence of eosinophils to newly excysted juveniles. Biochem Soc Trans 20, 85S.

Chai, J. Y. (2005). Intestinal trematode infections in Korea. Asian Parasitology Vol.1, Food-borne helminthiasis in Asia. Pp.79–102, AAA/FAP, Chiba, Japan.

Chai, J. Y., Murrell, K. D., and Lymbery, A. J. (2005). Fishborne parasitic zoonoses: Status and issues. Int J Parasitol 35, 1233–1254.

Chang, N. C., Hung, S. I., Hwa, K. Y., et al. (2001) A macrophage protein, Ym1, transiently expressed during inflammation is a novel mammalian lectin. J Biol Chem 276, 17497–17506.

Chen, H., Nara, T., Zeng, X., et al. (2000). Vaccination of domestic pig with recombinant paramyosin. against *Schistosoma japonicum* in China. Vaccine 18, 2142–2146.

Chen, M., Lu, Y., Hua, X., and Mott, K. E. (1994). Progress in assessment of morbidity due to *Clonorchis sinensis* infection: a review of recent literature. Trop Dis Bull 9, R7–R65.

Chen, S. N. (1986). Enzyme-linked immunosorbent assay (ELISA) for the detection of antibodies to *Angiostrongylus cantonensis*. Trans R Soc Trop Med Hyg 80, 398–405.

Chieppa, M., Rescigno, M., Huang, A. Y., and Germain, R. N. (2006). Dynamic imaging of dendritic cell extension into the small bowel lumen in response to epithelial cell TLR engagement. J Exp Med 203, 2841–2852.

Cho, S. Y., Chung, Y. B., and Kong, Y. (1992). Component proteins and protease activities in excretory-secretory product of sparganum. Korean J Parasitol 26, 163–167.

Choi, J. H., Lee, J. H., Yu, H. S., et al. (2006). Molecular and biochemical characterization of hemoglobinase, a cysteine proteinase, in *Paragonimus westermani*. Korean J Parasitol 44, 187–196.

Choi, M. H., Chai, J. Y., and Lee, S. H. (1998a). Purification and characterization of a 16–kDa cysteine proteinase of *Gymnophalloides seoi* (Gymnophallidae) metacercariae. J Parasitol 84, 350–355.

Choi, M. H., Choe, S. C., and Lee, S. H. (1999). A 54 kDa cysteine protease purified from the crude extract of Neodiplostomum seoulense adult worms. Korean J Parasitol 37, 39–46.

Choi, M. H., Park, W. J., Park, Y. K., Chai, J. Y., and Lee, S. H. (1998b). Isolation and characterization of a 40 kDa cysteine proteinase from *Gymnophalloides seoi* adult worms. Korean J Parasitol 36, 133–141.

Choi, M. H., Ryu, J. S., Lee, M., et al. (2003). Specific and common antigens of *Clonorchis sinensis* and *Opisthorchis viverrini* (Opisthorchidae, Trematoda). Korean J Parasitol 41, 155–163.

Chung, B. S., Zhang, H., Choi, M. H., et al. (2004). Development of resistance to reinfection by *Clonorchis sinensis* in rats. Korean J Parasitol 42, 19–26.

Chung, Y. B., Chung, B. S., Choi, M. H., Chai, J. Y., and Hong, S.T. (2000). Partial characterization of a 17 kDa protein of *Clonorchis sinensis* Korean J Parasitol 38, 95–97.

Chung, Y. B., Kong, Y., Joo, I. J., Cho, S. Y., and Kang, S. Y. (1995). Excystment of *Paragonimus westermani* metacercariae by endogenous cysteine protease. J Parasitol 81, 137–142.

Chung, Y. B., Kong, Y., Yang, H. J., Kang, S. Y., and Cho, S. Y. (1997a). Cysteine protease activities during maturation stages of *Paragonimus westermani*. J Parasitol 83, 902–907.

Chung, Y. B., Yang, H. J., Kang, S. Y., Kong, Y., and Cho, S. Y. (1997b). Activities of different cysteine proteases of *Paragonimus westermani* in cleaving human IgG. Korean J Parasitol 35, 139–142.

Chye, S. M., Chang, J. H., and Yen, C. M. (2000). Immunodiagnosis of human eosinophilic meningitis using an antigen of *Angiostrongylus cantonensis* L5 with molecular weight 204 kD. Acta Trop 75, 9–17.

Cliffe, L. J., Humphreys, N. E., Lane, T. E., Potten, C. S., Booth, C., and Grencis, R. K. (2005). Accelerated intestinal epithelial cell turnover: a new mechanism of parasite expulsion. Science 308, 1463–1465.

Collins, P. R., Stack, C. M., O'Neill, S. M., et al. (2004). Cathepsin L1, the major protease involved in liver fluke (*Fasciola hepatica*) virulence: propeptide cleavage sites and autoactivation of the zymogen secreted from gastrodermal cells. J Biol Chem 279, 17038–17046.

Cordova, M., Herrera, P., Nopo, L., et al. (1997). *Fasciola hepatica* cysteine proteinases: immunodominant antigens in human fascioliasis. Am J Trop Med Hyg 57, 660–666.

Crompton, D. (1999). How much human helminthiasis is there in the world? J Parasitol 85, 397–403.

Cross, J. H., Banzon, T., and Singson, C. (1978). Further studies on *Capillaria philippinensis*: Development of the parasite in the Mongolian gerbil. J Parasitol 64, 208–213.

Cross, J. H., and Basaca-Sevilla, V. (1983). Experimental transmission of *Capillaria philippinensis* to birds. Trans R Trop Med Hyg 77, 511–514.

Cross, J. H., and Chi, J. C. (1982). ELISA for the detection of *Angiostrongylus cantonensis* antibodies in patients with eosinophilic meningitis. Southeast Asian J Trop Med Public Health 13, 73–76.

Daengsvang, S. (1980). A Monograph on the Genus *Gnathostoma* and Gnathostomosis in Thailand. Tokyo: Southeast Asian Medical Information Center.

Dalton, J. P., Brindley, P. J., Knox, D. P., et al. (2003a). Helminth vaccines: from mining genomic information for vaccine targets to systems used for protein expression. Int J Parastol 33, 621–640.

Dalton, J. P., McGonigle, S., Rolph, T. P., and Andrews, S. J. (1996). Induction of protective immunity in cattle against infection with *Fasciola hepatica* by vaccination with cathepsin L proteinases and with hemoglobin. Infect Immun 64, 5066–5074.

Dalton, J. P., and Mulcahy, G. (2001). Parasite vaccines—a reality? Vet Parasitol 98, 149–167.

Dalton, J. P., Neill, S. O., Stack, C., et al. (2003b). *Fasciola hepatica* cathepsin L-like proteases: biology, function, and potential in the development of first generation liver fluke vaccines. Int J Parasitol 33, 1173–1181.

Daly, C. M., Mayrhofer, G., and Dent, L. A. (1999). Trapping and immobilization of *Nippostrongylus brasiliensis* larvae at the site of inoculation in primary infections of interleukin-5 transgenic mice. Infect Immun 67, 5315–5323.

Dantrakool, A., Morakote, N., and Uparanukraw, P. (2001). Production of monoclonal antibodies to the cuticle of advanced third-stage larva of *Gnathostoma spinigerum*. Southeast Asian J Trop Med Public Health, 32, 562–569.

Daschner, A., Alonso-Gomez, A., Cabanas, R., Suarez-de-Parga, J. M., and Lopez-Serrano, M. C. (2000). Gastroallergic anisakiasis: borderline between food allergy and parasitic disease-clinical and allergologic evaluation of 20 patients with confirmed acute parasitism by *Anisakis simplex*. J Allergy Clin Immunol 105, 176–181.

Daschner, A., and Pascual, C. Y. (2005). *Anisakis simplex*: sensitization and clinical allergy. Curr Opin Allergy Clin Immunol 5, 281–285.

De Maere, V., Vercauteren, I., Geldhof, P., Gevaert, K., Vercruysse, J., and Claerebout, E. (2005). Molecular analysis of astacin-like metalloproteases of *Ostertagia ostertagi*. Parasitology 130, 89–98.

Diaz-Camacho, S. P., Willms, K., Ramos, M. Z., del Carmen de la Cruz Otero, M., Nawa, Y., and Akahane, H. (2002). Morphology of *Gnathostoma* spp. isolated from natural hosts in Sinaloa, Mexico. Parasitol Res 88, 639–645.

Dirgahayu, P., Fukumoto, S., Tademoto, S., Kina, Y., and Hirai, K. (2004). Excretory/secretory products from plerocercoids of *Spirometra erinaceieuropaei* suppress interleukin-1beta gene expression in murine macrophages. Int J Parasitol 34, 577–584.

Dixon, K. E. (1966). A morphological and histochemical study of the cystogenic cells of the cercariae of *Fasciola hepatica* L. Parasitology 56, 287–297.

Dixon, K. E., and Mercer, E. H. (1967). The formation of the cyst wall of the metacercariae of *Fasciola hepatica* L. Z Zellforsch Mikrosk Anat 77, 331–344.

Doanh, P. N., and Le, N. T. (2005). The DT. *Paragonimus* and paragonimiasis in Vietnam In: Arizono N, et al., eds. Asian Parasitology, Vol. 1. Food-Borne Helminthiasis in Asia, pp. 149–153. FAP, Chiba, Japan.

Don, T. A., Oksov, Y., Lustigman, S., and Loukas, A. (2007). Saposin-like proteins from the intestine of the blood-feeding hookworm, *Ancylostoma caninum*. Parasitology 134, 427–436.

El Dib, N. A., Sabry, M. A., Ahmed, J. A., El-Basiouni, S. O., and El-Badry, A. A. (2004). Evaluation of *Capillaria philippinensis* coproantigen in the diagnosis of infection. J Egypt Soc Parasitol 34, 97–106.

El-Malky, M., Maruyama, H., Hirabayashi, Y., et al. (2003). Intraepithelial infiltration of eosinophils and their contribution to the elimination of adult intestinal nematode, *Strongyloides venezuelensis* in mice. Parasitol Int 52, 71–79.

Emery, D. L. (1996). Vaccination against worm parasites of animals. Vet Parasitol 64, 31–45.

Espino, A. M., and Hillyer, G. V. (2004). A novel *Fascioloa hepatica* saposin-like recombinant protein with immuno-prophylactic potential. J Parasitol 90, 876–879.

Espino, A. M., Osuna, A., Gil, R., and Hillyer, G. V. (2005). *Fasciola hepatica*: humoral and cytokine responses to a member of the saposin-like protein family following delivery as a DNA vaccine in mice. Exp Parasitol 110, 374–383.

Falcao, H., Lunet, N., Neves, E., and Barros, H. (2002). Do only live larvae cause *Anisakis simplex* sensitization? Allergy 57, 44.
Fernandez de Corres, L., Audicana, M., Del Pozo, M. D., et al. (1996). *Anisakis simplex* induces not only anisakiasis: report on 28 cases of allergy caused by this nematode. J Invest Allergol Clin Immunol 6, 315–319.
Finkelman, F.D., Shea-Donohue, T., Goldhill, J., et al. (1997). Cytokine regulation of host defense against parasitic gastrointestinal nematodes: lessons from studies with rodent models. Annu Rev Immunol 15, 505–533.
Finkelman, F. D., Shea-Donahue, T., Morris, S. C., et al. (2004). Interleukin-4– and interleukin-13–mediated host protection against intestinal nematode parasites. Immunol Rev 201, 139–155.
Flavell, D. J. (1982). Acquired resistance to *Opisthorchis viverrini* in the hamster. Trans R Soc Trop Med Hyg 76, 800–802.
Flavell, D. J., and Flavell, S. U. (1986). *Opisthorchis viverrini*: pathogenesis of infection in immunodeprived hamsters. Parasite Immunol 8, 455–466.
Frazer, B. A., Fried, B., Fujino, T., and Sleckman, B. P. (1999). Host-parasite relationships between *Echinostoma caproni* and RAG-2–deficient mice. Parasitol Res 85, 337–342.
Fukao, T., Yamada, T., Tanabe, M., et al. (2002). Selective loss of gastrointestinal mast cells and impaired immunity in PI3K-deficient mice. Naute Immunol 3, 295–304.
Fujino, T., and Fried, B. (1993). Expulsion of *Echinostoma trivolvis* (Cort, 1914) Kanev, 1985 and retention of *E. caproni* Richard, 1964 (Trematoda) in C3H mice: pathological, ultrastructural, and cytochemical effects on the host intestine. Parasitol Res 79, 286–292.
Fukumoto, S., Hirai, K., Tanihata, T., Ohmori, Y., Stuehr, D. J., and Hamilton, T. A. (1997). Excretory/secretory products from plerocercoids of *Spirometra erinacei* reduce iNOS and chemokine mRNA levels in peritoneal macrophages stimulated with cytokines and/or LPS. Parasite Immunol 19, 325–332.
Garcia, F., Blanco, J. G., Garces, M., Juste, S., Fuentes, M., and Herrero, D. (2001). Freezing protects against allergy to *Anisakis simplex*. J Investig Allergol Clin Immunol 11, 49–52.
Garcia, L. S. (2001). *Anisakis* spp., *Phocanema* spp., *Contracaecum* spp., and *Pseudoterranova* spp. (Larval nematodes acquired from saltwater fish). In: Garcia, L.S., ed. Diagnostic Medical Parasitology, 4th ed. Herndon, VA: ASM Press, 322–324.
Garkavi, B.G. (1972). Species of *Trichinella* isolated from wild animals. Veteranariya 10, 90–91.
Gomez Gallego, S., Loukas, A., Slade, R. W., et al. (2005). Identification of an astacin-like metallo-proteinase taranscript from the infective larvae of *Strongyloides stercoralis*. Parasitol In, 54, 123–133.
Goto, Y., Tamura, A., Ishikawa, O., Miyachi, Y., Ishii, T., and Akao, N. (1998). Creeping eruption caused by a larva of the suborder Spirurina type X. Br J Dermatol 139, 315–318.
Haehling, E., Niederkorn, J. Y., and Stewart, G. L. (1995). *Trichinella spiralis* and *Trichinella pseudospiralis* induce collagen synthesis by host fibroblasts in vitro and in vivo. Int J Parasitol 25, 1393–400.
Hamajima, F., Fukuda, K., Yamakami, K., Oguma, T., Sakamoto, N., and Fujii, C. (1989). Infectivity of the lung fluke *Paragonimus westermani* in the snail *Semisulcospira libertina* and susceptibility of the snail to the fluke. Jpn J Parasitol 38, 241–256.
Hamajima, F., Yamamoto, M., Tsuru, S., et al. (1994). Immunosuppression by a neutral thiol protease from parasitic helminth larvae in mice. Parasite Immunol 16, 261–273.
Harada, M., and Suguri, S. (2001). A histochemical study of the secretory gland cells of *Cercaria shikokuensis* and their role during development from cercaria to metacercaria. Parasitol Int 50, 149–156.

Harmsen, M. M., Cornelissen, J. B., Buijs, H. E., Boersma, W. J., Jeurissen, S. H., and van Milligen, F. J. (2004). Identification of a novel *Fasciola hepatica* cathepsin L protease containing protective epitopes within the propeptide. Int J Parasitol 34, 675–682.

Harrison, L. J. S., and Parkhouse, R. M. E. (2001). Antigens of parasitic helminthes in protection and pathology. In: Chowdhury, N., and Tada, I., eds. Perspectives on Helminthology. Enfield, US: Science Publishers, 369–384.

Hillyer, G. V. (1986). Fascioliasis, Paragonimiasis, Clonorchiasis, and Opisthorchias is. In: Walls, K.W., and Schantz, P.M., eds. Immunodiagnosis of Parasitic Diseases. Vol. 1. Helminthic Diseases. Orlando: Academic Press, 39–68.

Hillyer, G. V. (1995). Comparison of purified 12 kDa and recombinant 15 kDa *Fasciola hepatica* antigens related to a *Schistosoma mansoni* fatty acid binding protein. Mem Inst Oswaldo Cruz 90, 249–253.

Hillyer, G. V., Garcia Rosa, M. I., Alicea, H., and Hernandez, A. (1988). Successful vaccination against murine *Schistosoma mansoni* infection with a purified 12 Kd *Fasciola hepatica* cross-reactive antigen. Am J Trop Med Hyg 38, 103–110.

Hillyer, G. V., and Serrano, A. E. (1982). Cross protection in infections due to *Schistosoma mansoni* using tegument antigens of *Fasciola hepatica*. J Infect Dis 145, 728–732.

Hong, S. J., Lee, Y. H., Chung, M. H., Lee, D. H., and Woo, H. C. (1994). Egg positive rates of Clonorchis sinensis and intestinal helminthes among residents in Kagye-ri, Saengbiryang-myon, Sanchong-gun, Kyongsangnam-do. Korean J Parasitol 32, 271–273.

Hong, S. T., and Cross, J. H. (2005). *Capillaria phillipinensis* infection in Asia. In: Arizono, N., et al., eds. Asian Parasitology, Vol. 1. Food-Borne Helminthiasis in Asia. Chiba, Japan: FAP, 225–229,

Horii, Y., Khan, A.I., and Nawa, Y. (1994). Persistent infectio of *Strongyloides venezuelensis* and normal expulsion of *Nippostrongylus brasiliensis* in Mongolian gerbils, *Meriones unguiculatus*, with reference to the cellular responses in the intestinal mucosa. Parasite Immunol 15, 175–179.

Iglesias, R., Leiro, J., Ubeira, F. M., Sanamarina, M. T., Navarrete, I., and Sanmartin, M. L. (1996). Antigenic cross reactivity in mice between third-stage larvae of Anisakis simplex and other nematodes. Parasitol Res 82, 378–381.

Ikeda, T. (1998). Cystatin capture enzyme-linked immunosorbent assay for immunodiagnosis of human paragonimiasis and fascioliasis. Am J Trop Med Hyg 59, 286–290.

Ikeda, T. (2003). Involvement of cysteine proteinases in excystment of Paragonimus ohirai metacercariae induced by sodium cholate and A23187. J Helminthol 77, 21–26.

Ikeda, T., Oikawa, Y., and Nishiyama, T. (1996). Enzyme-linked immunosorbent assay using cysteine proteinase antigens for immunodiagnosis of human paragonimiasis. Am J Trop Med Hyg 55, 435–437.

Infante-Duarte, C., and Kamradt, T. (1999). Th1/Th2 balance in infection. Springer Semin Immunopathol 21, 317–338.

Intapan, P.M., and Maleewong, W. (2001). In vitro excystation of Paragonimus heterotremus metacercariae. J Parasitol 87, 1184–1186.

Intapan, P. M., Maleewong, W., Sukeepaisarnjaroen, W., and Morakote, N. (2006). Potential use of *Trichinella spiralis* antigen for serodiagnosis of human capillariasis philippinensis by immunoblot analysis. Parasitol Res 98, 227–231.

Irving, J. A., Spithill, T. W., Pike, R. N., Whisstock, J. C., and Smooker, P. M. (2003). The evolution of enzyme specificity in *Fasciola* spp. J Mol Evol 57, 1–15.

Ishii, H., Mukae, H., Inoue, Y., et al. (2001). A rare case of eosinophilic pleuritis due to sparganosis. Intern Med 40, 783–785.

Ishikawa, N., Horii, Y., and Nawa, Y. (1993). Immune-mediated alteration of the terminal sugars of goblet cell mucins in the small intestine of *Nippostrongylus brasiliensis*-infected rats. Immunology 78, 303–307.

Ishikawa, N., Horii, Y., Oinuma, T., Suganuma, T., and Nawa, Y. (1994). Goblet cell mucins as the selective barrier for the intestinal helminthes: T-cell independent alteration of goblet cell mucins by immunologically "damaged" *Nippostrongylus brasiliensis* worms and its significance on the challenge infection with homologous and heterologous parasites. Immunology 81, 480–486.

Ishikawa, N., Shi, B-B., Khan, A. I., and Nawa, Y. (1995). Reserpine-induced sulphomucin production by goblet cells in the jejunum of rats and its significance in the establishment of intestinal helminths. Parasite Immunol 17, 581–586.

Ishiwata, K., Diaz-Camacho, S.P., Ogata, K., Nakamura-Uchiyama, F., Hiromatsu, K., and Nawa, Y. (2003). Evaluation of the antigenic similarities of adult-worm extracts from three Gnathostoma species, using sera from Mexican and Japanese patients with Gnathostoma infections. Ann Trop Med Parasitol 97, 629–637.

Ishiwata, K., Nakao, H., Nose, R., et al. (1997). Gnathostomiasis in frog-eating snakes from Japan. J Wildl Dis 33, 877–879.

Jankovic, D., Kullberg, M. C., Noben-Trauth, N., Caspar, P., Paul, W. E., and Sher, A. (2000). Single cell analysis reveals that IL-4 receptor/Stat6 signaling is not required for the in vivo or in vitro development of CD4+ lymphocytes with a Th2 cytokine profile. J Immunol 164, 3047–3055.

Japanese Ministry of Health, Labor, and Welfare, Information Service for Rhuematic Fever and Allergy. http://www.mhlw.go.jp/new-info/kobetu/kenkou/ryumachi/index.html (in Japanese, 2006 updated).

Jittimanee, J., Sermswan, R.W., Puapairoj, A., Maleewong, W., and Wongratanacheewin, S. (2007). Cytokine expression in hamsters experimentally infected with *Opisthorchis viverrini*. Parasitol Immunol 29(3), 159–167.

Johansson, E., Aponno, M., Lundberg, M., and van Hage-Hamsten, M. (2001). Allergenic cross-reactivity between the nematode *Anisakis simplex* and the dust mites *Acarus siro*, *Lepidoglyphus destructor*, *Tyrophagus putrescentiae*, and *Dermatophagoides pteronyssinus*. Allergy 56, 660–666.

Johnston, C. E., Bradley, J. E., Behnke, J.M., Matthews, K. R., and Else, K. J. (2005). Isolates of *Trichuris muris* elicit different adaptive immune responses in their murine host. Parasite Immunol 27, 69–78.

Jung, S.K., Mai, A., Iwamoto, M., et al. (2000). Purification and cloning of an apoptosis-inducing protein derived from fish infected with *Anisakis simplex*, a causative nematode of human anisakiasis. J Immunol 165, 1491–1497.

Kan, H., Ogata, T., Taniyama, A., Migita, M., Matsuda, I. and Nawa, Y. (1995). Extraordinarily high eosinophilia and elevated serum interleukin-5 level observed in a patient infected with *Paragonimus westermani*. Pediatrics 96, 351–354.

Kennedy, M. W., Tierney, J., Ye, P., et al. (1988). The secreted and somatic antigens of the third stage larva of *Anisakis simplex*, and antigenic relationship with *Ascaris suum*, *Ascaris lumbricoides*, and *Toxocara canis*. Mol Biochem Parasitol 31, 35–46.

Kearn, G. C. (1967). Experiments on host-finding and host-specificity in the monogeneal skin parasite *Entobdella soleae*. Parasitology 57, 585–605.

Kitagaki, K., Businga, T.R., Racila, D., Elliott, D.E., Weinstock, J. V., and Kline, J. N. (2006). Intestinal helminths protect in a murine model of asthma. J Immunol 177, 1628–1635.

Knox, D.P., Redmond, D.L., Skuce, P.J., and Newlands, G.F. (2001). The contribution of molecular biology to the development of vaccines against nematode and trematode parasites of domestic ruminants. Vet Parasitol 101, 311–335.

Koga, M., Akahane, H., Ogata, K., et al. (1999). Adult *Gnathostoma cf. binucleatum* obtained from dogs experimentally infected with larvae as an etiological agent in Mexican gnathostomiasis; external morphology. J Helminthol Soc Wash 66, 41–46.

Koga, M., and Ishii, Y. (1990). Persistence of *Gnathostoma hispidum* in chronically infected rats. J Helminthol 64, 46–50.

Kong, Y., Chung, Y. B., Cho, S. Y., Choi, S. H., and Kang, S. Y. (1994a). Characterization of three neutral proteases of *Spirometra mansoni* plerocercoid. Parasitology 108, 359–368.

Kong, Y., Chung, Y. B., Cho, S. Y., and Kang, S. Y. (1994b). Cleavage of immunoglobulin G by excretory-secretory cathepsin S-like protease of *Spirometra mansoni* plerocercoid. Parasitology 109, 611–621.

Kong, Y., Kang, S. Y., Kim, S. H., Chung, Y. B., and Cho, S. Y. (1997). A neutral cysteine protease of *Spirometra mansoni* plerocercoid invoking an IgE response. Parasitology 114, 263–271.

Kong, Y., Yun, D. H., Cho, S. Y., Sohn, W. M., Chung, Y. B., and Kang, S. Y. (2000). Differential expression of the 27 kDa cathepsin L-like cysteine protease in developmental stages of *Spirometra erinacei*. Korean J Parasitol 38, 195–199.

Korenaga, M., Hitoshi, Y., Yamaguchi, N., Sato, Y., Takatsu, K., and Tada, I. (1991). The role of interleukin-5 in protective immunity to *Strongyloides venezuelensis* infection in mice. Immunology 72, 502–507.

Kuramochi, T., Machida, M., Araki, J., Uchidea, A., Kishiro, T., and Nagasawa, K. (1996). Minke whales (*Balaenoptrera acutorostrata*) are one of the major final hosts of *Anisakis simplex* (Nematoda: Anisakidae) in the northwestern North Pacific Ocean. Rep Int Whale Commun 46, 415–419.

Kwiatkiewicz-Pauw, M., Blotna-Filipiak, M., Gabryel, P., and Gustowska, L. (1994). Contribution of muscle cell to formation of the capsule around *Trichinella spiralis* larva. Ultrastructural studies. Wiad Parazytol 40, 369–374.

Kwon, T. C., Kang, C. M., and Choi, D. W. (1987). Passive transfer of immunity against *Clonorchis sinensis* by peritoneal exudates cells in mice. Korean J Parasitol 25, 45–50

Lai, S. C., Jiang, S. T., Chen, K. M., and Lee, H. H. (2005). Matrix metalloproteinases activity demonstrated in the infective stage of the nematodes, *Angiostrongylus cantonensis*. Parasitol Res 97, 466–471.

Lantz, C. S., Boesiger, J., Song, C. H., et al. (1998). Role for interleukin-3 in mast cell and basophil development and in immunity to parasites. Nature 392, 90–93.

Lee, E.G., Na, B.K., Bae, Y.A., et al. (2006). Identification of immunodominant excretory-secretory cysteine proteases of adult *Paragonimus westermani* by proteome analysis. Proteomics 6, 1290–1300.

Lee, J. D., Yen, and C. M. (2005). Protease secreted by the infective larvae of *Angiostrongylus cantonensis* and its role in the penetration of mouse intestine. Am J Trop Med Hyg 72, 831–836.

Lee, J. S., Kim, I. S., Sohn, W. M., Lee, J., and Yong, T. S. (2006a). A DNA vaccine encoding a fatty acid-binding protein of *Clonorchis sinensis* induces protective immune response in Sprague-Dawley rats. Scand J Immunol 63, 169–176.

Lee, J. S., and Yong, T. S. (2004). Expression and cross-species reactivity of fatty acid-binding protein of *Clonorchis sinensis*. Parasitol Res 93, 339–343.

Lee, J. Y., Cho, P. Y., Kim, T. Y., Kang, S. Y., Song, K. Y., and Hong, S. J. (2002). Hemolytic activity and developmental expression of pore-forming peptide, clonorin. Biochem Biophys Res Commun 296, 1238–1244.

Li, R. S. (1980). Study on animal hosts of fasciolopsiasis in Shanxi Province. J Gansu Agriculture Univ 1, 16.

Lie, K. J., Jeong, K. H., and Heyneman, D. (1981). Selective interference with granulocyte function induced by *Echinostoma paraensei* (Trematode) laravae in *Biomphalaria glabrata* (Mollusca). J Parasitol 67, 790–796.

Lie, K. J. (1982). Survival of *Schistosoma mansoni* and other trematode larvae in the snail *Biomphalaria glabrata*. A discussion of the interference hypothesis. Trop Geogr Med 34, 111–122.

Lightowlers, M. W., and Rickard, M. D. (1988). Excretory-secretory products of helminth parasites: effects on host immune response. Parasitology 96, S123–S166.

Liu, J. X., and Cheng, Y. Z. (1996). Echiostomiasis. In: Yu, E. S., Lin, J. H., and Chen, G. J., eds. Zoonoses in China. Fuzhou, China: Fujian Science and Technology Publishing House, 863–874.

Lo Re, V. III, and Gluckman, S. J. (2003). Eosinophilic meningitis. Am J Med 114: 217–223.

Loker, E. S., and Bayne, C. J. (1986). Immunity to trematode larvae in the snail *Biomphalaria*. Symp Zool Soc Lond 56, 199–220.

Lopez-Serrano, M. C., Gomez, A. A., Daschner, A., et al. (2000). Gastroallergic anisakiasis: findings in 22 patients. J Gastroenterol Hepatol 15, 503–506.

Lorenzo, S., Iglesias, R., Leiro, J., et al. (2000). Usefulness of currently available methods for the diagnosis of *Anisakis simplex* allergy. Allergy 55, 627–633.

Maki, J., and Yanagisawa, T. (1986). Demonstration of carboxyl and thiol protease activities in adult *Schistosoma mansoni, Dirofilaria immitis, Angiostrongylus cantonensis* and *Ascaris suum*. J Helminthol 60, 31–37.

Mako, T., and Akahane, H. (1985). On the larval *Gnathostoma doloresi* found in a snake, *Dinodon semicarinatus* from Amami-Oshima Is., Japan. Jpn J Parasitol 34, 493–499 (in Japanese with English abstract).

Maleewong, W., Sombatsawat, P., Intapan, P.M., Wongkham, C., and Chotmongkol, V. (2001). Immunoblot evaluation of the specificity of the 29–kDa antigen from young adult female worms *Angiostrongylus cantonensis* for immunodiagnosis of human angiostrongyliasis. Asian Pac J Allergy Immunol 19, 267–273.

Maruyama, H., Aoki, M., Okamura, S., Yoshida, A., Itagaki, T., and Ohta, N. (2003). Adult worms of the rodent intestinal nematode, Strongyloides venezuelensis, successfully invade chick intestinal mucosa. Parasitol Int 52, 35–39.

Maruyama, H., Hirabayashi, Y., El-Malky, M., et al. (2002). *Strongyloides venezuelensis*: longitudinal distribution of adult worms in the host intestine is influenced by mucosal sulfated carbohydrates. Exp Parasitol 100, 179–185.

Maruyama, H., Nishimaki, A., Takuma, Y., et al. (2006). Successive changes in tissue migration capacity of developing larvae of an intestinal nematode, *Strongyloides venezuelensis*. Parasitology 132, 411–418.

Maruyama, H., Yabu, Y., Yoshida, A., Nawa, Y., and Ohta, N. (2000). A role of mast cell glycosaminoglycans for the immunological expulsion of intestinal nematode, *Strongyloides venezuelensis*. J Immunol 164, 3749–3754.

Matsumoto, N., Mukae, H., Nakamura-Uchiyama, F., et al. (2002). Elevated levels of thymus and activation-regulated chemokine (TARC) in pleural effusion samples from patients infested with *Paragonimus westermani*. Clin Exp Immunol 130, 314–318.

Matsuoka, H., Nakama, T., Kisanuki, H., et al. (1994). A case report of serologically diagnosed pulmonary anisakiasis with pleural effusion and multiple lesions. Am J Trop Med Hyg 51, 8198–22.
Metwali, A., Setiawan, T., Blum, A. M., et al. (2006). Induction of CD8+ regulatory T cells in the intestine by *Heligmosomoides polygyrus* infection. Am J Physiol Gastrointest Liver Physiol 291, G253–259.
Mimori, T., Tada, I., Kawabata, M., Ollaue, W. L., Calero, G. H., and De Chong, Y. F. (1987). Immunodiagnosis of human gnathostomiasis in Ecuador by skin test and ELISA using *Gnathostoma doloresi* antigen. Jpn J Trop Med Hyg 15, 191–196.
Min, D. Y. (1990). Cestode infections in Korea. Korean J Parasitol 28(suppl), 123–144.
Min, D. Y., Lee, Y. A., Ryu, J. S., et al. (2004). Caspase-3–mediated apoptosis of human eosinophils by the tissue-invading helminth *Paragonimus westermani*. Int Arch Allergy Immunol 133, 357–364.
Miura, K., Fukumoto, S., Dirgahayu, P., and Hirai, K. (2001). Excretory/secretory products from plerocercoids of *Spirometra erinaceieuropaei* suppress gene expressions and production of tumor necrosis factor-alpha in murine macrophages stimulated with lipopolysaccharide or lipoteichoic acid. Int J Parasitol 31, 39–47.
Miyazaki, I. (1991). Helminthic Zoonoses. Japan: International Medical Foundation of Japan. Paragonimiasis, pp. 76–146; Gnathostomiasis, pp. 368–409.
Mohanty, M.C., and Ravindran, B. (2002). Deficiency of antibody responses to T-independent antigens in gerbils—*Meriones unguiculatus*. Dev Comp Immunol 26, 385–391.
Mohrs, M., Shinkai, K., Mohrs, K. and Locksley R.M. (2001). Analysis of type 2 immunity in vivo with a bicistronic IL-4 reporter. Immunity 15, 303–311.
Montgomery, J., Feldman, A., Despommier, D. D., Stewart, G. L., and Haehling, E. (1995). A method for isolation and partial purification of *Trichinella spiralis* nurse cells. J Parasitol 81, 649–652.
Morakote, N., and Kong, Y. (1993). Antigen specificity of 36 and 31 kDa proteins of *Spirometra erinacei* plerocercoid in tissue invading nematodiasis. Korean J Parasitol 31, 169–171.
Morrison, C. A., Colin, T., Sexton, J. L., et al. (1996). Protection of cattle against *Fasciola hepatica* infection by vaccination with glutathione S-transferase. Vaccine 14(17–18), 1603–1612.
Mosmann, T. R., and Coffman, R. L. (1989). TH1 and TH2 cells: different patterns of lymphokine secretion lead to different functional properties. Annu Rev Immunol 7, 145–173.
Mulcahy, G., and Dalton, J. P. (2001). Cathepsin L proteinases as vaccines against infection with *Fasciola hepatica* (liver fluke) in ruminants. Res Vet Sci 70, 83–86.
Mulcahy, G., O'Connor, F., McGonigle, S., et al. (1998). Correlation of specific antibody titre and avidity with protection in cattle immunized against *Fasciola hepatica*. Vaccine 16(9–10), 932–939.
Muro, A., Ramajo, V., Lopez, J., Simon, F., and Hillyer, G. V. (1997). *Fasciola hepatica*: vaccination of rabbits with native and recombinant antigens related to fatty acid binding proteins. Vet Parasitol 69(3–4), 219–229.
Murphy, K. M., and Reiner, S. L. (2002). The lineage decisions of helper T cells. Nat Rev Immunol 2, 933.
Na, B. K., Kim, S. H., Lee, E. G., et al. (2006). Critical roles for excretory-secretory cysteine proteases during tissue invasion of *Paragonimus westermani* newly excysted metacercariae. Cell Microbiol 8, 1034–1046.
Na, B. K., Lee, H. J., Cho, S. H., et al. (2002). Expression of cysteine proteinase of *Clonorchis sinensis* and its use in serodiagnosis of clonorchiasis. J Parasitol 88, 1000–1006.

Nagano, I., Pei, F., Wu, Z., et al. (2004). Molecular expression of a cysteine proteinase of *Clonorchis sinensis* and its application to an enzyme-linked immunosorbent assay for immunodiagnosis of clonorchiasis. Clin Diagn Lab Immunol 11, 411–416.

Nagano, I., and Takahashi, Y. (2005). Morphology of *Trichinella*. In: Arizono, N. et al., eds. Asian Parasitology, Vol. 1. Food-Borne Helminthiases in Asia. AAA and FAP, 263–266.

Nair, M. G., Cochrane, D. W., and Allen, J. E. (2003). Macrophages in chronic type 2 inflammation have a novel phenotype characterized by the abundant expression of Ym1 and Fizz1 that can be partly replicated in vitro. Immunol Lett 85, 173–180.

Nair, M. G., Gallagher, I. J., Taylor, M. D., et al. (2005). Chitinase and Fizz family members are a generalized feature of nematode infection with selective upregulation of Ym1 and Fizz1 by antigen-presenting cells. Infect Immun 73, 385–394.

Nakamura-Uchiyama, F., Mukae, H., and Nawa, Y. (2002a). Paragonimiasis: a Japanese perspective. Clin Chest Med 23, 409–420.

Nakamura-Uchiyama, F., Yamasaki, E., and Nawa, Y. (2002b). One confirmed and six suspected cases of cutaneous larva migrans caused by overseas infection with dog hookworm larvae. J Dermatol 29, 104–11.

Nakanishi, K., Yoshimoto, T., Tsutsui, H., and Okamura, K. (2001). Interleukin-18 regulates both Th1 and Th2 responses. Ann Rev Immunol 19, 423–474.

Nawa, Y. (1991). Historical review and current status of gnathostomiasis in Asia. Southeast Asian J Trop Med Public Health 22(suppl), 217–219.

Nawa, Y. (2002). Host-Parasite Interplay-Messages from Parasites. In: Proceedings of the 10th Japan Korea Parasitologists' Seminar, Forum Cheju 10, pp. 17–20.

Nawa, Y., Hatz, C., and Blum, J. (2005). Sushi delights and parasites: The risk of fishborne and foodborne parasitic zoonoses in Asia. Clin Infect Dis 41, 1297–1303.

Nawa, Y., Ishikawa, N., Tsuchiya, K., et al. (1994). Selective effector mechanisms for the expulsion of intestinal helminths. Parasite Immunol 16, 333–338.

Nawa, Y., Kiyota, M., Korenaga, M., and Kotani, M. (1985). Defective protective capacity of *W/W^v* mice against *Strongyloides ratti* infection and its reconstitution with bone marrow cells. Parasite Immunol 7, 429–438.

Nawa, Y., and Korenaga, M. (1983). Mast and goblet cell responses in the small intestine of rats concurrently infected with *Nippostrongylus brasiliensis* and *Strongyloides ratti*. J Parasitol 69, 1168–1170.

Nawa, Y., and Nakamura-Uchiyama, F. (2004). An overview of gnathostomiasis in the world. Southeast Asian J Trop Med Public Health 35(suppl 1), 87–91.

Nishino, C., Asaishi, K., and Hayasaka, K. (1990). Immune response to *Anisakis* larvae in healthy humans. In: Ishikura, H., and Kikuchi, K., eds. Intestinal Anisakiasis in Japan. Tokyo: Springer-Verlag, 251–256.

Nobayashi, M., Hirabayashi, H., Sakaki, T., et al. (2006). Surgical removal of a live worm by stereotactic targeting in cerebral sparganosis. Case report. Neurol Med Chir (Tokyo) 46, 164–167.

Noben-Trauth, N., Shultz, L. D., Brombacher, F., Urban, J. F. Jr, Gu, H., and Paul, W. E. (1997). An interleukin 4 (IL-4)-independent pathway for CD4+ T cell IL-4 production is revealed in IL-4 receptor-deficient mice. Proc Natl Acad Sci USA 94, 10838–10843.

Noda, S. (1990). Hemocyte encapusulation of *Angiostrongylus cantonensis* in M-line *Biomphalaria glabrata* infected with *Echinostoma paraensei*. Jpn J Parasitol 635–627.

Nolan, T. J., Megyeri, Z., Bhopale, V. M., and Schad, G. A. (1993). *Strongyloides stercoralis*: the first rodent model for uncomplicated and hyperinfective strongyloidiasis, the Mongolian gerbil (*Meriones unguiculatus*). J Infect Dis 168, 1479–1484.

Nopparatana, C., Setasuban, P., Chaicumpa, W., and Tapchaisri, P. (1991). Purification of *Gnathostoma spinigerum* specific antigen and immunodiagnosis of human gnathostomiasis. Int J Parasitol 21, 677–687.

Nuamtanong, S. (1996). The evaluation of the 29 and 31 kDa antigens in female *Angiostrongylus cantonensis* for serodiagnosis of human angiostrongyliasis. Southeast Asian J Trop Med Public Health 27, 291–296.

Ogata, K., Nawa, Y., Akahane, H., et al. (1998). Short report: Gnathostomiasis in Mexico. Am J Trop Med Hyg 58, 316–318.

Ohtaki, H., and Ohtaki, R. (1989). Clinical manifestation of gastric anisakiasis. In: Ishikura, H., and Namiki, M., eds. Gastric Anisakiasis in Japan. Tokyo: Springer-Verlag, 37–46.

O'Neill, S. M., Mills, K. H., and Dalton, J. P. (2001). *Fasciola hepatica* cathepsin L cysteine proteinase suppresses *Bordetella pertussis*-specific interferon-gamma production in vivo. Parasite Immunol 23, 541–547.

Oshima, T. (1987). Anisakiasis – Is the sushi bar guilty? Parasitol Today 3, 44–48.

Ovington, K. S., and Behm, C. A. (1997). The enigmatic eosinophil: investigation of the biological role of eosinophils in parasitic helminth infection. Mem Inst Oswaldo Cruz Rio de Janeiro 92(suppl II), 93–104.

Park, H., Kim, S. I., Hong, K. M., et al. (2002). Characterization and classification of five cysteine proteinases expressed by *Paragonimus westermani* adult worm. Exp Parasitol 102, 143–149.

Park, Y. K., Hwang, M. K., and Chung, P. R. (2006). Encystment and metacercariae development of *Echinostoma cinetorchis* cercariae in an in vitro culture system. J Parasitol 92, 1010–1013.

Pascual, C.Y., Crespo, J.F., San Martin, S., et al. (1997). Cross-reactivity between IgE-binding proteins from Anisakis, German cockroach, and chironomids. Allergy 52, 514–520.

Paykari, H., Dalimi, A., and Madani, R. (2002). Immunization of sheep against *Fasciola gigantica* with glutathione S-transferase.Vet Parasitol 105, 153–159.

Phares, C. K. (1987). Plerocercoid growth factor: a homologue of human growth hormone. Parasitology Today 3, 346–349.

Phares, K., and Kubik, J. (1996). The growth factor from plerocercoids of *Spirometra mansonoides* is both a growth hormone agonist and a cysteine proteinase. J Parasitol 82, 210–215.

Pichler, W. J. (1999). Anisakiasis: immunity, allergy or both? Lessons on the natural role of immunoglobulin E from a nematode infestation. Clin Exp Allergy 29, 1161–1163.

Pinlaor, S., Tada-Oikawa, S., Hiraku, Y., et al. (2005). *Opisthorchis viverrini* antigen induces the expression of Toll-like receptor 2 in macrophage RAW cell line. Int J Parasitol 35, 591–596.

Pozio, E., La Rosa, G., Murrell, K. D., and Lichtenfels, J. R. (1992). Taxonomic revision of the genus *Trichinella*. J Parasitol 78, 654–659.

Raes, G., De Baetselier, P., Noel, W., Beschin, A., Brombacher, F., and Hassanzadeh Gh, G. (2002). Differential expression of FIZZ1 and Ym1 in alternatively versus classically activated macrophages. J Leukoc Biol 71, 597–602.

Rajasekariah, G. R., Mitchell, G. F., Chapman, C. B., and Montague, P. E. (1979). *Fasciola hepatica*: attempts to induce protection against infection in rats and mice by injection of excretory/secretory products of immature worms. Parasitology 79, 393–400.

Raybourne, R., Deardorff, T. L., and Bier, J. W. (1986). *Anisakis simplex*: Larval excretory secretory protein production and cytostatic action in mammalian cell cultures. Exp Parasitol 62, 92–97.

Reite, O. B. (2005). The rodlet cells of teleostean fish: their potential role in host defence in relation to the role of mast cells/eosinophilic granule cells. Fish Shellfish Immunol 19, 253–267.

Reite, O.B., and Evensen, O. (2006). Inflammatory cells of teleostean fish: a review focusing on mast cells/eosinophilic granule cells and rodlet cells. Fish Shellfish Immunol 20, 192–208.

Rescigno, M., Urbano, M., Valzasina, B., et al. (2001). Dendritic cells express tight junction proteins and penetrate gut epithelial monolayers to sample bacteria. Nat Immunol 2, 361–367.

Rimoldi, M., Chieppa, M., Salucci, V., et al. (2005). Intestinal immune homeostasis is regulated by the crosstalk between epithelial cells and dendritic cells. Nat Immunol 6, 507–514.

Ritterson, A. L. (1966). Nature of the cyst of *Trichinella spiralis*. J Parasitol 52, 157–161.

Rodriguez-Sosa, M., David, J. R., Bojalil, R., Satoskar, A. R., and Terrazas, L. I. (2002). Cutting edge: susceptibility to the larval stage of the helminth parasite *Taenia crassiceps* is mediated by Th2 response induced via STAT6 signaling. J Immunol 168, 3135–3139.

Rodriguez-Sosa, M., Saavedra, R., Tenorio, E. P., Rosas, L.E., Satoskar, A. R., and Terrazas, L. I. (2004). A STAT4–dependent Th1 response is required for resistance to the helminth parasite *Taenia crassiceps*. Infect Immun 72, 4552–4560.

Rojekittikhun, W., Yamashita, T., Saito, S., Watanabe, T., Azuma, T., and Sendo, F. (1993). Purification of *Gnathostoma spinigerum* larval antigens by monoclonal antibody affinity chromatography. Southeast Asian J Trop Med Public Health 24, 680–684.

Ross, J. G. (1967). Studies of immunity to *Fasciola hepatica*: acquired immunity in cattle, sheep and rabbits following natural infection and vaccine procedures. J Helminthol 41, 393–399.

Rossjohn, J., Feil, S. C., Wilce, M. C., Sexton, J. L., Spithill, T. W., and Parker, M. W. (1997). Crystallization, structural determination and analysis of a novel parasite vaccine candidate: *Fasciola hepatica* glutathione S-transferase. J Mol Biol 273, 857–872.

Sajid, M., and McKerrow, J.H. (2002). Cysteine proteases of parastic organisms. Mol Biochem Parasitol 120, 1–21.

Saksirisampant, W., Chawengkiattikul, R., Kraivichain, K., and Nuchprayoon, S. (2001). Specific IgE antibody responses to somatic and excretory-secretory antigens of third stage *G. spinigerum* larvae in human gnathostomiasis. J Med Assoc Thai 84(suppl 1), S173–181.

Sampaio Silva, M. L., Correia Da Casta, J. M., Viana Daa Costa, A. M. V., et al. (1996). Antigenic components of excretory-secretory products of adult *Fasciola* hepatic recognized in human infections. Am J Trop Med Hyg 54, 146–148.

Sasaki, Y., Yoshimoto, T., Maruyama, H., et al. (2005). IL-18 with IL-2 protects against *Strongyloides venezuelensis* infection by activating mucosal mast cell-dependent type 2 innate immunity. J Exp Med 202, 607–616.

Sastre, J., Lluch-Bernal, M., Quirce, S., et al. (2000). A double-blind, placebo-controlled oral challenge study with lyophilized larvae and antigen of the fish parasite, *Anisakis simplex*. Allergy 55, 560–564.

Selzer, P. M., Pingel, S., Hsieh, I., et al. (1999). Cysteine protease inhibitors as chemotherapy: lessons from a parasite target. Proc Natl Acad Sci USA 96, 11015–11022

Seo, B. S., Lee, S. H., Cho, S. Y., Chai, J. Y., Hong, S. T., Han, I. S., Sohn, J. S., Cho, B. H., Ahn, S. R., Lee, S. K., Chung, S. C., Kang, K. S., Shim, H. S., and Hwang, I. S.(1981). An epidemiologic study on clonorchiasis and metagonimiasis in riverside areas in Korea. Korean J Parasitol 19(2), 137–150.

Sexton, J. L., Wilce, M. C., Colin, T., Wijffels, G. L., et al. (1994). Vaccination of sheep against *Fasciola hepatica* with glutathione S-transferase. Identification and mapping of antibody epitopes on a three-dimensional model of the antigen. J Immunol 152, 1861–1872.

Sheung-To, C., and Gibson, J. B. (1970). The histochemistry of biliary mucins and the changes caused by infestation with *Clonorchis sinensis*. J Pathol 101, 185–197.

Shi, B. B., Ishikawa, N., Itoh, H., et al. (1994). Goblet cell mucins of four genera of the subfamily Cricetinae with reference to the protective activity against *Strongyloides venezuelensis*. Parasite Immunol 16, 553–559.

Shi, B. B., Ishikawa, N., Itoh, H., et al. (1995). Goblet cell hyperplasia induced by *Strongyloides venezuelensis*-infection in Syrian golden hamster, *Mesocricetus auratus*. Int J Parasitol 25, 399–402.

Shih, H. H., and Chen, S. N. (1991). Immunodiagnosis of angiostrongyliasis with monoclonal antibodies recognizing a circulating antigen of mol. wt 91,000 from *Angiostrongylus cantonensis*. Int J Parasitol 21, 171–177.

Shin, E. H., Osada, Y., Chai, J. Y., Matsumoto, N., Takatsu, K., and Kojima, S. (1997). Protective roles of eosinophils in Nippostrongylus brasiliensis infection. Int Arch Allergy Immunol 114(suppl 1), 45–50.

Shin, M. H., Chung, Y. B., and Kita, H. (2005). Degranulation of human eosinophils induced by *Paragonimus westermani*-secreted protease. Korean J Parasitol 43, 33–37.

Shinkai, K., Mohrs, M., and Locksley, R. M. (2002). Helper T cells regulate type-2 innate immunity in vivo. Nature 420, 825–829.

Shortman, K., and Liu, Y. J. (2002). Mouse and human dendritic cell subtypes. Nat Rev Immunol 2, 151–161.

Sithithaworn, P., Haswell-Elkins, M. R., Mairiang, P., et al. (1994). Parasite-associated morbidity: liver fluke infection and bile duct cancer in northeast Thailand. Int J Parasitol 24, 833–843.

Sithithaworn, P., and Haswell-Elkins, M. (2003). Epidemiology of *Opisthorchis viverrini*. Acta Trop 88, 187–194.

Smith, A. M., Dowd, A. J., Heffernan, M., Robertson, C. D., and Dalton, J. P. (1993a). *Fasciola hepatica*: a secreted cathepsin L-like proteinase cleaves host immunoglobulin. Int J Parasitol 23, 977–983.

Smith, A. M., Dowd, A. J., McGonigle, S., et al. (1993b). Purification of a cathepsin L-like proteinase secreted by adult *Fasciola hepatica*. Mol Biochem Parasitol 62, 1–8.

Smith, W. D. (1999). Prospects for vaccines of helminth parasites of grazing ruminants. Int J Parasitol 29, 17–24.

Smooker, P. M., Kennedy, N. J., Steeper, K. R., Christopoulos, H., and Spithill, T. W. (2001). *Fasciola*: kinetics and quality of humoral responses to fatty acid binding protein and cathepsin L following delivery as DNA vaccines in mice. Exp Parasitol 97, 154–160.

Smooker, P. M., Steeper, K. R., Drew, D. R., Strugnell, R. A., and Spithill, T. W. (1999). Humoral responses in mice following vaccination with DNA encoding glutathione S-transferase of Fasciola hepatica: effects of mode of vaccination and the cellular compartment of antigen expression. Parasite Immunol 21, 357–364.

Smythies, L. E., Coulson, P. S., and Wilson, R. A. (1992). Monoclonal antibody to IFN-γ modifies pulmonary inflammatory responses and abrogates immunity to *Schistosoma mansoni* in mice vaccinated with attenuated cercariae. J Immunol 149, 3654–3658.

Sohn, W. M., and Lee, S. H. (1998). The first discovery of larval *Gnathostoma hispidum* (Nematoda: Gnathostomidae) from a snake host, Agkistrodon brevicaudus. Korean J Parasitol 36, 81–89.

Song, C. Y., and Dresden, M. H. (1990). Partial purification and characterization of cysteine proteinases from various developmental stages of *Paragonimus westermani*. Comp Biochem Physiol [B] 95, 473–476.

Southgate, V. R. (1971). Observations on the fine structure of the cercariae of *Notocotylus attenuatus* and formation of the cyst wall of the metacercara. Z Zellforsch Mikrosk Anat 120, 420–449.

Spithill, T. W., and Dalton, J. P. (1998). Progress in development of liver fluke vaccines. Parasitol Today 14, 224–228.

Spithill, T. W., Piedrafita, D., and Smooker, P. M. (1997). Immunological approaches for the control of fasciolosis. Int J Parasitol 27, 1221–1235.

Sriveny, D., Raina, O. K., Yadav, S. C., et al. (2006). Cathepsin L cysteine proteinase in the diagnosis of bovine *Fasciola gigantica* infection. Vet Parasitol 135, 25–31.

Sugane, K., Kusama, Y., Takamoto, M., Tominaga, A., and Takatsu, K. (1996). Eosinophilia, IL-5 level and recovery of larvae in IL-5 transgenic mice infected with *Toxocara canis*. J Helminthol 70, 153–158.

Sugaya, H., Aoki, M., Abe, T., Ishida, K., and Yoshimura, K. (1997a). Cytokine responses in mice infected with *Angiostrongylus cantonensis*. Parasitol Res 83, 10–15.

Sugaya, H., Aoki, M., Yoshida, T., Takatsu, K., and Yoshimura, K. (1997b). Eosinophilia and intracranial worm recovery in interleukin-5 transgenic and interleukin-5 receptor alpha chain-knockout mice infected with *Angiostrongylus cantonensis*. Parasitol Res, 83, 583–590.

Sugaya, H., Graeff-Teixeira, C., Ishida, K., Matsuda, S., Katahira, K., and Yoshimura, K. (2002). Interleukin-5 transgenic mice show augmented resistance to *Angiostrongylus costaricensis* infection. Parasitol Res 88, 350–355.

Sugaya, H., and Yoshimura, K. (1988). T-cell-dependent eosinophilia in the cerebrospinal fluid of the mouse infected with *Angiostrongylus cantonensis*. Parasite Immunol 10, 127–138.

Suntharasamai, P., Desakorn, V., Migasena, S., Bunnag, D., and Harinasuta, T. (1985). ELISA for immunodiagnosis of human gnathostomiasis. Southeast Asian J Trop Med Public Health 16, 274–279.

Taira, N., Yoshifuji, H., and Boray, J. C. (1997). Zoonotic potential of infection with *Fascioloa* spp. by consumption of freshly prepared raw lier containing immature flukes. Int J Parasitol 27, 775–779.

Takahashi, S., Ishikura, H., Sato, M., and Iwasa, K. (1993). Seroimmunodiagnostics of anisakiosis. Annu Rev Immunol Cyugai-Medical, Tokyo, 211–217 (in Japanese).

Takeda, K., Tanaka, T., Shi, W., et al. (1996). Essential role of Stat6 in IL-4 signalling. Nature 380, 627–630.

Taniguchi, H., Mukae, H., Matsumoto, N., et al. (2001). Elevated IL-5 Levels in pleural fluid of patients with paragonimiasis westermani. Clin Exp Immunol 123, 94–98.

Taniguchi, Y., Ando, K., Shimizu, M., Nakamura, Y., and Yamazaki, S. (1994). Creeping eruption due to larvae of the suborder Spirurina—a newly recognized causative parasite. Int J Dermatol 33, 279.

Tantrawatpan, C., Maleewong, W., Wongkham, C., Wongkham, S., Intapan, P. M., and Nakashima, K. (2005). Serodiagnosis of human fascioliasis by a cystatin capture enzyme-linked immunosorbent assay with recombinant *Fasciola gigantica* cathepsin L antigen. Am J Trop Med Hyg 72, 82–86.

Taylor, M. D., LeGoff, L., Harris, A., Malone, E., Allen, J. E., and Maizels, R. M. (2005). Removal of regulatory T cell activity reverses hyporesponsiveness and leads to filarial parasite clearance in vivo. J Immunol 174, 4924–4933.

Tendler, M., Brito, C. A., Vilar, M.M., et al. (1996). A *Schistosoma mansoni* fatty acid-binding protein, Sm14, is the potential basis of a dual-purpose anti-helminth vaccine. Proc Natl Acad Sci U S A 93, 269–273.

Tendler, M., Vilar, M. M., Brito, C. A., Freire, N. M., Katz, N., and Simpson, A. (1995). Vaccination against schistosomiasis and fascioliasis with the new recombinant antigen Sm14: potential basis of a multi-valent anti-helminth vaccine? Mem Inst Oswaldo Cruz 90, 255–256.

Tiengkamol, Y., and Brockelman, C. R. (1982). *Angiostrongylus cantonensis:* biogenic amines in the lungs of infected rats. Exp Parasitol 54, 121–128.

Toledo, R., Esteban, J-G., and Fried, B. (2006). Immunology and pathology of intestinal trematodes in their definitive hosts. Adv Parasitol 63, 285–365.

Toledo, S., and Fried, B. (2005). Echinostomes as experimental models in adult parasite-vertebrate host interactions. Trend Parasitol 21, 251–254.

Torres, D., and Espino, A. M. (2006). Mapping of B-cell epitopes on a novel 11.5–kilodalton *Fasciola hepatica-Schistosoma mansoni* cross-reactive antigen belonging to a member of the *F. hepatica* saposin-like protein family. Infect Immun 74, 4932–4938.

Tort, J., Brindley, P. J., Knox, D., Wolfe, K. H., and Dalton, J. P. (1999). Proteinases and associated genes of parasitic helminthes. Adv Parasitol 43, 161–266.

Uchida, A., Kawakami, Y., Yuzu, S., et al. (1998). Prevalence and histopathology of parasitization in minke whales from the western North Pacific Ocean and the southern Sea of Okhotsk. Rep Int Whale Commun 48, 475–479.

Uparanukraw, P., Harnnoi, T., Dantrakool, A., Morakote, N., and Chaicumpa, W. (1999). Recognition of deglycosylated larval proteins of *Gnathostoma spinigerum* by a monoclonal antibody and human gnathostomiasis antiserum. J Helminthol 73, 273–276.

Uparanukraw, P., Morakote, N., Harnnoi, T., and Dantrakool, A. (2001). Molecular cloning of a gene encoding matrix metalloproteinase-like protein from Gnathostoma spinigerum. Parasitol Res 87, 751–757.

Urban, J. F. Jr, Noben-Trauth, N., Donaldson, D. D., et al. (1997). IL-13, IL-4Ralpha, and Stat6 are required for the expulsion of the gastrointestinal nematode parasite *Nippostrongylus brasiliensis*. Immunity 8, 255–264.

Valinas, B., Lorenzo, S., Eiras, A., Figueiras, A., Sanmartin, M. L., and Ubeira, F. M. (2001). Prevalence of and risk factors for IgE sensitization to *Anisakis simplex* in a Spanish population. Allergy 56, 667–671.

van Milligen, F. J., Cornelissen, J. B., and Bokhout, B. A. (2000). *Fasciola hepatica*: an antigen fraction derived from newly excysted juveniles, containing an immunoreactive 32–kDa protein, induces strong protective immunity in rats. Exp Parasitol 94, 163–171.

van Milligen, F. J., Cornelissen, J. B., and Bokhout, B. A. (1999). Protection against *Fasciola hepatica* in the intestine is highly correlated with eosinophil and immunoglobulin G1 responses against newly excysted juveniles. Parasite Immunol 21, 243–251.

van Rijt, L. S., Jung, S., Kleinjan, A., et al. (2005). In vivo depletion of lung CD11c+ dendritic cells during allergen challenge abrogates the characteristic features of asthma. J Exp Med 201, 981–991.

Voehringer, D., Reese, T.A., Huang, X., Shinkai, K., and Locksley, R.M. (2006). Type 2 immunity is controlled by IL-4/IL-13 expression in hematopoietic non-eosinophil cells of the innate immune system. J Exp Med 203, 1435–1446 (erratum in J Exp Med, 203, 1617).

Voehringer, D., Shinkai, K., and Locksley, R.M. (2004). Type 2 immunity reflects orchestrated recruitment of cells committed to IL-4 production. Immunity 20, 267–277.

Waikagul, J., and Radomyos, P. (2005). Intestinal trematode infections in Thailand. In: Asian Parasitology, Vol. 1. Food-Borne Helminthiasis in Asia. Chiba, Japan: AAA/FAP, 79–102.

Weaver, C. T., Harrington, L. E., Mangan, P. R., Gavrieli, M., and Murphy, K. M. (2006). Th17: an effector CD4 T cell lineage with regulatory T cell ties. Immunity 24, 677–688.

Welch, J.S., Escoubet-Lozach, L., Sykes, D.B., Liddiard, K., Greaves, D.R., and Glass, C.K. (2002) TH2 cytokines and allergic challenge induce Ym1 expression in macrophages by a STAT6–dependent mechanism. J Biol Chem 277, 42821–42829.

Williamson, A.L., Lustigman, S., Oksov, Y., et al. (2006). *Ancylostoma caninum* MTP-1, an astacin-like metalloprotease secreted by infective hookworm larvae, is involved in tissue migration. Infect Immun 74, 961–967.

Wilson, M. S., Taylor, M. D., Balic, A., Finney, C. A., Lamb, J. R., and Maizels, R. M. (2005). Suppression of allergic airway inflammation by helminth-induced regulatory T cells. J Exp Med 202, 1199–1212.

Wiwanitkit, V. (2005). A review of human sparganosis in Thailand. Int J Infect Dis 9, 312–316.

Wongkham, C., Tantrawatpan, C., Intapan, P. M., Maleewong, W., Wongkham, S., and Nakashima, K. (2005). Evaluation of immunoglobulin G subclass antibodies against recombinant Fasciola gigantica cathepsin L1 in an enzyme-linked immunosorbent assay for serodiagnosis of human fasciolosis. Clin Diagn Lab Immunol 12, 1152–1156.

Wongratanacheewin, S., Rattanasiriwilai, W., Priwan, R., and Sirisinha, S. (1987). Immunodepression in hamsters experimentally infected with Opisthorchis viverrini. J Helminthol 61, 151–156.

World Health Organization. (1995). Control of foodborne trematode infections. Report of WHO Study group. WHO, Geneva, Switzerland.

Wranicz, M.J., Gustowska, L., Gabryel, P., Kucharska, E., and Cabaj, W. (1998). *Trichinella spiralis*: induction of the basophilic transformation of muscle cells by synchronous newborn larvae. Parasitol Res 84, 403–407.

Wu, G. (1947). Development of *Fasciolopsis buski* in rabbits. Natl Med J China, 33, 175.

Wynn, T.A., Jankovic, D., Hieny, S., Cheever, A.W., and Sher, A. (1995). IL-12 enhances vaccine-induced immunity to Schistosoma mansoni in mice and decreases T helper 2 cytokine expression, IgE production, and tissue eosinophilia. J Immunol 154, 4701–4709.

Wynn, T. A., Reynolds, A., James, S., et al. (1996). IL-12 enhances vaccine-induced immunity to schistosomes by augmenting both humoral and cell-mediated immune responses against the parasite. J Immunol 157:4068–4078.

Yadav, S. C., Saini, M., Raina, O. K., Nambi, P. A., Jadav, K., and Sriveny, D. (2005). *Fasciola gigantica* cathepsin-L cysteine proteinase in the detection of early experimental fasciolosis in ruminants. Parasitol Res 97, 527–534.

Yamakami, K. (1986). Purification and properties of a thiol protease from lung fluke adult *Paragonimus ohirai*. Comp Biochem Physiol [B] 83, 501–506.

Yamasaki, H., Toki, T., and Oya, H. (1989). A cysteine proteinase from the liver fluke *Fasciola* spp: purification, characterization, localization and application to immunodiagnosis. Jpn J Parasitol 38, 373–384.

Yang, H. J. (2004). Modification of carbohydrate compositions of 31/36 kDa proteins of plerocercoids (sparganum) of *Spirometra mansoni* grown in different intermediate hosts. Korean J Parasitol 42, 77–79.

Yang, S. H., Park, J. O., Lee, J. H., et al. (2004). Cloning and characterization of a new cysteine proteinase secreted by *Paragonimus westermani* adult worms. Am J Trop Med Hyg 71, 87–92.

Yoshida, Y. (2006). Sparganosis. In: Illustrated Human Parasitology, 7th ed. Tokyo: Nanzan-do, 194–195.

Yoshikawa, M., and Ishizaka, S. (2005). Anisakidosis in Japan. In: Arizono, N. et al., eds. Asian Parasitology, Vol. 1. Food-Borne Helminthiasis in Asia. Chiba, Japan: FAP, 217–224.

Yoshimito, T., and Nakanishi, K. (2006). Role of IL-18 in basophils and mast cells. Allergol Int 55, 105–113.

Yoshimura, H. (1990). Clinical patho-parasitology of extra-gastrointestinal anisakiasis. In: Ishikura, H., and Kikuchi, K., eds. Intestinal Anisakiasis in Japan. Tokyo: Springer-Verlag, 145–154.

Yoshimura, K., Aiba, H., Hirayama, N., Yosida, T. H. (1979). Acquired resistance and immune responses of eight strains of inbred rats to infection with *Angiostrongylus cantonensis*. Jpn J Vet Sci 41, 245–259.

Yoshimura, K., Sato, K., Oya, H., and Nakagawa, M. (1980). Transfer of young adult *Angiostrongylus cantonensis* from mastomys and mice to the pulmonary arteries of rats immunized against the donor hosts. Jpn J Vet Sci 42, 567–572.

Yoshimura, K., Sugaya, H., and Ishida, K. (1994). The role of eosinophils in *Angiostrongylus cantonensis* infection. Parasitol Today 10, 231–233.

Yoshimura, K., and Yamagishi, T. (1976). Reaginic antibody productions in rabbits and rats infected with *Angiostrongylus cantonensis*. Jpn J Vete Sci 38, 33–40.

Yu, S.H., and Mott, K.E. (1994). Epidemiology and morbidity of food-borne intestinal nematode infections. Trop Dis Bull 91, R125–R152.

11 Molecular Epidemiology of Food-Borne Parasitic Zoonoses

R.C. Andrew Thompson, Rebecca J. Traub, and Nevi Parameswaran

An increasing number of parasites are being added to the list of those that can be transmitted via food or water and that pose a risk to human health if ingested. These zoonotic infections usually have complicated life cycles requiring a number of hosts for completion or a diversity of cycles of transmission that may interact. The challenge in all control efforts is to break the cycle of transmission that may lead to human infection, which requires the ability to detect and characterize the relevant parasite life cycle stage in food or water. This requires tools that are both sensitive and specific, and often beyond the limitations of conventional techniques such as microscopy. This is not only because of the insensitivity of many microscopy-based procedures but principally due to the need to characterize the infectious agent in terms of its potential health risk. Many parasites produce infective stages—larvae, eggs, cysts, oocysts—that often do not possess species- or strain-specific morphological features that can be used to differentiate between stages from parasites that may be infective to humans and those that are not. Taeniid eggs and many protozoan oocysts provide excellent examples of this, as do many helminth larvae in tissue. This is where the ability to detect genetic variability using molecular techniques has proved to be so valuable in epidemiological investigations and has provided the basis for the establishment of molecular epidemiology as a rapidly growing field of study.

This chapter describes what underpins molecular epidemiological studies, demonstrates their value with a variety of food- and waterborne parasitic zoonoses, and using two case studies, the fishborne trematodes and *Toxoplasma*, indicate progress with molecular epidemiological approaches and future research needs.

What Is Molecular Epidemiology?

Molecular epidemiology provides the tools (both laboratory and analytical) that have predictive significance and that epidemiologists can use to better define the etiology of specific diseases and work toward their control (Thompson et al., 1999; Thompson, 2002). The scope of molecular epidemiological investigations is greatest with infectious diseases, particularly parasites due to the complexity of the ecological interactions involved. Molecular epidemiology contributes significantly to the characterization, detection, and diagnosis of infectious diseases, and to an understanding of the processes involved in their emergence by studying the origins and ecology of emerging infections. As such, molecular epidemiological investigations require not only the development and application of appropriate molecular tools, but also the application of both evolutionary biology and population genetics for a meaningful interpretation of the data obtained by molecular epidemiology. Ideally, molecular epidemiological techniques should also facilitate determinations of why infections occur and if they are likely to occur in the future by identifying risk factors and obstacles that may limit prevention and control.

The Role of Molecular Epidemiology in Infectious Disease Control

Controlling infectious diseases depends on the ability to rapidly detect and characterize the etiological agents, and on the establishment of adequate surveillance systems to monitor the effects of control programs, and warn of outbreaks or incursions so that appropriate intervention strategies can be implemented.

From an epidemiological viewpoint, we need to understand the ecology of the etiological agents (Poulin, 1998; Hudson, 2005). This will enable us to better understand and predict their transmission dynamics, how life cycles may interact, and the nature of interactions within the host. This requires an input from both population and evolutionary biology particularly to understanding the genetic structure and evolution of infectious agents, their population biology and ecology, and the evolutionary consequences of medical and public health interventions (Levin et al., 1999). In recent years the information generated as a result of molecular epidemiological investigations has produced much new data requiring rigorous population genetic analysis and modeling in order for it to be interpreted in an epidemiological context (Constantine, 2003). In this context emphasis has been given to the importance of appropriate analysis and the value of characterizing the genetic diversity of infectious agents at different levels of specificity (Hall, 1996; Thompson and Lymbery, 1996; Monis et al., 2005). The latter requires choosing molecular tools that are capable of discriminating genetic variants at different hierarchical levels, and the region of DNA examined must be appropriate to the issue being addressed (Thompson et al., 1999; Monis et al., 2005), for example, taxonomy, diagnosis, population genetics, evolutionary relationships, and isolate tracking (Table 11.1).

Our ability, using molecular techniques, to detect and characterize the genetic variability of infectious agents, particularly at the intraspecific level (often resulting in the recognition and description of new species), can be seen as the foundation for the majority of molecular epidemiological studies. The application of appropriate molecular tools will aid in the identification and surveillance of infectious agents and in determining sources of infection. The availability of such tools, particularly

TABLE 11.1. Characterizing genetic diversity in parasites.

Function	Purpose	Tools
Discrimination above species level	Taxonomy/phylogeny	Highly conserved coding regions, e.g., SSU rDNA, certain mitochondrial genes
Discrimination between species	Taxonomy/diagnosis/epidemiology	Moderately conserved regions, e.g., coding mitochondrial genes, ITS rDNA, and other loci (e.g., house-keeping genes such as GDH, TPI, HSP, Actin, etc.); LAMP
Discrimination between intraspecific variants/strains/ genotypes	Population genetics/breeding systems (e.g., cross *vs.* self fertilization) / host specificity/ molecular epidemiology/ conservation (e.g., predicting susceptibility to pathogens)/ biosecurity (exotic and emerging pathogens)	Variable regions, e.g., allozymes, RAPD, AFLP, PFGE, PCR-RFLP, pyrosequencing, LAmp
Discrimination between individual isolates/clonal lineages/ subgenotypes/ ecological interactions within host	Fingerprinting/molecular epidemiology—tracking transmission of subgenotypes/ sources of infection and risk factors/competitive interactions and course of infection	Fingerprinting techniques, e.g., mini/microsatellites, SSCP
Genetic markers/ linking phenotype and genotype	Identifying phenotypic traits of clinical and epidemiological significance, e.g., virulence, infectivity, drug sensitivity	Genotype linked to phenotype via (1) genetic map; (2) RDA; (3) sequencing and/or RT-PCR of genes thought to be linked to phenotypic traits

In some cases, there may be overlap between the tools (regions of DNA) used and function. This will depend on the group of parasites being studied and the level of variation detectable by a particular approach (modified from Monis et al., 2005).

AFLP, amplified fragment length polymorphism; ITS, internal transcribed spacer; LAMP, loop-mediated isothermal DNA amplification; PCR, polymerase chain reaction; PCR-RFLP, PCR-coupled restriction fragment length polymorphism; PFGE, pulse field gel electrophoresis; RAPD, random amplified polymorphic DNA; RDA, representational difference analysis; RDNA, ribosomal DNA.

those based on the polymerase chain reaction (PCR), which obviate the need for laboratory amplification of infectious agents and enable direct examination of clinical or environmental isolates, has had an enormous impact on the genetic characterization, diagnosis, and taxonomy of infectious disease agents. Molecular epidemiology is also very useful in the individual host not only for diagnosis but also for monitoring the effects of treatment and competitive interactions, for example in malaria (Walliker, 2000), and for discriminating between treatment failures and recurrent infection.

From a practical perspective, genotyping parasites should not become a dominant aim of molecular epidemiological investigations since the existence of different genotypes does not imply that they necessarily have some phenotypic importance. The scope and potential of molecular epidemiology is much greater, and in this regard the search for genetic markers for medically important traits such as infectivity, drug resistance, and virulence present important challenges for molecular epidemiology. The availability of appropriate phenotypic markers facilitates more accurate predictions to be made about the clinical course of any resultant disease and the most appropriate therapy. Such research will be made much easier through the availability of an increasing number of fully sequenced genomes of parasites and their vectors (Coppel and Black, 2005).

The Applications of Molecular Epidemiology to the Control Food- and Waterborne Parasite Zoonoses

The control of food- and waterborne parasitic zoonoses depends on the rapid and accurate detection of the etiological agents so that cycles of transmission can be identified, and the potential for interaction between cycles evaluated. Effective control also requires the ability to characterize parasites from different stages in their life cycles in tissues, blood, feces, or the environment, on the basis of epidemiologically useful features. The latter include host specificity, public health significance in terms of zoonotic potential, virulence, and drug sensitivity. Traditional diagnostic techniques involving microscopy have thus been complemented by a variety of molecular tools that provide additional information about the causative agents. The application of such tools has also helped to resolve taxonomic issues that resulted in much controversy in the past, when new species or strains were described on the basis of phenotypic characteristics or epidemiological observations in particular endemic areas. A formal nomenclature is essential for effective communication and provides the stability that underpins epidemiological investigations. However, the lack of morphological differences between many such variants only compounded an often confusing taxonomic picture that has in many cases taken decades to resolve. Such was the situation with *Trichinella* and *Cryptopsoridium*, but as a result of the application of molecular tools many taxonomic issues have been resolved and, as a consequence, communication has been markedly enhanced.

Characterization—The Tools

As indicated above, the application of molecular biological methods to parasite taxonomy and epidemiology has much to do with detecting and analyzing variation between parasites. Thus the major advances in recent years have been in defining the most appropriate regions of DNA to use for detecting variation at different levels of specificity. This requires choosing molecular tools that are capable of discriminating genetic variants at different hierarchical levels, and the region of DNA examined must be appropriate to the level of questions being asked (Monis et al., 2005; Table 10.1). Analyzing such genetic variation is also dependent on appropriate and rigorous analysis (Constantine, 2003), the reliability of which is enhanced when a number of genetic loci are used. The value of such tools is greatest if they can be applied directly to fecal or tissue specimens, as well as environmental samples and food, and if there is the potential to automate such procedures. In this respect, PCR-based techniques have provided very powerful epidemiological tools that obviate the need for laboratory amplification. The latter imposed a major limiting factor in characterizing parasites refractory to in vitro culture, which was exacerbated by the selective factors associated with the culture of parasites. In the future, emphasis will be given to the establishment of high throughput molecular assays such as pyrosequencing, as well as their field applicability. For example, pyrosequencing techniques have the added advantage of allowing the simultaneous detection of multiple species/genotypes in a single sample (Sreekumar et al., 2005; Ahmadian et al., 2006; Stensvold et al., 2006). Loop-mediated isothermal DNA amplification (LAMP) is a newly developed, rapid, quantitative, highly sensitive and specific nucleic acid–based, non-PCR diagnostic tool (Notomi et al., 2000; Mori et al., 2004; Yano et al., 2007), applicable to low-cost laboratory settings. This simple molecular test can be carried out on a bench with a heating block instead of a thermal cycler and may prove to be an invaluable "field-friendly" tool for screening and quantifying infections in host populations while providing important genotypic information.

Evolutionary Relationships

Defining evolutionary relationships between closely related infectious agents can give important insights into their origins, host associations, and ecology. The development of molecular tools for the genetic characterization of parasite species has provided the opportunity for phylogenetic studies, which, by demonstrating affinities within phyla, will provide a better understanding of biology and ecology, which are fundamental to control.

A wealth of molecular data has been produced for the different encapsulated species of *Trichinella* over the last 15 years, but there is still no consensus as to whether *T. spiralis* or *T. nelsoni* is the origin of the encapsulated species (Pozio and Zarlenga, 2005). This determination requires further additional in-depth analyses. The non-encapsulated species are genetically distinct from encapsulated species, but the three non-encapsulated species are genetically quite distinct from each other, with unique host ranges and broadening geographical ranges serving to raise questions about the term "*Trichinella*-free" as it applies to geographic localities and trade restrictions. With *Fasciola hepatica*, molecular analyses of European and Asian populations support separate lineages for these populations and two independent domestications (Semyenova et al., 2006). Whether the genetic diversity between these two lineages reflects phenotypic differences of practical significance in terms of definitive host preference remains to be determined.

Phylogenetic analysis of molecular data from *Cryptosporidium* has confirmed that *Cryptosporidium* is not as closely related to coccidian parasites as originally suspected (Morrison and Ellis, 1997) but rather to the gregarines (Carreno et al., 1999; Leander et al., 2003a,b; Leander and Keeling 2004). Thus, within the digestive tract of vertebrates, there are at least two distinct lineages of apicomplexan parasites that have exploited this ecological niche (Barta and Thompson, 2006). The first lineage are the classically recognized coccidia including eimeriid and isosporoid coccidia that are each well-supported natural groups of taxa that share a common ancestry. The second lineage appears to be restricted to only *Cryptosporidium* species that have apparently evolved from gregarine or protococcidian ancestors. A gregarine ancestry for cryptosporidians will have a significant impact on how we understand and deal with the basic biology of cryptosporidians (Cavalier-Smith and Chao, 2004; Barta and Thompson, 2006; Leander et al., 2006), and the epidemiology of infections with species of *Cryptosporidium*. Such studies not only will provide clues about the evolution of intracellular parasitism, but also, in the case of *Cryptosporidium*, will provide a better understanding of the host parasite relationship, as well as the development of the parasite in the environment.

Taxonomy

Of all the food-borne helmith zoonoses, *Trichinella* provides the best example of the value of molecular characterization in determining the taxonomy of species in the genus. All recognized species and genotypes of the genus are morphologically indistinguishable at all developmental stages, and the application of molecular tools has been instrumental in defining the taxonomy of both encapsulated and non-encapsulated forms. This has provided a firm foundation on which to interpret the wealth of epidemiological data obtained over a period of more than 50 years since the discovery of what we now know as *T. nelsoni*. Excellent progress is also being made in resolving taxonomic issues with *Paragonimus* using molecular techniques (see below). In contrast, there is an urgent need to develop and apply appropriate molecular tools to numerous other food-borne helminths in order to determine the extent of genetic diversity and importance of reported phenotypic differences. For example, with the diphyllobothriids, molecular studies have shown that the morphological variation seen within *D. dendriticum* populations has a genetic basis (De Vos et al., 1990), but the epidemiological significance of this variation remains to be investigated. Preliminary molecular studies of anisakids have provided evidence for new species in European and North American waters and have emphasized the urgent need for a thorough, geographically widespread systematic study of the genera *Anisakis, Pseudoterranova*, and related genera (Chai et al., 2005). The public health significance of food-borne trematode infections is an increasing problem, particularly in Southeast Asia, and, as will be discussed below, there is great scope for the application of molecular epidemiological tools in their control.

As with many protozoan parasites with few morphological characteristic for species discrimination, early workers relied largely on host occurrence in describing species of *Giardia* and *Cryptosporidium*. This resulted in the description of a large number of species and a history of taxonomic confusion and controversy (O'Donoghue, 1995; Thompson, 2002). Today we recognize that *Giardia* and *Cryptosporidium* are phenotypically and genotypically heterogeneous assemblages of largely morphologically identical genotypes and species (Morgan et al., 1999; Monis and Thompson, 2003; Thompson et al., 2003a). With both *Giardia* and *Cryptosporidium*, a large number of species and genotypes are now recognized that differ principally in their host ranges. Some species and genotypes appear to be restricted to particular species of hosts (e.g., *G. psittaci*; *C. canis*) or closely related host assemblages (e.g., *G. bovis*; *C. baileyi*), whereas others have broad host ranges including humans (e.g., *G. duodenalis*; *C. parvum*) and are therefore of zoonotic significance.

It is not surprising, therefore, that the lack of morphological characteristics to discriminate variants led to much debate over whether phenotypic differences were real and reflected genetic differences or were the result of environment/host-induced changes (Fall et al., 2003). Clearly, the application of PCR-based molecular tools has had an enormous impact on our understanding of the nature of variation in these two protozoan genera. The direct characterization of oocysts recovered from fecal or environmental samples using PCR-based procedures has had a major impact not only on resolving the taxonomy of *Giardia* and *Cryptosporidium* at the species level but also on the molecular epidemiology of infections. The recent molecular and biological evidence that has demonstrated *Cryptosporidium*'s closer relationship to gregarine protozoa than to coccidia (Carreno et al., 1999; Hijjawi, et al., 2002) not only helps to explain the increasing numbers of novel genotypes that are being discovered, but also emphasizes that the specificity of environmental detection procedures for *Cryptosporidium* could be compromised by cross-reactivity with gregarine protozoa that are ubiquitous in fresh water environments (Bull et al., 1998; Hijjawi et al., 2002; Tenter et al., 2002; Barta and Thompson, 2006).

With *Toxoplasma*, what was considered to be a genetically conserved genus with just one species, *T. gondii*, may now be questioned in light of an increasing number of diverse isolates being examined using more discriminatory molecular tools (see below).

Diagnosis and Detection

In any epidemiological investigation, there is a need for sensitive and specific diagnostic procedures for detecting the etiological agents of infectious disease. The PCR-based procedures have proved to have greater sensitivity and specificity than conventional diagnostics that are reliant on microscopy or immunodiagnosis. For example, sensitive molecular techniques have been developed for both *Giardia* and *Cryptosporidium* that not only are more sensitive than conventional techniques, but also can provide information on the genotype or species present by combining PCR with restriction fragment length polymorphism (RFLP) analysis, without having to resort to costly and time-consuming sequencing (Caccio et al., 2002, 2005). Although one of the major advantages of such PCR-based procedures is the ease of interpretation that usually involves the visualization of a small number of bands on a gel, the very high sensitivity of most PCR-based procedures can also present problems of interpretation. For example, a recent survey of parasites of domestic cats using microscopy found that 5% were infected with *Giardia*, whereas PCR revealed that 80% were positive, a level supported by the detection of fecal antigen (McGlade et al., 2003). Such a result raises questions concerning both the clinical and epidemiological significance of such presumably low-level infections with *Giardia* that result in minimal excretion of infective stages.

Apart from *Trichinella*, the application of molecular tools for the successful diagnosis or detection of food-borne helminths is limited. This is discussed in more detail with fish-borne trematodes, below, but with other helminth species there is an urgent need for concerted research efforts. Evidence from a number of reports has demonstrated that appropriate molecular tools may be available, but they need to be applied in epidemiological studies. For example, several studies have been undertaken on the molecular characterization of species of *Gnathostoma* (Martinez-Salazar and Leon-Regagnon, 2005; Ngarmamonpirat et al., 2005; Ando et al., 2006), but further studies are required to determine the epidemiological value of the regions of DNA examined. Although there is a clear need for molecular tools for the species-specific detection of *Fasciola* and *Angiostrongylus* in snail hosts, despite a number of reported studies there has been little success with *Fasciola* (see Mas-Coma et al., 2005), and promising results with *Angiostrongylus* (Caldeira et al., 2003) need to be repeated and applied in epidemiological studies.

Epidemiology and Transmission

From an epidemiological perspective, the requirement of molecular tools is that they provide some predictive ability with respect to the etiology of an infection or disease outbreak and the characteristics of the causative agent(s). Such information is essential for control, especially in determining zoonotic potential. There is a need to apply molecular tools to provide basic ecological data on food-borne helminths, their cycle of transmission particularly for those zoonotic species with low definitive host specificity. This is certainly the case with food-borne trematodes (see below) but also applies to most other food-borne helmiths, even *Trichinella*. For example, there are still many countries where epidemiological information on animal and human infections is lacking, for example, India, many African countries, and countries in Central and South America and Asia (Chai et al., 2005). Even with well-described species of anisakid nematodes, there is limited information on the full extent of their geographical distribution, host range, and prevalence rates in definitive, intermediate, or paratenic hosts (Chai et al., 2005).

In contrast, the application of molecular tools to food-borne protozoa has proved very successful when applied to clearly defined epidemiological data or endemic foci. An important and unresolved question with many parasites relates to their zoonotic potential. *G. duodenalis* and *C. parvum* are maintained in a variety of transmission cycles that can be maintained independently and do not require interaction between them. Thus *G. duodenalis* can be maintained in wild animal cycles or cycles involving domestic animals. Similarly, *C. parvum* can be maintained in cycles involving livestock, especially cattle. What is not understood are the circumstances under which such cycles may interact, resulting in zoonotic transfer. Numerous studies have characterized isolates of *Giardia* and *Cryptosporidium* collected from different hosts and have demonstrated the occurrence of the same species/genotype in humans and other animals (Monis and Thompson, 2003). Such data is indicative of zoonotic potential but gives no information on the frequency of zoonotic transmission. Such information has been obtained from molecular epidemiological studies that genotype isolates of the parasites from susceptible hosts in localized, well-defined foci of transmission or as a result of longitudinal surveillance and genotyping of positive cases (Hunter and Thompson, 2005). In the former, recent research in a localized endemic focus of transmission has provided convincing data on the zoonotic transmission of *G. duodenalis* between dogs and humans (Traub et al., 2004). With *Cryptosporidium*, there is considerable epidemiological data demonstrating strong links between contact with infected livestock and human infections (Fayer et al., 2000; Stantic-Pavlinic et al., 2003; Hunter and Thompson, 2005). This is not the case with *Giardia*, but with both organisms, infected livestock have long been incriminated as sources for the waterborne transmission of cryptosporidiosis and giardiasis (Fayer et al., 2000; Thompson, 2004). Interestingly, the application of genotyping procedures to the contaminating isolate(s) has often incriminated human effluent as the source. However, in a study undertaken of cryptosporidiosis patients in Scotland, *C. parvum* was shown to be the causative agent in 84% of 67 cases, supporting livestock fecal pollution of water sources as the leading cause of human sporadic cryptosporidiosis (Goh et al., 2004).

Fundamental questions about the transmission of parasites will also increasingly be resolved with the application of appropriate molecular tools. For example, the ability to detect *Toxoplasma* in the tissues of domestic and wild animals is raising questions about the maintenance and transmission of *Toxoplasma* in areas where cats are not abundant. In this respect, the application of PCR-based diagnostic procedures in studies on sheep in Europe and native mammals in Austaralia (Duncanson et al., 2001; Parameswaran et al., 2006) suggests that the role of vertical transmission may be more important than previously envisaged for the maintenance of the parasite in some populations.

Surveillance and Biosecurity

Although technological advances such as pasteurization and proper canning have controlled or eliminated some food-borne diseases, new causes are being identified and may increase in importance in the future. In the past, the consumption of raw or undercooked meat and fish was associated with specific cultures and practices, but with shifting consumer fashions, increased international travel, globalization of food supplies, and cosmopolitan eating habits, what once were regarded as rare diseases are now becoming increasingly more recognized (WHO, 1999; Cohen, 2000; Slifko et al., 2000). An estimated 2.5 million food-borne illnesses in the United States were caused by parasitic diseases (300,000, 2,000,000, and 225,000, for *Cryptosporidium parvum, Giardia duodenalis*, and *Toxoplasma gondii*, respectively), all with zoonotic implications. In addition to these protozoa, foodstuffs have been recognized as an emerging etiological source of infections with *Cyclospora* and *Blastocystis* (Sunnotel et al., 2006). *Toxoplasma* serves as an excellent example of a parasite that continues to exploit novel routes of transmission that become available as a result of societal changes. It remains one of the most common parasitic infections of humans and other animals, one of the top three causes of food-borne disease and one of five infectious diseases causing more than 90% of food-related deaths (Mead et al., 1999). In an estimated 50% of cases, *Toxoplasma* is transmitted by food (United States Department of Agriculture, 1995). Clearly, the nature of food-borne illness is changing, and there is also growing awareness of the role of wildlife in the food-borne transmission of *Toxoplasma* infection (Tenter et al., 2000). Molecular epidemiology will play an increasing role in surveillance of new and emerging food- and waterborne zoonotic diseases.

International food-borne disease outbreaks have become more common and have introduced new parasites to an area, such as the outbreaks in the U.S. of *Cyclospora* associated with imported raspberries (Herwaldt and Ackers, 1997; Herwaldt and Beach, 1999; Thompson et al., 2000). Such outbreaks have highlighted the public health significance of the increasing usage of surface water for irrigation, which can indirectly cause human infection with *Cyclospora* and other enteric protozoan infections via the consumption of contaminated fresh produce (Sunnotel et al., 2006). As such, there is a need for rapid detection and characterization of the contaminating agent, in order to determine the source of contamination and public health risk. Molecular epidemiological tools have proven very valuable in determining sources of infection in food- and waterborne outbreaks of disease caused by protozoan zoonoses, and greatly improve our understanding of contamination routes (reviewed in Dawson, 2005). Furthermore, retrospective genotypic characterization of parasite isolates in longitudinal surveys has helped to identify risk factors for infection. For example, such studies have shown that direct or indirect association with cattle is the main driver for infections with *C. parvum*, whereas direct or indirect association via contaminated food is the biggest risk for infection with *C. hominis* (Hunter and Thompson, 2005). Bivalve molluscan shellfish have the ability to filter large amounts of water and concentrate oocysts of *Cryptosporidium* and *Giardia* within their gills, increasing the health risk of consuming raw or undercooked shellfish. The genotyping of oocysts in shellfish has helped to determine sources of contamination and public health risk.

Changes in dietary practice have also been implicated as a reason for the emergence of several helminth zoonoses including capillariasis, anisakiasis, and gnathostomiasis (McCarthy and Moore, 2000). There is scope for such approaches with food-borne helminths, but there has been little application to date. For example, the illegal importation of pork products from *Trichinella* endemic areas has resulted in several human outbreaks in Germany, Italy, and the United Kingdom (reviewed in Pozio and Zarlenga, 2005). Apart from the public health significance of such outbreaks, such illegal importation has biosecurity implications since a country's trade can be affected by the presence of a parasite like *Trichinella*. For example, crocodile meat for export now has to be tested for infection with *T. papuae* because of the discovery of a new closely related species, *T. zimbabwensis*, infecting crocodiles in Zimbabwe (Owen et al., 2001; Pozio et al., 2002). The demonstrated occurrence of *T. spiralis* in Australia would have an enormous impact on trade because Australia could no longer be considered to be "*Trichinella* free." Clearly, there is scope for the application of molecular tools for the detection and characterization of *Trichinella* in meat in order to determine sources of contamination.

Fish-Borne Trematodes

To date, the contribution of molecular tools to the study of the fish-borne trematode (FBT) zoonoses has been limited with the exception of the information it has provided on the population structure and taxonomy of *Paragonimus* spp. and, more recently, the diagnostic tools for the detection and species-specific differentiation of the liver and intestinal flukes from clinical and environmental samples.

Despite efforts aimed at control the liver flukes *Opisthorchis* spp., *Clonorchis sinensis*, lung flukes *Paragonimus* spp., and intestinal flukes *Echinostoma* spp., heterophyids continue to remain a significant public health problem, especially in remote rural communities in Southeast Asia (WHO, 1995). Many factors contribute to this problem including a lack of public education, the continued popularity of consuming raw, undercooked, or insufficiently processed foods, expansion of aquiculture and its distribution networks, widespread zoonotic reservoirs, poor socioeconomic conditions, lack of improved sanitation, and poor food inspection measures (Keiser and Utzinger, 2004). These liver, intestinal, and lung flukes share a common feature and life cycle involving a definitive mammalian host and two obligate intermediate hosts, a snail and a freshwater crustacean for *Paragonimus*, and one of many freshwater fish for *Opisthorchis* and *Clonorchis*. *Paragonimus* spp. also utilize paratenic hosts such as wild pigs or rodents. Due to the similarity of their overall epidemiological features, the control of these trematode zoonoses have common features that to date have consisted of mainly human intervention with varying degrees of success. One of the key strategies for control advised by the World Health Organization is integrating mass drug treatment of people at risk together with education and improved sanitation in endemic areas, which aims to reduce overall morbidity and reduce the contamination of the environment with eggs from human excreta (WHO, 2004). However, the extent to which eggs from infected humans contribute to the continuation of the cycle remains unclear, and the role of reservoir hosts, especially semidomesticated and feral dogs, cats, pigs, and rats, in maintaining the natural cycle of infection remains poorly explored. Moreover, the provision of accurate and high-throughput diagnostic tests for the detection and differentiation of the liver from the minute intestinal flukes from clinical and environmental samples is urgently required for accurate prevalence data to be generated in humans, animals, and intermediate hosts and for monitoring the success of control measures (Chai et al., 2005).

There is limited literature available describing targeted control programs for *Paragonimus*; however, one such study in Yongjia County in China saw a reduction in the prevalence and intensity of *Paragonimus* among cats, dogs, and crustaceans following public education and mass chemotherapy in humans, suggesting that humans are the principal hosts there (Chen et al., 2001). In Thailand, an integrated and long-term liver fluke control program implemented by the Ministry of Public Health has also led to significant reduction in the prevalence of *O. viverrini* in many parts of the country, but in some parts of northeastern Thailand the prevalence of *O. viverrini* continues to remain at a highly endemic level with an average prevalence of 24% and a prevalence of up to 72% in some districts (Sriamporn et al., 2004), as humans continue traditional practices of eating undercooked fish. Moreover, the discrepancies of prevalence observed among humans and reservoir host populations from different community settings within Asia still remain largely unexplained. In some areas, the prevalence of *O. viverrini*–like eggs may be as high as 9% (range 5% to 19% per village) among human populations but as low as 4% among domestic animals as observed in a remote rural community in Chacherngsao Province in southeast Thailand (Traub et al., unpublished). On the other hand, a number of studies have reported highly endemic levels of potentially zoonotic species of fish-borne trematodes in dogs in India (Sahai, 1969; Traub et al., 2005) and in dogs, cats, pigs, and fish intermediate hosts in Nghe An province, Vietnam (Nguyen et al., 2006a; Tran et al., 2006) despite human infection being negligible (Nguyen et al., 2006b). In other areas such as Guangdong and Guangxi provinces in China (Lin et al., 2005) and in northern Vietnam (Queuche et al., 1997; De et al., 2000), highly endemic levels of *C. sinensis* and *P. heterotremus*, respectively, were reported among both human and reservoir hosts (cats, dogs, pigs). There is, therefore, a lack of fundamental knowledge and understanding of the specific transmission patterns and ecology of these parasites among and within different populations in Asia (Lun et al., 2005). It is unclear whether these differences in prevalence are caused by the varying dietary habits of human and reservoir hosts or by a biological or genetic variation within trematode populations associated with the maintenance of the life cycle, much like *P. westermani,* where it is known that dogs and cats are good hosts for both diploid and triploid forms of *P. westermani* in most places, except in Malaysia, where they were found to be poor hosts for the parasite (Blair et al., 1999).

Knowledge gaps also occur with respect to ascertaining the molluscan host for diploid species of *P. westermani*. Apart from Malaysia and the Philippines, east and northeast China, Japan, Korea, and Taiwan, nothing is known about the molluscan hosts in any other endemic country (Blair et al., 1999). Also relatively little is known on the distribution and specificity of crustacean hosts of *P. westermani* (Blair et al., 1999). It is imperative that these knowledge gaps be filled for the success of future control programs. However, an integral part of prevention and control of these emerging FBT zoonoses will be based on early detection, simultaneous epidemiological investigation in humans, reservoir and intermediate host populations, multidisciplinary collaboration, and the development of advanced diagnosis and surveillance tools. Research into the development and testing under field and clinical conditions appropriate methods to improve the reliability of diagnosis of FBT infections in both humans and domesticated animals (reservoir hosts), including utilizing molecular based methods was one of the key recommendations made by the World Health Organization for the future understanding and control of these FBT infections (WHO, 2004). The Hazard Analysis Critical Control Point (HACCP) approach to commercial fish pond management aimed at minimizing contamination of the ponds with trematode eggs and snail infections shows potential as a means for control of fish-borne trematodiasis (Khamboonruang et al., 1997). The development and classification of hazards and the verification procedures for ensuring that the HACCP system is functioning optimally, however, necessitate the provision of accurate and high-throughput diagnostic tests.

Adult and metacercaria characteristics have been widely used to distinguish among species of *Paragonimus* in liver and intestinal flukes from crustacean and fish hosts, respectively. However, apart from their identification being laborious and requiring highly skilled personnel, the cyst morphology is not always constant within a species; for example, metacercarial polymorphisms occur within *P. ohirai* (Blair et al., 1999) and the *P. skrjabini* complex (Blair et al., 2005). Furthermore, identifying liver and intestinal fluke metacercaria in fish is notoriously difficult, especially when there are multiple species of infections present. Microscopic examination of fecal samples (and sputum in the case of *Paragonimus* spp.) is the most widely employed technique for diagnosis of an active infection of food-borne trematodes with prevalence estimates usually based on the formalin ethyl acetate concentration technique (FECT) or the Kato Katz (KK) method. The FECT it is a highly laborious and time-consuming technique compared to KK, but can be utilized to examine feces from both humans as well as reservoir hosts such as dogs and cats (Bowman, 2003; Hong et al., 2003). Kato Katz is a suitable technique to utilize in large surveys but is limited to its use for human fecal samples only. Furthermore, the diagnostic sensitivity of KK is relatively low and may miss light infections (Hong et al., 2003). Also, the eggs of the liver and Hetrophyidae family of flukes are difficult to distinguish morphologically from *Opisthorchis* and *C. sinensis*, and it is suggested that epidemiological screening surveys in the past have been largely inaccurate in determining the relative prevalence of these flukes (Chai et al., 2005). Moreover, eggs of *O. viverrini* and *C. sinensis* are indistinguishable by light microscopy, and differential diagnoses of the two species are necessary in endemic regions such as central Vietnam, where both species may overlap. Compounding difficulties exist with the diagnosis of *Paragonimus* spp., as their eggs must be differentiated from other trematodes and pseudophyllidean cestodes, and the dimension of some species of *Paragonimus* overlap. Furthermore, the eggs are frequently not found in the sputum and stool of infected individuals, especially in the cases of extrapulmonary manifestations that can occur with diploid *P. westermani* as well as *P. miyazakii* and *P. skrjabini* that do not normally mature in humans (Blair et al., 1999). Also, clinical and radiographic features of the pulmonary forms often resemble that of tuberculosis, and extrapulmonary forms may be misdiagnosed as malignancies.

A monoclonal antibody enzyme-linked immunosorbert assay (ELISA) based on the detection of an 89-kd *O. viverrini* metabolic antigen in feces (Sirisinha et al., 1995) and serum specimens (Wongsaroj et al., 2001) as well as a recombinant cysteine-proteinase–based ELISA of *C. sinensis* (Nagano et al., 2004) have the potential for providing a sensitive, specific, and suitable alternative for mass screening for *O. viverrini* and *C. sinensis* in human populations. Immunodiagnostic assays are also useful for the diagnosis of paragonimiasis and the only option for prepatent and extrapulmonary infections. They could provide a simple, rapid, and cheap diagnostic option for large-scale screening in humans, usually with high sensitivity but low specificity due to cross-reactions with other trematodes (reviewed by Blair et al., 1999). More recent tests such as recombinant cysteine proteases (Yang et al., 2004) and yolk ferritin form cross-reactions with other trematodes, but have yet to be tested for cross-reactions in human populations harboring other species of *Paragonimus*. Furthermore, the aforementioned immunodiagnostic assays are not useful for surveys of fish and crustacean intermediate hosts or snails and have yet to be tested on naturally infected animal reservoir hosts.

A number of species-specific PCR-based methods have recently been developed that are capable of detecting and differentiating trematode species from clinical and environmental samples obtained from definitive and intermediate hosts. The key advantage of these techniques are speed, increased discriminatory power, and the ability to analyze small amounts of sample, which is particularly important for detecting and characterizing etiological agents directly from host or clinical samples. More importantly, they negate the need for laborious morphological examination of the individual adult flukes following anthelmintic purging (in humans) or necropsy in reservoir host populations or morphological examination of the intermediate hosts for the presence of cercariae and metacercariae. This may be even more important if mixed species of parasites are present and may be overlooked if the burden of one species or genotype of parasite is low compared to another.

An *O. viverrini*–specific PCR test based on amplification of a highly repetitive DNA sequence in the parasite genome capable of detecting *O. viverrini* eggs directly from hamster feces was developed (Wongratanacheewin et al., 2001). In evaluations using human fecal samples, the test was shown to have an analytical sensitivity of 100% and analytical specificity of 97.8% in moderate to severe infections (more than 1000 eggs per gram [epg] feces) and an analytical sensitivity of 68.2% in the detection of light infections (epg $<$1000) with a detection limit of 200 eggs (Wongratanacheewin et al., 2002). In a separate field study this same PCR detected significantly less positive cases compared to both the FECT and KK methods with an overall diagnostic sensitivity of 45% in fecal eggs counts >1000, which was attributed to technical problems (Stensvold et al., 2006). This same PCR was also utilized to detect *O. viverrini* from experimentally infected snails and fish with high analytical sensitivity and specificity (Maleewong et al., 2003). More recently, a mitochondrial-based multiplex PCR for the identification and discrimination of *C. sinensis* and *O. viverrini* from different life stages (adults, metacercarea, and eggs) from fish intermediate hosts and infected human patients was developed (Le et al., 2006) but its sensitivity to detect eggs compared to the FECT and KK methods has yet to be evaluated in a field study. Similarly, a species-specific PCR assay has been developed to detect and differentiate six species of heterophyid intestinal flukes based on their 18S recombinant DNA (rDNA) gene from cercariae, metacercariae, and adults, from snails, fish, and birds, respectively (Dzikowski et al., 2004). Ideally, a high throughput PCR-based diagnostic test to differentiate both the liver-fluke species from the minute intestinal flukes is urgently required. However, its design and development is currently hampered by a lack of sequence information for a number of FBT species at the appropriate genetic loci (http://www.ncbi.nlm.nih.gov/).

Polymerase chain reaction RFLP as well as multiplex species specific PCR-based on the ITS2 region of *Paragonimus* spp. have also been developed to detect and differentiate metacercariae of *P. westernmani* and *P. heterotremus* (Sugiyama et al., 2005) and *P.westermani* and *P. miyazakii* (Sugiyama et al., 2002) whose distributions overlap in intermediate hosts from Thailand and Japan, respectively. The PCR methods have also been attempted to amplify a species-specific probe of *P. heterotremus* directly from eggs in stool samples of experimentally infected cats (Intapan et al., 2005). This technique had an analytical sensitivity of five eggs in 0.6 g of feces and it cross-reacted with *P. westermani* and *P. siamensis* but not with other parasites and bacteria. More recently, a PCR capable of amplifying the *cox1* and *ITS2* genes directly from *P. heterotremus* eggs in human feces was also described (Le et al., 2006); however, both these PCR techniques are yet to be proven suitable for a large-scale epidemiological study. Even so, their use may be limited in human surveys due to the possible absence of eggs from infected individuals. Polymerase chain reaction is likely in the short term to be utilized for confirmation of clinical cases of Paragonimiasis, especially in returning overseas travelers and immigrants in countries where this is an affordable practice (Schuster et al., 2006).

Apart from their ability as diagnostic tools, molecular-based tools have provided valuable information on the population structure, taxonomy, host specificity, and biology of *Paragonimus* spp. found in Southeast Asia. Techniques such as allozyme electrophoresis (Agatsuma et al., 1993), random amplified polymorphic DNA (RAPD) markers (Intapan et al., 2004), and more commonly phylogenetic analysis of nuclear ribosomal DNA, ITS, and especially mitochondrial sequences of *Paragonimus* spp. have shown, for example, that the Philippine and Malaysian populations of *P. westermani* are genetically very different from those from diploid and triploid populations from northeast Asia (that form a single clade) (Agatsuma et al., 1993; Iwagami et al., 2000) and genetically different from each other, which have supported biological differences in their life cycles. *Paragonimus westermani* from South Asia utilize thiarid snails, while those from the northeast utilize pleurocercid snails. *Paragonimus westermani* in the northeast group are usually found in dogs, cats, or foxes, which are relatively poor hosts in Malaysia, but in the Philippines, the species is naturally parasitic in rats (reviewed by Blair et al., 1999). It was therefore proposed that the taxa currently known as *P. westermani* probably represents a complex species that awaits proper characterization (Blair, 2000). In another example, subadult *Paragonimus* that migrate through tissues of humans, *P. skrjabini* (China) and *P. miyazakii* (Japan), utilize two different snail hosts and differ noticeably in adult morphology. However, molecular phylogenetic studies have shown that the two species are extremely closely related to each other, and it has been proposed that both taxa be referred to as the same subspecies *P. skrjabini miyazakii* from various geographical locations in China and Japan (Blair, 2000; Blair et al., 2005). Molecular phylogenetic studies have also been extremely useful in resolving the taxonomy and nomenclature of a number of previous isolates of *Paragonimus*, some described as new species but that are most probably invalid and belong to the *P. skrjabini* complex (Blair et al., 2005). A similar situation also probably exists with a third species complex *P. iloktsuenensis, P. ohirai*, and *P. sadoensis*, which were long distinguished on their cyst morphology and snail host, despite their similar adult morphologies but shown to belong to the same species complex using phylogentic analysis of their cox1 sequences (Blair et al., 1997; Blair, 2000). Studies of the genomes of *Paragonimus* species are likely to continue due to the complex and poorly understood nature of the genus. Further studies on the population structure of the liver and intestinal flukes are also warranted in different hosts from various geographical locations in Southeast Asia.

Of primary importance is the recognition of the public health significance of these fish- and invertebrate-borne trematode zoonoses, especially in Southeast Asia and to invest in research that will provide tools for their control. Such tools include the development of molecular diagnostic techniques for high-throughput detection and accurate identification of trematode stages from various clinical and environmental samples and studies to further explore the genetic diversity, population biology, and host-specificity of these fish-borne trematodes in both human and animal hosts.

Toxoplasma

Toxoplasma gondii is a ubiquitous protozoan parasite that infects up to one third of the world's population (Montoya and Liesenfeld, 2004). Molecular techniques to date have identified three major strains of *Toxoplasma* that differ in virulence and epidemiological pattern of occurrence. The zoonotic potential of *Toxoplasma* lies in its ability to infect animal tissue (and therefore meat) and its ability to be shed in the feces of acutely infected felids. *Toxoplasma* has a predilection to infect nervous and muscular tissue; however, it can be widely dispersed in the body, particularly in acute infections. A majority of infections with *Toxoplasma* go unnoticed due to the ability of the parasite to form chronic cysts in tissues without causing clinical signs. Manifestations of *Toxoplasma* infection that are known to cause significant clinical signs in humans are congenital infection, adult eye infection, and reactivated infection in severely immunocompromised patients.

It has been widely accepted that *Toxoplasma* has a low genetic diversity due to the common finding of strains that can be grouped into three highly clonal but closely related lineages (Johnson, 1997, Howe and Sibley, 1995, Su et al., 2003). However, it is increasingly being proposed that the genetic diversity among *T. gondii* strains is greater than current estimates due to the sampling bias that has resulted from the study of strains from humans and domestic animals primarily originating from North America and Europe. In addition, the conserved nature of the molecular tools that have been used in the past to investigate genetic diversity may have exacerbated the situation.

Genetic Diversity

A number of molecular studies have been undertaken that have brought many to the conclusion that *Toxoplasma* comprises three clonal lineages. Studies on *Toxoplasma* lineages began with isoenzyme analysis and antigenic analysis and then progressed to using molecular tools such as PCR combined with RFLP (PCR-RFLP), random amplified polymorphic DNA PCR (RAPD-PCR), DNA sequencing, and microsatellite DNA analysis. Upon using these analyses on stocks of *Toxoplasma* isolates found in humans and domestic animals, it was concluded that the genome of *Toxoplasma* is highly conserved. Previous studies have characterized *Toxoplasma* into virulent and avirulent lineages. In an initial genetic study by Sibley and Boothroyd (1992), PCR-RFLP analysis at the SAG-1, 850, and BS loci of 10 mouse virulent strains revealed an essentially identical genotype among the isolates, and it was concluded that the virulent strains of *T. gondii* comprise a single clonal lineage. This genetically homogeneous virulent lineage was also found on isoenzyme analysis, which showed that most virulent isolates fell into a single zymodeme (Darde et al., 1992). The RFLP analysis of DNA polymerase α genes (Binas and Johnson, 1998) in addition to gene sequence data from HSP70 (Lyons and Johnson, 1998) and reverse-transcriptase PCR of SAG1 (Windeck and Gross, 1996) all confirm the dichotomy of a virulent and avirulent lineage. In addition, a study by Ajzenberg et al. (2002a) used eight microsatellite markers to type 84 *Toxoplasma* isolates and applied evolutionary genetic analysis to show the *Toxoplasma* population structure consists of two clonal lineages. Similarly, Guo et al. (1997) used RAPD-PCR to separate 35 *Toxoplasma* strains into virulent and avirulent strains and a genotype of avirulent genotypes.

The existence of three clonal lineages was considered after a number of studies using multilocus PCR-RFLP. One of the first studies to suggest the existence of three clonal lineages was Parmley et al. (1994), who used RFLP analysis at three loci (P22, SAG1, and 850). Virulent strains were genetically identical and comprised a single lineage (group A), consistent with previous studies. However, the heterogeneity seen among the 21 avirulent strains was categorized into two genetically identical clonal lineages (group B and C). A subsequent study by Howe and Sibley (1995) had a larger number of isolates and produced similar findings. The RFLP analysis was employed at six loci with 106 strains. It was found that virulent strains were represented in one clonal lineage (type I), whereas avirulent strains were represented in two clonal lineages (type II and type III). It was concluded that *T. gondii* has a clonal population structure in that >95% of isolates fall clearly into one of three distinct lineages. The theory of a highly clonal genetic structure for *T. gondii* has been confirmed by many studies subsequent to Howe and Sibley (1995), including Boothroyd and Grigg (2002). However, it was also noted that an increasing number of isolates have been found that do not fit into the three distinct genotypes. These novel isolates fall into two general classes: recombinant strains, which have genotypes that are clearly related to the three dominant types, and exotic strains, which have a significant level of polymorphism and often originate from nondomesticated animals or remote areas.

A majority of the novel strains isolated to date have been found from areas outside Europe and North America or in nondomesticated animals. In contrast, many of the *Toxoplasma* isolates that have been used to identify three clonal lineages have been collected from human patients and domestic animals in Europe and North America (Ajzenberg et al., 2004; Lehmann et al., 2006). In light of this knowledge, it is possible that the genetic diversity found in *T. gondii* to date is highly underestimated. The collection of isolates tested for the clonal theory may not reflect the true status of *T. gondii* in remote geographical areas or in tropical regions where the ecological system is very different from that in temperate developed areas. Although several studies have attempted to identify *Toxoplasma* genotypes in chickens from exotic countries such as Egypt, Argentina, India, and Brazil (Dubey et al., 2002, 2003a–d), many were limited by their use of only one PCR-RFLP marker and because chickens in farms are simply indicators of strain prevalence in a domestic environment. Rare atypical strains have been sampled mainly from tropical regions such as French Guiana (Bossi et al., 1998, Darde et al., 1998, Carme et al., 2002) and Brazil (Ferreira Ade et al., 2006; Khan et al., 2006) or in unusual host species such as deer, bear, cougar, or sea otter (Howe and Sibley, 1995; Darde et al., 1998; Lehmann et al., 2000, Miller et al., 2004).

In a study to investigate the hidden genetic diversity of *Toxoplasma*, Ajzenberg et al., (2004) analysed 43 parasite strains isolated from remote geographical regions using multilocus microsatellite sequencing and phylogenetic analysis. Microsatellite markers were used in this study because of their high discriminatory power, their usefulness in assessing molecular epidemiological studies, and their ability to deduce phylogenetic relationships at the intraspecific level or for recently diverged species (Ajzenberg et al., 2004). The study showed that while a majority of the isolates fell into the three clonal lineages as expected, all nine strains from French Guiana were clearly atypical and not only very different from the three major lineages, but also from atypical strains from France and Uruguay.

This is in marked contrast to the domestic samples tested from France where 95% of strains belonged to the three main types. Overall, the data from Ajzenberg et al. (2004), which included sequencing data and phylogram hierarchy, were not in agreement with a strictly clonal model. It was concluded that the most parsimonious hypothesis is that *T. gondii* presents a complex population structure with a mix of clonal and sexual propagation. When the proportion of typical strains is compared to the strains with recombined genotypes in different regions, it can be seen that the highest proportion of genetic exchanges may occur in the wild cycle in countries where domestic breeding has a short history. It is probable that typical strains have primarily been used in previous studies to describe the global clonal structure of *Toxoplasma*.

Analysis of a small number of *Toxoplasma* isolates from wild animals in North America showed genotypes with different combinations of types I, II, and III, and atypical alleles (Howe and Sibley, 1995, Ajzenberg et al., 2004). One of the most common novel genotypes isolated from wildlife to date is type X, isolated from southern sea otters along the Californian coast. The novel genotype was first identified in otter brain tissue collected between 1998 and 2002. Multilocus PCR-RFLP of 35 isolates, in addition to DNA sequencing at conserved genes *(18S rDNA, ITS-1)* and polymorphic genes *(B1, SAG1, SAG3*, and *GRA6)* was used to identify two distinct genotypes infecting the otters: type II and a type X. Sixty percent of sea otter isolates tested in the study were of *Toxoplasma* genotype X, with the remaining 40% being of genotype II (Miller et al., 2004). The type X genotype possessed distinct alleles at three of the four polymorphic loci sequenced. An expanded analysis of otter isolates characterized 15 additional isolates obtained from otters in 2004, and two other marine species (Conrad et al., 2005). Multi-locus PCR-RFLP identified all 15 otter isolates as type X. Combining both sets of otter genotyping results brings the total of type X isolates to 38/50 or 76% of all otter isolates examined. *Toxoplasma* isolates from two other species of marine mammal (Pacific harbor seal and Californian sea lion) were also identified as being of the type X strain using the same procedure as in the otters in addition to direct DNA sequencing of GRA6 PCR amplification products. The result of 76% of *Toxoplasma* isolates from otters being of a novel strain is in marked contrast to the study of Howe and Sibley (1995), who genotyped 106 primarily domestic strains in which the vast majority had one of three identical or highly similar genotypes. Only four of 106 strains tested had extensively mixed genotypes, three of which appeared to be natural recombinants of type II and III, while one was a recombinant of types I and II (Howe and Sibley, 1995). The otter study is further evidence to suggest the genetic diversity of *Toxoplasma* in wildlife and geographically isolated areas is underestimated.

A number of studies in addition to those in otters, which have genotyped *Toxoplasma* isolates from wildlife (Dubey et al., 2004a) and from geographically isolated locations (Dubey et al., 2002, 2003a–d) have not found novel strains. Many of these studies have used one PCR-RFLP marker to identify isolates as either type I, II, or III. Studies that have identified novel or recombinant strains of *Toxoplasma* (Howe and Sibley, 1995; Bossi et al., 1998; Darde et al., 1998;

Lehmann et al., 2000; Carme et al., 2002) have used more discriminatory techniques including isoenzyme analysis, microsatellite analysis, multilocus PCR-RFLP, and gene sequencing. Therefore, it can be speculated that misidentification of unusual recombinants or novel strains can result from the use of a single genetic marker in genotype analysis.

Studies in Brazil

Several studies have shown an unusual *T. gondii* population structure in the country of Brazil. Polymerase chain reaction RFLP at eight independent loci was used to determine the clonal lineage of 20 *Toxoplasma* stains isolated from humans and animals in Brazil (Ferreira Ade et al., 2006). The finding that 100% of strains in the analyzed *T. gondii* population were natural recombinants was particularly different from the expected frequency. Previous studies have reported that regardless of the host and geographical origin, approximately 95% of *Toxoplasma* isolated belong to one of three genetically distinct lineages (Darde et al., 1992; Howe and Sibley, 1995). Several studies that have used single-locus PCR-RFLP of Brazillian *Toxoplasma* strains have identified a high frequency of types I and III and an absence of type II in Brazil (Dubey et al., 2003a,c, 2004b; de A. dos Santos et al., 2005). The study by Ferreira Ade et al. (2006) illustrates the usefulness of multilocus PCR-RFLP in identifying hybrid strains of *Toxoplasma*. The authors concluded that even the analysis of two loci may lead to the misidentification of the genotype of Brazilian strains. For example, if all Brazilian isolates in the study were analysed using the SAG1 and B1 loci, they would have been identified as being the type I lineage.

A recent study in Brazil used a combination of two techniques to genotype human and animal samples from the area (Khan et al., 2006). Nested PCR amplification of four different RFLP markers (SAG2, GRA6, SA3, and BTUB) was combined with sequencing of the UPRT-1 intron in order to classify strains as having clonal, recombinant, or exotic genotypes. It was concluded that most Brazilian *T. gondii* strains were clustered into two new groups that were intermediate between types I and III. The results suggested the presence of at least two additional haplotypes that are present in Brazil and differ from North America and European lineages. Thirteen of the 22 Brazilian strains shared a new allele that was distinguished by six additional polymorphisms not seen in the clonal lineages. Other Brazilian strains analyzed contained equally divergent but unique alleles that in some cases formed smaller groups. Only one strain from Brazil contained a haplotype characteristic of one of the clonal lineages. The abundance of genotypes that do not fit into clonal lineages shows the global *T. gondii* population structure to be more complex than previously recognized. When analyzed by multilocus PCR-RFLP, the new South American genotypes initially appeared to be composed of different combinations of alleles seen in the clonal types, similar to findings of a previous report from Brazil (Ferreira Ade et al., 2004). It was suggested that direct sequencing of introns from housekeeping genes (such as UPRT-1) provides a more accurate picture of sequence divergence as seen in Khan et al. (2006) and Su et al. (2003). The investigation showed the presence of a low genetic diversity of *Toxoplasma* in Brazil, although the strains included genotypes uncommon in North America and Europe. Overall, it was found that most strains from Brazil do not fit the clonal pattern seen in North America and that using only one to two loci to genotype strains can underestimate the genetic diversity of *T. gondii*, particularly in regions outside North America and Europe that may not have a clonal pattern. Further strain comparisons based on a wider set of markers will be necessary to define the global population structure of *T. gondii* and to determine the relationships between major strain types seen in different regions.

Meat Studies

The ingestion of meat has long been known to be a risk factor in the transmission of *Toxoplasma*. Meat from an infected animal can contain *Toxoplasma* bradyzoites and is infective when eaten rare or undercooked. A study of Seventh-Day Adventists, who as a group follow a diet containing no meat, found a significantly lower proportion of people in this group to be infected with *Toxoplasma* compared to a control group (Roghmann et al., 1999). In addition, inhabitants of France, particularly Parisians, who have a high rate of eating rare meat, also have among the highest rate of *Toxoplasma* infection in the world (Papoz et al., 1986).

Only a small number of studies to date have looked for *Toxoplasma* DNA in meat and meat products. Aspinall et al. (2002) found 27 out of 77 meat products in the United Kingdom tested to be positive by PCR analysis. Upon PCR-RFLP of SAG2 and DNA sequencing, 21 samples were genotyped as type I, while six samples contained parasites of both types I and II. Meat products tested came from pork, lamb, and beef, but were predominately pork based and ranged from sausages to minced meat and bacon. In a different study of cured meats, all samples tested were negative by PCR, but it is probable that the high salt content of some cured meats limited the sensitivity of the PCR assay, resulting in some false negatives (Warnekulasuriya et al., 1998). It is unknown how many *Toxoplasma* PCR-positive meat samples are infectious to humans, as PCR only detects DNA and does not give an indication of parasite viability. Relatively few studies published have tested the prevalence of viable *Toxoplasma* in commercial meat. One large scale study by (Dubey et al., 2005) found a very low prevalence of viable *Toxoplasma* cysts in pork (0.38%) and a zero prevalence in chicken and beef in a survey of the United States using cat and mouse bioassays.

Significance of Toxoplasma Genotype in Human Disease Manifestations

The identification of virulence-associated loci by molecular methods has served as a tool to investigate the relationship between *Toxoplasma* genotype and disease manifestations. A number of studies have attempted to link the strain of

Toxoplasma with a particular manifestation of infection. Although the vast majority of infections are asymptomatic, congenital infection, ocular disease, and reactivated toxoplasmosis are three commonly described disease manifestations of *Toxoplasma* infection in humans.

It has been known for some time that certain strains of *Toxoplasma* (type I) are highly virulent in mice, whereas others are avirulent. Strains generally fit into two extremes in mice: highly virulent, with an LD_{100} (the dose at which 100% of animals die) of one parasite, or avirulent, with an LD_{100} of several thousand parasites (Boothroyd and Grigg, 2002). There are, however, no guarantees that the differences in virulence seen in mice will also be seen in other animals. One indication that the differences seen in mice may extend to other animals came from the observation that type I strains develop approximately one third faster in human foreskin fibroblasts than type II and III strains (Boothroyd and Grigg, 2002). It is unknown what strains are responsible for the bulk of human infections, as most human infections do not exhibit overt disease and form chronic cysts that cannot be genotyped by testing bodily fluids. Tachyzoites present in severe disease are present at the site of disease, and depending on the disease location, can subsequently occupy the amniotic fluid, aqueous humor, or cerebrospinal fluid. It is therefore possible to access tachyzoites in severe infections, which subsequently enables molecular analysis and strain typing.

There are some conflicting views on which genotypes are more common in different manifestations of *Toxoplasma* infection. A number of studies have reported that type II strains predominate in AIDS and congenital infections (Boothroyd and Grigg, 2002). Analyses in France indicate that out of 45 *Toxoplasma* infections in AIDS patients, 34 were with type II strains (Howe et al., 1997). The same study also reported all 13 congenital infections analyzed to be type II. A limited analysis of samples from the U.S. supports the trend of type II strains being most common in congenital infection (Howe and Sibley, 1995). Another investigation in France looked primarily at congenital infection using isoenzyme analysis combined with microsatellite markers and found that of 40 samples, 22 were type II, eight type III, seven type I, and three samples were not of the three dominant types (Ajzenberg et al., 2002a). An additional study genotyped 86 samples, primarily from France and Belgium, which were examined by both mouse inoculation and microsatellite analysis (Ajzenberg et al., 2002b). Isolates were characterized using eight microsatellite markers, and it was found that a majority (85%) were type II, 8% were type I, 3% were type III, and 4% were atypical genotypes. This study also analyzed the relationship between *Toxoplasma* genotype and clinical manifestations of patients with congenital toxoplasmosis. Type II isolates were predominant among both the severe toxoplasmosis group as well as the group of patients with benign and asymptomatic toxoplasmosis. In contrast, no type I *Toxoplasma* DNA was isolated from the benign and asymptomatic group of patients (Ajzenberg et al., 2002b). Although the time the fetus is initially infected with *Toxoplasma* plays a large role in the severity of infection in the fetus, the data from this study suggest the strain of *Toxoplasma* may also play a role in the severity of disease. A conflicting study in Spain (Fuentes et al., 2001) identified type I *Toxoplasma* to predominate in congenital toxoplasmosis, although several factors may be involved in the difference in results of this study (Ajzenberg et al., 2002b). Not only was there a geographical difference between studies, but also mice were not used to amplify *Toxoplasma* numbers in the Spanish study. The growth of parasites in mice and cell culture has been associated with artificial selection of virulent strains of *Toxoplasma* and so may have contributed to the conflicting results.

Animals have also been found to commonly be infected with type II strains of *Toxoplasma* in a number of studies including those in France (Ajzenberg et al., 2002a), the U.S. (Howe et al., 1997), and the U.K. (Owen and Trees, 1999).

Brazil seems to be an outlier in that a majority of isolates originating from Brazil have been genotyped as type I, recombinants of type I, or novel strains. The fact that a majority Brazilian isolates genotyped are closely related to the type I lineage may be a significant finding (Ferreira Ade et al., 2006). From a recent study of 20 Brazilian isolates, 85% showed a significant degree of virulence (Ferreira Ade et al., 2006). These findings are in contrast to studies performed in the U.S. and Europe in which most strains were avirulent types II or III. Previous studies in Brazil have shown a comparatively high seropositivity of humans to *T. gondii* (Neto et al., 2000; Petersen et al., 2001, Silveira et al., 2001), and a high prevalence of *T. gondii* in food animals such as pigs (de A. dos Santos et al., 2005) and chickens (Dubey et al., 2002), as well as a high prevalence in dogs (da Silva et al., 2005) and cats (Dubey et al., 2004b). Overall, there seems to be a high level of transmission of *Toxoplasma* in Brazil. It has been suggested that the high frequency of virulent strains found in Brazil may be in part responsible for the high frequency of acquired ocular toxoplasmosis in humans in Brazil (Ferreira Ade et al., 2006; Khan et al., 2006). Cases of ocular toxoplasmosis in Brazil are often recurrent and serious in nature (Glasner et al., 1992; Silveira et al., 2001).

Type I *Toxoplasma* is strongly associated with ocular toxoplasmosis as deduced by a number of investigations in human patients. In contrast to other epidemiological studies of *Toxoplasma* disease manifestations, data concerning ocular disease and its association with type I strains is consistent and not conflicting. Vallochi et al. (2005) showed that parasite DNA isolated from all 11 ocular toxoplasmosis patients in Brazil were from virulent *Toxoplasma* strains. A study of U.S. patients (Grigg et al., 2001) observed that in rare occurrences of ocular toxoplasmosis in otherwise healthy adults, type I and novel virulent strains were found in all six patients. Conversely, type II and III strains were found to cause ocular disease only in immunosuppressed patients. Similar results were seen in Canada (Burnett et al., 1998) and Brazil (Glasner et al., 1992), where isolates from ocular toxoplasmosis outbreaks were found to be type I strains (Boothroyd and Grigg, 2002). A recent study from Brazil (Khan et al., 2006) found a large majority (seven of 11) of strains isolated from patients with ocular toxoplasmosis to be of a novel genotype intermediate between type I and II. Other isolates consisted of genotype I (two of 11) and combinations of type I and II (one of 11) or combinations of type I, II, and III (one of 11) genotypes.

The relationship between *Toxoplasma* genotype and severity of infection has also been investigated in sea otters in the Californian coast. Infection with *Toxoplasma* and associated meningoencephalitis was recognized as a major cause of death in subadults and prime-aged adult sea otters, accounting for 16% of total otter mortality (Miller et al., 2004). A novel type X strain was identified to predominate in all infected otters, being present in 72% of all beach-cast otters examined by genotypic analysis (Conrad et al., 2005). It was found that type X–infected otters tended to have moderate to severe meningoencephalitis on histopathology more frequently than type II infected otters (Miller et al., 2004). In addition, more otters infected with type X *Toxoplasma* had *T. gondii* associated meningoencephalitis as a primary cause of death when compared with type II infected otters (Miller et al., 2004). It is clear that further analysis is needed to ascertain the virulence of the type X genotype, particularly in Californian sea otters.

Conclusion

Molecular epidemiological research on food-borne protozoan zoonoses has demonstrated the value of such approaches in resolving taxonomic issues, investigating transmission patterns and identifying traits of epidemiological significance such as infectivity and virulence. In contrast, there has been limited application of molecular epidemiology to the food-borne helminths. Numerous studies have described regions of DNA with potential value as diagnostics or for detecting variation but they have yet to be applied in epidemiological studies. This in part may be due to the fact that the food-borne helminths are principally problems in the developing world in contrast to parasites such as *Toxoplasma* and *Cryptosporidium*. There is thus a need for more comprehensive collaborative networks and training programs to ensure that the benefits of molecular epidemiology are translated to the regions where they can benefit the efforts to control these neglected diseases.

References

Agatsuma, T., Lai, K., Ow-Yang, C.K., Habe, S., Sugiyama, H., Hirai, H., and Kawashima, K. (1993). Genetic differentiation between Malaysian and other Asian *Paragonimus westermani*. Trop Biomed 10, 45–52.

Ahmadian, A., Ehn, M., and Hober, S., (2006). Pyrosequencing: history, biochemistry and future. Clin Chim Acta 363, 83–94.

Ajzenberg, D., Banuls, A.L., Su, C., Dumetre, A., Demar, M., Carme, B., and Darde, M.L. (2004). Genetic diversity, clonality and sexuality in *Toxoplasma gondii*. Int J Parasitol 34, 1185–96.

Ajzenberg, D., Banuls, A.L., Tibayrenc, M., and Darde, M.L. (2002a). Microsatellite analysis of *Toxoplasma gondii* shows considerable polymorphism structured into two main clonal groups. Int J Parasitol 32, 27–38.

Ajzenberg, D., Cogne, N., Paris, L., Bessieres, M.H., Thulliez, P., Filisetti, D., Pelloux, H., Marty, P., and Darde, M.L. (2002b). Genotype of 86 *Toxoplasma gondii* isolates associated with human congenital toxoplasmosis, and correlation with clinical findings. J Infect Dis 186, 684–9.

Ando, K., Tsunemori, M., Akahane, H., Tesana, S., Hasegawa, H., and Chinzei, Y. (2006). Comparative study on DNA sequences of ribosomal DNA and cytochrome c oxidase subunit 1 of mitochondrial DNA among five species of gnathostomes. J Helminthol 80, 7–13.

Aspinall, T.V., Marlee, D., Hyde, J.E., and Sims, P.F. (2002). Prevalence of *Toxoplasma gondii* in commercial meat products as monitored by polymerase chain reaction–food for thought? Int J Parasitol. 32, 1193–1199.

Barta, J.R., and Thompson, R.C.A. (2006). What is *Cryptosporidium*? Reappraising its biology and phylogenetic affinities. Trends Parasitol 22, 463–468.

Binas, M., and Johnson, A.M. (1998). A polymorphism in a DNA polymerase alpha gene intron differentiates between murine virulent and avirulent strains of *Toxoplasma gondii*. Int J Parasitol 28, 1033–40.

Blair, D., Agatsuma, T., and Watanobe, T. (1997). Molecular evidence for the synonymy of three species of *Paragonimus, P. ohirai miyazaki*, 1939, *P. iloktsuenensis chen*, 1940 and *P. sadoensis miyazaki et al.*, 1968. J Helminthol 71, 305–310.

Blair, D., Xu, Z.B., and Agatsuma, T. (1999). Paragonimiasis and the genus *Paragonimus*. Adv Parasitol 42, 113–222.

Blair, D. (2000). Genomes of *Paragonimus westermani* and related species: current state of knowledge. Int J Parasitol 30, 421–426.

Blair, D., Chang, Z.S., Chen, M.G., Cui, A.L., Wu, B., Agatsuma, T., Iwagami, M., Corlis, D., Fu, C.B., and Zhan, X.M. (2005). *Paragonimus skrjabini* Chen, 1959 (Digenea: Paragonimidae) and related species in eastern Asia: a combined molecular and morphological approach to identification and taxonomy. Syst Parasitol 60, 1–21.

Boothroyd, J.C., and Grigg, M.E. (2002). Population biology of *Toxoplasma gondii* and its relevance to human infection: do different strains cause different disease? Curr Opin Microbiol 5, 438–42.

Bossi, P., Caumes, E., Paris, L., Darde, M.L., and Bricaire, F. (1998). *Toxoplasma gondii*–associated Guillain-Barre syndrome in an immunocompetent patient. J Clin Microbiol 36, 3724–5.

Bowman, D.D. (2003). Diagnostic parasitology. In: Bowman, D.D., Lynn, R.C., and Eberhard, M.L., eds. Georgi's Parasitology for Veterinarians. St. Louis: Saunders, 287–358.

Bull, S., Chalmers, R., Sturdee, A.P., Curry, A., and Kennaugh, J. (1998). Cross-reaction of an anti-*Cryptosporidium* monoclonal antibody with sporocysts of *Monocystis* species. Vet Parasitol 77, 195–197.

Burnett, A.J., Shortt, S.G., Isaac-Renton, J., King, A., Werker, D., and Bowie, W.R. (1998). Multiple cases of acquired toxoplasmosis retinitis presenting in an outbreak. Ophthalmology 105, 1032–1037.

Caccio, S., De Giacomo, M., and Pozio, E., (2002). Sequence analysis of the b-*giardin* gene and development of a polymerase chain reaction-restriction fragment length polymorphism assay to genotype *Giardia duodenalis* cysts from human faecal samples. Int J Parasitol 32, 1023–1030.

Cacciò, S.M., et al. (2005). Unravelling *Cryptosporidium* and *Giardia* epidemiology. Trends Parasitol 21, 429–437.

Caldeira, R.L., Carvalho, O.S., Mendonca, C.L.F.G., Graeff-Teixeira, C., Silva, M.C.F., Ben, R., Maurer, R., Lima, W.S., and Lenzi, H.L. (2003). Molecular differentiation of *Angiostrongylus costaricensis, A. cantonensis,* and *A. vasorum* by polymerase chain reaction-restriction fragment length polymorphism. Mem Inst Oswaldo Cruz 98, no.8.

Carme, B., Bissuel, F., Ajzenberg, D., Bouyne, R., Aznar, C., Demar, M., Bichat, S., Louvel, D., Bourbigot, A.M., Peneau, C., Neron, P., and Darde, M.L. (2002). Severe

acquired toxoplasmosis in immunocompetent adult patients in French Guiana. J Clin Microbiol 40, 4037–44.
Carreno, R.A., Martin, D.S., and Barta, J.R. (1999). *Cryptosporidium* is more closely related to the gregarines than to coccidia as shown by phylogenetic analysis of apicomplexan parasites inferred using small-subunit ribosomal RNA gene sequences. Parasitol Res 85, 899–904
Cavalier-Smith, T., and Chao, E.E. (2004). Protalveolate phylogeny and systematics and the origins of sporozoa and dinoflagellates (Phylum Myzozoa nom. nov.). Eur J Protistol 40, 185–212.
Chen, M.G., Chang, Z.S., Shao, X.Y., Liu, M.D., Blair, D., Chen, S.H., Zhang, Y.N., Hong, J.L., Shen, B.G., and Feng, Z. (2001). Paragonimiasis in Yongjia County, Zhejiang Province, China: clinical, parasitological and karyotypic studies on Paragonimus westermani. Southeast Asian J Trop Med Public Health 32, 760–769.
Chai, J-Y., Murrell, K.D., and Lymbery, A.J. (2005). Fish-borne parasitic zoonoses: status and issues. Int J Parasitol 35, 1233–1254.
Cohen, M.L. (2000). Changing patterns of infectious disease. Nature 406,762–767.
Conrad, P.A., Miller, M.A., Kreuder, C., James, E.R., Mazet, J., Dabritz, H., Jessup, D.A., Gulland, F., and Grigg, M.E. (2005). Transmission of *Toxoplasma*: clues from the study of sea otters as sentinels of *Toxoplasma gondii* flow into the marine environment. Int J Parasitol 35, 1155–68, 2005.
Constantine, C.C. (2003). Importance and pitfalls of molecular analysis to parasite epidemiology. Trends Parasitol 19, 346–348.
Coppel, R.L., and Black, C.G. (2005). Parasite genomes. Int J Parasitol 35, 465–479.
da Silva, A.V., Pezerico, S.B., Lima, V.Y.d., Moretti, L.d.A., Pinheiro, J.P., Tanaka, E.M., Ribeiro, M.G., and Langoni, H. (2005). Genotyping of *Toxoplasma gondii* strains isolated from dogs with neurological signs. Vet Parasitol 127, 23–27.
Darde, M.L., Bouteille, B., and Pestre-Alexandre, M. (1992). Isoenzyme analysis of 35 Toxoplasma gondii isolates and the biological and epidemiological implications. J Parasitol 78, 786–94.
Darde, M.L., Villena, I., Pinon, J.M., and Beguinot, I. (1998). Severe toxoplasmosis caused by a *Toxoplasma gondii* strain with a new isoenzyme type acquired in French Guyana. J Clin Microbiol 36, 324, 1998.
Dawson, D. (2005). Foodborne protozoan parasites. Int J Food Microbiol 103, 207–227.
de A. dos Santos, C.B., Carvalho, A.C.F.B.d., Ragozo, A.M.A., Soares, R.M., Amaku, M., Yai, L.E.O., Dubey, J.P., and Gennari, S.M. (2005). First isolation and molecular characterization of Toxoplasma gondii from finishing pigs from Sao Paulo State, Brazil. Vet Parasitol 131, 207–211.
De, N.V., Cong, L.D., Kino, H., Son, D.T., and Vien, H.V. (2000). Epidemiology, symptoms and treatment of paragonimiasis in Sin Ho district, Lai Chau province, Vietnam. Southeast Asian J Trop Med Public Health 31(suppl 1), 26–30.
De Vos, T., Szalai, A.J., and Dick, T.A. (1990). Genetic and morphological variability in a population of *Diphyllobothrium dendriticum* (Nitzsch, 1824). Syst Parasitol 16, 99–105.
Dubey, J.P., Graham, D.H., Blackston, C.R., Lehmann, T., Gennari, S.M., Ragozo, A.M., Nishi, S.M., Shen, S.K., Kwok, O.C., Hill, D.E., and Thulliez, P. (2002). Biological and genetic characterisation of *Toxoplasma gondii* isolates from chickens (Gallus domesticus) from Sao Paulo, Brazil: unexpected findings. Int J Parasitol 32, 99–105.
Dubey, J.P., Graham, D.H., da Silva, D.S., Lehmann, T., and Bahia-Oliveira, L.M. (2003a). *Toxoplasma gondii* isolates of free-ranging chickens from Rio de Janeiro, Brazil: mouse mortality, genotype, and oocyst shedding by cats. J Parasitol 89, 851–3.

Dubey, J.P., Graham, D.H., Dahl, E., Hilali, M., El-Ghaysh, A., Sreekumar, C., Kwok, O.C., Shen, S.K., and Lehmann, T. (2003b). Isolation and molecular characterization of *Toxoplasma gondii* from chickens and ducks from Egypt. Vet Parasitol 114, 89–95.

Dubey, J.P., Graham, D.H., De Young, R.W., Dahl, E., Eberhard, M.L., Nace, E.K., Won, K., Bishop, H., Punkosdy, G., Sreekumar, C., Vianna, M.C., Shen, S.K., Kwok, O.C., Sumners, J.A., Demarais, S., Humphreys, J.G., and Lehmann, T. (2004a). Molecular and biologic characteristics of *Toxoplasma gondii* isolates from wildlife in the United States. J Parasitol 90, 67–71.

Dubey, J.P., Hill, D.E., Jones, J.L., Hightower, A.W., Kirkland, E., Roberts, J.M., Marcet, P.L., Lehmann, T., Vianna, M.C., Miska, K., Sreekumar, C., Kwok, O.C., Shen, S.K., and Gamble, H.R. (2005). Prevalence of viable *Toxoplasma gondii* in beef, chicken, and pork from retail meat stores in the United States: risk assessment to consumers. J Parasitol 91, 1082–93.

Dubey, J.P., Navarro, I.T., Graham, D.H., Dahl, E., Freire, R.L., Prudencio, L.B., Sreekumar, C., Vianna, M.C., and Lehmann, T. (2003c). Characterization of *Toxoplasma gondii* isolates from free range chickens from Parana, Brazil. Vet Parasitol 117, 229–34.

Dubey, J.P., Navarro, I.T., Sreekumar, C., Dahl, E., Freire, R.L., Kawabata, H.H., Vianna, M.C., Kwok, O.C., Shen, S.K., Thulliez, P., and Lehmann, T. (2004b). *Toxoplasma gondii* infections in cats from Parana, Brazil: seroprevalence, tissue distribution, and biologic and genetic characterization of isolates. J Parasitol 90, 721–6.

Dubey, J.P., Venturini, M.C., Venturini, L., Piscopo, M., Graham, D.H., Dahl, E., Sreekumar, C., Vianna, M.C., and Lehmann, T. (2003d). Isolation and genotyping of *Toxoplasma gondii* from free-ranging chickens from Argentina. J Parasitol 89, 1063–4.

Duncanson, P., Terry, R.S., Smith, J.E., and Hide, G. (2001). High levels of congenital transmission of *Toxoplasma gondii* in a commercial sheep flock. Int J Parasitol 31, 1699–1703.

Dzikowski, R., Levy, M.G., Poore, M.F., Flowers, J.R., and Paperna, I. (2004). Use of rDNA polymorphism for identification of Heterophyidae infecting freshwater fishes. Dis Aquatic Organisms 59, 35–41.

Fall, A., Thompson, R.C.A., Hobbs, R.P., and Morgan-Ryan U. (2003). Morphology is not a reliable tool for delineating species within *Cryptosporidium*. J Parasitol 89, 399–402.

Fayer, R., Morgan, U., and Upton, S.J. (2000). Epidemiology of Cryptosporidium: transmission, detection and identification. Int J Parasitol 30, 1305–1322.

Ferreira Ade, M., Vitor, R.W., Gazzinelli, R.T., and Melo, M.N. (2006). Genetic analysis of natural recombinant Brazilian *Toxoplasma gondii* strains by multilocus PCR-RFLP. Infect Genet Evol 6, 22–31.

Ferreira, Ade, M., Vitor, R.W.A., Carneiro, A.C.A.V., Brandao, G.P., and Melo, M.N. (2004). Genetic variability of Brazilian *Toxoplasma gondii* strains detected by random amplified polymorphic DNA-polymerase chain reaction (RAPD-PCR) and simple sequence repeat anchored-PCR (SSR-PCR). Infect Genet Evol 4, 131–142.

Fuentes, I., Rubio, J.M., Ramirez, C., and Alvar, J. (2001). Genotypic characterization of *Toxoplasma gondii* strains associated with human toxoplasmosis in Spain: direct analysis from clinical samples. J Clin Microbiol 39, 1566–70.

Glasner, P.D., Silveira, C., Kruszon-Moran, D., Martins, M.C., Burnier Junior, M., Silveira, S., Camargo, M.E., Nussenblatt, R.B., Kaslow, R.A., and Belfort, R., Jr. (1992). An unusually high prevalence of ocular toxoplasmosis in southern Brazil. Am J Ophthalmol 114, 136–44.

Goh, S., Reacher, M., Casemore, D.P., Verlander, N.Q., Chalmers, R., Knowles, M., Williams, J., Osborn, K., and Richards, S. (2004). Sporadic cryptosporidiosis, North Cumbria, England, 1996–2000. Emerg Infect Dis 10, 1007–1015.

Grigg, M.E., Ganatra, J., Boothroyd, J.C., and Margolis, T.P. (2001). Unusual abundance of atypical strains associated with human ocular toxoplasmosis. J Infect Dis 184, 633–639.

Hall, A. (1996). Editorial: What is molecular epidemiology? Trop Med Int Hlth 1, 407–408.

Herwaldt, B.L., and Ackers, ML. (1997). An outbreak in 1996 of cyclosporiasis associated with imported raspberries. The *Cyclospora* Working Group. N Engl J Med 336, 1548–1556.

Herwaldt, B.L., and Beach, M.J. (1999). The return of *Cyclospora* in 1997: another outbreak of cyclosporiasis in North America associated with imported raspberries. Cyclospora Working Group. Ann Intern Med 130, 210–220.

Hijjawi, N.S. Meloni, B.P., Ryan, U.M., Olson, M.E., and Thompson, R.C. (2002). Successful in vitro cultivation of *Cryptosporidium andersoni* with evidence for the existence of novel extracellular stages in the *Cryptosporidium* life cycle. Int J Parasitol 32, 1719–1726.

Hong, S.T., Choi, M.H., Kim, C.H., Chung, B.S., and Ji, Z. (2003). The Kato-Katz method is reliable for diagnosis of *Clonorchis sinensis* infection. Diagn Microbiol Infect Dis 47, 345–347.

Howe, D.K., Honore, S., Derouin, F., and Sibley, L.D. (1997). Determination of genotypes of *Toxoplasma gondii* strains isolated from patients with toxoplasmosis. J Clin Microbiol 35, 1411–1414.

Howe, D.K., and Sibley, L.D. (1995). *Toxoplasma gondii* comprises three clonal lineages: correlation of parasite genotype with human disease. J Infect Dis 172, 1561–1566.

Hudson, P. (2005). Parasites, diversity and the ecosystem. In: Thomas, F., Renaud, F., and Guégan, J.-F, eds. Parasitism and ecosystems. New York: Oxford University Press, 1–12.

Hunter, P.R., and Thompson, R.C.A. (2005). The zoonotic transmission of *Giardia* and *Cryptosporidium*. Int J Parasitol 35, 1181–1190.

Intapan, P.M., Kosuwan, T., Wongkham, C., and Maleewong, W. (2004). Genomic characterization of lung flukes *Paragonimus heterotremus, P, siamensis, P. harinasutai, P. westermani* and *P. bangkokensis* by RAPD markers. Vet Parasitol 124, 55–64.

Intapan, P.M., Wongkham, C., Imtawil, K.J., Pumidonming, W., Prasongdee, T.K., Miwa, M., and Maleewong, W. (2005). Detection of Paragonimus heterotremus eggs in experimentally infected cats by a polymerase chain reaction-based method. J Parasitol 91, 195–198.

Iwagami, M., Ho, L., Su, K., Lai, P., Fukushima, M., Nakano, M., Blair, D., Kawashima, K., and Agatsuma, T. (2000). Molecular phylogeographic studies on *Paragonimus westermani* in Asia. J Helminthol 74, 315–322.

Johnson, A.M. (1997). Speculation on possible life cycles for the clonal lineages in the genus toxoplasma. Parasitol Today 13, 393–397.

Keiser, J., and Utzinger, J. (2004). Chemotherapy for major food-borne trematodes: a review. Expert Opin Pharmacother 5, 1711–1726.

Khamboonruang, C., Keawvichit, R., Wongworapat, K., Suwanrangsi, S., Hongpromyart, M., Sukhawat, K., Tonguthai, K., and Lima dos Santos, C.A. (1997). Application of hazard analysis critical control point (HACCP) as a possible control measure for *Opisthorchis viverrini* infection in cultured carp (Puntius gonionotus). Southeast Asian J Trop Med Public Health 28 Suppl 1, 65–72.

Khan, A., Jordan, C., Muccioli, C., Vallochi, A.L., Rizzo, L.V., Belfort, R., Jr., Vitor, R.W., Silveira, C., and Sibley, L.D. (2006). Genetic divergence of *Toxoplasma gondii* strains associated with ocular toxoplasmosis, Brazil. Emerg Infect Dis 12, 942–949.

Le, T.H., Van De, N., Blair, D., McManus, D.P., Kino, H., and Agatsuma, T. (2006). *Paragonimus heterotremus* Chen and Hsia (1964), in Vietnam: a molecular identification and relationships of isolates from different hosts and geographical origins. Acta Trop 98, 25–33.

Leander, B.S., Harper, J.T., and Keeling, P.J. (2003a). Molecular phylogeny and surface morphology of marine aseptate gregarines (Apicomplexa): *Selenidium* spp. and *Lecudina* spp. J Parasitol 89, 1191–1205.

Leander, B.S., Clopton, R.E., and Keeling, P.J. (2003b). Phylogeny of gregarines (Apicomplexa) as inferred from small-subunit rDNA and b-tubulin. Int J Syst Evol Microbiol 53, 345–354.

Leander, B.S., and Keeling, P.J. (2004). Morphostasis in alveolate evolution. Trends Ecol Evol 18, 395–402.

Leander, B.S., Lloyd, S.A., Marshall, W., and Landers, S.C. (2006). Phylogeny of marine gregarines (Apicomplexa)—*Pterospora, Lithocystis* and *Lankesteria*—and the origin(s) of coelomic parasitism. Protist 157, 45–60.

Lehmann, T., Blackston, C.R., Parmley, S.F., Remington, J.S., and Dubey, J.P. (2000). Strain typing of *Toxoplasma gondii*: comparison of antigen-coding and housekeeping genes. J Parasitol 86, 960–971.

Lehmann, T., Marcet, P.L., Graham, D.H., Dahl, E.R., and Dubey, J.P. (2006). Globalization and the population structure of *Toxoplasma gondii*. Proc Nat Acad Sci U S A 103, 11423–1148.

Levin, B.R., Lipsitch, M., and Bonhoeffer, S. (1999). Population biology, evolution, and infectious disease: convergence and synthesis. Science 283, 806–809.

Lin, R., Li, X., Lan, C., Yu, S., and Kawanaka, M. (2005). Investigation on the epidemiological factors of *Clonorchis sinensis* infection in an area of south China. Southeast Asian J Trop Med Public Health 36, 1114–1117.

Lun, Z.R., Gasser, R.B., Lai, D.H., Li, A.X., Zhu, X.Q., Yu, X.B., and Fang, Y.Y. (2005). Clonorchiasis: a key foodborne zoonosis in China. Lancet Infect Dis 5, 31–41.

Lyons, R.E., and Johnson, A.M. (1998). Gene sequence and transcription differences in 70 kDa heat shock protein correlate with murine virulence of *Toxoplasma gondii*. Int J Parasitol 28, 1041–1051.

Maleewong, W., Intapan, P.M., Wongkham, C., Wongsaroj, T., Kowsuwan, T., Pumidonming, W., Pongsaskulchoti, P., and Kitikoon, V. (2003). Detection of *Opisthorchis viverrini* in experimentally infected bithynid snails and cyprinoid fishes by a PCR-based method. Parasitology 126, 63–67.

Martinez-Salazar, E.A., and Leon-Regagnon, V. (2005). Confirmation of *Gnathostoma binucleatum* Almeyda-Artigas, 1991, advances third-stage larvae in Tres Palos Lagoon, Mexico, by morphological and molecular data. J Parasitol 91, 962–965.

Mas-Coma, S., Bargues, M.D., and Valero, M.A. (2005). Fascioliasis and other plant-borne trematode zoonoses. Int J Parasitol 35, 1255–1278.

McCarthy, J., and Moore, T.A. (2000). Emerging helminth zoonoses. Int J Parasitol 30, 1351–60.

McGlade, T.R., Robertson, I.D., Elliott, A.D., and Thompson, R.C.A. (2003). High prevalence of *Giardia* detected in cats by PCR. Vet Parasitol 110, 197–205.

Mead, P.S., Slutsker, L., Dietz, V., McCaig, L.F., Bresee, J.S., Shapiro, C., Griffin, P.M., and Tauxe, R.V. (1999). Food-related illness and death in the United States. Emerg Infect Dis 5, 607–625.

Miller, M.A., Grigg, M.E., Kreuder, C., James, E.R., Melli, A.C., Crosbie, P.R., Jessup, D.A., Boothroyd, J.C., Brownstein, D., and Conrad, P.A. (2004). An unusual genotype of *Toxoplasma gondii* is common in California sea otters (Enhydra lutris nereis) and is a cause of mortality. Int J Parasitol 34, 275–284.

Monis, P.T., and Thompson, R.C.A. (2003). *Cryptosporidium* and *Giardia*-zoonoses: fact or fiction? Infect Genet Evol 3, 233–244.

Monis, P.T., Giglio, S., Keegan, A.R., and Thompson, R.C.A. (2005). Emerging technologies for the detection and genetic characterisation of parasites. Trends Parasitol 21, 340–346.

Montoya, J.G., and Liesenfeld, O. (2004). Toxoplasmosis. Lancet 363, 1965–76.

Morgan, U.M., Xiao, L., Fayer, R., Lal, A.A., and Thompson, R.C.A. (1999). Variation in *Cryptosporidium*: towards a taxonomic revision of the genus. Int J Parasitol 29, 1733–1751.

Mori, Y., Kitao, M., Tomita, N., and Notomi, T. (2004). Real-time turbidimetry of LAMP reaction for quantifying template DNA. J Biochem Biophys Methods 59, 145–157.

Morrison, D.A., and Ellis, J.T. (1997). Effects of nucleotide sequence alignment on phylogeny estimation: a case study of 18 rDNA of Apicomplexa. Mol Biochem Parasitol 14, 428–441.

Nagano, I., Pei, F., Wu, Z., Wu, J., Cui, H., Boonmars, T., and Takahashi, Y. (2004). Molecular expression of a cysteine proteinase of Clonorchis sinensis and its application to an enzyme-linked immunosorbent assay for immunodiagnosis of clonorchiasis. Clin Diagn Lab Immunol 11, 411–416.

Neto, E.C., Anele, E., Rubim, R., Brites, A., Schulte, J., Becker, D., and Tuuminen, T. (2000). High prevalence of congenital toxoplasmosis in Brazil estimated in a 3-year prospective neonatal screening study. Int J Epidemiol 29, 941–947.

Ngarmamonpirat, C., Waikagul, J., Petmitr, S., Dekumyoy, P., Rojekittikhun, W., and Anantapruti, M.T. (2005). Analysis of sequence variation in *Gnathostoma spinigerum* mitochondrial DNA by single-strand conformation polymorphism analysis and DNA sequence. Parasitol Int 54, 65–68.

Nguyen, T.L.A., Murrell, K.D., Nguyen, C.T., Giang, H.H., Johansen, V.M., and Thamsborg, M.S. (2006a). Domestic animals serve as reservoir hosts for fishborne treamatodes in Nghe An, Vietnam. In: Waikagul, J., ed 5th Seminar on Food- and Water-borne Parasitic Zoonoses. Book of abstracts. Mahidol University, Bangkok: Faculty of Tropical Medicine, 78.

Nguyen, V.D., Do, T.D., Nguyen, T.H., Pham, A.T., Bui, V.T., Chai, J.Y., Sohn, W.M., Kim, J.L., Murrell, K.D., and Dalsgaad, A. (2006b). Human prevalence survey of fishborne trematodes in Nghe An, Nam Dinh and Giang province, Vietnam, 2004–2005. In: Waikagul, J., ed 5th Seminar on Food- and Water-Borne Parasitic Zoonoses. Book of Abstracts, Mahidol University, Bangkok: Faculty of Tropical Medicine, 79.

Notomi, T., Okayama, H., Masubuchi, H., Yonekawa, T., Watanabe, K., Amino, N., and Hase, T. (2000). Loop-mediated isothermal amplification of DNA. Nucleic Acids Res 28:E63.

O'Donoghue, P.J. (1995). *Cryptosporidium* and Cryptosporidiosis in man and animals. Int J Parasitol 85, 525–530.

Owen, I.L., Pozio, E., Tamburrini, A., Danaya, R.T., Bruschi, F., and Gomez Morales, M.A. (2001). Focus of human trichinellosis in Papua New Guinea. Am J Trop Med Hyg 65, 553–557.

Owen, M.R., and Trees, A.J. (1999). Genotyping of *Toxoplasma gondii* associated with abortion in sheep. J Parasitol 85, 382–384.

Papoz, L., Simondon, F., Saurin, W., and Sarmini, H. (1986). A simple model relevant to toxoplasmosis applied to epidemiologic results in France. Am J Epidemiol 123, 154–161.

Parameswaran N., O'Hanbdley, R.M., Vitali, S., Walton, S, Fenwick, S.G., and Thompson, R.C.A. (2006). The investigation of vertical transmission of *Toxoplasma gondii* in zoo bred marsupials. Perth Zoo Arazpa Conference 2006, 1–5, Australasian Regional Zoos Association of Zoological Parks and Aquaria. http://www.arazpa.org.au/2006%20ARAZPA%20Conference%20Proceedings.htm.

Parmley, S.F., Gross, U., Sucharczak, A., Windeck, T. Sgarlato, G.D., and Remington, J.S. (1994). Two alleles of the gene encoding surface antigen P22 in 25 strains of *Toxoplasma gondii*. J Parasitol 80, 293–301.

Petersen, E., Pollak, A., and Reiter-Owona, I. (2001). Recent trends in research on congenital toxoplasmosis. Int J Parasitol 31, 115–144.

Poulin, R. (1998). Evolutionary Ecology of Parasites: From Individuals to Communities. London: Chapman & Hall.

Pozio, E., Foggin, C.M., Marucci, G., La Rosa, G., Sacchi, L., Corona, S., Rossi P., and Mukaratirwa, S. (2002). *Trichinella zimbabwensis* n.sp. (Nematoda), a new non-encapsulated species from crocodiles *(Crocodylus niloticus)* in Zimbabwe also infecting mammals. Int J Parasitol 32, 1787–1799.

Pozio, E., and Zarlenga, D.S. (2005). Recent advances on the taxonomy, systematics and epidemiology of *Trichinella*. Int J Parasitol 35, 1191–1204.

Queuche, F., Cao Van, V., and Le Dang, H. (1997). [Endemic area of paragonimiasis in Vietnam.] Sante 7, 155–159.

Roghmann, M.C., Faulkner, C.T., Lefkowitz, A., Patton, S., Zimmerman, J., and Morris, J.G., Jr. (1999). Decreased seroprevalence for *Toxoplasma gondii* in Seventh Day Adventists in Maryland. Am J Trop Med Hyg 60, 790–2.

Sahai, B.N. (1969). A survey of the helminth parasites of stray dogs in and around Bareilly, Uttar Pradesh. Indian Vet J 46, 734.

Schuster, H., Agada, F.O., Anderson, A.R., Jackson, R.S., Blair, D., McGann, H., and Kelly, G. (2006). Otitis media and a neck lump—current diagnostic challenges for *Paragonimus*-like trematode infections. J Infect 54, 103–106.

Semyenova, S.K., Morozova, E.V., Chrisanfova, G.G., Gorokhov, V.V., Arkhipov, I.A., Moskvin, A.S., Movsessyan, S.O., and Ryskov A.P. (2006). Genetic differentiation in Eastern European and Western Asian populations of the liver fluke, *Fasciola hepatica*, as revealed by mitochondrial *NAD1* and *COX1* genes. J Parasitol 92, 525–530.

Sibley, D.L. (2003). Recent origins among ancient parasites. Vet Parasitol 115, 185–98.

Sibley, L.D., and Boothroyd, J.C. (1992). Virulent strains of *Toxoplasma gondii* comprise a single clonal lineage. Nature 359, 82–85.

Silveira, C., Belfort, R., Jr., Muccioli, C., Abreu, M.T., Martins, M.C., Victora, C., Nussenblatt, R.B., and Holland, G.N. (2001). A follow-up study of *Toxoplasma gondii* infection in southern Brazil. Am J Ophthalmol 131, 351–354.

Sirisinha, S., Chawengkirttikul, R., Haswell-Elkins, M.R., Elkins, D.B., Kaewkes, S., and Sithithaworn, P. (1995). Evaluation of a monoclonal antibody-based enzyme linked immunosorbent assay for the diagnosis of *Opisthorchis viverrini* infection in an endemic area. Am J Trop Med Hyg 52, 521–524.

Slifko, T.R., Smith, H.V., and Rose, J.B. (2000). Emerging parasite zoonoses associated with water and food. Int J Parasitol 30, 1379–1394.

Sreekumar, C., Hill, D.E., Miska, K.B., Vianna, M.C.B., Yan, L., Myers, R.L., and Dubey, J.P. (2005). Genotyping and detection of multiple infections of *Toxoplasma gondii* using Pyrosequencing. Int J Parasitol 35, 991–999.

Sriamporn, S., Pisani, P., Pipitgool, V., Suwanrungruang, K., Kamsa-ard, S., and Parkin, D.M. (2004). Prevalence of *Opisthorchis viverrini* infection and incidence of cholangiocarcinoma in Khon Kaen, Northeast Thailand. Trop Med Int Health 9, 588–594.

Stantic-Pavlinic M., Xiao L., Glaberman S., Lal, A.A., Orazen T., Rataj-Verglez A., Logar J., and Berce I. (2003). Cryptosporidiosis associated with animal contacts. Wien Klin Wochensch 115, 125–127.

Stensvold, C.R., Saijuntha, W., Sithithaworn, P., Wongratanacheewin, S., Strandgaard, H., Ornbjerg, N., and Johansen, M.V. (2006). Evaluation of PCR based coprodiagnosis of human opisthorchiasis. Acta Trop 97, 26–30.

Stensvold, C.R., Traub, R.J., von Samson-Himmelstjerna, G., Jespersgaard, C., Nielsen, H.V., and Thompson, R.C.A. (2007). *Blastocystis*: subtyping isolates using Pyrosequencing 116, 111–119.

Su, C., Evans, D., Cole, R.H., Kissinger, J.C., Ajioka, J.W., and Sibley, L.D. (2003). Recent expansion of *Toxoplasma* through enhanced oral transmission. Science 299, 414–416, 2003

Sugiyama, H., Morishima, Y., Kameoka, Y., and Kawanaka, M. (2002). Polymerase chain reaction (PCR)-based molecular discrimination between *Paragonimus westermani* and *P. miyazakii* at the metacercarial stage. Mol Cell Probes 16, 231–236.

Sugiyama, H., Morishima, Y., Rangsiruji, A., Binchai, S., Ketudat, P., Kameoka, Y., and Kawanaka, M. (2005). Molecular discrimination between individual metacercariae of *Paragonimus heterotremus* and *P. westermani* occurring in Thailand. Southeast Asian J Trop Med Public Health 36 Suppl 4, 102–106.

Sunnotel, O., Lowery, C.J., Moore, J.E., Dooley, J.S.G., Xiao, L., Millar, B.C., Rooney, P.J., and Snelling, W.J. (2006). *Cryptosporidium*. Lett Appl Microbiol 43, 7–16.

Tenter, A.M., Heckeroth, A.R., and Weiss, L.M. (2000). *Toxoplasma gondii* from animals to humans. Int J Parasitol 30, 1217–1258.

Tenter, A.M., Barta, J.R., Beveridge, I., Duszynski, D.W., Mehlhorn, H., Morrison, D.A., Thompson, R.C.A., and Conrad, P.A. (2002). The conceptual basis for a new classification of the coccidia. Int J Parasitol 32, 595–616.

Thompson, R.C.A. (2002). Presidential address: rediscovering parasites using molecular tools—towards revising the taxonomy of *Echinococcus, Giardia* and *Cryptosporidium*. Int J Parasitol 32, 493–496.

Thompson, R.C.A. (2004). The zoonotic significance and molecular epidemiology of *Giardia* and giardiasis. Vet Parasitol 126, 15–35.

Thompson, R.C.A., and Lymbery, A.J. (1996). Genetic variability in parasites and host-parasite interactions. *Parasitology* 112, S7–S22.

Thompson, R.C.A., Constantine, C.C., and Morgan U.M. (1999). Overview and significance of molecular methods: what role for molecular epidemiology? Parasitology 117: s161–s175.

Thompson, R.C.A., Deplazes, P., and Eckert, J. (2003a). Observations on the development of *Echinococcus multilocularis* in cats. J Parasitol 89, 1086–1088.

Thompson, R.C.A., Morgan, U.M., Hopkins, R.M., and Pallant, L.J. (2000). Enteric protozoan infections. In: Thompson, R.C.A., ed. Molecular Epidemiology of Infectious Diseases. London: Arnold, 194–209.

Thompson, R.C.A., Owen, I.L., Puana, I., Banks, D., Davis, T.M.E., and Reid, S.A. (2003b). Parasites and biosecurity—the example of Australia. Trends Parasitol 19, 410–416.

Tran, T.K.C., Dalsgaard, A., Pham, A.T., and Murrell, K.D. (2006). Cross-sectional prevalence study of zoonotic trematodes in fish from a north Vietnamese fish farming community. In: Waikagul, J., ed. 5th Seminar on Food- and Water-Borne Parasitic Zoonoses. Book of abstracts. Mahidol University, Bangkok: Faculty of Tropical Medicine, 77.

Traub, R. (2004). Dogs, humans and gastrointestinal parasites: unravelling epidemiological and zoonotic relationships in an endemic tea-growing community in Northeast India. PhD, Murdoch University, Murdoch, Western Australia.

Traub, R.J., Monis, P.T., Robertson, I., Irwin, P., Mencke, N., and Thompson, R.C.A. (2004). Epidemiological and molecular evidence supports the zoonotic transmission of *Giardia* among humans and dogs living in the same community. Parasitology 128, 253–262.

Traub, R.J., Robertson, I.D., Irwin, P., Mencke, N., and Thompson, R.C.A. (2005). Canine gastrointestinal parasitic zoonoses in India. Trends Parasitol 21, 42–48.

United States Department of Agriculture. (1995). Pathogen reduction; hazard analysis and critical control point (HACCP) systems; proposed rule, vol 60. Washington, DC: Federal Register, 6774–6889.

Vallochi, A.L., Muccioli, C., Martins, M.C., Silveira, C., Belfort, R., Jr, and Rizzo, L.V. (2005). The genotype of *Toxoplasma gondii* stains causing ocular toxoplasmosis in humans in Brazil. Am J Ophthalmol 139, 350–351.

Walliker, D. (2000). Malaria. In: Thompson, R.C.A., ed. Molecular Epidemiology of Infectious Diseases. London: Arnold, 93–112.

Warnekulasuriya, M.R., Johnson, J.D., and Holliman, R.E. (1998). Detection of *Toxoplasma gondii* in cured meats. Int J Food Microbiol 45, 211–215.

WHO. (1995). Control of Foodborne Trematode Infections. WHO Technical Report Series no. 849.

WHO. (1999). Report of the WHO informal Consultation on Monitoring Drug Efficacy in the Control of Schistosomiasis and Intestinal Nematodes. WHO/CPC/SIP/99.1.

WHO. (2002). Joint WHO/FAO Workshop on Food-borne Trematode in fections in Asia RS/2002/GE/40(VTN), Hanoi, Vietnam, pp. 1–62.

WHO. (2004). Joint WHO/FAO Workshop on Food-borne Trematode in fections in Asia RS/2002/GE/40(VTN), Hanoi, Vietnam, 1–62.

Windeck, T., and Gross, U. (1996). *Toxoplasma gondii* strain-specific transcript levels of SAG1 and their association with virulence. Parasitol Res 82, 715–719.

Wongratanacheewin, S., Pumidonming, W., Sermswan, R.W., and Maleewong, W. (2001). Development of a PCR-based method for the detection of *Opisthorchis viverrini* in experimentally infected hamsters. Parasitol 122, 175–180.

Wongratanacheewin, S., Pumidonming, W., Sermswan, R.W., Pipitgool, V., and Maleewong, W. (2002). Detection of *Opisthorchis viverrini* in human stool specimens by PCR. J Clin Microbiol 40, 3879–3880.

Wongsaroj, T., Sakolvaree, Y., Chaicumpa, W., Maleewong, W., Kitikoon, V., Tapchaisri, P., Chongsa-nguan, M., and Cross, J.H. (2001). Affinity purified oval antigen for diagnosis of *Opisthorchiasis viverrini*. Asian Pacific J Allergy Immunol 19, 245–258.

Yang, S.H, Park, J.O., Lee, J.H., Jeon, B.H., Kim, W.S., Kim, S.I., Yun, K.J., Jeong, E.T., Lee, K.W., Kim, Y.M., and Lee, H., P. (2004). Cloning and characterization of a new cysteine proteinase secreted by *Paragonimus westermani* adult worms. Am J Trop Med Hyg 71, 87–92.

Yano, A., Ishimartu, R. and Hujikata, R. (2007). Rapid and sensitive detection of heat-labile I and heat-stable I enterotoxin genes of enterotoxigenic *Escherichia coli* by loop-mediated isothermal amplification. J Microbiol Methods 68, 414–420.

Index

Printed in the United States of America

Printed in the United States
128625LV00005B/7/P

9 780387 713571